谨以此书献给中国哺乳动物学创始人

夏武平教授（1918~2009）

———————————

This book is dedicated to the founder of Mammalogy in China

Professor Xia Wuping（1918-2009）

"十三五"国家重点出版物出版规划项目

中国哺乳动物多样性

编目、分布与保护

1————

蒋志刚 / 主编

EDITOR-IN-CHIEF
JIANG ZHIGANG

DIVERSITY OF
CHINA'S MAMMALS
INVENTORY,
DISTRIBUTION AND
CONSERVATION

海峡出版发行集团
海峡书局

图书在版编目（CIP）数据

中国哺乳动物多样性：编目、分布与保护 / 蒋志刚
主编. — 福州：海峡书局，2024.6
　　ISBN 978-7-5567-1063-8

　　Ⅰ．①中… Ⅱ．①蒋… Ⅲ．①哺乳动物纲—生物多样
性—生物资源保护—工作概况—中国 Ⅳ．①Q959.8

中国国家版本馆CIP数据核字(2023)第005621号

审图号：GS（2024）1541号

出 版 人：林前汐
策 划 人：曲利明　李长青
主　　编：蒋志刚
责任编辑：林洁如　张　帆　陈　尽
营销编辑：邓凌艳　陈洁蕾
责任校对：卢佳颖
装帧设计：黄舒埼　李　晔　董玲芝　林晓莉

ZHŌNGGUÓ BŬRŬ DÒNGWÙ DUŌYÀNGXÌNG： BIĀNMÙ、 FĒNBÙ YŬ BĂOHÙ
中国哺乳动物多样性：编目、分布与保护

出版发行：海峡书局
地　　址：福州市台江区白马中路15号
邮　　编：350004
发行电话：0591-88600690
印　　刷：深圳市泰和精品印刷有限公司
开　　本：889毫米×1194毫米　1/16
印　　张：106.5
图　　文：1704码
版　　次：2024年6月第1版
印　　次：2024年6月第1次印刷
书　　号：ISBN 978-7-5567-1063-8
定　　价：1200元（全三册）

也许是因为我们智人是哺乳动物，哺乳动物是地球上最易识别的动物群体之一。事实上，我们人类已经成为地球上的主导物种，带来了广泛的影响，这些影响有好有坏。尽管地球上自有生命历史以来，已经有成千上万的物种自然灭绝，但随着人类的增加，灭绝的速度正在加快。国际自然保护联盟(IUCN)濒危物种红色名录(Red List of Threatened Species)最近的评估记录了大约四分之一的哺乳动物物种处于危险之中。显然，我们有理由担心这些濒临灭绝的哺乳动物的命运。它们的命运对我们人类的命运至关重要。

我们对哺乳动物的了解继续稳步增加，新的分析技术使我们对当今地球上哺乳动物的物种限制和物种数量有了更好的了解。今天，继续记录世界哺乳动物物种的多样性和分布比以往任何时候都更加重要。中国是一个哺乳动物种类丰富的国家，其特点是秦岭和横断山的大熊猫、金羚牛、仰鼻猴，以及青藏高原的藏羚羊和野牦牛。中国哺乳动物的科学研究和描述始于15世纪的大发现时代。中国大部分早期的哺乳动物研究工作是由来自欧洲国家和北美的探险家进行的，他们经常与中国同事合作。中国第一份哺乳动物名录是由美国哺乳动物学家格洛弗·艾伦和他的同事在20世纪初编制的。

近几十年来，中国与世界其他地区的科学交流变得越来越普遍和重要。在我的研究生涯的早期阶段，我很幸运地与中国同事开始了合作野外考察和博物馆研究。20世纪80年代，我在美国史密森学会考察青藏高原时，参观了中国科学院西北高原生物研究所，蒋志刚是该所生态学的年轻研究人员。1998年，我和志刚在马里兰大学的校园里重新联系上了，当时我们正在参加保护生物学会的年会。志刚在2003年和2012年两次来到史密森尼国家自然历史博物馆研究哺乳动物标本。美国博物馆收藏了中国早期的珍贵标本。这些学生和研究人员的持续交流促进了中国的动物学研究，大大增加了我们对世界哺乳动物的认识。

我高兴地看到《中国哺乳动物多样性：编目、分布与保护》一书在《生物多样性公约》第十五次缔约方大会之际成稿。作为一个积极编写关于哺乳动物的分类、分布和保护状况的著作的人，我自己的研究很大程度上依赖于这样的著作。我们对哺乳动物的了解现在已经大幅增加，我们计划和管理保护项目的能力将继续快速增长。

D. E. 威尔逊博士

2021年8月29日

于华盛顿特区

Foreword /

Mammals are one of the most recognizable groups of animals on Earth, perhaps because we Homo sapiens are mammals. In fact, we have become the dominant species on Earth, with wide-reaching consequences, both good and bad. Although thousands of species have gone extinct naturally during the history of life on Earth, the pace of extinction is quickening as humankind increases. Recent assessments by the IUCN Red List of Threatened Species have documented that roughly a quarter of all mammalian species are in peril. Clearly we have reason to worry about the fate of those threatened mammals. Their fate is critically important to the fate of our own species.

Our knowledge of mammals continues to grow steadily, and new analytical techniques have led to a far better understanding of species limits and of the number of species of mammals on Earth today. Today, more than ever, it is vital to continue recording the diversity and distribution of mammal species worldwide. China is a country with an impressive mammalian fauna, characterized by the Giant Panda, Golden Takin, and the snub-nosed monkeys in the Qinling and Hengduanshan Mountains along with Chiru (Tibetan Antelope) and Wild Yak on the Qinghai –Tibetan Plateau. The mammals of China have been scientifically studied and described beginning with the era of Great Discovery in the 15th century. Much of the early work on mammalogy in the country was conducted by explorers from European countries and North America, frequently working with Chinese colleagues. The first checklist of China's mammals was compiled by Glover Allen and colleagues early in the 20th century.

Scientific exchanges between China and the rest of the world have become increasingly common and important in recent decades. I was fortunate to begin field and museum studies with Chinese colleagues at an early stage in my own research career. I visited the Northwest Institute of Biology, Chinese Academy of Sciences during a Smithsonian Institution expedition to the Qinghai-Tibetan Plateau in the 1980s, Zhigang Jiang was a young researcher in the Ecology Department of the institute. I reconnected with Zhigang on the University of Maryland Campus in 1998, when we were attending the Annual Conference of the Society for Conservation Biology. Zhigang came to study specimens in the mammal collection of the Smithsonian Institution's National Museum of Natural History twice, in 2003 and 2012. U.S. museums harbor valuable collections of Chinese specimens from the early days. These continuing exchanges of students and researchers improve zoological research in China and add greatly to our knowledge of mammals of the world.

I am delighted to see this book, *Diversity of China's Mammals: Inventory, Distribution and Conservation* finished on the occasion of the 15th Conference of Parties of Convention on Biological Diversity. As an active compiler of works on the taxonomy, distribution, and conservation status of mammals, my own research depends heavily on just such works as this. Our knowledge of the mammals has now grown substantially and our ability to plan and manage conservation programs will continue apace.

D. E. Wilson
Washington, D.C.
August 29, 2021

生物多样性是人类赖以生存的物质基础，是人类美学、艺术、文化的源泉，是地球生物圈的功能组分。生物多样性是客观世界的属性之一，从基因、功能蛋白、个体、种群、生态系统都存在多样性。生物学家林奈建立的自然系统，为认识生物类群的层次结构和种类多样性奠定了基础。达尔文发现的"生存竞争、适者生存"自然规律是理解生物多样性的关键（Darwin，1859）。大自然是一只无形之手，甄别了不同基因组合、不同个体的生存力，孕育了生机勃勃的万千生物世界。生物世界是如此的复杂，生命类型是如此的多样，从列文虎克在显微镜下发现细菌至今，人们已经花费了3个多世纪的时间，为认识生物多样性的组成和结构而努力，许多极富天赋的生物学家为此而穷尽毕生的精力。

作为人类，我们与其他哺乳动物最接近，但我们对哺乳动物进化和生物地理学的理解本身仍在进化。目前已知的最古老的真兽化石是在中国发现的、生活在1.6亿年前的侏罗纪晚期的中华侏罗兽 *Juramaia sinensis*（Luo et al.，2011）。那些原始哺乳类体型小，与现代食虫类同源。那些弱小的、胆怯的原始哺乳类仅夜间在恐龙世界的边缘生境中活动，它们在寻找食物时，探头探脑、胆颤心惊、不时打量四周，生怕成为丛林中其他猛兽的腹中之食。若不是那些史前巨兽——恐龙突然从地球上消失，无法预料哺乳类能够繁衍分化为今日如多样的类型，占据从海洋、极地、雪原、戈壁、草原、森林、冰缘、悬崖、湿地、河流、湖泊、地下、洞穴乃至人造生境等一系列生境，进而跃居地球生态系统食物链的顶端，智人在自然界的地位发生质变（Wallace，1903）。

从食虫类进化而来的灵长类，早期是一类树栖动物，是热带雨林中的杂食者（Wilson and Mittermeier，2012）。森林树冠既为灵长类提供给了丰富的食物，包括水果、叶片、芽苞、昆虫，也为它们提供了躲避猛兽、遮风避雨的栖息场所。地球环境变迁驱动着生物进化。非洲大陆的森林面积随着气候的寒暖干湿变化而变化。当森林消失后，早期的类人猿不得不下树，开始在草原上寻找生计。它们摇摇晃晃，步履蹒跚，在地面迈开了艰难的步子。为了获得开阔的视野，类人猿不得不站立起来，四下瞭望，慢慢地进化到双足直立行走方式。这种哺乳类中唯一的双足行走方式，改变了人类的身体结构，将我们与其他灵长类动物和大多数哺乳动物区分开来。古人类诞生了。

以色列历史学家哈纳雷博士（2011）在畅销书《人类简史》中，以历史学家的视角审视了人类进化的历程。他认为人类语言与人类想象力的出现促成了人类社会的形成、人类社会组织结构和人类社会生产力的发展。人类语言与人类想象力都离不开人类新脑皮层的形成，人类对周边环境的探究、认知和改造。通过畜养动物、种植庄稼，人类改变了自己获得能量与蛋白质的途径，摆脱了靠天吃饭的采集狩猎方式，增长了人类的体质和寿命。通过20世纪的绿色革命，人类的农业生产力得到空前发展，提高了地球的人类负载量。人类已经成为地球上最成功的生物。

人类的足迹已经遍布地球表面每一个角落。人类潜入深海、横跨沙漠、登上雪峰、踏上极地。人类还改变了地球表面，从太空观察地球，可以看到人类活动在地球表面留下的痕迹。人类已经成为一种重要的地质营力，自19世纪中期的大加速以来，这种影响甚至在地质记录中都是可见的。农田、牧场、人工林、养殖水面、城市等人造面状景观和道路、渠道、管线等人造线性景观取代、分割了原始植被，改变了区域的生物区系。随着生活方式的改变，随着人口日益增加，人类产生的生活垃圾，人类生产排出的废水、废气和废渣污染环境，改变大气圈，改变气候，危及地球上生物的生存，导致了生物多样性危机，许多物种由于人类活动导致的生境丧失、生境改变而濒临灭绝。

有识之士认识到人类在欢庆生命、制造空前繁荣的同时，也制造了这场环境污染、生物多样性危机和全球气候变化危机，同时也造成了人类自身的生存危机，因为目前人类生存离不开地球生物圈。生态系统服务不仅对我们的生存至关重要，对地球上所有生命都至关重要。于是，在联合国的主导下，30年前，《生物多样性公约》在巴西里约热内卢开放签署。生物多样性一词开始登堂入室，走入千家万户，成为社会生活中的高频词。保护生物多样性成为国际社会的共识。人们认识到生物多样性是人类生存的基础，通过生物多样性保育、生物多样性可持续利用以及遗传资源惠益共享的途径，才能实现人类社会的福祉，保障人类的可持续发展。

生物多样性保育、生物多样性可持续利用以及遗传资源惠益共享的前提条件之一是对生物多样性的认知。即地球上究竟有多少生物多样性？生物多样性在哪里？物种是生物分类的基本单元，其多样性是生物多样性的表征和测度。20世纪80年代，英国理论生态学家罗伯特·梅（1988）曾经提出了一个著名的问题：地球上有多少物种？近半个世纪过去了，尽管人们对这一问题进行了孜孜不倦地探索，然而，目前我们仍仅能给出这一问题的初步答案。于是，《生物多样性公约》要求各缔约国进行生物多样性编目，查清生物资源的种类、数量与分布，查明生物多样性濒危的原因，制定并实施保护措施。在寻找这些问题答案的过程中，我们开始了解物种在哪里，并绘制出了全球多样性的热点。这些热点地区不仅分布在多样的热带地区，也分布在诸如中国南部和西部等过渡带、生态交错带和异质带，这些地区独特的地质历史留下了丰富的生物多样性遗产。

中国哺乳动物是一个有地域特色的区系。许多动物在第四纪冰川的避难所中生存至今。大熊猫*Ailuropoda melanoleuca*是中国特有的哺乳动物，作为中国"国宝"，大熊猫受到世界人们的喜爱，成为世界自然基金会的会徽，是全球野生动物保护的象征。中国还有白鱀豚*Lipotes vexillifer*、藏羚*Pantholops hodgsonii*、野牦牛*Bos mutus*、普氏原羚*Procapra przewalskii*、伊犁鼠兔*Ochotona iliensis*、海南大鼯鼠*Petaurista hainana*、长江江豚*Neophocaena asiaeorientalis*、黔金丝猴*Rhinopithecus brelichi*、台湾鼠耳蝠*Myotis taiwanensis*和高黎贡林猬*Mesechinus wangi*等特有动物。对中国哺乳动物种类、数量与分布进行编目，为其保育提供基础资料是本书的目的。

随着考察范围扩大、研究手段的更新、数据库的建设，人们记录的世界哺乳动物种数不断增加。哺乳动物编目在不断更新。20世纪90年代，Wilson and Reeder（1993）报道全球有26目4629种现生哺乳动物种。12年后，他们将这一数目更新为29目5416种（Wilson and Reeder，2005）。2018年，国际自然保护联盟受威胁物种红色名录评估了5488种哺乳动物。Burgin et al.（2018）报道已知哺乳动物有6495种，其中96种是最近灭绝的。美国哺乳动物学会哺乳动物数据库（2021）收录27目167科1335属6554种，其中现生野生哺乳动物6432种，灭绝哺乳动物103种，家养哺乳动物19种。同样值得注意的是，在哺乳动物中，大多数是啮齿动物和蝙蝠，它们的新种被描述率仍然很高，但是，自然界存在许多肉眼无法区分的隐种。例如Chornelia et al.（2022）发现在11种亚洲菊头蝠中都存在隐种。隐种的存在意味着我们需要进一步的工作来记录包括隐种在内的所有物种。

为了加快识别和描述新物种的能力，特别是研究小型和隐蔽的物种，开发新的工具是必要的。PCR技术为生物学研究手段给生物学研究带来了革命性变化，将生物多样性研究带入一个崭新的层次。通过DNA测序，人们能够识别原来肉眼看不出的DNA系列的差异，并能够利用核DNA、线粒体DNA的差异识别生物种，甚至识别生物个体，产生了谱系系统学。第二代测序技术加快了测序过程，降低了成本，扩大了应用范围。因此，许多以前未被认识的物种被添加到物种编目之中。谱系系统与传统分类结合，成为生物分类学的主流手段。于是，现代生物编目中增加了许多以前未能识别的种。然而，物种概念也是导致地球上有多少物种这一问题没有确切答案的原因之一。

物种是一个颇有争议的概念，是一个假说概念框架（Sandall et al.，2023；Groves and Grunn，2011）。物种的概念一直在演化。从达尔文时代至今，生物学界仍没有一致的物种定义。生物以种存在。种是生命科学的基石，是生物分类的基本阶元，是生物进化的基本单元，是生态系统的基本功能单元，也是生物多样性的基本单位。然而，从柏拉图和亚里士多德的模式种概念、达尔文的唯名论种概念（达尔文在《物种起源》一书中说他认为物种这个名词完全是为了方便起见，任意用来表示一群相互密切类似的个体的），到群体种概念（生物种是一些具有形态和遗传相似的个体组成）、表型种概念（生物种是表型上能识别的生物个体的集合）、生态种概念（物种是生态系统的功能单元，每个物种占据一个生态位）、时间种概念（当一个物种后代的表型差异足以与其祖先区别时，即形成了新种）、分支种概念（分支种以生物进化每个分支事件，即两个线系的衍征产生，作为物种的识别标准），Emily L. Sandall et al.（2023）则指出："物种可被认作一个有公认名称的属于最低分类等级分类单元的一组生物。"她们将物种作为是生物分类系统的最小单元，进一步发展了物种的概念。

Mayden（1997）总结了25种进化种概念，发现这些概念之间有交叉，甚至有包含关系，但很多概念都有独特的标准。过去分类学家多应用形态特征分类，后来出现了数值分类方法和支序分类方法。现代分子生物学技术的飞速发展，打开了一扇认识生物起源分化的新窗口，为分类学提供了新的手段。Apagow et al.（2004）对7个哺乳动物类群使用谱系种（Phylogenetic species）概念比应用非谱系种概念多发现了71%的物种。在他综述的91例涵盖所有生物类群的研究中，使用谱系种概念发现的物种数比使用非谱系种概念发现的平均多121%。随着谱系种概念的广泛被接受与应用，可以期待会在今后的研究中人们会报道更多的新种。

分类学研究是分歧较大的一门基础学科，也是一门经验学科。不同生物类群、不同分类学家对物种有不同认识，对于一个种是否有效，一个分类单元的分合，学者们往往仁者见仁、智者见智，无法取得统一意见。不仅不同类群分类学家的物种标准不同，同一类群的分类学家的物种标准有时也不一致。中国哺乳类中啮类（Glires，包括啮齿目和兔形目）即是一个争议较多的类群。另一个争议大的类群是偶蹄类。人们对鹿科和牛科的分类存在不同的观点（Groves and Grunn，2011；Wilson and Mittermeiar，2011；IUCN，2021）。尽管过去分类学家强调生物种的概念，但是在实践中常常难以以生殖隔离来区分物种。许多物种间隔离机制事实上是空间隔离，当两个种的个体相遇时常常产生杂交。然而，若不以生殖隔离来区分物种，物种分类多依靠专家的经验标准。学者在研究中不断质疑，标新立异，在实践中无法统一物种的标准，是目前生物分类的意见分歧的根源。一个类群往往合久必分，分久必合，于是，物种数目一直在波动，物种编目老是在变化。

尽管著名的国际公约，如濒危野生动植物种国际贸易公约、保护野生动物迁徙物种公约，以及IUCN受威胁物种红色名录等，所采用的物种分类系统都不相同，但人们在进行生物多样性编目时采取了包涵兼容的态度。编目时，海纳百川、博采各家的观点，收纳各家的研究成果，是务实开放的治学态度。在进行生物多样性大数据的建设时，人们发现不同类群与不同地区的研究深度存在差异。对一些类群已经深入开展研究，而另一些类群则缺乏研究。尽管如此，现有的知识是我们的宝贵财富。科学知识是不同领域、不同类群专家知识的集合。百家争鸣、求同存异，正是本书的宗旨，我们收集了截至2023年10月的文献资料，建立了中国哺乳动物编目，为后续的中国哺乳动物的研究与保育提供资料。众多的特有哺乳类仅分布在中国，更增加了这一编目的紧迫性和重要性。

中国的季风带（包括南亚季风和东亚季风带）、干旱区和高原区等三大自然地理带的日照与降水差异悬殊。中国境内有冲积平原、海岸盐沼、黄土高原、喀斯特地貌、丘陵山地、极旱荒漠、永久冻土、内蒙古草原、青藏高原等地理单元，中国地质环境经历了海陆变迁、造山运动、高原隆升等一系列重大地质事件。高山和峡谷折叠和伸展了地表，创造了三维气候和垂直植被带，为植物和动物提供了多种多样的生态位，栖息着不同的哺乳动物，其中很大一部分是特有种。

生物地理学是研究动植物地理分布的学科。人们在收集生物标本时，即开始了生物地理分布的研究。在地理大发现与生物学大发现中，人们开始了生物地理学研究。A. R. Wallace是世界动物地理的先驱，1876年他提出了世界生物地理区理论。他的生物地理学理论后来发展为研究植物和动物分布及其制约因素的地理生物（Geobiological）途径，以及基于相似物种、类群、生态系统、生物群落、植被或地表被覆研究生物地理的生物地理（Biogeological）途径。现在，还产生了研究生物地理的第3条途径，即谱系地理学途径。

人类活动已经改变了地球上的原始植被、改变了生物地理版图。我们迫切需要了解生物多样性格局，确定生物多样性热点地区，解决濒危物种地理分布与人类活动的空间规律这一复杂而又相互关联的保护问题，于是，产生了保护生物地理学（Conservation Biogeography）这一门新兴的分支学科。保护生物地理学利用生物地理学概念、工具和方法，预测关键物种、濒危物种和生态系统的现实分布、形成机制和未来版图，解决现实世界的生物多样性保护问题。本书的哺乳动物种分布和生物地理研究为探索中国哺乳动物多样性、特有性空间分布，为规划自然保护区、国家公园，识别自然遗产地，为研究哺乳动物多样性保护提供了基础资料。

中国对珍稀濒危物种的保护、对自然生态系统的保护，曾经停滞了相当长一段时间。当中国再次向世界打开国门，珍稀濒危物种的保护、对自然生态环境的保护再次起步。在培养动物学研究人才，开展动物学科学基础研究的同时，中国制定了野生动物保护法，颁布了国家重点保护野生动物名录，使得中国野生哺乳动物保护走向法治化轨道。同时，在全国范围内评估保护区和保护区的有效性。中国还签署了濒危野生植物种国际贸易公约（CITES）、拉姆萨公约、生物多样性公约，并参加了迁徙物种公约赛加羚羊保护议定书、人与生物圈保护区网络，成为履行保护濒危物种、生态环境和生物多样性国际责任的大国之一。承办联合国《生物多样性公约》第十五次缔约方大会表明中国在保护生物多样性方面的带头作用。

2021年10月11日至24日，联合国《生物多样性公约》第十五次缔约方大会（CBD COP15）在中国昆明召开。这是联合国首次以生态文明为主题召开的全球性会议，也是生物多样性公约缔约国大会首次在中国举行。与会各方分享生物多样性保育和生态文明建设经验，共同商讨生物多样性保护大计。2022年12月8日至20日，联合国《生物多样性公约》第十五次缔约方大会第二阶段会议在加拿大蒙特利尔举行，通过了"昆明-蒙特利尔全球生物多样性框架"，为今后直至2030年乃至更长一段时间的全球生物多样性治理擘画新蓝图。与会国承诺到2030年"采取紧急管理行动，阻止人类造成的已知受威胁物种的灭绝"。该文件是继生物多样性"爱知目标"后，全球新的生物多样性保护行动计划，将为未来生物多样性带来新希望。

昆明市是云南省省会，是此次会议会址的理想选择，因为云南省气候生境多样，生物多样性丰富，拥有超过半数的中国已知物种，被誉为"植物王国、动物王国、世界花园"。云南省南北跨度900多千米，从西双版纳的热带风光到香格里拉皑皑雪峰，相当于将中国海南岛到青藏高原的气候带压缩到了云南省境内。云南省仅占全国4.1%的国土面积，但有除沙漠外的所有地球生态系统类型，拥有众多的生物，是中国生物多样性宝库，也是全球生物多样性保护的关键区和全球34个物种最丰富的生物多样性热点地区之一。所以选择昆明作为联合国《生物多样性公约》第十五次缔约方大会会址是名实所归。

为纪念这次生物多样性盛会，我们编写了这本《中国哺乳动物多样性：编目、分布与保护》，更新了中国哺乳动物多样性，展示了中国哺乳动物研究与保育，提供了一份全新的中国哺乳动物多样性编目，再次证实中国是世界哺乳动物多样性最丰富的国家。本书收集中国研究者的研究成果，也是我们多年工作的汇集，包括近年来我们在云南开展的野外调查工作。当代世界著名哺乳动物学家Don E. Wilson博士欣然命笔，为本书作序。本书的写作得到马建章院士、张亚平院士和广大同仁的大力支持。由于时间仓促，写作中难免挂一漏万，欢迎各位读者指出遗漏和错误，以便在今后的再版中修订。

动物学是一门综合科学，带有公众科学的特征。动物学研究不仅需要专门的分类学家和保护生物学家参与，还需要公民科学家参与。动物多样性是生物多样性的组成部分。生物多样性科学也带有公众科学的特征。以观鸟、观兽、动物摄影、考察野生动物为目的的野外考察、探险历来是公民喜爱的亲近自然，了解自然的户外活动，这些活动为生物多样性研究积累了海量的数据。近年来，数码摄影技术的普及、红外自动摄影的广泛应用，专业与业余作者的野生动物影像、视频、科普作品通过网络普及传播，增加了公众对野生动物的认识，还提供了野生动物活动的季节性细节，更激发了公众对野生动物的热爱。动物学在中国再次成为公众科学。本书既为研究人员提供一本中国哺乳动物学参考书，也为公众提供一本深入浅出、图文并茂的中国哺乳动物学读本。期待越来越多的公民探索野生哺乳动物、保育濒危哺乳动物。

2023年6月8日
北京市门头沟区永定镇 "远洋新天地"

Biodiversity is the material basis for human survival, the source of human art and culture, and the functional component of the Earth's biosphere. Biodiversity is an attribute of the objective biological world, including diversity at gene, protein, individual, population, ecosystem and biome levels. *The Nature System* established by the biologist Linnaeus in the 18th century laid the foundation for understanding the hierarchy and diversity of biological groups. Charles Darwin discovered the rule of "Struggle for Existence, Survival of the Fittest" in nature is key to understanding biodiversity (Darwin, 1859). Nature is an invisible hand that selects the gene combinations, of different individuals in different environments, and produces a vibrant world of millions of organisms. The biological world is so complex and the types of life so diverse, Sine Leeuwenhoek discovered bacteria under a microscope it has taken more than three centuries for the biologists to understand composition and structure of biodiversity, and many gifted biologists have devoted their whole lives to the study.

As humans we identify most closely with our fellow mammals, yet our understanding of mammalian evolution and biogeography is in itself still evolving. The oldest eutherian fossil known today was *Juramaia sinensis*, was discovered in China and lived 160 million years ago in the late Jurassic period (Luo et al., 2011). The earliest mammals were small, rodent-sized animals, yet still had the distinctive features shared by modern mammals of warm blood, and thus high energy demands. Primitive mammals diverged from insectivores. Those timid, nocturnal animals only hung on the fringes of the world of dinosaurs. These creatures poked around and jittered, from time to time when foraging, fearing of being eaten by other predators in the landscapes of the day. If dinosaurs were not the suddenly disappearance from the Earth, it is hard to predict the mammals could have diversified to the scale today, occupying such a variety of types of habitats from the ocean, polar, snowfield, cliff, desert, grassland, forest, periglacial, wetlands, rivers, lakes, underground, caves and man-made habitat, and mammals finally sit on the top of the food chain in today's global ecological system(Wallace, 1903).

Primates evolved from insectivores, and were the arboreal omnivores in the rainforest (Wilson and Mittermeier, 2012). Forest canopies provided primates with abundant food, including fruit, leaves, buds and insects, as well as sheltering them from the rain and predators. Environmental changes on Earth drive the evolution of living things, including primates. When forests were gone in some parts of the African continent as the climate changed, early Hominids had to descend from trees and begin to seek a livelihood on the ground in Savannahs, they staggered, hobbled, made difficult strides on the ground and had to stand up to get a better view around. Slowly, slowly they evolved to a bipedal upright walking mode, though as our skeletal architecture shows we are not well adapted to be bipeds, and still have the vestigial relics of our ancestral primate tails. Bipedalism, changed the body structure of humans and gave birth to hominids, differentiating us from all other primates, and the majority of mammals.

In his best-selling book, *Sapiens: A Brief History of Humankind*, Dr. Yuval Noah Harari (2011), an Israeli historian, examines human evolution from a historian's perspective. In his opinion, the emergence of human language and fabrication contributed to the formation of human society, promoted the organizational structure of human society and the development of the productivity of human societies. Human language and human imagination are inseparable from the human neocortex and from the process of human's exploring, conceptualizing and transforming the surrounding environment. By raising animals and growing crops, human beings changed the way we extract energy and protein from nature, and got rid of the hunter-gatherer way of obtaining foods, and thus improving human nutrition and increasing human lifespan by orders of magnitude. Through the Green Revolution of the 20th century, human agricultural productivity further increased the human population carrying capacity of Earth. Humans have become the most successful mammal on Earth.

Humans have left footprints on every corner of the Earth's surface. Humans have dived deep into the oceans, marched crossed deserts, climbed snow-capped peaks, and set foot in the polar regions. Humans have also changed the surface of the Earth. Viewed from space, we can see the traces that human society has left on the Earth's surface. Humans have become an important geological agent, and since the great acceleration in the mid-19th century that impact has been visible even in the geological record. Anthropogenic landscapes such as farmland, pasture, forest plantations, aquaculture, towns and cities, and artificial linear structures such as roads, channels, pipelines and power lines have replaced or divided the original vegetation and have changed the flora and fauna in many parts of the world. With ever-growing human population and the change in lifestyle, people produce more household waste, humans pollute the environment with waste water, waste gas and solid pollutants, change the atmosphere, change the climate, and threaten the survival of life on Earth. This has led to a biodiversity crisis, with many species facing extinction due to habitat loss and environmental change as a consequence of human activities.

People start to realize that while celebrating life and creating unprecedented prosperity, mankind has also created the crisis of environmental pollution, biodiversity loss and global climate change, as well as the crisis of human survival. Therefore, led by the UNEP, United Nations 30 years ago, the Convention on Biological Diversity was opened for signature in Rio de Janeiro, and the word "biodiversity" enters every household and becomes a common word in social life. The protection of biodiversity has become the consensus of the international community, because human still cannot survive without the biosphere on Earth, and ecosystem services are fundamental not only to our existence, but to all life on Earth. It is recognized that biodiversity is the basis of human existence, and the well-being of human society can only be realized through biodiversity conservation, sustainable utilization of biodiversity and benefit-sharing of genetic resources between nations and individuals.

One of the preconditions for biodiversity conservation is the sustainable utilization of biodiversity and benefit-sharing of genetic resources. Yet to share benefits from biodiversity, and to ensure it is effectively conserved there are fundamental questions that must first be answered; How many species is there on Earth? Where is this biodiversity?

To answer these questions, we must first identify and map species, as species are the basic unit of taxonomy and measure of biodiversity. In the 1980s, British theoretical ecologist Robert J. May (1988) once wisely asked: How many species are there on Earth? Half a century later, despite much exploration, we still only have finite answers to the question, and our estimates still range by several million. Therefore, the Convention on Biological Diversity requires State parties to inventory biodiversity, to identify the types, quantities and distribution of biological resources, to identify the causes of biodiversity endangerment, and to formulate and implement conservation measures. In the quest for the answers to these questions we have begun to understand where species are, and mapped out global hotspots of diversity. These hotspots fall not only within the diverse tropics, but also in transitional and heterogeneous zones such as the South and West of China, where a unique geological history has left a rich legacy of biodiversity.

The mammals of China have a range of distinct characteristics. The Giant Panda *Ailuropoda melanoleuca* is regarded as a "national treasure" of the country, it is also loved by people all over the world. Giant Panda has become the emblem of WWF and a symbol of global wildlife conservation, and a conservation success story for the 21st century. There are also other endemic species like Baji *Lipotes vexillifer*, Tibetan antelope *Pantholops hodgsonii*, wild yak *Bos mutus*, Przewalski's gazelle *Procapra przewalskii*, Ili Pika *Ochotona iliensis*, Hainan Flying Squirrel *Petaurista hainana*, Guizhou Snub-nosed Monkey *Rhinopithecus brelichi*, and Taiwan Mouse-eared Bat *Myotis taiwanensis*, Wang's Hedgehog *Mesechinus wangi* and a long list of other species. The purpose of this book is to catalog the species, number and distribution of mammals in China for streamlining conservation strategy.

With the expansion of the scope of field investigation, the updating of research methods, and the construction of databases, the number of recorded mammal species in the world is increasing. The mammal inventory is constantly being renewed. In the 1990s, Wilson and Reeder (1993) reported 4,629 living mammal species in 26 orders worldwide. They updated this number to 29 orders, 5,416 species in 2005. The International Union for Conservation of Nature Red List of Threatened Species evaluated 5,488 mammals in 2008. Burgin et al. (2018) reported that there are 6,495 known species of mammals, 96 of which are recently extinct. The Mammal Database of the American Society of Mammals (2021) includes 6,554 species of 1,335 genera, 167 families, and 27 orders, including 6,432 living wild mammals, 103 extinct mammals, and 19 domestic mammals. It is also important to note that of mammals, the majority are rodents and bats, for which description of *Species nova* remain high, and the existence of cryptic species means much further work is needed to document all species. For example, Chornelia et al. (2022) found cryptic species in 11 species of Asian horseshoe bats.

To improve our ability to identify and describe new species, especially smaller and more cryptic species, new tools are necessary. Giving biodiversity research a completely new approach, molecular approaches and PCR (Polymerase Chain Reaction) technology have brought revolutionary changes to biological research. Phylogenic systematics, which combines traditional classification, has become the mainstream means of taxonomy. Now, through DNA sequencing, people can identify differences in DNA series that were previously invisible to the naked eye, and can use differences in nuclear DNA, and mitochondrial DNA to identify species, and even individuals. The second-generation sequencing speeds up the process, and reduces costs, thereby widening its application. Consequently, many species previously unrecognized have been added to modern inventories, increasing the rate new species are discovered. However, the puzzling concept of species is also what has left the question, of how many species are there on Earth, unanswered.

Species are a controversial concept, which is the conceptual framework for forming hypotheses that represent species (Sandall et al., 2023; Groves and Grunn,2011). Since Darwin's time, there has been no agreed definition of species in biology. Living things exist as species. Species are the cornerstone of life science, the basic taxon of taxonomy, the basic unit of biological evolution, and the basic functional unit in ecosystems, as well as the basic unit of biodiversity. However, from Plato and Aristotle's Type Species Concept, Darwin's Nominalism Concept (Darwin said in his book *Origin of Species*: the term, species, I think that is purely for convenience, any used to represent a group of individuals closely similar to each other), to Population Species Concepts (biological species are populations have morphological and genetic similar individuals). Phenotypical Species Concept is a collection of individual organisms with identifiable biological phenotype, Ecological Species Concept (species are ecological functional units, each species occupies a niche), Temperate Species Concept (new species forms as the phenotypic differences of offspring accumulated over time), Cladistic Species Concept (a species is a lineage of populations between two phylogenetic branch points (or speciation events). Emily L. Sandall et al.(2023) give a definition of species as "a group of organisms that can be considered one taxonomic unit, typically as the lowest taxonomic rank that has an accepted name." They regard species as the smallest unit in the taxonomy, thus further they develop the concept of species.

Mayden(1997) summarized 25 concepts of evolutionary species and found that they overlapped, but many had unique criteria. In the past, taxonomists used morphological features in classification. Later, the numerical taxonomy method and cladistics method appeared. The rapid development of modern molecular biology technology has opened a new window to understand the origin and differentiation of organisms and provided a new means for taxonomy. Apagow et al. (2004) found that 71% more species were found using the Phylogenetic Species Concept than using the non-phylogenetic species concept in seven mammalian groups. In the 91 studies he reviewed, the number of species discovered using the Phylogenetic Species Concept was on average 121% higher than using the non-phylogenetic species concept. With the acceptance and application of the Phylogenetic Species Concept, it is expected that more new species will be reported in future studies.

Taxonomic research is a basic and empirical science with disputes. Different taxonomists of different taxa may hold a different understanding of what constitutes a species, and the degree of differences needed. As for whether a species is valid and whether taxa shall be split or lumped, different people often cannot reach an agreement. Not only do taxonomists of different groups may have different standards for species, but taxonomists of the same group may also have a different opinion on species, for example in Glires (including rodents and lagomorphs) among Chinese mammals. Another controversial taxonomic group is the artiodactylus. There are different views on the classification of Cervicidae and Bovidae (Groves & Grunn, 2011; Wilson & Mittermeiar, 2011; IUCN, 2021). Although taxonomists have emphasized the Biological Species Concept in the past, in practice it is often impracticable to distinguish species by means of reproductive isolation. The mechanism of interspecies isolation is actually isolation in space, and interbreeding often occurs when individuals of two species meet in nature. However, without reproductive isolation to distinguish species, taxonomy relies on the empirical criteria of experts, which is the root of the divergence of estimated species diversity.

Despite the fact that even well-known international conventions such as the Convention on International Trade in Endangered Species of Wild Fauna and Flora (CITES), the Convention on the Conservation of Migratory Species of Wild Animals (CMS), and the IUCN Red List of Threatened Species have different taxonomy systems, people adopt an open-minded attitude, absorbing the viewpoints of different origins when inventorying biodiversity, which is a good practice. In the construction of biodiversity big data, the depth of research on different groups is not the same. Some taxa are well studied, while others are less well studied. Nonetheless, our knowledge is a valuable asset. Scientific knowledge is a collection of expert knowledge in different fields and groups. This is exactly the purpose of this book. We collected the data from literature up to October 2023 to establish the inventory of mammals in China for the subsequent research and conservation of Chinese mammals, of which a large portion are endemic.

The physical geography of China is divided into the Monsoon Zone (which includes the South Asian monsoon and East Asian monsoon zones), Arid Zone and Plateau Zone which differ in sunshine and precipitation. Geographical units like alluvial plain, coastal salt marsh, the loess plateau, karst landforms, hilly mountains, extremely arid deserts, steppe, and the Qinghai-Tibet Plateau permafrost are found across the country. China's geological environment has experienced a series of major geological events like transgression and recession, orogeny, and plateau uplifting. High mountains and deep canyons fold and stretch across the surface of the Earth, creating a three-dimensional climate that provides a wide range of ecological niches for plants and animals, and complex geological history with lasting impacts on modern biodiversity patterns.

Biogeography is the study of the geographical distribution of plants and animals. People began to study biogeographic distribution when they collected biological specimens. The Great Discovery marked nascent biogeography. A. R. Wallace was a pioneer of world zoogeography. In 1876 he proposed the biogeographic regions of the world. A.R. Wallace's theory of biogeography later developed into the geobiological approach to study the distribution of plants and animals and their constraints, as well as the biogeographic approach to study biogeography based on similar species, ecosystems, biomes or land cover. Now, there is a third approach to study biogeography, the genealogical geographic approach.

Human activity has altered the original vegetation and biogeographic patterns across the Earth. We urgently need to understand, and delineate the biodiversity hotspots, geographic distribution of endangered species and the spatial pattern of human activity to develop conservation blueprint, therefore it leads to the emergence of a new branch of Biogeography -- Conservation Biogeography. It uses biogeography concepts, tools and methods to predict the distribution and fate of keystone species, endangered species and ecosystems and to solve real-world biodiversity conservation problems. In this book, the distribution of mammal species provides information for the study of mammal diversity conservation, provides data for planning nature reserves, national parks and identifying natural heritage sites.

The protection of rare and endangered species and natural ecological systems in China could be viewed to have stagnated for a long period of time. Yet since China opened its door to the world again, the protection of rare and endangered species and the natural ecological environment have started again, with nationwide assessments of protected areas and protected area effectiveness. While training zoologists and carrying out field surveys and basic scientific research in zoology, China has formulated Wildlife Protection Law and issued a List of State Key Protected Wild Animals. Protection of wild mammals in China now includes clear legal provisions. China has signed the Convention on International Trade in Endangered Species of Wild Fauna and Flora (CITES), Convention on Biological Diversity, Ramsar Convention, and participated in the Saiga Conservation Protection Protocol of Conservation of Migratory Species, Man and Biosphere Reserve Network. The country takes up the international responsibility of the protection of endangered species, biodiversity and the ecological environment, and in hosting the CBD COP15 demonstrates a commitment to leading in the field of biodiversity conservation.

From 11 to 24 October 2021, the 15th Conference of the Parties to the United Nations Convention on Biological Diversity (CBD COP15) was held in Kunming, China. This was the first time for the UN to hold a global conference on ecological civilization, and the first time for the Conference of the Parties to the Convention on Biological Diversity held in China. The participants shared their experience in biodiversity conservation and ecological civilization building, and discussed plans for biodiversity conservation. The second phase of the 15th Conference of the Parties to the United Nations Convention on Biological Diversity was held in Montreal, Canada, from December 8 to 20, 2022, and adopted the "Kunming-Montreal Global Biodiversity Framework" to draw a new blueprint for global biodiversity governance up to 2030 and beyond. Participants pledged to "take urgent management action to halt the human-caused extinction of known threatened species" by 2030. The document is a new global action plan for biodiversity conservation after the Biodiversity Aichi Goals, which will bring new hope for the future of biodiversity.

Kunming is the capital of Yunnan Province, which has many climatic types, diverse habitats and rich biodiversity, spanning more than 900 kilometers from north to south. Kunming is a natural choice for such a conference, as Yunnan hosts over half of China's known species, with the tropic climate in Xishuangbanna to highland climate in the snow-peaked mountain in Xianggelila, climate types equivalent to the tropic climate on Hainan Island to the highland climate on the Qinghai-Tibetan Plateau are all compressed into the province. Yunnan is known as "the Plant Kingdom, Animal Kingdom, the Garden of the

World ". Accounted for only 4.1% of the country's land area, the province has all terrestrial biomes on Earth except desert biomes. Yunnan Province is also one of the world's top 34 biodiversity hot spots and a key region for global biodiversity conservation.

CBD COP15 was a historic event in biodiversity conservation, and consequently we wrote the book, Mammals of China: Diversity, Distribution and Conservation, to integrate research and conservation of mammals in China. This book is a collection of research findings from researchers and a compilation of our work over the years, including our fieldwork in Yunnan in recent years. This book confirms that China is a global mammal diversity hotspot. We are proud of and grateful to Dr. Don Wilson, the well-known contemporary mammologist of the world for writing the Preface for the book. We thank Academician Ma Jianzhang and Academician Zhang Yaping for their support to us. Due to the short time limit, there may be unavoidable omissions and mistakes, we welcome readers to point out the omissions and mistakes for the revision in future reprints.

Zoology is an integrative science, involving not only dedicated taxonomists and systematists, but citizen scientists. Animal diversity is an integral part of biodiversity. Biodiversity science thus also has characteristics of Citizen Science. In recent years, with the popularization of digital photography technology and the wide application of infrared automatic photography (camera traps) in the field, wildlife images and videos of professional and amateur authors have been widely spread through the Internet, enriching the zoological knowledge of public and inspired a public love of wildlife, as well as providing seasonal detail on wild animal movements. In China, Zoology appeared once again as a Citizen Science. The compilation of this book not only provides a reference book for researchers, but also provides a popular science book for the public. We are looking forward to seeing more and more citizens explore and protect threatened mammals in wild.

Jiang Zhigang
June 8, 2023
"Sino Ocean", Yongding Town, Mentougou
District, Beijing

物种序号

依本书采用的分类系统排列的物种各论序号。

The serial number of the species chapters which is arranged according to the taxonomy system of the book.

物种属性 / Species Character

· 学名 / Scientific name

物种的科学名称，亦称拉丁名。采用拉丁文二名法(Binomial Nomenclature)，由"属名"+"种名"组成。学名随着分类系统变化而变化，本书采用的*Handbook of World Mammals* (Wilson et al.，2007~2019)的分类系统，*Handbook of World Mammals* (Wilson et al.，2007~2019)尚未收录的新种采用命名人制定的名称。

Scientific name of a species, also known as the Latin name, using the Latin Binomial Nomenclature, consisting of "genus name" + "species name". The scientific name changes with the taxonomy system. This book adopts the taxonomy of *Handbook of World Mammals* (Wilson et al., 2007-2019), and for these new species that are not included in the *Handbook of World Mammals* (Wilson et al., 2007-2019). We adopt the name by the species authority.

· 中文名 / Chinese name

种的中文名称。以科学文献，如《中国动物志》采用的中文名称为准，若有关类群的动物志尚未出版，则采用目前权威科学文献中最常用的中文名称。对于那些已经被分列多个种的广布种，若该种分列后，只有一个种分布在中国，则在采用新的学名的同时，继续使用该种的原中文名。例如*Gazella yarkandensis*仍沿用鹅喉羚（蒋志刚等，2017）。

The Chinese name of the species. The Chinese names used in the scientific literature, such as *Fauna Sinica*, shall prevail. If the volume of *Fauna Sinica* of the relevant group has not been published, the most commonly used Chinese names in the scientific literature shall be used. For those widely distributed species that have been split into several species, if only one species is found in China, the original Chinese name of the species will continue to be used along with the new scientific name. For example, "鹅喉羚" is still used as Chinese name for the Goittered Gazelle, *Gazella yarkandensis*(Jiang et al., 2017)

· 英文名 / English name

种的英文名称。本书采用的*Handbook of World Mammals* (Wilson et al.，2007~2019)哺乳动物英文名称。中国研究者近年来发现的新种，则采用命名人使用的英文名称，若尚无英文名称，则参照有关规则定名，或请专家制定英文名称。

The English name of the species. Handbook of *World Mammals* (Wilson et al., 2007-2019) is adopted as the reference of English names of species in this book. For the new species discovered by Chinese researchers in recent years, we use the English name of the authority. If no English names are available, we refer to relevant rules for English name or ask experts to give an English name to the species.

· 亚种 / Subspecies

该种的种下分类单元，亚种是动物命名法规认可的唯一种下阶元(Rosenberg et al.，2012)。亚种在该种分布区内有相对固定的分布区，并与该种其他亚种存在形态差异(Mayr，1982；Monroe，1982)。亚种以拉丁文三名法由"属名"+"种名"+"亚种名"组成。

The subspecies is a taxon of this species which has a relatively fixed distribution area within the distribution range of this species and it is morphologically different from other subspecies of this species (Mayr, 1982; Monroe, 1982). Subspecies are the only such rank recognized in the zoological code (Rosenberg et al., 2012). The subspecies is a Latin trinomial nomenclature consisting of "genus name" + "species name" + "subspecies name".

- **模式标本产地** / Type Locality

种的模式标本发现地点，以标本标签、数据库记录以及科学文献记载为准，保留原文。由于20世纪前，研究者多采用威妥玛拼音标注采集地名。二战后，世界上许多国家的版图和地名发生了变化。许多模式标本产地已经难以查出，作者尽可能查找、核对了原模式标本产地所在的现代国家。

The location of where the type specimens was collected is determined by the specimen labels, database records and scientific literature quote the original text. Researchers used Wade-Giles system to spell the location names in 18th and 19th centuries, furthermore, after World War II, the map of the world changed; it is rather difficult to find out the origin of many type specimens. The authors tried their best to find out and check the modern countries where the original type specimens were collected.

分类地位 / Taxonomy

物种的分类地位包括如下内容（Rosenberg et al.，2012；Vaughan et al.，2013）：

Taxonomy of the species includes the following information (Rosenberg et al., 2012; Vaughan et al., 2013):

- **目** Order

- **科** Family

- **该科建立者及其文献** Family Authority

- **属** Genus

- **该属建立者及其文献** Genus Authority

其他名称 / Other Name(s)

物种的其他名称，包括其他中文、英文和曾用拉丁名。

- **其他中文名** / Other Chinese Name(s)

为了中文名的标准化，避免术语混乱，并考虑中文名在科学文献中的使用频次，本书收入的物种其他中文名一般不超过三个。

In order to standardize Chinese names and to avoid terminology confusion, we consider the frequency of use of those Chinese names, generally not taking more than three other Chinese names for a species included in this book.

- **其他英文名** / Other English Name(s)

科学文献中常用的其他英文名，主要依据美国哺乳动物学会哺乳动物多样性数据库（www.mammaldiversity.org）。

Other commonly used English names of the species in scientific literature cited in the ASM Mammal Diversity Database(www.mammaldiversity.org) are used as other English name of the species.

形态及生境 / Morphology and Habitat

· 照片、形态图 / Photo、Sketch

物种的生态照片或形态图。

An ecological photograph or morphology sketch of the species.

· 形态特征 / Morphological Characteristics

引用文献或根据作者的测量结果对该物种形态特征的描述。

A description of morphological characters of the species, adopted from the cited literature or based on the measurement of the authors.

· 生境 / Habitat

引用文献或根据作者的研究结果对该物种生存的栖息生境类型的划分。

The type of habitat the species inhabits based on the literature or the author's findings.

地理分布 / Geographic Distribution

· 国内分布 / Domestic Distribution

该种在中国的分布情况。

The distribution of the species in China.

· 全球分布 / World Distribution

该种在世界各国家和地区的分布。

The distribution of the species in countries and regions of the world.

· 生物地理界 / Biogeographic Realm

过去动物地理分布采用Darlington（1957）的动物地理区分类：古北界（Palearctic Region）、新北界（Nearctic Region）、东方界（东洋界）（Oriental Region）、埃塞俄比亚界（Ethiopian Region）、新热带界（Neotropical Region）和澳大利亚界（Australian Region）。现在生物地理界的划分依据Udvardy（1975）以及Olson and Dinerstein（1998）的系统，将地球陆地表面划分为8个生物地理界：古北界（Palearctic）、新北界（Nearctic）、非洲热带界（Afrotropic）、新热带界（Neotropic）、澳大拉西亚界（Australasia）、印度马来界（Indomalaya）、大洋洲界（Oceanian）和南极界（Antarctic）。中国横跨古北界与印度马来界。本书根据美国哺乳动物学会哺乳动物多样性数据库以及本书作者的研究结果，确定了每个物种的生物地理界（Biogeographic Realm）。注意到尽管Udvardy（1975）与Olson and Dinerstein（1998）明确了全球生物地理界大致分界，不同动物的专家对各自研究类群的生物地理区划并不相同，有关讨论见概论第2章。

Darlington's (1957) system was adopted for zoogeographic region classification for an extended period, which divided the world zoogeographic region into the Palearctic Region, the Nearctic Region, the Oriental Region, the Ethiopian Region, the Neotropical Region and Australia Australian Region. Now, according to Udvardy (1975) and Olson and Dinerstein (1998), the current biogeographic realm system divides Earth's land surface into 8 biogeographic realms: Palearctic, Nearctic,

Afrotropic, Neotropic, Australasia, Indomalaya, Oceanian, and Antarctic. China straddles the Palearctic and Indomalaya borders. The attribute of the Biogeographic Realm of each species is identified based on the Mammalian Diversity Database of the American Mammalian Society and the authors' research. Note that although Udvardy (1975) and Olson and Dinerstein (1998) define global biogeographic boundaries, the standard of biogeographic realm used by different taxa experts for classifying their respective study taxa are not the same, as discussed in Chapter 2.

· WWF生物群系 / WWF Biome

WWF生物群系根据气候、动物区系和植物区系来区分的地球上的不同野生生物栖息生境类型。WWF将地球陆地表面划分为8个生物地理界，14种主要生物群系，包含867个生态区。我们依据物种的分布区，确定了其所属WWF生物群系及生物地理界。中国涉及的WWF生物群系及其所属生物地理界详见附录3。

WWF Biomes are the various regions of the planet that are distinguished by their climate, fauna and flora. The WWF terrestrial scheme divides the Earth's land surface into eight biogeographic realms, and 14 major biomes types containing 867 ecoregions . According to the species' distribution area, we identified the WWF biome and biogeographic kingdom the species belongs to. For those WWF biomes in China and its biogeographic realms used in the book, see Appendix 3.

· 动物地理分布型 / Zoogeographic Distribution Type

动物地理分布型代码采用扩充的张荣祖（1999）的动物地理分布型代码的研究结果, 详见附录4。Darlington（1957）与Udvardy（1975）、Olson and Dinerstein（1998）的动物地理区分类基本不影响生物地理界以下生物地理分布区划。

Codes for Zoogeographic Distribution Type of Zhang Rongzu (1990) are expanded (Appendix 4) and applied to classify the Zoogeographic Distribution Type of each species by the authors. Basically, realms recognized by Darlington (1957), Udvardy (1975) and Olson and Dinerstein (1998) do not affect the Zoogeographic Distribution Type of each species.

· 分布标注 / Distribution Note

该种是否是中华人民共和国疆域的特有种。

Whether the species is endemic to the People's Republic of China.

· 分布图 / Distribution Map

该种在中华人民共和国分布区地图。海洋哺乳动物的分布图则图示了该种在中华人民共和国周边海域的分布。

以《中国哺乳动物多样性及地理分布》（蒋志刚等，2015）一书和IUCN网站（https://www.iucnredlist.org/）中国物种空间分布图（Extent-of-occurrence distribution maps）以及本书作者的野外考察为基础，通过查阅中国知网（https://chn.oversea.cnki.net/）和谷歌学术（Google Scholar，https://scholar.google.com/）在2015至2020年国内外发表的物种分类分布文献，确定新物种增加分布地，删除现已被认为消失的原分布地，从而获得该种的初步分布图。初步分布图成稿后，然后经过本书主编与编委会专家的审查修改定稿，有些分布图经过多次修改才定稿。

Distribution map of the species in the People's Republic of China. The distribution map of a marine mammal shows the distribution of the species in the saewaters the People's Republic of China.

Based on the distribution maps of the mammals in China (Jiang Zhigang et al., 2015) and the IUCN Red List website (https://www.iucnredlist.org/), and the field survey data of the authors, initial distribution maps of mammal species native to China were drawn. Then consulted the China National Knowledge Infrastructure (CNKI, https://chn.oversea.cnki.net/) and Google Scholar (https://scholar.google.com/) for the literature published from 2015 to 2020, at home and abroad, The initial distribution maps were obtained by ascertaining the addition of new ranges of the species and deleting the original distribution that the species is now thought to have disappeared. After the preliminary distribution map is finalized, it is reviewed and revised by the Editor-in--Chief and the experts of the Editorial Board of the book, and some distribution maps were finalized after several revisions.

濒危状况 / Threatened Status

· 中国生物多样性红色名录等级 / CB RL Category (2021)

中国研究人员依据IUCN受威胁物种红色名录等级标准确定的该种的受威胁等级（蒋志刚等，2021）。部分种在本书成书之前依据新信息确定了濒危等级。

Threatened category of the species assessed by Chinese researchers based on the IUCN Criteria of Red List of Threatened Species (Jiang et al., 2021). New information was used to determine the categories of some species before the book was finalized.

· IUCN红色名录 / IUCN Red List (2021)

2021年，IUCN红色名录网站发布的该种的IUCN受威胁物种红色名录等级。本书成书前，IUCN红色名录网站发布的该种的IUCN受威胁物种红色名录等级有变化者，特别标注年份。

The IUCN Red List of Threatened Species status for this species which is published on the IUCN Red List website in 2021. Prior to the completion of this book, the IUCN Red List of Threatened Species was updated with a special year.

· 威胁因子 / Threats

威胁物种生存的主要威胁因子。

Major threats to the survival of a species.

法律保护地位 / Legal Protection Status

· 国家重点保护野生动物等级 / Category of National Key Protected Wild Animals (2021)

该种是否列入国家林业和草原局、农业农村部公告（2021年第3号）颁布的《国家重点保护野生动物名录》及其保护级别。

Whether the species is listed on the List of National Key Protected Wild Animal promulgated by the Official Decree of National Forestry and Grassland Administration and the Ministry of Agriculture and Rural Affairs (No. 3 of 2021) and its protection category.

· "三有" 名录 / TWIESSV (2023)

2016年修订的《中华人民共和国野生动物保护法》将"国家保护的有益的或者有重要经济、科学研究价值的陆生野生动物名录"更名为"有重要生态、科学、社会价值的陆生野生动物名录"。简称"三有"名录。2023年6月30日，国家林业和草原局颁布新调整的《有重要生态、科学、社会价值的陆生野生动物名录》，共收录野生动物1924种，其中哺乳类91种。

The Wildlife Protection Law of the People's Republic of China, amended in 2016, changed the name of the *List of Important Beneficial, Economic and Scientific Valuable Terrestrial Wildlife under State Protection* to the *List of Terrestrial Wildlife of Important Ecological, Scientific and Social Values under State Protection*, TWIESSV for short. On June 30, 2023, the National Forestry and Grassland Administration issued the newly adjusted TWIESSV, which included 1,924 species of wild animals, including 91 species of mammals.

· CITES附录等级 / CITES Appendix (2023)

该种是否列入濒危野生动植物种国际贸易公约的附录I、 II或III。

Whether the species is listed in the appendixes of Convention on International Trade in Endangered Species of Wild Fauna and Flora （CITES).

· 迁徙物种公约附录 / CMS Appendix (2020)

该种是否列入保护迁徙野生动物种，亦称之为迁徙物种公约的附录。

Whether the species is listed in the appendixes of Convention on the Conservation of Migratory Species of Wild Animals, also known as the Convention on Migratory Species (CMS).

· 保护行动 / Conservation Action

针对该种采取的保护行动，如建立自然保护区、开展人工保护繁育、放归自然、禁止贸易等。

Conservation actions are taken for protecting this species, including establishing nature reserves, conservation breeding, re-wilding, and prohibition of trade, etc.

参考文献 / References

发现、报道、收录、研究该种的重要科学文献。文献引用格式基于*Forntier in Ecology and Evolution*格式。

Important literature that discovered, reported, inventoried and studied the species. The document citation format is based on the *Forntier in Ecology and Evolution* format.

编目物种收录原则 / Principle for species inventory in the book

本书的物种编目基于如下原则取舍收录物种：

1. 国际权威哺乳动物名录、专著和数据库收录的中国哺乳动物种。

2. 在中国有分布记录的种或标本采集地为中国的种。

3. 20世纪在中国有分布记录的，但近期未发现、可能在中国局部灭绝的种。

4. 尽管在中国可能局部灭绝，但是国家已经在该种原分布区建立保护地、繁育中心、开始重引入的种。

5. 正在被人类驯化，但尚未形成家养品种的哺乳动物。

Species selected for inclusion in this inventory of the book are based on the following principles:
1. Species of China that are included in international authoritative mammal checklists, monographs, and databases.
2. Species or species with their type specimen collection sites recorded in China.
3. Species that have been recorded in China in the 20th century but have not been discovered recently and may be extirpated in China.
4. Although the species may be extirpated in China, breeding centers or protected areas have been established in their original ranges, and reintroduction of the species to its original range has been launched.
5. Mammals that are being domesticated by humans but have not yet formed domestic breeds.

总目录 / General Contents

第1卷目录 / Contents of the First Volume

哺乳动物学：动物学中研究哺乳动物的一个分支学科

哺乳动物学家：拥有世界上最好工作的人！

——美国哺乳动物学会（www.mammalsociety.org）

Mammalogy / a branch of zoology dealing with mammals

Mammalogist / a person with the best job in the world!

——American Society of Mammalogists (www.mammalsociety.org)

第1章 多样性

哺乳动物是地球上适应能力最强的动物类群。从戈壁到深海，从"世界屋脊"的雪山高原到海滨湿地，到处都有哺乳动物的分布。然而，哺乳动物却是地球上出现较晚的高等生物类群，属于温血羊膜类（Endothermic Amniotes）动物。最早的哺乳动物从下孔类（Synapsids）演化而来，在3亿年前的石炭纪（Carboniferous Period）末才出现（Kemp，2005）。早期的哺乳动物是地球上一个占据边缘生态位的生物类群。在恐龙灭绝之后，哺乳动物迅速辐射进化，占据了地球上恐龙腾出的中型与大型生态位，其中一些种类演化成地球陆地上和海洋中体型最大动物（Smith et al.，2010），成为世界上最成功的动物类群之一。

1 Diversity

Mammals are among the most adaptive groups of animals on Earth. Mammals are found everywhere, from the Gobi to the deep sea, from the snow-capped mountains on the "Roof of the World" to coastal wetlands. Mammals are a relatively recent group of higher organisms in evolution, belonging to the endothermic amniotes. The earliest mammals evolved from Synapsids and only appeared at the end of the Carboniferous Period 300 million years ago (Kemp, 2005). Early mammals were a group of organisms that occupied marginal ecological niches on Earth. After the extinction of dinosaurs, mammals evolved rapidly and radially, occupying the medium and large ecological niches that dinosaurs had left behind on Earth. Some of these species evolved into the largest animals on Earth's land and sea (Smith et al., 2010), becoming one of the most successful animal groups in the world.

· 哺乳类多样性

地球陆地与海洋生境中有多少物种，半个世纪以来，生物学家对这一问题进行了研究，不同研究者用不同方法得到的物种估计数不同，估计地球物种总数在50万～1亿。直到今天，生物学家们没有得到一个确切的地球物种数目（蒋志刚，2016）。编目是联合国《生物多样性公约》规定一项生物多样性研究的基础工作。哺乳动物是相对研究得比较透彻的生物类群，然而，随着研究的深入，人们记录的世界哺乳动物种数仍在不断增加。1982年，Wilson and Reeder（1982）报道全球有哺乳类20目135科1033属4170种；20世纪90年代，Corbet and Hill（1991）报道全球有哺乳类21目4327种，而Wilson and Reeder（1993）则报道全球有26目4631种现生哺乳动物种（表1.1）。这两个数据的差异主要来自各自分类系统在有袋类分类上的差异。在进化中，许多演化支没有留下后代，很多哺乳类属和物种已经灭绝。McKenna and Bell（1997）发现了4000多个已灭绝的哺乳动物属，大多数已灭绝的哺乳类是在新生代（Cenozoic），在地球上哺乳动物演化史的最后三分之一阶段灭绝。这些已经灭绝的物种是什么样子的，是什么导致了它们的灭亡，目前我们仍知之有限（Rose，2006）。

· Mammalian diversity

How many species exist in terrestrial and marine habitats on Earth? Biologists have studied this question for half a century. Different researchers have come up with different estimates of species using different methods, with estimates ranging from 500,000 to 100 million. Until today, biologists still do not have an exact number of species on Earth (Jiang, 2016). Inventory is the basic work of biodiversity research under the United Nations Convention on Biological Diversity. Mammals are a relatively well-studied group of living things, but the number of mammals worldwide continues to increase as research continues. Wilson and Reeder (1982) reported 4,170 living mammal species in 20 orders, 135 families and 1,033 genera. In the 1990s, Corbet and Hill (1991) reported 4,327 species in 21 orders, while Wilson and Reeder (1993) reported 4,631 living mammal species in 26 orders (Table 1.1). The differences between Corbet and Hill (1991) and Wilson and Reeder (1993) records were due to the differences in the classification of marsupials in their respective classification systems. During evolution, many clades left no offspring, and many mammalian genera and species became extinct. McKenna and Bell (1997) found more than 4,000 extinct mammal genera, most of which became extinct in the Cenozoic, the last third of mammalian evolution on Earth. What did these extinct species look like, what led to their demise, we still know very little (Rose, 2006).

当今社会，交通和通信日益发达，生命支撑系统日臻完善，使得人类能够在较短时间内到达地球上任何地点，例如，我们基本上实现了一日达到全国范围内的任何地点的目标。随着人类考察自然范围的扩大，人们有机会收集原来没有发现的生物种类。环境DNA手段、红外摄影技术、数码摄影技术、无人机技术以及蜂窝无线通信网络的出现，扩大了人类的视野，为在野外自动记录野生动物影像和行为提供了技术手段，揭示了过去鲜为人知的野生动物世界。全球卫星定位系统、地理信息系统又为野生动物的定位、跟踪、空间制图、管理提供了新的技术手段，为野生动物的多样性、地理分布研究与保护管理积累了丰富的素材。

Transportation and communication have become more and more developed, and logistics support systems have been improved day by day, enabling humans to reach any place on the planet within a short time in modern time. For instance, we can reach any place in China in a day. As the scope of human exploration of nature expanded, people had the opportunity to collect species that had not been found before. The advances of environmental DNA, infrared photography, digital photography, drones, and cellular wireless communication networks has expanded our horizons and provided the technology to automatically record wildlife images and their behavior in the wild, revealing a previously unknown wildlife world. Global Positioning System (GPS) and Geographic Information System (GIS) have provided new technical means for locating, tracking, mapping, and managing wild animals, and accumulated rich materials for the diversity, geographical distribution studies and conservation management of wild animals.

随着新的分类学方法的应用，如支序系统学（Schuh，2000）的出现，特别是第二代测序技术的应用，分子系统学（Molecular Phylogenetics，Suárez-Díaz & Anaya-Muñoz，2008）方法使得人们发现了新的物种，提出了新分类系统。通过新种描述或由于原有物种的分裂或合并，世界上的哺乳动物数目总在不断增加，哺乳动物编目不断被刷新。Wilson and Reeder（2005）记录了世界哺乳动物1229属153科29目5416种。国际自然保护联盟（IUCN，2008）完成了一份为期5年的全球动物红色名单评估，其中评估了5488种哺乳动物。Burgin et al.（2018）在《哺乳动物学杂志》上报道目前已知的哺乳动物种类有6495种，其中96种是最近灭绝的。美国哺乳动物学会（American Society of Mammalogists）在Burgin et al.（2018）的基础上，补充了新发现的种，该会哺乳动物数据库（2021）收录27目167科1335属6554种，其中现生野生哺乳动物6447种，灭绝哺乳动物103种，家养哺乳动物19种（表1.1）。

With the application of new taxonomic methods, such as cladistic systematics (Schuh, 2000) the application of the second-generation sequencing technology, Molecular Phylogenetics (Suarez-Diaz & Anaya-Munoz, 2008) method, new species have been discovered from *species nova* description or specie splits/lumping and new taxonomic systems have been proposed. Generally, the number of mammal species in the world is increasing; the inventory of mammals is constantly being updated. Wilson and Reeder (2005) classified the world's mammals into 1,229 genera, 153 families, 29 orders and 5,416 species. The International Union for Conservation of Nature (IUCN, 2008) completed a five-year global Red List assessment of animals, which assessed 5,488 mammal species. Burgin et al. (2018) reported in the Journal of Mammalogy that there are currently 6,495 known mammal species, of which 96 are recently extinct. In addition to Burgin et al. (2018), the American Society of Mammalogists (2021) included 6,554 species in 1,335 genera, 167 families, and 27 orders, including 6,447 living wild mammal species, 103 recent extinct mammal species, and 19 domestic mammals (Table 1.1).

除了单孔目鸭嘴兽科和针鼹科3属5种哺乳类为卵生外，6490种哺乳动物是胎生的，其中有91属379种有袋类（Burgin et al.，2018）。胎盘类哺乳动物是胎生哺乳类中物种最多的一类，占哺乳类的绝大多数，共1220属，6111种。哺乳动物中，啮齿目、翼手目、食虫目、灵长目、鲸偶蹄目、食肉目是物种数目最丰富的6个目，都属于胎盘类。啮齿目包括仓鼠、豪猪、河狸、水豚，翼手目包括蝙蝠和果蝠，灵长目包括长臂猿、猴和狐猴，鲸偶蹄目包括鲸和偶蹄类，食肉目动物包括猫、狗、鼬、熊、海豹和水獭。食虫目包括鼩鼱、鼹鼠。Beck et al.（2006）的研究更新了人们对原来的食虫目Insectivora的认识，将Family Erinaceidae、Family Soricidae、Family Talpidae、Family Solenodontidae和Family Nesophontidae（已经灭绝的西印度洋群岛鼩鼱）归入劳亚食虫目Eulipotyphla。Beck et al.（2006）的研究还表明劳亚食虫目是单系的。

In addition to the five species of mammals belonging to three genera of monotreme (Ornithorhynchidae (platypus) and Echidnidae), which are oviparous, 6,490 species of mammals are viviparous, among which there are 379 species of marsupials belonging to 91 genera (Burgin et al., 2018). Placental mammals are the largest group of viviparous mammals, accounting for most mammals, with 6,111 species in 1,220 genera. Among mammals, Rodentia, Chiroptera, Insectivora, Primata, Cetartiodactyla, and Carnivora are the six orders with the most abundant species; all belonging to the placenta. Order Rodentia includes hamsters, porcupines, beavers, and capybaras. Order Chiroptera includes bats and fruit bats. Order Primate includes gibbons, monkeys, and lemurs. Order Cetartiodactyla includes whales and even-toed ungulates and Order Carnivora includes cats, dogs, ferrets, bears, seals, and otters. Order Insectivora includes shrews, and moles. Beck et al.

(2006) updated the understanding of the original Order Insectivora, added Family Erinaceidae, Family Soricidae, Family Talpidae and Family Solenodontidae and Family Nesnophontidae (extinct West Indian Shrews) into Order Eulipotyphla, and they demonstrated that Order Eulipotyphla is monophyletic.

表1.1 从1982年至今，世界哺乳动物多样性的变化*

Table 1.1 Changes in world mammal diversity from 1982 to the present

年份 Year	目数 Orders	科数 Families	属数 Genera	现生种数 Extant Species
1982	20	135	1,033	4,170
1993	26	132	1,135	4,631
2005	29	153	1,230	5,341
2017	27	159	1,267	5,475
2018	27	167	1,314	6,399
2021	27	167	1,343	6,447
2022	27	167	1,347	6,495

* 数据来源：Mammal Species of the World (Wilson and Reeder,1982, 1995, 2005), the IUCN Red List of Threatened Species (2017), and the Mammal Diversity Database (America Society of Mammalogists, 2018, 2021, 2022)

· 对人类进化的意义

人类脱胎于哺乳动物。现代人类离不开哺乳类。人类在新石器时代驯化了家畜，使得人类社会实现了从采集狩猎文明到农耕文明的转化。家畜生产方式从早期游牧业过渡到定居圈养。驯养哺乳动物在新石器时代农业和文明发展中的重要性早已经被人们认识（Darwin，1868）。尽管人类仅仅驯化了极少数哺乳动物种，哺乳动物的驯养使得世界各地的农业文明取代了狩猎-采集文明。动物驯养反映了人与动物的相互关系及其对动物饲养和繁殖产生影响（Zeder，2015）。通过人工选择，驯化哺乳动物产生了数以千计的品种，也培育了新的动物，如Bennett et al.（2022）在叙利亚阿勒颇郊外的Umm el-Marra，一处有着4500年历史的墓地里发现了罕见的、可能是已知的人类创造的第一个杂交动物——家驴与叙利亚野驴杂交的"kunga"的骨骼（图1.1）。以驯养哺乳动物为基础的新型农业经济导致了人类社会重组，全球生物多样性改变，以及地球地貌和大气层重大变化（Zeder，2008），使得人类在历史上迈出的重要一步（Larson & Burger，2013）。

随着人类社会生产力的提高，人类社会的组织化程度进一步提高，组织层次增加，人类合作的群体增大，人类改造自然的能力增强。在这一过程中，哺乳动物一直是人类生活资料、生产资料和文化艺术源泉。家养哺乳动物为人类提供肉类、奶制品、毛皮。猫狗是人们的传统伴侣动物，现在，许多其他哺乳动物，如豚鼠、仓鼠、兔，甚至猪，已经成为人类新的伴侣动物。马、牛和骆驼参与人类的体育运动。实验动物并被用作科学研究、医药实验的模式生物。工业革命前，家畜还曾经是人类的重要运输、牵引和耕作的动力。

• Implications to human evolution

Humans originated from mammals, and modern humans cannot survive without mammals. In the Neolithic Age, human beings domesticated wild animals, which led to the transformation of human society from a gatherer and hunter-gatherer civilization to an agricultural civilization, and changed the living style of people from nomadism to settlement. Livestock production mode changed from early nomadic to captive livestock breeding. The importance of domesticated mammals in the development of neolithic agriculture and civilization has long been recognized (Darwin, 1868). Though human beings only domesticated several species of mammals. New animals were also bred, such as Bennett et al. (2022) reported the skeleton of "Kunga", possibly the first known hybrid animal created by humans, a cross between a domestic donkey and a Syrian wild donkey, was found in a 4,500-year-old cemetery in Umm El-Marra, outside Aleppo, Syria (Figure 1.1). The domestication of mammals replaced hunter-gatherer civilizations around the world with agricultural civilizations. The new agricultural economy, based on domesticating mammals, led to a radical restructuring of human society, changes in global biodiversity, and major changes in the Earth's landscape and atmosphere (Zeder, 2008). Through artificial selection, human beings successfully bred thousands of breeds of domestic animals. It was an important stage in the history of mankind.

With improvement of productivity, the organizational level of human society is further improved, the scale of human cooperation is greatly enlarged, and the power of humans to transform nature is enhanced. During the process of human evolution, mammals are

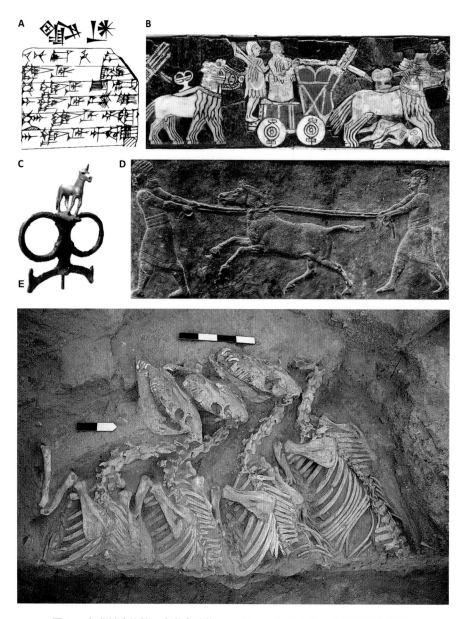

图1.1 人类创造的第一个杂交动物——"kunga"的图像、文字描述和骨骼

Figure 1.1 Painting, text description and skeleton of "Kunga", the first hybrid animal species created by man

（A)在UrIII Girsu/Lagaš（大英博物馆BM23836）泥板上刻有公元前3000年的kunga楔形符号(ANŠE.BARxAN)，出现了多次，并在并列的图画中突出。（B）《乌尔旗》的细节展示了一支在战斗中拉着四轮马车的马队（图片来源：大英博物馆图片）。（C）乌尔皇家墓穴中的缰绳环和装饰马驹的图像。（D）尼尼微刻板："猎野驴"（公元前645－635年）（大英博物馆，伦敦）。照片来源：E. Andrew Bennett。Photo credit：E. Andrew Bennett。（E）叙利亚阿勒颇郊外的Umm el-Marra，一处有着4500年历史的墓地里发现了罕见已知的人类创造的第一个杂交动物的骨骼。Photo credit：Glenn Schwartz。Science Advances CC BY-NC license.

（A）Third millennium BCE cuneiform signs for the kunga (ANŠE.BARxAN) above a photo and drawing of a clay tablet from UrIII Girsu/Lagaš (British Museum BM23836). (B) Detail from the Standard of Ur shows an equid team pulling a four-wheeled wagon in battle (photo credit: The British Museum Images). (C) Image of a rein ring with decorative equid from a royal grave at Ur, contemporary and similar to those visible in the Standard of Ur. (D) Nineveh panel: "hunting wild asses" (645 to 635 BCE) (British Museum, London). Photo credit: E. Andrew Bennett. (E) Rare donkey bones were buried in a 4500-year-old cemetery at Umm el-Marra outside Aleppo, Syria. Photo credit: Glenn Schwartz. Science Advances CC BY-NC license.

indispensable to human life, production, culture, and art. Domestic mammals provide meat, dairy products, and fur for humans. Dogs and cats are traditional companion animals. Now many other mammals such as guinea pigs, hamsters, rabbits and even pigs have become our new companion animals. Horses, cattle, and camels participate in human sports. Laboratory animals are used as model organisms for scientific research and medical experiments. Before the Industry Revolution, livestock used to be an important transport, pulling and farming power before the Industrial Revolution.

全球人类饲养的家养哺乳动物品种数为5769个，其中，中国有533个家养哺乳动物品种，占全球家养哺乳动物品种的9%。全球驯养哺乳动物占世界人类饲养的肉食牲畜的很大部分。据联合国粮农组织（http://www.fao.org/faostat，2021）的统计资料，2019年全世界饲养了15亿头牛、12亿只羊、10.9亿只山羊、8.5亿头猪、2亿头水牛、29亿只兔子、5900万匹马、5000万头驴、3700万峰双峰驼以及近2000万只啮齿动物（非实验用的水豚等啮齿类，在南美用于肉食）。除了肉用、乳用之外，哺乳动物的皮毛可用于制鞋、服装和室内装潢，绵羊、山羊和羊驼等哺乳动物的绒毛一直被人类用来做衣服和保暖材料。

There are 5,769 domestic mammal breeds raised by humans in the world, of which 533 are in China, accounting for 9% of the global domestic mammal breeds. Global domestic mammals comprise a large portion of the world's meat industry. According to the newest statistics of the Food and Agriculture Organization of the United Nations (http://www.fao.org/faostat, 2021), in 2019, the world raised 1.5 billion cattle, 1.2 billion sheep, 1.09 billion goats and 850 million pigs, 200 million water buffalo, 2.9 billion rabbit, 59 million horses, 50 million donkeys, 37 million humped Bactrian camels and nearly 20 million rodents (non-laboratory rodents like capybara which is farmed for meat). In addition to meat and milk, mammal skins are used in making shoes, clothing and for interior decoration. The fur or wool of mammals such as sheep, goats and alpacas has long been used by humans for clothing.

最初的家畜由野生哺乳动物通过人工选育驯养而来（Irving-Pease et al., 2018）。狗 *Canis familiaris* 是最早被驯化的动物。狗曾被认为是灰狼 *Canis lupus* 的一个亚种 *C. l. familiaris*。狗是由狩猎采集者驯养的，最早的证据可追溯到伊拉克一个洞穴中发现的1.2万年前的一块狗下颚骨。最近，意大利佛罗伦萨大学的Bartolini-Lucenti et al.（2021）分析了一只177万～176万年前欧洲大型犬的遗骸，早于猎犬在更新世卡拉布里期（180万至80万年前）从亚洲向欧洲和非洲的广泛迁移。随着人类社会的演化，狗的职责已经从狩猎扩展到放牧、保护、拉车、充当人类的伙伴。人们还在不断地选育新品种，美国养犬俱乐部（https://www.akc.org/）目前在其网站列出了全球175种犬种，其中大多数犬种只有几百年的驯养历史，是人类近期选育培养的。Larson and Burger（2013）回顾了最近的理论进展，提出了一个家养动物驯化起源过程的观点，强调过去10000年中种群混合在影响家养动物基因组中发挥了关键作用。

The first domestic animals were tamed from wild mammals through artificial selection. Dogs *(Canis familiaris)* were the earliest domesticated animals. Dogs were once thought to be a subspecies *(Canis lupus familiaris)* of the Gray Wolf *(Canis lupus)*. The earliest dogs were domesticated by hunters and gatherers, with the earliest evidence dating back to a 12,000-year-old dog jaw found in a cave in Iraq. However, Bartolini-Lucenti et al. (2021) of the University of Florence, Italy, analyzed the remains of a large European dog from 1.77 to 1.76 million years ago, which predated the widespread migration of hounds from Asia to Europe and Africa during the Calabria Period of the Pleistocene (1.8 million to 800000 years ago). As human society changes, modern dogs' duties have expanded from hunting to herding, guarding, and pulling carts, to serve as companions to humans. People still breed new breeds, the American Kennel Club (https://www.akc.org) currently lists 175 breeds of dogs, most of which are only a few hundred years old and recently bred by humans. Larson & Burger (2013) reviewed recent theoretical advances and proposed a view of the origin of domestication in domestic animals, emphasizing that population mixing has played a key role in influencing the genomes of domestic animals over the past 10000 years.

大约1万年前，为获得稳定的食物、皮革和羊毛来源，人们在西南亚驯化了绵羊和山羊（Chessa et al., 2009）。家马 *Equus ferus caballus* 的野生祖先最早出现在16万年前，现在已经灭绝。2012年，基于考古学和遗传学证据，研究人员得出结论，它们的驯化可以追溯到6000年前的欧亚西部哈萨克斯坦草原。

Sheep and goats were domesticated in Southwest Asia about 10,000 years ago to obtain a stable source of food, leather, and wool (Chessa et al., 2009). The wild ancestor of the domestic horse *(Equus ferus caballus)* first appeared 160,000 years ago and is now extinct. In 2012, based on archaeological and genetic evidence, researchers concluded that the domestication of sheep could be traced back 6,000 years ago to the Kazakh steppes of western Eurasia.

无论是在基础生物学研究（如遗传学、生理学或生物化学）中，还是在新药物的开发中，都利用哺乳动物作为科学实验动物。每年有数百万哺乳动物被用于实验，特别是老鼠和大鼠（Council of the European Parliament，2013）。基因敲除鼠是研究人员用人工DNA片段替换或破坏现有基因，使基因失活或"敲除"的实验室老鼠（www.genome.gov/about-genomics）。基因敲除鼠使研究功能未知的基因成为可能（Zeder，2008）。哺乳动物中有一小部分是非人类灵长类动物，因其遗传结构与人类高度相似而被用于研究（European Biomedical Research Association, 1996）。

Some mammals are used as scientific laboratory animals, whether in basic biological research (such as genetics, psychology, and biochemistry) or in developing of new drugs in the pharmacy. Millions of mammals are used in experiments each year, especially mice and rats (Council of the European Parliament, 2013). A knockout mouse is a laboratory mouse in which researchers have inactivated, or "knocked out," an existing gene by replacing it or disrupting it with an artificial piece of DNA (www.genome.gov/about-genomics). Knockout mouse makes it possible to study gene sequences whose function is unknown (Zeder, 2008). A small percentage of mammals are non-human primates, used for research because of their similarity to humans (European Biomedical Research Association, 1996).

从农业起源开始，包括牛和马在内的家畜就被用于工作和运输，随着机械化运输和农业机械的出现，它们的数量不断减少。2004年，它们仍然为第三世界的小型农场提供了80%的电力，为世界运输提供了20%的动力，同样主要是在农村地区。在不适合使用轮式车辆的山区，驮畜继续运输货物（Larson，2014）。

Working livestock, and packing animals, including cattle and horses, have been used for farming and transportation since the origins of agriculture. The numbers of such livestock have declined with the advent of mechanized transport and farm machinery. Nevertheless, livestock still supplied 80 percent of the electricity for small farms in the third world and 20 percent of the power for world transportation, again mainly in rural areas in 2004. In mountainous areas where wheeled vehicles are not suitable, pack animals continue to transport goods (Larson, 2014).

非人类哺乳动物在人类文化中扮演着角色。哺乳动物被广泛用于体育比赛。非人类哺乳动物是受欢迎的宠物，猫狗是最早作为人类伴侣动物的动物，世界各地的家庭饲养了数千万只狗、猫和其他动物，包括兔类和啮齿类。从旧石器时代开始，哺乳动物即是人类艺术的源泉，出现在神话、宗教、文学和电影中。在旧石器时代晚期史前岩画中，猛犸象、狮、犀牛、马和鹿等哺乳动物是最早的艺术题材。在印度尼西亚苏拉威西岛的更新世洞穴中发现了出现在岩画的马和狗（Aubert et al.，2014）。

Non-human mammals play a role in human culture. Mammals such as horses and dogs are widely used in sports. Non-human mammals are the most popular pets. Dogs and cats were the first animals to serve as companions to humans. Tens of millions of dogs, cats, and other animals, including rabbits and rodents, are kept in homes around the world. Since the Paleolithic age, mammals have been the source of human art, appearing in myth, religion, literature, and film. Mammals such as mammoths, lions, rhinoceroses, horses, and deer are among the earliest art subjects in the upper Paleolithic prehistoric cave paintings, e. g. the Pleistocene cave art of animal painting from Sulawesi, Indonesia was reported by Aubert et al. (2014).

在世界许多地方，许多种类的哺乳动物被猎取作为运动狩猎和食物，在北美、欧洲和世界其他地方，鹿类和野猪是受欢迎的狩猎动物（Mayor et al.，2021；Fa et al.，2014）。在非洲，狩猎作为野生动物管理与保护的手段（Baker，1997；Lindsey et al.，2006）。现在野生哺乳动物的价值是生态价值、科学价值和社会价值超越了其传统价值。哺乳动物成为生态系统的构件，哺乳动物是采食者、捕食者、种子扩散传播者。草食哺乳动物通过采食植物，摄取被绿色植物固定的太阳能，草食动物被食肉哺乳动物捕食，这样，太阳能从生态系统的生产者传递到消费者。哺乳动物是自然生态系统食物链中的重要一环，维系着地球生物圈的能流和物质流。一种关键哺乳动物的灭绝可能影响生态系统的结构与功能（蒋志刚，2001）。

In many parts of the world, many kinds of mammals are hunted for sport and food. In North America and Europe and many other places of the world, deer and wild boar are popular game animals (Mayor et al., 2021; Fa et al., 2014). In Africa, hunting is regarded as a means of wildlife management and conservation (Baker, 1997; Lindsey et al., 2006). At present, however, the traditional value of wild mammals is outweighed by their ecological value, scientific value, and social value. Mammals are the building blocks of the ecosystem, and mammals are foragers, predators, and seed dispersers in nature. Herbivorous mammals absorb the solar energy fixed by green plants by grazing or browsing on plants; herbivorous mammals are predated by carnivorous mammals. In this way, solar energy is transferred from producers of the ecosystem to consumers of the ecosystem. Mammals are important links in the food chain of the natural ecosystem, maintaining the energy flow and material flow of the Earth's biosphere. Extinction of a key mammalian species may affect the structure and function of an ecosystem (Jiang, 2001).

· 研究简史

中国哺乳动物区系的形成可以追溯到第三纪末（周明镇，1964）。已知最古老的真兽类化石是1.6亿年前的侏罗纪晚期小鼩鼱状的中华侏罗纪兽 *Juramaia sinensis*，被称为"来自中国的侏罗纪母亲"（Luo et al.，2011，图1.2）。已知最早的后兽亚纲（Metatherian）是在中国辽宁省1.25亿年前早白垩世页岩中发现的中国袋兽（Sinodelphys）（Luo et al.，2007）。由于新生代地球板块运动，古地中海的形成与消失，影响了欧亚大陆动物的进化、扩散和灭绝历程，极大地影响了欧亚大陆动物的多样性格局（Zhao et al.，2022）。由于大地板块运动的原因，中国没有原始的单孔类、有袋类动物。尽管在中国古籍《山海经》中即有动物的描述，然而，囿于人们接触自然的机会，古代学者缺乏接触深山老林野生动物的机会，遑论对野生动物开展科学研究，建立系统认识，动物学没有成为一门科学。中国哺乳动物的发现与定名研究始于19世纪的科学大发现时代，外国传教士、使团成员和探险考察人员在中国采集了大量的植物标本。然而，对中国哺乳动物的系统记录和研究则在20世纪后才开始。关于中国哺乳动物的第一部系统记述是20世纪初Dr. G. M. Allen（1940）完成的*The Mammals of China and Mongolia*一书。在该书中记录了中国（不包括新疆、西藏、台湾和东北）哺乳动物8目30科97属314种。

- The research history

The formation of mammalian fauna in China can be traced back to the end of the Tertiary (Zhou, 1964). The oldest known eutherian fossil is the late Jurassic small shrew-like *Juramaia sinensis* of 160 million years ago, known as the "Jurassic mother from China" (Luo et al., 2011, Figure 1.2). The earliest known subclass Metatherian was the Sinodelphys found 125 million years ago in the Early Cretaceous shales of Liaoning Province, China (Luo et al., 2007). Due to the Cenozoic tectonic movement, the formation and disappearance of the closure of the Tethys Sea, affected the evolution, dispersal, and extinction of animals in the Eurasia continent, and greatly affected the diversity pattern of animals in Eurasia (Zhao et al., 2022). There are no primitive monotreme marsupials in China because of the tectonic plate movement. Although there are descriptions of animals in the ancient Chinese book *The Classic of Mountains and Rivers*, however, ancient Chinese scholars lacked chance to watch wild animals in deep mountains and high sea, not to mention to conduct the scientific research and to draw a systematic understanding of wild animals, consequently Zoology had not become a science in ancient China. The discovery and scientific description of mammals in China was not started until the Golden Age of Discovery in the 17th century, when foreign missionaries, members of diplomatic missions and explorers collected a large number of plant and animal specimens from China. However, the systematic recording and study of Chinese mammals was yet to wait until arriving of the 20th century. The first systematic account of China's mammals was the book, *The Mammals of China and Mongolia*, by Dr. G. M. Allen and his colleagues, completed in the early 20th century. In that book, Dr. Allen (1940) described 314 species of mammals in 97 genera, 30 families, and 8 orders recorded in China (excluding Xinjiang, Tibet, Northeast China, and Taiwan).

图1.2 复原的已知最古老的真兽类——中华侏罗纪兽 *Juramaia sinensis*（160万年前），体长70-100mm ［©N Namura］

Figure. 1.2 Restoration of the oldest known eutheria, *Juramaia sinensis* (1.6 million years ago), about 70-100mm in length ［©N Namura］

随后，一批学者从西方留学归国，开始了中国动物区系的研究。而后在中国科学院建立了专门动物学研究机构，一些省份也相继建立了动物学研究机构，高等院校设立了相关专业，组织开展了专项动物学考察和区域动物学考察。逐步建立了中国现代动物学。由于中国哺乳动物学研究起步较晚。迄今《中国动物志》的兽纲编研工作仅完成了《兽纲 第八卷 食肉目》（高耀亭等，1987）、《兽纲 第六卷 啮齿目（下）仓鼠科》（罗泽珣等，2000）和《兽纲 第九卷 鲸目 食肉目 海豹总科 海牛目》（周开亚，2004），目前对中国哺乳动物的分类与编目工作仍在继续。寿振黄（1962）记录了中国哺乳动物12目52科180属405种。夏武平（1963）在《中国经济动物志——兽类》中记载"我国共有兽类13目390余种"。张荣祖（1979）记录了12目44科183属414种。郑昌琳（1986）

报道中国哺乳动物13目51科210属509种，其中包括对少数非原产于我国的引入种，如豚鼠（*Cavia cobava*）、毛丝鼠（*Chinchilla lanigera*）、北美水貂（*Neovison vison*，现更名为*Neogale vison*）和美洲早獭（*Marmota monax*）等。Nowak（1999）在 *Mammals of the World*（《世界哺乳动物》）第6版中记录了中国哺乳动物共12目52科240属560种。王应祥（2003）将中国哺乳动物编目更新为13目55科235属607种。Wilson and Reeder（2005）在 *Mammalian Species of the World* 第3版中记录了中国哺乳动物13目54科245属572种。Smith（2009，2013）报道中国哺乳动物14目53科240属585种，其中陆生哺乳动物556种，海洋哺乳动物29种。

Later, a group of scholars returned from the West and began studying the fauna of China. zoological research institutes have set up at the *Academia Sinica* (Chinese Academy of Sciences) and in some provinces, and zoology relevant majors were established at universities. Special and regional field surveys have been organized since then. Modern zoology is gradually established in the country. Therefore, mammalian research started rather late in China. So far, only the eighth volume of Fauna Sinica, *Carnivora* (Gao et al., 1987), the sixth volume of *Fauna Sinica, Rotonidae* (Part II): *Cricetidae* (Luo et al., 2000), and the ninth volume of *Fauna Sinica, Cetacean Carnivora, Sealozoidae, Hyacinidae* (Zhou, 2004) have been compiled and published. Work on the classification and inventory of mammals in China is still ongoing. Shou Zhenhuang (1962) recorded 405 species, 180 genera, 52 families, and 12 orders of mammals in China. Xia Wuping (1963) recorded that "There were over 390 species of mammals in 13 orders in China". Zhang Rongzu (1979) recorded 414 species, 183 genera, 44 families, and 12 orders. Zheng Changlin (1986) reported 509 species, 210 genera, 51 families, and 13 orders of mammals in China, including a few introduced alien species in the inventory, such as *Cavia cobava, Chinchilla lanigera, Neovison vison* (*Neogale vison*) and *Marmota monax*. Nowak (1999) recorded 560 species of 240 genera, 52 families and 12 orders of Chinese Mammals in the 6th edition of *Mammals of the World*. Wang Yingxiang (2003) updated the number of Chinese mammals to 607 species, 235 genera, 55 families, and 13 orders. Wilson & Reeder (2005) recorded 572 Species, 245 genera, 13 orders, and 54 families *in the Mammalian Species of the World* (3rd edition). Smith (2009, 2013) reported 585 species of 240 genera, 53 families, and 14 orders of mammals in China, including 556 terrestrial mammal species and 29 species of marine mammals.

中国新哺乳动物记录不断增加、新的哺乳动物种不断被发现。动物学是一门地域特色鲜明的科学。近年来，由于新一代哺乳动物学家的成长、野外考察范围的扩大、网络数据库的建设与运行，新技术手段在野外考察中的运用，分子生物学技术在分类鉴定与谱系地理学的广泛应用，人们发现了中国哺乳动物的新种、新记录种。同时，随着人们动物学知识的积累，人们还对哺乳动物分类系统进行了修订，提出了新的哺乳动物分类系统。例如IUCN受威胁物种红色名录将鲸目Cetacea与偶蹄目Artiodactyla合并为鲸偶蹄目Cetartiodactyla（IUCN，2014），因为这两个目的相同进化起源（Graur & Higgins，1994；Gatesy et al.，1999）。

Zoology is a science branch with distinct regional characteristics. In recent years, thanks to the growth of a new generation of mammalian scientists, the expansion of field survey scope, the construction and operation of internet databases, with the application of new techniques in field survey, and the wide application of molecular biology techniques in taxonomy, new species, and new species records of mammals in China have been discovered. At the same time, with the accumulation of zoological knowledge, people also revised the mammal taxonomy and put forward a new one. For example, the IUCN Red List of Threatened Species merged Cetacea and Artiodactyla into Cetartiodactyla (IUCN, 2014), because these two orders share the same evolutionary origin (Graur & Higgins, 1994; Gatesy et al., 1999).

从2008年起，我们收集了正式出版文献中的中国哺乳动物资料。以Wilson & Reeder（2005）*Mammal Species of the World*和王应祥（2003）《中国哺乳动物种与亚种分类名录与分布大全》的分类系统为基础，前者是目前世界公认的哺乳动物分类权威著作，由IUCN、濒危野生动植物种国际贸易公约（Convention on International Trade in Endangered Species of Wild Fauna and Flora，CITES）推荐的世界哺乳动物分类体系，后者是一部集20世纪中国哺乳动物分类和分布之大全的分类著作。此外，还参考了IUCN Red List of Threatened Species（2018），潘清华、王应祥等（2007）的《中国哺乳动物彩色图鉴》，Smith et al.（2009）《中国兽类野外手册》，刘瑞玉（2008）《中国海洋生物名录》，盛和林等（1998）《中国野生哺乳动物》以及张荣祖（1997）的《中国哺乳动物分布》，灵长类分类还参考了夏武平和张荣祖（1995）的分类系统，海兽类参考了周开亚（2004）的分类系统，获得了一份全新的中国哺乳动物编目。最终收集整理了中国（包括台湾地区）所有哺乳动物种类，包括发表的中国新哺乳动物种和新记录种，对中国哺乳动物物种多样性编目进行了全面增补与修订，发表了《中国哺乳动物多样性》，记录了中国哺乳动物12目55科246属673种（蒋志刚等，2015）。

Since 2008, we have collected data on China's mammals in the peer-referred literature. Based on the taxonomic systems of *Mammal Species of the World* (3rd edition)(Wilson & Reeder, 2005) and the *Complete Catalogue and Distribution of Mammal Species and Subspecies in China* (Wang, 2003), the former is the most recognized works on mammal taxonomy in the world, which is recommended by the Convention on International Trade in Endangered Species of Wild Fauna and Flora (CITES), the latter is the first bibliography of the classification and distribution of mammals in China in the 21st century. In addition, we refer to Xia Wuping and Zhang Rongzu (1995) for the taxonomy and distribution of primates in China, and Zhou Kaiya (2004) for marine mammals in China. *IUCN Red List of Threatened*

Species (2018), *Color Guide of Chinese Mammals* by Pan Qinghua, Wang Yingxiang et al. (2007). *Field Manual of Chinese Mammals* by Smith et al. (2009), *Catalogue of Marine Life in China* (Liu, 2008), *Wild Mammals in China* (Sheng et al., 1998), and *Distribution of Mammals in China* (Zhang, 1997) were also referenced. A new inventory of Chinese mammals was compiled then. We also collected the publication about mammal species, including new species records and new species in the country from peer-referred journals. In 2015, we published *China's Mammal Diversity*, which reported 673 species belonging to 246 genera and 55 families were recorded (Jiang et al., 2015).

Wilson等编著的*Handbook of World Mammals*丛书共9卷，从2009年开始出版第1卷食肉目，到2019年出版第9卷翼手目，共收录全球哺乳类5771种。*Handbook of World Mammals*出版周期长达11年。在该丛书编纂期间发现的一些新种未被收录进该丛书中较早出版的卷中。Burgin et al.（2018）报道了世界哺乳动物1314属6496种。Burgin et al.（2018）的哺乳动物分类系统基本与IUCN红色名录采用的哺乳类系统一致。2020版IUCN红色名录工作组估计全球哺乳动物种（除人类外），实际评估了5899种哺乳类的生存状况。依据美国哺乳动物学会全球哺乳动物数据库（2022），截至2022年底，全球共有哺乳动物27目167科1347属6615种；在这些物种中，101种是最近灭绝的种、6514种是现生物种、20种是家养种、6494种是野生种。

The series of *Handbook of World Mammals*, edited by Dr. Wilson et al. consists of nine volumes. The first volume of the book series, *Carnivora*, was published in 2009, and the ninth volume of the book series, *Chiroptera*, was published in 2019, covering a total of 5,771 mammal species worldwide. The duration of publishing of the *Handbook of World Mammals* extends to 11 years, so some new species discovered later than the compilation of early volumes of the series are not included. Burgin et al. (2018) reported that there are 6,496 species of mammals in 1,314 genera in the world. The mammalian taxonomy of Burgin et al. (2018) is basically consistent with the mammalian taxonomy adopted by the IUCN Red List of Threatened Species. The 2020 edition of the IUCN Red List estimates there were 6,495 mammal species (excluding *Homo sapiens*) worldwide, IUCN Red List Working Group assessed the living status of 5,899 mammal species. According to the Global Mammal Database of the American Society of Mammalogists (2021), there were 27 orders, 167 families, 1,347 genera, and 6,615 species of mammals worldwide at the end of 2022; of those species, 101 were recently extinct, 6,514 were living species; 20 were domestic species, 6,494 were living wild species.

- **中国哺乳类新发现**

进入21世纪后，动物学家们仍不断发现新种。以灵长类为例，如Sinha et al.（2005）报道了分布在藏南的猕猴实为一个猕猴新种，命名为藏南猕猴*Macaca munzala*。范朋飞团队发现分布在高黎贡山的白眉长臂猿与东白眉长臂猿的形态与DNA有明显差别，于是，将高黎贡山的东白眉长臂猿命名为一个新种——高黎贡白眉长臂猿（天行长臂猿 *H. tianxing*，Fan et al.，2017）。然而，Wilson and Mittermeier（2012）报道中国藏南有东白眉长臂猿分布。人们还不断报道哺乳动物的新分布种记录。如胡一鸣等（2017）在西藏山南地区考察时记录到原分布在喜马拉雅山脉南麓的叶猴（*Trachypithecus pileatus*）在中国西藏山南地区的新分布，并建议将Shortridge's langur（*T. shortirdgei*）更名为"灰叶猴"，将*T. pileatus*的中文名翻译为"戴帽叶猴"。

- **New discovery in mammal fauna**

Well into the 21st century, zoologists continue to discover new species. Taking primates for example, Sinha et al. (2005) reported that the macaque distributed in southern Tibet was a new species of macaque, and named it *Macaca munzala*. Fan Pengfei's team found differences in morphology and DNA sequences in the *Hoolock* in Gaoligong Mountains. Therefore, the *Hoolock* in Gaoligong Mountains was named a new species–Sky-walking Gibbon (*H. tianxing*, Fan et al., 2017). However, Wilson and Mittermeier (2012) reported *Hoolock leuconedys* marginally distributed in Zangnan Region. Hu Yiming et al. (2017) recorded a new distribution of *Trachypithecus pileatus*, formerly only distributed in the southern foothills of the Himalayas, in the Shannan Region of Tibet during a field expedition. They also suggested changing Shortridge's Langur (*T. shortirdgei*) to "灰叶猴 (Gray Langur)" and translating the Chinese name of *T. pileatus* to "戴帽叶猴 (Hooded Langur)".

这期间，刘少英团队完成的对中国鼠兔科的重新厘定（刘少英等，2017），降级了5个鼠兔种，提升了4个鼠兔亚种为种，增加了5个新种。中国有29种鼠兔分布，北美鼠兔*O. princeps*、斑颈鼠兔*O. collaris*、荷氏鼠兔*O. hoffinanni*、阿富汗鼠兔*O. rufescens*和草原鼠兔*O. pusilla*在中国没有分布。刘少英等（2017）还对田鼠亚科Arvicolini族啮齿动物进行了系统发育分析并发表了2个新种；He et al.（2015）发表了梵净山管鼻蝠*Murina fanjingshanensis*。党飞红等（2017）订正了渡濑氏鼠耳蝠*Myotis rufoniger*，Chen et al.（2017）研究了缺齿鼩属*Chodsigoa*的系统发育并发表了1个新种，Cheng等（2017）通过分子系统学澄清了猪尾鼠属*Typhlomys*的系统发育关系并发表了1个新种，蒋学龙等（2017）重新厘定了壮鼠属*Hadromys*物种的分类问题。刘丽等（2018）基于线粒体基因、形态学和栖息地指标厘清了高原鼢鼠和甘肃鼢鼠的分类。蒋志刚团队先后分别完成了第2次全国陆栖野生动物考察的普氏原羚专项（平晓鸥等，2018）、白肢野牛与爪哇野牛专项（丁晨晨等，2018）、麋鹿专项（蒋志刚等，2017）以及豚鹿专项调查（Ding et al.，2021）。蒋志刚等（2017）补充了以前知之甚少的藏南地区哺乳动物信息，还按照Wilson & Mittermeier（2012）的 *Handbook of World Mammals*, Groves & Grubb（2011）的 *Ungulate Taxonomy* 对中国有蹄类进行了分类，再次更新了中国哺乳动物多样性编目（蒋志刚等，2017）。

During the period, Liu Shaoying et al. (2017) added five new species, downgraded five species of pika, and upgraded 4 subspecies to species of pika when they reclassified pika in China. They reported that there are 29 species of pika in China, of which *Ocurtona princeps*, *O. collaris*, *O. hoffinanni*, *O. rufescens* and *O. pusilla* are not found in the country. Liu et al. (2017) also conducted phylogenetic analysis on rodents of the Subfamily Arvicolini and reported two new species. He et al. (2015) published the Fanjingshan tubular-nosed bat *(Murina fanjingshanensis)*. Dang Feihong et al. (2017) revised *Myotis rufoniger*; Chen et al. (2017) studied the phylogeny of and published a new species of the genus. Cheng et al. (2017) clarified the phylogenetic relationship of *Typhlomys* through molecular phylogeny and published new species. Jiang Xuelong et al. (2017) redefined the classification of *Hadromys* species. Liu et al. (2018) re-classified Plateau Zokor and Gansu Zokor based on mitochondrial DNA, morphology, and habitat indicators. Jiang Zhigang's team completed the fieldwork of the Second National Survey on Wild Animals, such as Przewalski's Gazelle (Ping, et al., 2018), Gaur and Banteng (Ding, et al., 2018), Hog Deer (Ding et al., 2021), and the mouse deer (Jiang et al., 2021). Jiang Zhigang et al. (2017) also added previously unknown information about Mammals in Southern Tibet to the checklist. Taxonomy in *Handbook of World Mammals* (Wilson & Mittermeier, 2012) and *Ungulate Taxonomy* (Groves & Grubb, 2011) were applied to classify ungulates of China. The inventory of mammal diversity in China was updated again (Jiang et al., 2017).

· **中国哺乳类多样性**

蒋志刚等（2017）在《中国哺乳动物多样性（第2版）》一书中将中国哺乳动物编目更新为1总目13目56科248属693种，比《中国哺乳动物多样性》第1版多1总目1目1科3属20种，将鲸目与偶蹄目合并为鲸蹄总目。人们对其中18种中国哺乳动物的分类地位尚存在争议。蒋志刚等（2021）在《中国生物多样性红色名录·哺乳动物卷》再次更新了物种编目，收录了中国藏南的哺乳动物、新记录种，如道氏东京鼠（成市等，2018）等，增加了省级分布新记录

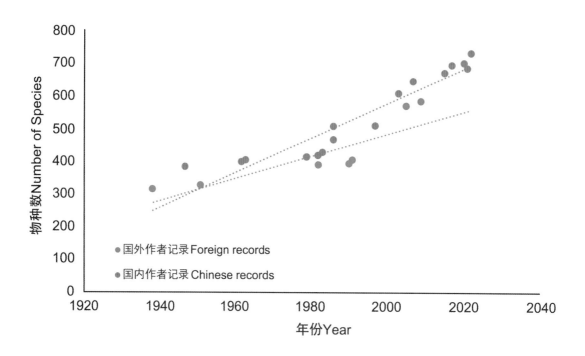

图1.3 中国哺乳动物物种多样性历史记录*

Figure 1.3 Historial records of mammal diversity in China

* 参考文献来源 References: Allen, 1938, 1940; Zheng（郑作新），1947; Elleman & Morrison, 1951; Xia et al.（夏武平等），1962; Zheng（郑作新），1963; Zhang（张荣祖），1979; Zheng（郑作新），1982; Hancki et al., 1982; Sheng et al.（盛和林等），1983; Institute of Zoology, Chinese Academy of Sciences（动物研究所），1986; Zheng（郑昌琳），1986; McNeely et al., 1990; Corbet & Hill, 1991; Wilson & Reeder, 1993; Zhang（张荣祖），1997; Wang（王应祥），2003; Wilson & Reeder（2005）; Pan et al.（潘清华等），2007; Smith et al.（2010）; Jiang et al.（蒋志刚等），2015; Jiang et al.（蒋志刚等），2017; Jiang et al.（蒋志刚等），2020; Wei et al.（魏辅文等），2021; Jiang et al.（蒋志刚等），2022

（程峰等，2018；黄太福等，2018；黄湘元等，2018；李伟东等，2019；罗娟娟等，2019；周智鑫，2019；曹慧等，2020）和考察结果（张璐，范朋飞，2020），更新中国哺乳动物种数为701种（包括智人在内）（图1.3）。中国已经成为世界哺乳动物多样性最丰富的国家之一。其中，中国兔形目46种，占世界兔形目种类的43%；灵长目30种，占33%；鲸偶蹄目113种，占19%；鳞甲目2种，占22%。但是，大陆板块起源的缘故，古冈瓦纳大陆的哺乳类，如原哺乳类亚纲Subclass Prototheria单孔目Order Monotremata、有袋类下纲Infraclass Marsupialia、负鼠目Order Didelphimorphia等7目在中国缺如，除海牛目Order Sirenia外，非洲兽总目Superorder Afrotheria其余各目，贫齿总目Superorder Xenarthra各目亦在中国缺如。

- **Diversity of mammals in China**

Jiang Zhigang et al. (2017) in *China's Mammal Diversity* (2nd Edition) updated the Chinese mammal inventory to one super order, 13 orders, 56 families, 248 genera and 693 species, one super order, 1 family, 3 genera and 20 species more than those in *China's Mammal Diversity* (1st edition). The taxonomic status of 18 Chinese mammals was still in debate at that time. Jiang Zhigang et al. (2021) updated the mammalian species inventory again in the *Red List of Biodiversity in China, Mammal Volume*, and recorded 701 mammal species in China (including *Homo sapiens*, Figure 1.3). China is one of the countries with the world's richest mammal diversity. Among them, 46 species of lagomorphs are recorded in China, accounting for 43% of the world total, 30 species of primates, accounting for 33% of the world total; Cetartiodactyla 113 species, accounting for 19% of the world total. 2 species of Squamata, account for 22% of the world's total (Table 1.1). But because of the origin of the continental plates, the ancient Gondwana mammals, for example, seven orders such as Order Monotremat of Subclass Prototheria, Order Didelphimorphia of Infraclass Marsupialia are missing in the country. In addition, except for the Order Sirenia, other orders of Superorder Afrotheria and Superorder Xenarthra are also absent from the country.

最近，中国兽类学家更新了部分哺乳动物类群的分类与编目。裴枭鑫等（2021）讨论了中国巢鼠属分类与分布。程继龙等（2021）报道了中国跳鼠总科物种的系统分类学研究进展。唐明坤等（2021）报道了中国森林田鼠族系统分类研究进展。刘少英等（2020）报道了中国䶄亚科田鼠族Microtini分类学研究进展与中国已知种类。江廷磊等（2020）报道了中国蝙蝠生物学研究进展及其保护对策。杨光等（2020）报道了中国海兽研究进展。中国研究人员还报道了新种与新记录种，如李飞虹等（2020）报道了安氏白腹鼠的形态分化，修订了其分布范围。刘少英等（2020）报道了中国兽类新记录——白尾高山䶄。陈中正等（2021）报道了侯氏猬的新分布。这些研究成果均在本书中收录。

Recently, Chinese mammalogists have updated the taxonomy and inventory of some mammal groups. Pei Xiaoxin et al. (2021) discussed the classification and distribution of the genus *Nestrat* in China. Cheng Jilong et al. (2021) reported the progress in the systematics of Dipodoidea in China. Tang Mingkun et al. (2021) studied systematic classification of voles in China. Liu Shaoying et al. (2020) reported on the taxonomic studies of Microtini in China and described the known species in China. Jiang Tinglei et al. (2020) reported advances in bat biology and bat conservation strategies in China. Yang Guang et al. (2020) reported the research progress on marine mammals in China. Chinese researchers have also reported new species and new records of species, such as Li Feihong et al. (2020), who reported morphological differentiation and revised distribution range of *P. anseri*. Liu Shaoying et al. (2020) reported a new record of Arvicoline in China. Chen et al. (2021) reported a new distribution of Hou's Hedgehog. The results of these studies are all collected in this book.

本书增加了《中国哺乳动物多样性（第2版）》发表以来的发现的新种，如鼹科Talpidae的*Uropsilus dabieshanensis* (Hu et al., 2021)、*Alpiscaptulus medogensis* (Chen et al., 2020)，鼩鼱科Soricidae的*Crocidura anhuiensis* (Zhang, Zhang, Li, 2020)、*Crocidura donyangjiangensis* (Liu, Chen, Liu, 2020)、*Crocidura huangshanensis* (Yang, Zhang, Li, 2020)；蝙蝠科Vespertilionidae的*Murina jinchui* (Yu et al., 2020)、*Murina liboensis* (Zeng et al., 2018)、*Murina rongjiangensis* (Zeng et al., 2018)、*Kerivoula furva* (Kuo et al., 2017)、*Petaurista mishmiensis* (Krishna et al., 2016)、*Eupetaurus nivamons* (Li et al., 2021)、仓鼠科Cricetidae的*Eothenomys shimianensis* (Liu et al., 2018)、*Eothenomys bialuojishanensis* (Liu et al., 2019)、*Eothenomys meiguensis* (Liu et al., 2019)、*Eothenomys jinyangensis* (Liu et al., 2019)、*Eupetaurus tibetensis* (Jackson et al., 2021)，鼠科Muridae的*Niviventer fengi* (Ge et al., 2020)、*Niviventer gladiusmaculus* (Ge et al., 2018)、*Niviventer mekongis* (Ge et al., 2020)、*Niviventer sacer* (Li et al., 2020)、*Typhlomys huangshanensis* (Hu et al., 2021)，刺山鼠科Platacanthomyidae的*Typhlomys huangshanensis* (Hu et al., 2021)以及跳鼠科Dipodidae的*Chimaerodipus auritus* (Shenbrot et al., 2017)等20多种。其中2021年是中国哺乳动物研究新发现众多的一年。

This book adds new species discovered since the publication of *Mammalian Diversity in China* (2nd Edition), such as *Uropsilus dabieshanensis* of *Talpidae* (Hu et al., 2021), *Alpiscaptulus medogensis* (Chen et al., 2020), *Crocidura anhuiensis* (Zhang, Zhang, Li, 2020) and *Crocidura donyangjiangensis* (Liu, Chen, Liu, 2020), *C. huangshanensis* (Yang, Zhang, Li, 2020); *Murina jinchui* of Vespertilionidae (Yu et al., 2020), *M. rongjiangensis* (Zeng et al., 2018), *Kerivoula furva* (Kuo et al., 2017), *Petaurista mishmiensis* (Krishna et al., 2016), *Eupetaurus nivamons* (Li et al., 2021), *Eothenomys shimianensis* (Liu et al., 2018), *E. bialuojishanensis* (Liu et al., 2019) and *E. meiguensis* in Cricetidae (Liu et al., 2019), *E. jinyangensis* (Liu et al., 2019), *Eupetaurus tibetensis* (Jackson et al., 2021), *Niviventer fengi* (Ge et al., 2020), *N. gladiusmaculus* (Ge et al., 2018), *N. mekongis* (Ge et al., 2018), *N. sacer* (Li et al., 2020) of the family Muridae, *Typhlomys huangshanensis* from Platacanthomyidae (Hu et al., 2021) and *Chimaerodipus auritus* from Dipodidae (Shenbrot et al., 2017), and so on. The year 2021 is a year of numerous discoveries in mammal research in China.

在本书的写作过程中, 在综合研究的基础之上, 我们全面整理了中国哺乳动物编目, 将中国哺乳动物更新为61 科257属740种, 此外, 收录家养哺乳动物13种 (附录1), 外来定殖哺乳动物3种 (附录2), 还有分布与分类存在争议的哺乳动物9种 (附录6)。中国现生哺乳动物属数约占全球哺乳类属数23%; 种数约占全球哺乳类属数13% (表1.2)。

Based on the comprehensive research, we inventoried the mammals in China, and updated the inventory to 61 families, 257 genera, and 740 species (+ *Homo sapiens* + 3 Habituated alien species + 15 Domesticated mammals=758 species). The number of existing mammal genera in China accounts for 23% of the global number of mammal genera. The number of species accounts for 13% of the mammalian genera in the world (Table 1.2).

表1.2 本书的哺乳动物编目与《中国生物多样性红色名录哺乳动物卷》(2021) 的比较

Table 1.2 A comparison of the mammal inventory in this book and the Red List of China's Biodiversity: Mammals (2021)

目 Order	科数 No. Families (2021)	科数 No. Families (2023)	属数 Genera (2021)	属数 Genera (2023)	种数 Species (2021)	种数 Species (2023)	特有种 No. Endemics (2021)	特有率 Enden =mic % (2021)	特有种 No. Endemics (2023)	特有率 Enden =mic % (2023)
劳亚食虫目 EULIPOTYPHLA	3	3	24	26	89	100	32	36%	37	37.9%
攀鼩目 SCANDENTIA	1	1	1	1	1	1	0	0%	0	0.0%
翼手目 CHIROPTERA	7	7	33	35	138	144	29	21%	28	19.4%
灵长目 PRIMATES	3	3	8	9	30	30	8	27%	8	27.6%
鳞甲目 PHOLIDOTA	1	1	1	1	3	2	0	0%	0	0.0%
食肉目 CARNIVORA	10	11	40	41	64	66	2	3%	4	6.1%
鲸目 CETACEA	9	15	26	54	39	103	2	5%	15	14.6%
偶蹄目 ARTIODACTYLA	6		29		64		16	25%		
海牛目 SIRENIA	1	1	1	1	1	1	0	0%	0	0.0%
长鼻目 PROBOSCIDEA	1	1	1	1	1	1	0	0%	0	0.0%
奇蹄目 PERISSODACTYLA	2	2	3	3	6	6	0	0%	0	0.0%
啮齿目 RODENTIA	9	12	78	82	222	247	52	23%	75	30.4%
兔形目 LAGOMORPHA	2	2	3	3	42	41	14	33%	15	36.6%
总计 Total	55	59	248	257	700	740	155	22.1%	182	24.6%

中国哺乳动物在世界哺乳动物区系占有重要的地位（表1.3）。中国哺乳动物种占世界种数10%～30%的科有猴科Cercopithecidae（19种，占世界种数12%）、懒猴科Lorisidae（2种，13%）、兔科Leporidae（12种，18%）、豪猪科Hystricidae（2种，18%）、松鼠科Sciuridae（34种，11%）、鼩鼱科Soricidae（68种，15%）、蹄蝠科Hipposideridae（9种，10%）、菊头蝠科Rhinolophidae（21种，21%）、蝙蝠科Vespertilionidae（96种，19%）、犬科Canidae（8种，21%）、海狮科Otariidae（2种，13%）、海豹科Phocidae（3种，16%）、灵猫科Viverridae（8种，22%）、马科Equidae（3种，25%）、露脊鲸科 Balaenidae（1种，25%）、牛科Bovidae（33种，11%）、骆驼科Camelidae（1种，14%）、鹿科Cervidae（22种，24%）、鼷鹿科Tragulidae（1种，10%）、喙鲸科Ziphiidae（6种，27%）等21个科。相当中国哺乳动物种数约占世界种数的三分之一以上的科有象科Elephantidae（1种，占世界种数33%）、儒艮科Dugongidae（1种，50%）、长臂猿科Hylobatidae（8种，40%）、河狸科Castoridae（1种，50%）、跳鼠科Dipodidae（16种，43%）、鼹型鼠科Spalacidae（15种，54%）、猬科Erinaceidae（9种，38%）、鼹科Talpidae（23种，43%）、假吸血蝠科Megadermatidae（2种，33%）、猫科Felidae（13种，31%）、鼬科Mustelidae（21种，33%）、熊科Ursidae（4种，50%）、犀科Rhinocerotidae（3种，60%）、须鲸科 Balaenopteridae（7种，88%）、牛科Bovidae（33种，11%）、海豚科Delphinidae（16种，40%）、麝科Moschidae（6种，86%）、鼠海豚科Phocoenidae（3种，43%），加上中国特有单种科，中国哺乳动物种数占世界哺乳动物在中国有分布的目的物种总数（5989种）的12%。

China's mammals occupy an important position in the world's mammalian fauna (Table 1.3). There are 19 species of Cercopithecidae (12% of the world total in the family), 2 species of Lorisidae (13%), and 12 species of Leporidae (18%). Number of species in mammal families in China accounts for 10%-30% of the world's mammal species in the family, including Hystricidae (2 species, 27%), Sciuridae (34species, 11%), Soricidae (68 species, 15%), Hipposideridae (9 species, 10%), Rhinolophidae (21 species, 21%), Vespertilionidae (96 species, 19%), Canidae (8 species, 21%), Otariidae (2 species, 13%), Phocidae (3 species, 16%), Viverridae (8 species, 22%) and Equidae (3 species, 25%), Balaenidae (1 species, 25%), Bovidae (34 species, 11%), Camelidae (1 species, 14%), Cervidae (22 species, 24%), and Chevroidae. There were 21 families, including Tragulidae (1 species, 10%) and Ziphiidae (6 species, 27%). The number of mammal species in China which accounts for more than one-third of the world's species includes Elephantidae (1 species, 33%), Dugongidae (1 species, 50%), Hylobatidae (8 species, 40%), Castoridae (1 species, 50%), Dipodidae (16 species, 43%), Spalacidae (15 species, 54%), Erinaceidaeidae (9 species, 38%), Talpidae (23 species, 43%), Megadermatidae (2 species, 33%), Felidae (13 species, 31%), and Mustelidae (21 species, 33%), Ursidae (4 species, 50%), Manidae (3 species, 38%), Rhinocerotidae (3 species, 60%), and Balaenopteridae (7 species, 88%), Bovidae (34 species, 11%), Delphinidae (16 species, 40%), Moschidae (6 species, 86%) and Phocoenidae (3 species, 43%), With the addition of monospecies families endemic to China, the number of mammal species in China accounts for 12% of the world total of those mammal orders distributed in the country.

表1.3 中国与世界哺乳动物种属比较

Table 1.3 Comparison of the mammalian species and genera of China and the world

目 科	世界 World		中国 China(2021)			
	属数 N. Genera	种数 N. Species	属数 N. Genera	世界属数占比 % of World	种数 N. Species	世界种数占比 % of World
哺乳纲 Mammalia	1314	6496	250	19	739	11
长鼻目 Proboscidea	2	3	1	50	1	33
象科 Elephantidae	2	3	1	50	1	33
海牛目 Sirenia	3	5	1	33	1	20
儒艮科 Dugongidae	2	2	1	50	1	50
攀鼩目 Scandentia	4	24	1	25	1	4
树鼩科 Tupaiidae	3	23	1	33	1	4

目 科	世界 World		中国 China(2021)			
	属数 N. Genera	种数 N. Species	属数 N. Genera	世界属数占比 % of World	种数 N. Species	世界种数占比 % of World
灵长目 Primates	84	518	8	10	29	6
猴科 Cercopithecidae	23	160	4	17	19	12
长臂猿科 Hylobatidae	4	20	2	50	8	40
懒猴科 Lorisidae	4	15	1	25	2	13
兔形目 Lagomorpha	13	98	3	23	41	42
兔科 Leporidae	11	67	2	18	12	18
鼠兔科 Ochotonidae	1	30	1	100	29	97
啮齿目 Rodentia	513	2552	85	17	247	10
河狸科 Castoridae	1	2	1	100	1	50
仓鼠科 Cricetidae	145	792	20	14	74	9
跳鼠科 Dipodidae	13	37	8	62	16	43
林跳鼠科 Zopodidae	1	2	1	100	1	50
睡鼠科 Gliridae	9	29	2	22	2	7
豪猪科 Hystricidae	3	11	2	67	2	18
蹶鼠科 Siciatidae	19	1	1	100	5	26
鼠科 Muridae	157	834	21	13	69	8
刺山鼠科 Platacanthomyidae	2	5	1	50	5	100
松鼠科 Sciuridae	62	298	17	27	34	11
鼯鼠科 Pteromyidae	13	34	7	54	23	68
鼹型鼠科 Spalacidae	7	28	4	57	15	54
劳亚食虫目 Eulipotyphlak	54	518	24	44	100	18
猬科 Erinaceidae	10	24	6	60	9	38
鼩鼱科 Soricidae	26	440	11	42	68	14
鼹科 Talpidae	18	54	7	39	23	43
翼手目 Chiroptera	227	1386	38	17	144	10

目 科	世界 World		中国 China(2021)			
	属数 N. Genera	种数 N. Species	属数 N. Genera	世界属数占比 % of World	种数 N. Species	世界种数占比 % of World
鞘尾蝠科 Emballonuridae	14	54	1	7	2	4
蹄蝠科 Hipposideridae	7	88	3	43	9	10
假吸血蝠科 Megadermatidae	5	6	1	20	2	33
犬吻蝠科 Molossidae	19	122	2	11	3	2
狐蝠科 Pteropodidae	45	197	7	16	11	6
菊头蝠科 Rhinolophidae	1	102	1	100	21	21
蝙蝠科 Vespertilionidae	54	493	23	43	96	19
食肉目 Carnivora	130	305	40	31	66	22
小熊猫科 Ailuridae	1	2	1	100	2	100
犬科 Canidae	13	39	4	31	8	21
猫科 Felidae	14	42	7	50	13	31
獴科 Herpestidae	16	36	1	6	3	8
鼬科 Mustelidae	23	64	10	43	21	33
海狮科 Otariidae	7	16	2	29	2	13
海豹科 Phocidae	14	19	3	21	3	16
林狸科 Procyonidae	6	14	1	17	1	7
熊科 Ursidae	5	8	3	60	4	50
灵猫科 Viverridae	14	37	7	50	8	22
大熊猫科 Ailuropodidae[a]	1	1	1	100	1	100
鳞甲目 Pholidota	3	8	1	33	2	25
鲮鲤科 Manidae	3	8	1	33	2	25
奇蹄目 Perissodactyla	8	21	3	38	6	29
马科 Equidae	1	12	1	100	3	25
犀科 Rhinocerotidae	4	5	2	50	3	60
鲸偶蹄目 Cetartiodactyla	132	551	54	41	102	19

目 科	世界 World		中国 China(2021)			
	属数 N. Genera	种数 N. Species	属数 N. Genera	世界属数占比 % of World	种数 N. Species	世界种数占比 % of World
露脊鲸科 Balaenidae	2	4	1	50	1	25
须鲸科 Balaenopteridae	2	8	2	100	7	88
牛科 Bovidae	54	297	13	24	33	11
骆驼科 Camelidae	2	7	1	50	1	14
鹿科 Cervidae	18	93	11	61	22	24
海豚科 Delphinidae	17	40	13	76	16	40
灰鲸科 Eschrichtiidae	1	1	1	100	1	100
白鱀豚科 Lipotidae[b]	1	1	1	100	1	100
麝科 Moschidae	1	7	1	100	6	86
鼠海豚科 Phocoenidae	3	7	1	33	3	43
抹香鲸科 Physeteridae[c]	1	1	1	100	3	
恒河豚科 Platanistidae	1	1	1	100	1	100
猪科 Suidae	6	28	1	17	1	4
鼷鹿科 Tragulidae	3	10	1	33	1	10
喙鲸科 Ziphiidae	6	22	4	67	5	27
总计 Total	1119	5989	259	23	739[d]	12

a 从 Ursidae 分裂 Split from Ursidae
b 从 Iniidae 分裂 Split from Iniidae
c 种数无共识 No consensus on species number
d 不包括智人 Not include *Homo sapiens*

· 中国特有哺乳类

特有现象是一个分类单元仅在一个地理区域分布的现象。特有种是仅在一定地理区域发现的物种，具有代表性生物地理学意义。

由于高山峡谷和江河海洋地理隔离产生的生态位，中国有哺乳动物特有种182种，占中国哺乳动物总数的25%。其中，有举世闻名的特有动物，如大熊猫*Ailuropoda melanoleuca*、川金丝猴*Rhinopithecus roxellanae*、白唇鹿*Przewalskium albirostris*、白鱀豚*Lipotes vexillifer*等。中国哺乳动物中，特有种比例最高的类群是劳亚食虫目和兔形目，分别达36%和34%。中国啮齿目Rodentia特有种比例约占其总数的30%，中国灵长目多样性亦高，其中，27%为中国特有种。此外，扩散能力较弱的、需要特殊生境的，如主要生活在地下生境的鼹科Talpidae的23种中约61%为中国特有种、猬科Erinaceidae的9种中有30%为中国特有种。此外，中国拥有世界上最丰富的偶蹄类，共有63种，栖息于森林、草原、荒漠和高山悬崖生境，其中中国特有种比例为25%。此外，翼手目Chiroptera特有种比例约占其总数的五分之一。

- **China's endemic mammals**

Endemic phenomenon is that one taxon is only distributed in one geographical region. An endemic species is a species found only in a certain geographical area and is the representative of biogeography.

Due to the ecological niche created by the geographical isolation of high mountains and valleys, rivers and oceans, China has 182 endemic mammal species, accounting for 25% of the total mammal population in the country. Among them, there are the world's known unique animals, such as Giant Panda *(Ailuropoda melanoleuca)*, Snub-nosed Monkey *(Rhinopithecus roxellanae)*, White-lipped Deer *(Przewalskium albirostris)*, Baiji *(Lipotes vexillifer)* and so on. Among the mammals in China, the highest proportion of endemic species are found in the Order Eulipotyphla and the Order Lagomorpha, which reached 37% and 34%, respectively. The primate diversity in China is also high, among which 27% are endemic to China. In addition, those with weak dispersal capacity and requiring special habitats, such as Talpidae (23 species, 61% endemic to China) and Erinaceidae (30% of 9 species endemic to China), mainly live in underground habitats. In addition, China has the most abundant even-hoofed ungulates in the world, with a total of 63 species inhabiting forests, grasslands, deserts, and alpine cliff habitats, of which 25% are endemic to China. In addition, about one-fifth of species in Rodentia and Chiroptera are endemic to China.

中国哺乳类特有科有大熊猫科Ailuropodidae与白鱀豚科Lipotidae（Zhou et al.，1984）。白鱀豚科被全球分类学家普遍接受，然而，分类学家们对大熊猫*Ailuropoda melanoleuca*是归入大熊猫科还是熊科Ursidae的意见却不统一（朱靖，1974；胡锦矗，2000；Wilson & Reeder,2005；Pastor et al.，2008）。人们对科的标准尚缺乏统一的认识（Myer,1990）。Hendey（1980a, b）将大熊猫作为Agriotheriinae亚科的唯一种，朱靖（1974）、Thenius（1979）将大熊猫列入大熊猫科Ailuropodidae。O'Brien et al.（1985）发现在现代熊的辐射发生之前，大熊猫的祖先从黑熊进化。随着染色体的重组，大熊猫和熊在染色体和解剖形态上产生明显区别，是分子、染色体和形态进化不一致的一个例子。到20世纪90年代，多数西方学者认可大熊猫隶属于熊科的观点。中国多数学者则认为，仅从一个方面解决大熊猫分类问题失之偏颇，主张从演化特征、时序、空间、生态行为、功能形态、生理和古生物等各生物学分支学科进行分析论证（朱靖，1974）。大熊猫的染色体数目、出齿顺序、白齿结构、头骨结构、掌骨结构、食性等与熊类的都不相同，更重要的是大熊猫的祖先曾与熊类的祖先平行进化。继邱占祥等（1989）研究了云南中新世大熊猫的祖先化石以后，黄万波（1993）研究了大熊猫的颅骨、下颌骨及牙齿，进一步论证了早在中新世晚期，始熊猫（Ailurarctos）与始熊（Ursavus）就开始并行发展，前者发展为现生的大熊猫，后者发展为真熊，即今熊科。多数中国学者从生物学的不同学科和水平，尤其在从古生物学方面的研究结果，坚持把大熊猫独立为一科。《中国动物志·食肉目》（高耀亭，1987）、《中国哺乳动物分布》（张荣祖，1997）、《中国哺乳动物种、亚种与分布大全》（王应祥，2003）均将大熊猫独立为一科。《中国重点保护野生动物名录》也将大熊猫独立成科。张亚平等（1997）提出了熊超科，下列大熊猫科，提出了解决大熊猫分类难题的另一条途径。

Ailuropodidae (Zhu,1974; Thenius,1979) and Lipotidae (Zhou et al.,1984) are two mammal families endemic to China. Lipotidae is widely accepted by zoologists, however, taxonomists disagree on whether the giant panda belongs to the Family Ursidae or the Family Ailuropodidae (Zhu, 1974; Hu, 2000; Wilson & Reeder, 2005). There is no unified understanding of the standard of the family concept in taxonomy (Myer, 1969; 1990). Hendey (1980a, b) listed the giant panda as the only species of the subfamily Agriotheriinae, whereas Zhu (1974), Thenius (1979), and Hu (2000) listed the giant panda in Ailuropodidae. O'Brien et al. (1985) found that the ancestor of the giant panda evolved from the black bear before the occurrence of radiation in modern bears. With the recombination of chromosomes, pandas and bears have obvious differences in chromosome and anatomical morphology, they thought which is an example of molecular, chromosomal, and morphological evolution inconsistency. By the 1990s, most western scholars believed Giant Panda belonged to the bear family. However, most Chinese scholars think that the Giant Panda is a separate family from the bears based on the research results of different disciplines of biology, especially Paleontology. Zhu (1974) advocated it is biased in solving the taxonomy of Giant Panda only based on one aspect instead of these evidences from various biological branches such as ecological behavior, functional morphology, physiology, and paleontology. Qi Zhanjiang et al. (1989) studied the fossils of the giant panda at the Miocene relic site in Yunnan, China in 1993, Huang Wando studied the skull, mandible, and teeth of the Giant Panda, which further demonstrated that Ailurarctos and Ursavus began to develop in parallel as early as the late Miocene. The former developed into the living giant panda, and the latter evolved into the real bear, which is now the Family Ursidae. The chromosome number, dentate sequence, molar structure, skull structure, metacaracteric structure and feeding habits of giant pandas are all different from those of bears. More importantly, the ancestor of giant pandas once evolved in parallel with the ancestor of bears. Therefore, Giant Panda is listed in the Family Ailuropodidae in Fauna Sinica, Carnivora (Gao, 1987), in Distribution of Mammals in China (Zhang, 1997), and in Species, Subspecies and Distribution of Mammals in China (Wang, 2003) and in the List of China's Key Protected Wild Animals (2021). Zhang Yaping et al. (1997) put forward the Superfamily Ursidae over the Family Ailuropodidae, giving another way of solving the dispute.

第2章 分布

哺乳动物分布在一定地域空间。动物分布反映了动物与气候、植被的关系，也反映了一定区域内动物区系的历史演化。哺乳动物是生态系统中的消费者，由地球表面太阳能输入和降水及其所决定植被的空间分布是决定哺乳动物分布的第一性因素。地理隔离、哺乳动物的迁移扩散能力与人类活动，最终决定了现代哺乳动物的地理分布。气候变化、灾害、地质灾害是决定哺乳动物地理分布的随机因素。要研究生物地理，对一个区域生物区系的深入了解是必要的前提。生物地理研究生物地理分布规律与生物分类和生物的地理空间分布特征有关。现代生物分类体系在18世纪由瑞典生物学家林奈建立。人们在采集生物标本的同时，开始了有关生物分布的研究——生物地理学研究。

2 Distribution

Mammals are distributed in certain geographical spaces. The distribution of animals reflects the relationship between animals and climate and vegetation, and reflects the historical evolution of the fauna in a certain region. Mammals are the consumers in the ecosystem. Spatial distribution of vegetation determined by solar energy input and precipitation on the Earth's surface is the primary factor determining the distribution of mammals. Geographical isolation, the ability of mammals to disperse and migrate, and human activities ultimately determine the geographical distribution of modern mammals. Climate change, disasters and geological hazards are random factors that determine the geographical distribution of mammals. A thorough understanding of the flora of an area is a prerequisite for the study of biogeography. The distribution of biogeography is related to the classification of organisms and the geographical spatial distribution of organisms. The modern taxonomy system was established in the 18th century by the Swedish biologist Linnaeus. While collecting specimens, people began to record and study the distribution of organisms – biogeography as the pioneers Charles Darwin and Wallace did.

地球上的生命存在从生物个体、种群、生物群落、生态系统、生物地理区（生物群系，Biome）[1] 到生物圈的空间结构（图2.1）。种群是一定地域空间中分布个体的集合，这些个体之间存在基因交流，是一个基因库，当一个种群存在不能与种群内其他局部种群充分交流基因的局部种群时，这样的种群称之为复合种群（metapopulation）。一定空间中植物种群、动物种群以及微生物种群构成生物群落。生物群落与土壤、水分等非生物环境元素一道构成生态系统。通过绿色植物固定的太阳能，驱动生态系统的能量流和物质循环，使得生态系统成为一个功能整体。人类通过开垦、放牧、人工种植、施肥改造了自然生态系统，改变了生态系统能流的流向与分配，使得自然生态系统演变为自然经济社会生态系统，人类文明进程提升了人类改造自然的能力，改变了地球生物圈。生物群系是植被类型相同的一类生态系统的集合。生物圈包括地球表面所有生态系统，包括生物个体、非

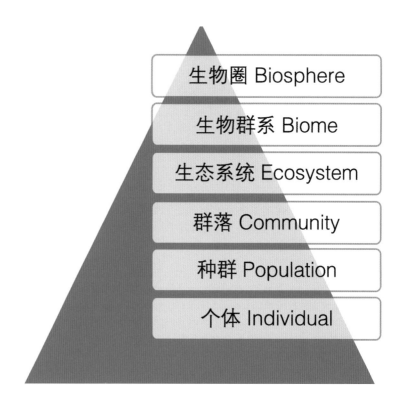

图2.1 地球生命的空间结构

Figure 2.1 The spatial structure of life on Earth

---

[1] Biome 一词在 1916 年由克莱门茨提出的，最初是 Möbius(1877) 的生物群落的同义词 (Clements，1917)。现在 Biome 的定义是基于一个空间地貌、植被、动物，排除了物种组成分类学信息的概念。国际生物学计划 (1964-74) 项目推广了生物群系的概念。Biome 一词亦被用作 "生物群系带" "生物地理省"。

The word Biome was coined by Clements in 1916 and was originally a synonym for the biological community of Möbius (1877) (Clements,1917). The current definition of the Biome is based on a spatial topography, vegetation, and animals, excluding the concept of taxonomic information for species composition. The International Programme of Biology (1964-74) popularized the concept of biome. The word Biome is also used as "Biome" or "Biogeographical Province".

生物环境组分，通过生态系统能流与物质流形成的功能整体。地球上的岩石圈、水圈、大气圈与生物圈构成生态圈（Möller，2010）。

The spatial hierarchical structure of life on Earth from individual organisms, populations, communities, ecosystems, and biogeographic areas (Biome) to biospheres (Figure 2.1). A population is a collection of individuals distributed in a certain geographical space. There are gene exchanges among these individuals, which is a gene pool. When a population has local populations that cannot fully exchange genes with other local populations in the population, such a population is called a metapopulation. (Levins, 1969). In certain space, plant population, animal population and microbial population constitute a biological community. Together with abiotic environmental.

哺乳动物分布受限于所在的生态系统、生物群系的分布范围。生物保护有两个目标：生物区系成员(个体、种群、物种等)的保存和功能生态系统的保存。对两者进行分类是一项生物地理学的任务（Udvarty，1975）。由此出发，研究哺乳动物保护生物地理学。

The distribution of mammals is also limited by the distribution of the ecosystem and biome. Conservation has two goals: the conservation of members of the biota (individuals, populations, species, etc.) and the conservation of functional ecosystems. The classification of the two is a biogeographical task (Udvarty,1975). From this point of view, we study mammal conservation biogeography.

· 生物地理区划

Alfred Russel Wallace是世界动物地理学研究的先驱。1876年，Wallace提出了世界生物地理区由澳大利亚界（Australian Realm）、埃塞俄比亚界（Ethiopian Realm）、新北界（Neoarctic Realm）、古北界（Palearctic Realm）和东方界（Oriental Realm，通常翻译为东洋界）组成。Wallace的生物地理学理论后来沿着两条线发展：前一个分支为地理生物（Geographical Biology，Geobiological）途径，研究植物和动物分布及其制约因素，以及研究种、高等类群或生态系统的空间发生及其制约因素；第二个分支为生物地理（Biological Geography，Biogeographic）途径，基于相似和不同的物种、高等类群、生态系统将地球表面划分为生物地理单元，在生物地理途径中，植被或地表被覆是分类的基础。

· Biogeographical regionalization

Alfred Russel Wallace was the pioneer of world zoogeography. In 1876, Wallace proposed that the world biogeographic regions include the Australian Realm, the Ethiopian Realm, the Neoarctic and the Palearctic Realm, and the Oriental Realm (Wallace, 1876). The biogeography theory proposed by A. R. Wallace later developed along two lines: the first branch is the Geographical Biology (Geobiological) Approach, which studies the distribution through the spatial occurrence of plant and animal species and limiting factors of taxa or ecosystem distributions. The second branch is called the Biological Geography (Biogeological) Approach, which divides the Earth's surface into biogeographic units based on similar and different species, higher groups, and ecosystems. In the biogeographic approach, Earth's surface vegetation or land cover is the basis of classification.

· 植被分类系统

植被指地球表面的植物被覆。地球植被是生态系统的第一性生产者，通过叶绿色固定能量，合成蛋白质、碳水化合物等生物质，为草食动物提供了食物与隐蔽所，太阳能通过肉食动物捕食草食动物进入生态系统顶级消费者，植被构成了动物的栖息地。

依据植被外观对植被分类，地球植被分为森林、灌丛、草原、湿地和沙漠等植被类型；可以按照植被分布纬度，将植被分为热带、亚热带、温带、亚极地和极地植被。可以按照植被分布区的气候类型分为多雨、多云、季节性干旱、干旱类型植被；可以根据植被分布的海拔高度，将植被划分为山地、亚山地、低地和海岸植被；还可以根据植物群落建群种的叶子形态，将植被划分为叶形针叶，阔叶林植被；根据植物群落建群种的周期叶片洞落性质分为落叶、半落叶、常绿色植被、多年生叶。还可以根据生境基底，将植被划分为陆生植被、水生植被、河岸红树林、湿地植被等（郭柯等，2020）。

· Vegetation classification system

Vegetation refers to the cover of plants on Earth's surface. The Earth's vegetation is the primary producer of the ecosystem. It provides food through the energy fixed and biomass such as proteins and carbohydrates synthesized by green plants. Solar energy enters the top consumer of the ecosystem through the carnivore preys on herbivores, and the vegetation also provides shelter to animals and constitutes their habitat.

Vegetation can be classified according to its appearance. Earth's vegetation types can be divided into forest, shrub, grassland, wetland, and desert vegetation. According to the latitude of vegetation distribution, vegetation is divided into tropical, subtropical, temperate, subpolar, and polar vegetation. According to the climate type of vegetation distribution area, vegetation can be divided into rainy, cloudy, seasonal arid and arid types. Vegetation can be further divided into mountain, sub-mountain, lowland, and coastal vegetation according to the altitude of vegetation distribution. The vegetation can also be divided into coniferous vegetation and broad-leaved forest vegetation according to the leaf morphology of the dominant species of a plant community. According to the nature of periodic leaf litter of the constructive species (edificato) of a plant community, it can be divided into deciduous, semi-deciduous, evergreen and perennial leaves. Vegetation can also be divided into terrestrial vegetation, aquatic vegetation, riverbank mangrove and wetland vegetation according to the habitat substrates (Guo et al., 2020).

不同学派提出了不同的植被分类系统，然而目前没有一个植被分类系统被广泛接受和应用，即使联合国教科文组织提出的国际植被分类系统（UNESCO，1973），目前也没有被广泛接受。为了方便学术交流和植被分类学研究，植被生态学界的主流做法是通过比较来寻求分类系统间的对应性（DeCáceres et al.，2015，2018），由此建立起不同植被分类系统的联系，郭柯等（2020）指出目前还很难看到建立起全球统一的植被分类系统的日期。

Different vegetation classification systems have been proposed by different schools in different regions, but no vegetation classification system has been widely accepted and applied, even the International Vegetation Classification Scheme proposed by UNESCO (UNESCO, 1973) has yet to be widely accepted at present. To facilitate academic exchanges and taxonomic studies of vegetation, the mainstream practice in the field of vegetation ecology is to seek interlinks between classification systems through comparison (DeCaceres et al., 2015, 2018), thus establishing the connection between different vegetation classification systems. Guo et al. (2020) point out that it is hard to see a date for establishing global vegetation classification system.

《中国植被》以"植物群落学－生态学"方法为基础进行中国植被分类，最近中国学者根据植物群落外貌和结构、植物群落生态地理特征和植物生态特性、植物群落种类组成以及植物群落动态特征划分中国植被（郭柯等，2020）。郭柯等（2020）提出中国植被分类修订方案沿用的三级主要分类单位是植被型（Vegetation Formation，早期使用Vegetation Type）、群系（Alliance，早期使用Formation）和群丛（Association）。

Vegetation in China is based on the "phytocoenology - ecology" approach to delinear China's vegetation. Recently Chinese vegetation scientists classified China's vegetation according to the appearance and structure of plant community, ecological and geographical characteristics of plant community and plant ecology, species composition of plant community, and the dynamic characteristic of the plant community of China's vegetation (Guo et al., 2020). Guo et al. (2020) proposed the three main classification units used in the revised vegetation classification scheme in China: Vegetation Formation (Vegetation Type was used in the early days) and Alliance (Formation was used in the early days) and Association.

植被由气温、降水所决定。植物学家基于Köppen（1846~1940）提出的气候分类系统与植被分类系统，从生物地理途径研究生物分布（Ashton and Zhu，2020）。植被和动物地理区划是基于地理空间划分的科学问题，两者都涉及空间边界，具有同质性。笔者曾经试图通过植被分类来确定中国哺乳动物地理分布型，现在看来，因为目前没有普遍接受的植被分类系统，这样做仍存在相当的困难。此外，地球表面人工植被已经取代天然植被。稳妥的做法是分别依据地理生物学途径研究中国哺乳动物分布，或依据大尺度植被分类框架建立生物群系内生物地理学途径，来确定中国哺乳动物分布。前者是一种由下而上的途径，如动物学家从模式标本采集地，野外考察确定动物的分布区，由动物的分布由点及面，确定动物的地理分布型；后者是一种由上而下的途径，如植被学家确定了地球的植被带、保护生物地理学家已经确定了地球的生物群系，人们对这些大尺度植被带、生物群系的边界没有或者很少争议，于是，可以根据动物分布点的经纬度来确定其在大尺度植被带、生物群系边界内的分布。也就是说，动物地理分布研究的尺度宜粗不宜细。

Vegetation is determined by temperature and precipitation. Based on the climate and vegetation classification systems proposed by Köppen (1846-1940), botanists study the distribution of organisms from the biogeographical approach (Ashton and Zhu, 2020). Both vegetation and zoogeographic subdivision are based on the scientific problem of geospatial division; thus, the two issues are homogenous because they all deal with geospace boundaries. The authors have attempted to determine the biogeographic distribution patterns of mammals in China through a vegetation classification scheme, but it is still difficult, because there is no universally accepted vegetation classification system, furthermore a large portion of the Earth's surface has been modified and replaced with anthropogenic land cover. A reliable approach is to determine the geographic distribution of mammals in China based on the Geobiological Approach or based on the large-scale vegetation classification and the biome framework (the Biogeographical Approach). The former is a bottom-up approach. Zoologists determine the geographic distribution area of animals from the sites of type specimen collection and field investigation on the species, thus determining the geographical distribution type of animals from linking distribution points to the distribution area of animals. The latter is a top-down approach, vegetation scientists determine the vegetation zone of the Earth, and conservation biogeographers have identified the Earth's biome. Generally, the boundary of these large-scale vegetation zones and biomes are without or a little controversy; therefore, the distribution of animals within the boundaries of large-scale vegetation zones and biomes can be determined by the latitude and longitude of their distribution points. In other words, the scale of the animal geographical distribution study should be coarse rather than fine.

· WWF生物群系与世界陆地生态区域

Udvarty（1975）比较了植物地理学家提出的北方界（Boreal Kingdom）、古北界（Palaeotropical Kingdom）、澳大利亚界（Australian Kingdom）、新北界（Neotropic Kingdom）与动物地理学家提出的生物地理学区划系统间的对应性，为IUCN提出了一个生物地理系统。这一系统被Cox（2001）、WWF（2021）继承。

· The WWF biome and the world's terrestrial ecological regions

Udvarty (1975) compared the Boreal Kingdom, Palaeotropical Kingdom, Australian Kingdom, and Neotropic Kingdom proposed by plant geographers, corresponding to the zoogeographical regionalization system proposed by zoogeographers, and recommended a biogeographical system for the IUCN. This system is inherited by Cox (2001) and WWF (2021).

生物群系（Biome）是以地球上不同气候区植被（更准确地说是地表覆盖，因为冰雪带亦是一个Biome，笔者更倾向于将Biome视为生物地理区）为基础划分的生态地理区域，是一个与生态气候区相关的概念。生物群系是由适应特定气候带的植被和野生动物组成的大型群落。地球上五种主要生物群落类型分别是水生、草地、森林、沙漠和冻土带（www.nationalgeographic.org/encyclopedia/biome/）。由于植被有多种划分方式，生物群系亦有不同的分类方法。人们对生物群系如何更细划分、不同类型生物群系的确切数量争论不休。然而，大的生物群系由适应气候特征植被类型划分，是一类生物气候区，在其中产生了适应特定气候、特定食物、特定栖息地的动物，是一个自组织的自然世界。因为生活在不同生物群系中的植物和动物存在明显的区别。

A biome is an ecogeographic region based on the vegetation (or, more accurately, the land cover, because the snow and ice belt is also a Biome, the authors prefer to regard a biome as a biogeographic region) of different climatic zones on Earth. It is a concept related to the ecoclimatic region. A biome is a large community of vegetation and wildlife adapted to a particular climate zone. Earth's five main biome types are aquatic, grassland, forest, desert, and tundra (www.nationalgeographic.org/encyclopedia/biome/). Because vegetation can be classified in different ways, biome can also be classified in different ways. There is much debate about the principles of finer division of biome and the exact number of different types of biomes on Earth. Large biomes, however, are divided by climate-adapted vegetation types, are a type of bioclimatic zones in which animals adapted to a particular climate, a particular food, and habitat, and are a self-organizing natural world because plants and animals that live in different biomes differ markedly.

生物群系是相对大的陆地或水体单元，包含一个独特自然群落的集合。相同生物群系中具有相似的环境条件，大多数物种相同，物种以及生态组分之间存在相互作用。生物群系对物种的长期存在至关重要（www.wwf.org）。目前使用的生物群系的边界接近于原始生物群系受到重大破坏或改变之前的分布范围。生物群系可视作在"界"或"世界生物地理区"之下的生物地理区域分类阶元。

A biome is a relatively large unit of land space or water space that contains a unique collection of natural communities. In a biome type, with similar environmental conditions, most species are the same, and there are interactions between species and ecological components within the biome. Thus, a biome is critical to the long-term existence of species (www.wwf.org). The boundaries of biome currently used are close to the original distribution range of the biome before it had been significantly damaged or altered. A biome could be taken as a classification element of a biogeographic region under a "Realm(Kingdom)" or "World Biogeographic Region".

生物群系共同的要素是特定气候下的植物生境以及动植物对环境的适应，目前生物群系其他要素也包括生物多样性和人类活动。一个生物群系的所有要素彼此之间存在着某种有意义的关系，生物群系要素中一个要素的变化会直接或间接地影响另一个要素。比如植被的变化，会导致动物栖息地的变化，会导致生物群系的变化。而人类活动则可能强化或弱化这一过程。

The common elements of a biome are the habitat of plants in a particular climate and the adaptation of plants and animals to the environment. Other elements of a biome now include biodiversity and human activities. All the elements of a biome are in some meaningful relationship to each other, and changes in one of the elements of a biome directly or indirectly affect the other. Changes in vegetation, for example, can lead to changes in the habitat of animals, consequently leading to changes in the biome. Human activity can either enforce or weaken this process.

世界自然基金会（WWF）将陆地生态区分为14个不同的生物群系，如森林、草原或沙漠。以生物群系为生物地理单元划分了世界陆地生态区域（Terrestrial Ecoregion of the World，TEOW），确定了地球上825个陆地生态区域和大约450个淡水生态区域（图2.2）。世界陆地生态区描述了每个生态区域的生物多样性特征、现状和种类以及对自然生境的威胁的严重程度，为全球保护战略的生物地理框架（worldbiomes.com）。

The World-Wide Fund for Nature (WWF) divides Earth's terrestrial ecosystems into 14 different biomes, such as forests, grasslands and deserts. Based on biomes, the Terrestrial Ecoregion of the World (TEOW) divides biogeographical units into 825 Terrestrial Ecological Regions and about 450 Freshwater Ecological Regions (Figure 2.2), also giving the characteristics, status, and species of biodiversity in each Ecological Region, as well as the severity of threats to natural habitats, as a biogeographical framework for global conservation strategies (worldbiomes.com).

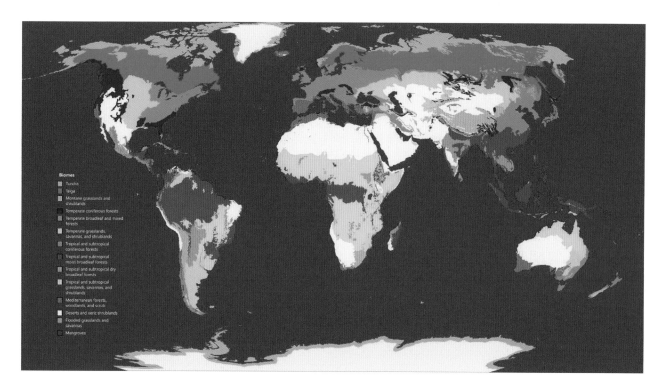

图2.2 基于生物群落的世界陆地生态区域(worldbiomes.com)

Figure 2.2 World Terrestrial Ecological Regions based on biomes (worldbiomes.com)

· 中国哺乳动物分布格局

中国幅宽辽阔，南北跨越近50个纬度，东西跨越60多个经度。按领陆面积排名，中国是世界第三大国。季风带、干旱区和高原地带的温度、日照、降水和地质历史的巨大差异决定了中国脊椎动物的分布格局（Luo et al., 2012）。中国地质环境经历了海进海退、造山、高原隆升等一系列重大地质事件。中国景观多样，海拔和气候差异很大。三江流域的高山和峡谷折叠，拉伸着地表，形成了三维气候和植被分区，为动植物物种创造了生态位，该地区的生物多样性在全国最高。来自印度洋的湿热气流沿雅鲁藏布江大峡谷移动，向北深入云南和西藏，形成局部热带环境。在中国，有冲积平原、海滨盐沼、黄土高原、喀斯特地貌、丘陵山地、极干旱沙漠、多年冻土、内蒙古草原、青藏高原等地理单元。哺乳动物栖息地多种多样，如溶洞、热带雨林、亚热带常绿阔叶林、暖温带针叶林混交林、冷针叶林、沼泽、红树林湿地、草地、草原、高寒草甸、高山苔原和砾石、冰川、沙漠和岩壁，栖息着高度多样化的哺乳动物区系（Jiang et al., 2015）。

· Mammalian distribution patterns in China

With a vast territory spanning nearly 50 latitudes from south to north and more than 60 longitudes from east to west. In terms of territorial areas China is the third largest country in the world. Large differences in temperature, sunshine and precipitation and geological history in the monsoon zone, arid zone and plateau zone shaped the distribution pattern of vertebrates in China (Luo et al., 2012). China's geological environment has experienced a series of major geological events, such as the transgression and regression of sea, mountain building, and the plateau's uplift. There is a diverse landscape and a big difference in altitude and climate. High mountains and canyons in the Three Parallel Rivers area fold and stretch the Earth's surface, creating a three-dimensional climate and vegetation zonation, and creating niches for biological species, both plants and animals, marking the area with the highest biodiversity in the country. Hot and humid air flows from the Indian Ocean moves along the Yarlung Zangbo River Grand Canyon, deep into Yunnan and Tibet northward, forming a local tropical environment. In China, there are alluvial plains, coastal salt marshes, Loess Plateau, karst landforms, hills and mountains, extremely arid deserts, permafrost, Inner Mongolia steppe, Qinghai-Tibet Plateau, and other geographical units. There are a variety of mammal habitats, such as karst caves, tropical rain forests, subtropical evergreen broad-leaved forests, mixed warm temperate coniferous forests, cool coniferous forests, marshes, mangrove wetland, grassland, steppe, alpine meadow, alpine tundra and scree, glacier, desert, and rock cliff habitats, Inhabit a highly diverse mammalian fauna (Jiang et al., 2015).

与高等植物、两栖爬行类、鸟类的物种分布格局相似，中国哺乳动物分布格局与温度与降水有关。例如，Hu et al.（2022）报道，喜马拉雅山脉降水增加驱动着脊椎动物种积累，而温度下降驱动着脊椎动物种减少。第三纪青藏高原隆升这一地质事件打破了中国哺乳动物分布的纬度格局。以50千米×50千米的栅格统计，西南部横断山区和云南南部哺乳动物物种密度是中国最高的，特别是西南地区，哺乳动物密度最高地区物种丰富度可达每2500平方千米164种。云贵高原、秦巴山区、东南沿海及海南、台湾的密度也高（图2.3）。这些地区分布的灵长类、翼手目、食虫目、啮齿目的多样性亦较高。中国北部，特别是青藏高原与南疆地区哺乳动物物种密度低（图2.3）。青藏高原，特别是高原面分布的种类以适应高寒生境的物种为主，哺乳动物产生了属级的分化，如藏羚属*Pantholops*。中国内陆干旱区则生存着耐受干旱、耐受环境温度极度波动的种类，如塔里木兔*Lepus yarkandensis*、郑氏沙鼠*Meriones chengi*等。

Like the distribution patterns of higher plants, amphibians, reptiles and birds, the distribution patterns of mammals in China are related to temperature and precipitation. For instance, Hu et al. (2022) reported that increasing precipitation drives species accumulation, whereas decreasing in temperature drives species decline in Himalayan vertebrates. The uplift of the Tibetan Plateau in the Tertiary broke the latitudinal pattern of mammal distribution in China. Based on a grid system of 50×50 km, the Hengduan Mountains and southern Yunnan Province have the highest mammal species density in China. Especially in southwest China, the region with the highest mammal density has a species richness of 164 species per 2,500 square km. Yunnan-Guizhou Plateau, Qinling-Daba Mountains, southeast coastal areas, Hainan Island and Taiwan Island also have high mammal species density (Figure 2.3). These areas are inhabited by primates. The diversities of bats, insectivores, and rodents are also high in these areas. The mammal species density is low in northern China, especially on the Qinghai-Tibet Plateau and in southern Xinjiang (Figure 2.3). The distribution of mammals in China is extremely uneven. The Tibetan Plateau, especially on the top of the plateau, is dominated by species adapted to the alpine habitat. Mammals have differentiated at the genus level, such as the genus *Pantholops.* Yankand Hare *Lepus yarkandensis* and Cheng's gerbil *Meriones chengi*, which tolerate drought and extreme temperature fluctuations, are found in the arid areas of inland China.

图2.3 中国哺乳动物种密度分布

Figure 2.3 Distribution of mammal species density in China

中国灵长类物种主要分布在中国南部，在横断山脉、喜马拉雅南坡和中国喀斯特地区灵长类物种多样性较高。云贵高原、四川盆地外围山区以及中部地区、华南地区鼠类多样性较高，新疆北部鼠类多样性则较低，其次为东北地区和西北干旱区，青藏高原腹地鼠种多样性最低。偶蹄类物种密度以青藏高原东南部和横断山地区最高，其次为云贵高原、秦巴山区、青藏高原东部和南岭地区，而青藏高原中西部、新疆南部、华北和东北地区密度较低。食肉动物分布密度与偶蹄目分布密度相似，青藏高原东南部、云南西南部和横断山地区食肉物种密度最高，其次是云贵高原、秦巴山、青藏高原东部和南岭、天山以及中国东北部，华北、内蒙古和青藏高原西部食肉动物种密度较低。在青藏高原东部、帕米尔高原和横断山区，兔形目物种多样性较高，在青藏高原北部和西部、中国东北部和中国东部季风带，兔形目物种多样性较低。在中国，蝙蝠种类主要集中在南方喀斯特地区，云南、贵州、广西、重庆、海南、台湾及东南沿海地区种类多样性较高，长江流域以北和青藏高原蝙蝠种类较少（图2.4）。

　　The primate species in China are mainly distributed in the southern part of China, and the primate species diversity is high in the Hengduan Mountains, the south slope of the Himalayas and the karst areas of China. Rodent species diversity is high in Yunnan-Guizhou Plateau, the peripheral mountains and central regions of Sichuan Basin, south China and low in northern Xinjiang, followed by northeast China and the arid region of Northwest China, and the lowest in the hinterland of Qinghai-Tibet Plateau. The species density of even-toed ungulates is highest in the southeast of Qinghai-Tibet Plateau and Hengduan Mountains, followed by Yunnan-Guizhou Plateau, Qinling-Bashan Mountains, eastern Qinghai-Tibet Plateau and Nanling Mountains, but is low in the central and western parts of Qinghai-Tibet Plateau, southern Xinjiang, north China, and northeast China. Carnivore distribution density is similar to Artiodactyla, the density of carnivorous species is highest in the southeast of Qinghai-Tibet Plateau, southwest of Yunnan Province and Hengduan Mountains, followed by the Yunnan-Guizhou Plateau, Qinling-Bashan Mountains, eastern of Qinghai-Tibet Plateau and Nanling Mountains, Tianshan Mountains and northeast China, and is low in North China, Inner Mongolia, and western Qinghai-Tibet Plateau. The lagomorph species diversity is high on the eastern Tibetan Plateau, the Pamir Plateau and in the Hengduan Mountains, and is low on the northern and western Tibetan Plateau, in the northeast of China, and the eastern monsoon zone of China. In China, bat species are concentrated in karst areas in the south, with high species diversity in Yunnan, Guizhou, Guangxi, Chongqing, Hainan, Taiwan, and coastal areas of southeast China, and few bat species in the north of the Yangtze River drainage and the Qinghai-Tibet Plateau (Figure 2.4).

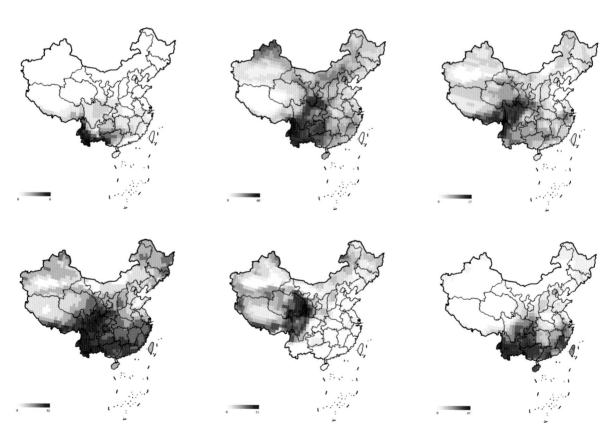

图2.4　哺乳纲灵长目（上左）、啮齿目（上中）、偶蹄类（上右）、食肉目（下左）、兔形目（下中）和翼手目（下右）在中国的物种数量密度分布

Figure 2.4 Species density distribution of Primates (upper left), Rodentia (upper middle), artiodactyls (upper right), Carnivora (lower left), Lagomorph (lower middle) and Chiroptera (lower right) in China

· 对中国哺乳动物地理的再认识

大陆哺乳动物可以向大陆内部或跨越大陆迁移扩散。动物分布受到地理隔离的影响。亚洲动物东亚-欧洲东西同纬度迁移扩散的主要地理屏障是青藏高原和亚洲中部干旱地区。稀薄的空气和高山陡坡沿着喜马拉雅山构成了难以逾越的障碍，形成了一道明显动物地理边界。亚洲中部干旱地区也限制了一些动物的扩散。除了适应北极地区的哺乳动物外，哺乳动物从亚洲到北美必须穿越寒带。第三纪以来，非洲-亚洲的连接是断断续续的，中国动物区系中仅有极少数古北界-非洲界元素。然而，在中国东部的东亚季风带，不存在明显的地理隔离，动物扩散使得古北界-印度马来界（古北界-东方界）在中国地理分界线变得模糊，同时动物地理分界也随着时间的推移、气候变化而改变。人们对中国南北动物地理分界的划分方式不一致。

• A new understanding of mammal geography in China

Continental mammals can migrate either inside or across continents. The distribution of animals is affected by geographical isolation. The Qinghai-Tibet Plateau and arid regions of central Asia are the main geographical barriers to the migration or diffusion of animals in the same latitude from East Asia to Europe. Thin air, high altitude and steep mountain slopes form an impenetrable barrier along the Himalayas, forming a clear zoogeographical boundary. Arid areas in central Asia also limit the spread of some animals in moist eastern China. Except for mammals adapted to cold climates, mammals must cross the cold zone to get from Asia to North America. Since the Tertiary, the Afro-Asian connection has been intermittent, and there are very few Palearctic - African elements in China's mammal faunas. However, in the East Asian Monsoon Zone of eastern China, there is no obvious geographical isolation, and the geographical boundary between Palearctic and Indomalaya (Palearctic and Oriental) in China is blurred. Furthermore, the zoogeographical boundary may change over time as the climate changes. It is difficult for people to agree with each other on the way of dividing the zoogeographical boundary between the north and south of China

人们对古北界与东方界（或中国-日本界）（Holt et al., 2013）、古北界与印度马来界（Udvardy，1975）在中国分界线的认识不同（图2.5）。有的依据地理生物途径的研究（Zhang 1999; Liu and Lu，2019），如Hoffmann（2001）试图根据物种分布范围和系统发生或生态渊源关系确定古北界在中国南界，他认为古北界和东方界之间存在过渡区，过渡区东段在中国东部地区中心约在30°N，南缘约在28°N，北缘约在33°N，过渡区在山地的海拔高度在1500～2500米，过渡区中心的海拔高度为2000米，西段的印度河作为东洋界和古北界两界的过渡区比作为东方界向西的延伸更为合适。植物地理学家的古北界与印度马来界分界是热带雨林在中国的分布区（Ashton and Zhu，2020），比Udvardy（1975）的生态地理区划中古北界的南缘更向南移。Olsen（1998）、WWF（2021）在WWF Biome-Ecoregion 的系统中（图2.2），对Udvarty（1975）系统的中国南北动物地理区划分界作了微调，将分界线北推到南岭一线（图2.5）。

People have different opinions on the boundary between Palearctic and Oriental Realm (or Sino-Japanese Realm) (Holt et al., 2013), or the Palearctic and IndoMalaya Realm (Udvardy, 1975) in China (Figure 2.5). There are studies based on Geobiological Approach (Zhang, 1999; Liu and Lu, 2019). For example, Hoffmann (2001) attempted to determine the southern boundary of the Palearctic Realm in China based on the distribution range and phylogenetic or ecological origin of species. He believed that there was a transition zone between the Palearctic and the Oriental realms. The center of the transition zone is about 30°N in eastern China, 28°N is the southern margin and 33°N is the northern margin. The altitude of the transition zone is between 1500 and 2500m in the mountains in central China, and the altitude of the center of the transition zone is 2,000 m. He also claimed that the Indus Region is more suitable as a transitional zone between the Oriental and Palearctic boundaries in the western section than a westward extension of the Oriental. According to vegetation ecologists, the northern edge of tropical rain forests is the boundary between the Palearctic and Indomalaya in China (Ashton and Zhu, 2020), which is more southerly than the southern margin of Palearctic in Udvardy's eco-geographical division (1975). Olsen (1998) and WWF (2021) in the WWF Biome-Ecoregion system (Figure 2.2) fine-tuned the Udvarty (1975) system for the delineation of zoogeographic divisions of north and south China. push the boundary north to the line north of the Nanling Mountains (Figure 2.5).

依据不同动物类群划分的动物地理区也可能不一致。一个例子是华莱士过渡带，印度尼西亚和马来西亚的脊椎动物地理分布划分问题。主要基于鸟类、昆虫和一些哺乳动物的分布绘制的华莱士线（Wallace's Line）与以哺乳动物分布为基础的莱德克线（Lydekker's Line）（Gunnell，2013）不同。由于中国东部东亚季风带的存在，古北界可能在中国东部存在辽阔的过渡带（图2.5）。Follows（2006）以"中国南方的蚂蚁属：对东方界-古北界界线的观察"为题，研究了东方界-古北界在中国的界线，确定了古北界和东方界蚂蚁属在中国的分布范围。他发现中国东部地区的67个东方蚂蚁属中，34个属（51%）的分布中位点达不到25°N。11个古北属中的6个（55%）的分布中位点向南不超过31°N，蚂蚁古北属分布的南界接近Udvardy（1975）的印度马来界，蚂蚁东方属分布的北界接近Smith（1983）的界线，这一研究是南方动物与北方动物在中国季风区存在一个宽阔过渡带的佐证。Holt et al.（2013）则利用动物系统发生和地理分布数据，在中国的古北界与东方界之间增加了中国-日本界。

Zoogeographic regions may also differ according to the studies in different animal groups. One example is the Wallace Transition Zone, the geographic division of vertebrate distribution in Indonesia and Malaysia. Wallace's Line is mainly drawn based on the distribution of birds, insects, and some mammals, which is different from Lydekker's Line based on the distribution of mammals (Gunnell, 2013). The Palearctic boundary may have a vast transition zone in eastern China due to the Eastern Asia Monsoon Belt (Figure 2.5). John Follows (2006) studied ant (Hymenoptera: Formicidae) genera in Southern China to assess the Oriental-Palearctic boundary. In determining the distribution range of Palearctic and Oriental ants in China, he found that 34 (51%) of the 67 genera of Oriental ants in eastern China are live in locations south of 25°N in their range. Six of the 11 Palearctic genera (55%) have locations south of 31°N. The southern boundary of the Palearctic genus is close to Udvardy's (1975) Indomalaya boundary, and the northern boundary of the Oriental genus is close to Smith's (1983) boundary. This study is evidence of a wide transition zone between southern and northern animals in China's monsoon region. Holt et al. (2013) used animals' phylogenetic and geographic distribution data and added a Sino-Japanese Realm between the Palearctic Realm and the Oriental Realm in China.

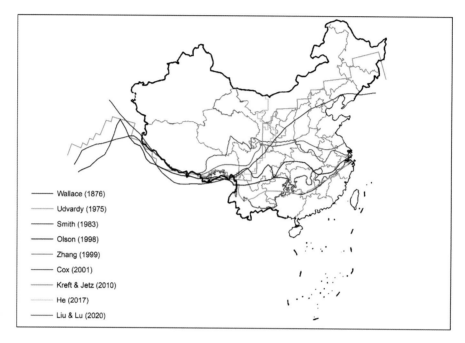

图2.5 各家对中国哺乳动物地理南北分界线的划分

Figure 2.5 The division of the north-south division of mammal geography in China by different schools

· 中国哺乳动物地理研究——地理生物途径

关于中国动物地理的划分存在不同的观点，如基于地理生物途径的传统动物地理的古北界与东方界（东洋界）分界与基于生物群系的生物地理途径的划分，两者显然不同。关于中国哺乳动物地理的研究始于郑作新、张荣祖（1956）。他们基于Wallace（1876）的动物地理系统，将中国动物地理划分为2界7区16亚区。后来，张荣祖（2011）将中国哺乳动物地理划分为2界7区19亚区。中国学者（郑作新和张荣祖，1956）认为，古北界与东方界（东洋界）在中国的分界以秦岭淮河为界，在中国西部以横断山的西部界线为界。刘嘉恒和路纪琪（2019）基于《中国哺乳动物多样性（第2版）》（蒋志刚等，2017）、《中国哺乳动物多样性》（蒋志刚等，2015）、《中国哺乳动物多样性及地理分布》（蒋志刚等，2015）以及《中国哺乳动物分布》（张荣祖，1997）的数据，进行多元相似性聚类分析，将中国的哺乳动物地理划分为2界7区10亚区。他们建议世界动物地理区划中古北界与东方界（东洋界）在中国境内的分界线为南迦巴瓦峰西部–玛卿岗–秦岭北部–黄土高原南缘–大别山–淮河南–长江中下游一线。

· Mammalian geography in China: The Geobiological Approach

There are different views on the zoogeographical division of mammals in China. The Palearctic and Oriental boundaries of traditional zoogeography based on the Geobiological Approach are obviously different from those of the Biogeographic Approach based on biomes. Zheng Zuoxin and Zhang Rongzu (1956) began to study the mammalian geography of China. Based on the zoogeographic system of Wallace (1876), divided the zoogeography of China into two kingdoms, seven regions and 16 subregions. Later, Zhang Rongzu (2011) divided the mammalian geography of China into two kingdoms, seven regions and 19 subregions. Chinese scholars (Zheng Zuoxin and Zhang Rongzu,

1956) believed that the eastern boundary between Palearctic Realm and Oriental Realm in China is the Qinling - Huaihe River, and the western boundary is Hengduan Mountain in western China. Liu Jiaheng and Lu Jiqi (2019) based on "China's Mammal Diversity (2nd edition)" (Jiang et al., 2017), "China's Mammalian Diversity and Geographical Distribution" (Jiang et al., 2015), and "Mammal Distribution in China" (Zhang, 1997), they conducted multivariate cluster analysis and divided the mammalian geography in China into two Kingdoms, seven Regions and 10 Subregions. They suggested that the dividing line between the boundary of Palearctic-Oriental in China is the line from the west of Nnamjagbarwa Peak to Maqing Hill to the north of the Qinling Mountains to the south of the Loess Plateau to the Dabie Mountains to the south of the Huaihe River to the middle and lower reaches of the Yangtze River.

我们利用本书中最新的中国哺乳动物编目和分布数据，依据物种的分布区标定一个物种的地理分布类型（依张荣祖，1999扩展），然后将中国哺乳动物按刘嘉恒和路纪琪（2019）生物地理分区统计。分布在古北界的中国陆生哺乳动物共计536种，其中以分布在内蒙古高原和东北区（Inner Mongolia Plateau and Northeast China Region）的哺乳动物种数最多（183种），东方界（东洋界）以西南内陆区（Southwest Inland Region）的陆生哺乳动物种数最多，共546种，其中该区的华南亚区（South China Sub-region）有315种，横断山亚区（Hengduan Mountain Sub-region）有231种（表2.1，一些物种跨古北界东方界分布）。

We use the latest Chinese mammal inventory and distribution data in this book to determine the biogeographic distribution type of mammals according to their zoogeographic distribution type expanded based on Zhang Rongzu(1999), we analyzed the biogeographic statistics of mammals in China in the zoogeographic system of Liu and Lu (2019). There are 536 species of terrestrial mammals of the Palearctic Realm in China, with the largest number (183) distributed in the Inner Mongolia Plateau and Northeast China Region. In the Oriental Realm, the Southwest Inland Region has the largest number of terrestrial mammal species (546 species), among which 315 species are found in the South China Sub-region, 231 species in Hengduan Mountain Sub-region (Table 2.1, some species extend to the two realms).

表2.1 中国古北界与东方界各地理亚区陆生哺乳动物地理分布

Table 2.1 Distribution of terrestrial mammals in the sub-regions of Palearctic Realm and Oriental Realm in China

界 realm	区 Region	亚区 Sub-region	亚区种数 [a] No. Species
古北界 Palearctic realm	Ⅰ新疆区 Xinjiang region	Ⅰ－1新疆区 Xinjiang region	130
	Ⅱ内蒙古高原和东北区 Inner Mongolia Plateau and Northeast China region	Ⅱ－1内蒙古高原亚区 Inner Mongolia Plateau sub-region	86
		Ⅱ－2东北亚区 Northeast China sub-region	97
	Ⅲ华北平原区 North China Plain region	Ⅲ－1黄海平原亚区 Huanghai Plain sub-region	64
		Ⅲ－2黄淮平原亚区 Huanghuai Plain sub-region	49
	Ⅳ西部和北部高原区 Western and Northern Plateau region	Ⅳ－1青藏高原和祁连－关中亚区 Qinghai-Tibet Plateau and Qilian Guanzhong sub-region	109
东方界（东洋界） Oriental realm	Ⅴ东南沿海区 Southeast coast region	Ⅴ－1 东南沿海亚区 South-east coast sub-region	192
	Ⅵ华中区 Central China region	Ⅵ－1华中亚区 Central China sub-region	152
	Ⅶ西南内陆区 Southwest inland region	Ⅶ－1华南亚区 South China sub-region	315
		Ⅶ－2横断山亚区 Heng-duan mountain sub-region	231

[a] 一个种的分布区可能跨越多个生物地理区 The distribution range of a species may extend across several biogeographic regions

根据WWF生态区200（Olsen et al.，2011，图2.2），将中国分为古北界（Palearctic Realm）和印度马来界（Indomalaya Realm）、10个生物群系和67个生态区。这些生物群系包括热带和亚热带湿润阔叶林，温带阔叶林和混交林，温带针叶林，北方森林／针叶林，温带草原，稀树草原和灌丛，洪水草原和稀树草原，山地草原和灌丛，岩石和冰原等。生态区包括雅鲁藏布江流域半常绿森林、海南岛季风雨林、喜马拉雅亚热带阔叶林等，其中海南岛季风雨林等35个生态区为WWF生态区200重点保护生态区（图2.6，附录3）。

- ## Studies on mammalian geography in China: The Biogeographical Approach

Based on the Ecoregion200 WWF (Olsen et al., 2011, Figure 2.2), China is divided into Palearctic Realm and Indomalaya Realm, 10 biomes and 67 ecological regions. These biomes include Tropical & Subtropical Moist Broadleaf Forests, Temperate Broadleaf & Mixed Forests, Temperate Conifer Forests, Boreal Forests/Taiga, Temperate Grasslands, Savannas & Shrublands, Flooded Grasslands & Savannas, Montane Grasslands & Shrublands, Rocks and Ice, etc. The ecological areas include Yarlung Zangbo Valley Semi-Evergreen Forests and Hainan Island Monsoon Rain Forests, Himalayan Subtropical Broadleaf Forests, etc., among which 35 ecoregions such as the Hainan Island Monsoon Rainforest are the priority protected ecoregions of Ecoregion200 WWF (Figure 2.6, Appendix 3).

在地理生物途径中通常视为古北界地理单元的贵州高原阔叶林和混交林(Guizhou Plateau Broadleaf and Mixed Forests)、云南高原亚热带常绿森林(Yunnan Plateau Subtropical Evergreen Forests)、长江平原常绿阔叶林（Changjiang Plain Evergreen Forests）、大巴山常绿森林（Daba Mountains Evergreen Forests）和四川盆地常绿阔叶林（Sichuan Basin Evergreen Broadleaf Forests）等生态区，Olsen et al.（2011）一并将这些生态区均列为古北界范围（图2.7）。

The Guizhou Plateau Broadleaf and Mixed Forests, the Yunnan Plateau Subtropical Evergreen Forests, Changjiang Plain Subtropical Evergreen Forests, Daba Mountains Evergreen Forests and Sichuan Basin Evergreen Broadleaf Forests, which are generally taken as Palearctic

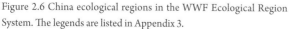

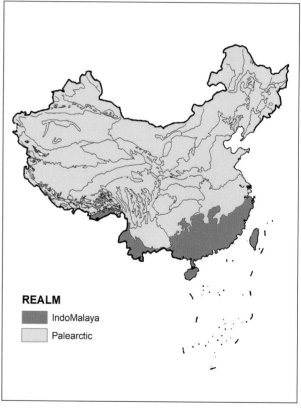

图2.6 WWF生态区中的中国生态区区划。图例参见附录3。

Figure 2.6 China ecological regions in the WWF Ecological Region System. The legends are listed in Appendix 3.

图2.7 根据WWF Global 200绘制的古北界与印度马来界在中国的分界线。

Figure 2.7 The boundary between the Palearctic Realm and the IndoMalaya Realm in China according to WWF Global 200.

geographical units in the Geobiological Approach, Olsen et al. (2011) merged these ecoregions into the range of Palearctic Realm (Figure 2.7).

依照Udvardy（1975）、Olsen et al.（2011）WWF Global 200的生物地理途径，Burgin et al.（2018）将世界哺乳动物分为新热带界（Neotropic）、非洲热带界（Afrotropic）、古北界（Palearctic）、印度马来界（Indomalaya）、新北界（Nearctic）、澳洲大洋洲界（Aust-Oceania）以及海洋（Ocean）七大哺乳类生物地理区。其中，海洋的哺乳类种密度最低，为每百万平方千米0.34种；陆地哺乳类种密度，以印度马来界为最高，每百万平方千米为127.2种，古北界最低，为21.5种。中国陆生哺乳类种密度为每百万平方千米75.6种，低于印度马来界、新热带界和非洲热带界，但高于澳洲大洋洲界、新北界和古北界（表2.2）。中国的哺乳动物种密度约为印度马来界平均种密度的一半，约为古北界平均种密度的3倍。

According to the Biogeographic Approach of Udvardy (1975), Olsen et al. (2011) WWF Global 200, Burgin et al. (2018) divided the zoogeographic zone of the world's mammals into the seven mammalian biogeographic regions: Neotropic, Afrotropic, Palearctic, Indomalaya, Nearctic, Aust-Oceania and the Ocean. The marine mammal species density is the lowest (0.34 species per million square kilometers). The terrestrial mammal species density in the Indomalaya Kingdom is the highest (127.2 species per million square kilometers) and the Palearctic Kingdom is the lowest (21.5 species per million square kilometers). The terrestrial mammal species density in China is 75.6 species per million square kilometers, lower than that of the Indomalaya, Neotropical and Afrotropic, but higher than that of the Aust-Oceania, Neoarctic and Palearctic (Table 2.3). The mammal species density in China is about half of that in the Indo- Malay Kingdom and about three times higher than that in the Palearctic Kingdom.

依据WWF的生物群系——生态地理区系统，在中国，生物群系热带和亚热带湿润阔叶林（Tropical & Subtropical Moist Broadleaf Forests）、温带阔叶和混交林（Temperate Broadleaf & Mixed Forests）与山地草原和灌丛（Montane Grasslands & Shrublands）分布的哺乳动物种数最多。生物群系水淹草原和稀树大草原（Flooded Grasslands & Savannas）与岩石和冰原（Rock & Ice）分布的哺乳动物种数最少。当然，水淹草原和稀树大草原和岩石和冰原的面积相对热带和亚热带湿润阔叶林、温带阔叶和混交林和温带针叶树森林的面积亦小。

In China, according to WWF's Biome – Ecoregion System, the biomes: Tropical & Subtropical Moist Broadleaf Forests, Temperate Broadleaf & Mixed Forests, and Montane Grasslands & Shrublands have the highest number of mammal species. The numbers of mammal species in the biomes: Flooded Grasslands & Savannas and Rock & Ice are among the lowest. Of course, the areas of Flooded Grasslands & Savannas and Rock and Ice are also smaller than those of Tropical & Subtropical Moist Broadleaf Forests, Temperate Broadleaf & Mixed Forests, and Temperate Conifer Forests.

表2.2 中国与世界生物地理区陆生哺乳类种密度比较

Table 2.3 Comparison of terrestrial mammal species density between China and world biogeographical regions

界 Realm	物种总数 Total species	面积 Area (million km²)	种密度 Density (species/ million km²)
新热带界 Neotropic	1,617	19	85.1
非洲热带界 Afrotropic	1,572	22.1	71.1
古北界 Palearctic	1,162	54.1	21.5
印度马来界 Indomalaya	954	7.5	127.2
新北界 Nearctic	697	22.9	30.4
澳洲大洋洲界 Aust-Oceania	527	8.6	61.4
海洋 Ocean	124	361.9	0.34
中国 China	726	9.6	75.6 [a]

[a] 不包括44种海洋哺乳类和淡水豚类

44 marine mammals and fresh water dolphins are excluded

依据生物地理途径分析，先确定中国哺乳动物地理分布型（附录4），然后确定中国哺乳动物的生物地理界属性。中国哺乳动物种以古北界（Palearctic）种类最多，约占中国哺乳类种数的一半；印度马来界（Indomalaya）种类约占四分之一强；古北界为主、印度马来界为辅（Palearctic，Indomalaya）的种类占9%，印度马来界为主，古北界为辅（Indomalaya、Palearctic）的种类占8%；这四种生物地理分布型物种占92%。其他兼有多个生物地理界属性的种，如兼有新北界、古北界（Nearctic，Palearctic）以及非洲热带界、澳大利西亚界、印度马来界、新北界、新热带界、大洋洲界（Afrotropical，Australasian，Indomalaya，Nearctic，Neotropical，Oceanian）的种，半数为海洋哺乳动物。都在13种以下（表2.3）。表2.1与表2.3的数据不同，因为前者基于地理生物途径，后者基于生物地理途径。

Based on Biogeographic Approach, we determined the geographical distribution types of mammals in China (Appendix 4), and then the biogeographic attributes of the mammals. The largest number of mammal species in China is the Palearctic, which accounts for about half of the mammal species in China. Indomalayan species account for roughly a forth. 9% of the species are mainly Palearctic, partial Indomalayan whereas 8% of the species are mainly Indomalayan, partial Palearctic. Species of the four biogeographical distribution types account for 92% of the total species. Other species that distribute across multiple biogeographic boundaries, such as Nearctic, Palearctic and Afrotropical, Australasian, Indomalaya, Neotropical, and Oceanian are all under 13 species (Table 2.3). The data in Table 2.1 differs from that in Table 2.3 because the former is based on the Geobiological Approach and the latter is based on the Biogeographic Approach.

表2.3 依据生物地理途径确定的中国哺乳动物生物地理界属性
Table 2.3 Zoo-geographic atributes of China's mammals according to the Biological- Geographic Approach

生物地理界 Biogeographic Realm	物种数 No. Species
古北界 Palearctic	343
印度马来界 Indomalaya	215
古北界、印度马来界 Palearctic, Indomalaya	66
印度马来界、古北界 Indomalaya, Palearctic	59
非洲热带界、澳大利西亚界、印度马来界、新北界、新热带界、大洋洲界、古北界 Afrotropical, Australasian, Indomalaya, Nearctic, Neotropical, Oceanian, Palearctic	13
新北界、古北界 Nearctic, Palearctic	11
非洲热带界、南极洲界、澳大利西亚界、印度马来界、新北界、新热带界、大洋洲界、古北界 Afrotropical, Antarctic, Australasian, Indomalaya, Nearctic, Neotropical, Oceanian, Palearctic	6
澳大利西亚界、印度马来界、新北界、新热带界、大洋洲界、古北界 Australasian, Indomalaya, Nearctic, Neotropical, Oceanian, Palearctic	3
古北界、非洲热带界、印度马来界 Palearctic, Afrotropic, Indomalaya	3
非洲热带界、澳大利西亚界、印度马来界 Afrotropical, Australasian, Indomalaya	2
非洲热带界、澳大利西亚界、印度马来界、古北界 Afrotropical, Australasian, Indomalaya, Palearctic	2
古北界、大洋洲界 Palearctic, Oceanian	2
古北界、非洲热带界 Palearctic, Afrotropic	2
古北界、新北界 Palearctic, Nearctic	2
全北界 Holartic	2
澳大利西亚界、印度马来界、古北界 Australasian, Indomalaya, Palearctic	1
非洲热带界、印度马来界、古北界 Afrotropic, IndoMalaya, Palearctic	1
非洲热带界、澳大利西亚界、印度马来界、新热带界、大洋洲界、古北界 Afrotropical, Australasian, Indomalaya, Nearctic, Oceanian, Palearctic	1

生物地理界 Biogeographic Realm	物种数 No. Species
新北界、古北界、印度马来界 Nearctic, Palearctic, Indomalaya	1
新热带界、印度马来界、新北界、古北界 Neotropical, Indomalaya, Nearctic, Palearctic	1
印度马来界、新北界、新热带界、古北界 Indomalaya, Nearctic, Neotropical, Palearctic	1
印度马来界、大洋洲界 Indomalaya, Oceanian	1
印度马来界、新热带界、大洋洲界、古北界 Indomalaya, Neotropical, Oceanian, Palearctic	1
全球分布 Global distribution	1
总计 Sum	740

谈到中国的动物地理，有几点需要记住。首先，中国是多山之国，山地生境对哺乳类成种有重要意义。山体具有立体气候，植被垂直分带，相对人类活动强度大的低海拔地区，是物种的避难所和"孤立的生态岛屿"。中国西南地区第四纪冰川活动微弱，横断山区峡谷切割，山体高耸，是我国重要的冰川动物避难地和物种发生中心（Zhang，1999）。黑姬鼠（*Apodemus nigrus* sp. nov., Ge et al., 2020）模式标本产于梵净山（海拔＞1984米)和重庆金佛山。黑姬鼠发现表明中国东南部高山作为温带物种的避难所和"孤立的生态岛屿"的重要性（Ge et al., 2020）。

Talking about the zoogeography of China, there are several points to remember. First, China is a mountainous country, and mountain habitats are important for mammal speciation. The mountain has a three-dimensional climate, and the vegetation is vertically zoned. Compared with the low altitude area with high intensity of human activities, it is the refuge of species and an "isolated ecological island". The Quaternary glaciation in southwest China is weak, and the Hengduan Mountain region is an important refuge for animals to survive the Ice Age and the center of speciation (Zhang, 1999). *Apodemus nigrus* sp. nov. (Ge et al., 2020), the type specimens were found in Fanjingshan Mountain (elevation > 1984 m), Guizhou and Jinfanshan Mountain, Chongqing City. The discovery of the *Apodemus nigrus* species nove suggests the importance of high mountains in southeast China as sanctuaries for temperate species and "isolated ecological islands" (Ge et al., 2020).

其二，东亚季风和南亚季风影响着中国的气候、植被和动物分布(Spicer，2017；Spicer et al.，2020)。雅鲁藏布江大峡谷是连接印度洋水汽从进入中国的天然通道(Zhang，1999)。来自印度洋的暖湿气流沿马拉布特拉河-雅鲁藏布江大峡谷形成雨带，将藏东南地区的降雨带北推3个纬度，推进到30°N，南亚季风将暖湿气流沿印缅地槽向东输送到102°E以东地区，使得这一地区低海拔地区发育的植物区系以印度马来界植物为主。东亚季风带给中国东部带来大量降水，形成了中国东部的森林带，特别是落叶阔叶带与常绿阔叶林带，为森林哺乳类提供了生境。

藏南地区位于喜马拉雅山南坡，这一地区以南方动物群的鸟类和哺乳动物为主。例如，懒猴 *Nycticebus pygmaeu*、亚洲象 *Elephas maximus* 和小竹鼠 *Cannomys badius*。此外，该地区还有许多特有种。藏南可以作为一个典型的例子，过去，我们对藏南的哺乳动物区系了解甚少，蒋志刚等（2017）报道了20多种藏南分布的"东方界（南方分布型）"哺乳动物，将进一步增加藏东南"南方分布型"（东方界/印度马来界）属性。

Second, South Asian, and East Asian monsoons affect the climate, vegetation, and animal distribution in China (Spicer, 2017; Spicer et al., 2020). The Yarlung Zangbo River Grand Canyon is a natural channel connecting water vapor from the Indian Ocean into China (Zhang, 1999). The warm and moist air from the Indian Ocean forms a rain belt along the Malaputla River-Yarlung Zangbo River Grand Canyon, pushing the rain belt into southeast Tibet three latitudinal degrees north to 30°N. The south Asian monsoon will transport the warm and moist air east along the Indo-Myanmar geosyncline to the region of 102°E. The flora developed in the lower elevations of this region is dominated by Indomalaya plants.

The southern Tibetan region is located on the southern slope of the Himalayas, with the Yarlung Zangbo River Grand Canyon a part of it. This region is dominated by birds and mammals of the southern fauna. Examples include the Slow Loris, *Nycticebus pygmaeu*, the Asian elephant, *Elephas maximus*, and the bamboo rat *Cannomys badius*. In addition, there are many endemic species in the area. Southern Tibet can be taken as a typical example. In the past, we had little understanding of the mammal fauna there. Jiang Zhigang et al. (2017) reported more than 20 Oriental mammals distributed in the Southern Tibet, which will further increase the "southern distribution type" (Oriental/Indomalaya) attributes of the Southeast Tibet.

第三，随着分子系统学的发展，经典的生物地理学得到了扩展，创造了一门新的学科，称为系统地理学，这一途径是研究生物地理的第三条途径。系统地理学的发展使得科学家们能够检验关于种群起源和扩散的理论，例如岛屿特有性。例如，经典的生物地理学家根据板块运动能够推测出夏威夷群岛的物种起源，而系统地理学让他们能够检验这些种群与亚洲和北美源种群之间的亲缘理论。Holt et al.（2012）结合21037种两栖类、鸟类和哺乳类动物的地理区域分布和系统发育关系数据，确定了11个界20个不同的动物地理区域。尽管由于数据完整性的缘故受到批评，但表明系统发育信息的整合提供了有价值的动物地理区域之间关系历史数据。Shuai et al.（2021）利用核酸系列数据库数据研究了中国哺乳动物，特别是受威胁哺乳动物的系统发生。我们期待在本书的出版后，看到更多的对中国哺乳动物系统地理学的类似工作，揭示中国哺乳动物起源和扩散和特有性。

Thirdly, phylogeographic approaches have been used to enhance our understanding of both biogeography and landscape genetics across a variety of spatial and temporal scales. Molecular systematics, created a new discipline called Phylogeography. This approach is the third approach to the study of biogeography. The development in Phylogeography has allowed scientists to test theories about the origin and spread of populations, such as island specificity. For example, whereas classical biogeographers could infer the origin of species in the Hawaiian Islands from plate movements, Phylogeography allows them to test theories of the kinship between these populations and populations of Asian and North American origin. Holt et al. (2012) identified 20 different zoogeographic regions in 11 kingdoms by combining the geographical distribution and phylogenetic relationship data of 21,037 species of amphibians, birds, and mammals. Although the reasons such as data integrity were criticized, it is suggested that the integration of phylogenetic information provides valuable historical data on relationships between zoogeographic regions. Shuai et al. (2021) studied the phylogeny of Chinese mammals, especially threatened mammals, using DNA database data. We expect to see more similar work on the phylogeography of Chinese mammals after the publication of this book, revealing the origin and spread and uniqueness of Chinese mammals.

第四，新种的发现也是生物地理的新发现。一个地区的新记录种也将影响生物地理分布格局研究。近年来，随着分子生物学、支序分类系统学、谱系系统学的发展，人们对种的认识发生了变化，分类系统发生了变化，大量的亚种被提升为种，人们还发现了形态上不易区别的隐种。Burgin et al.（2018）即报道了自2005年Wilson发表*Mammalian Species of the World*（3ʳᵈ ed.）以来，由于物种的拆分（Splits of species），全球增加了700个哺乳动物新种，其中食肉目增加了23个，灵长类增加了80个，翼手目增加了130个，啮齿类增加了180个，偶蹄目则增加了219个。通常依据物种的分布区标定一个物种的分布类型。当不可能获得所有物种的分布型数据时，所有生物地理学研究都是抽样研究。抽样研究的物种数目对研究有潜在影响：当一个标识为"古北界"的物种一分为二时，将增大数据中"古北界"的权重，相反地，当两个标识为"古北界"的物种合并为一个种时，将减少数据中"古北界"的权重。目前在研究中发生的绝大多数情形是前者——物种的拆分。

Fourthly, the discovery of new species is also a biogeographical discovery. New species recorded in an area will also influence the study of biogeographic distribution patterns. In recent years, with the development of molecular biology, Cladistics and Phylogenetics, people's understanding of species has changed, the taxonomic system has changed, and a large number of subspecies have been promoted to full species status. People also found the morphological indistinguishable cryptic species. Burgin et al. (2018) reported that since the publication of *Mammalian Species of the World* (3ʳᵈ ed.) by Wilson in 2005, 700 new mammal species were added to the Mammal Inventory of the World due to the splitting of species. Among them, 23 species were added to Carnivora, 80 to Primates, 130 to bats, 180 to rodents, and by 219 to Artiodactyla. The distribution type of a species is determined by its distribution area. When it is impossible to obtain distribution pattern data for all species, all biogeographical studies are sampling studies. The number of species sampled has a potential impact on the study: when a species labeled Palearctic is split, it increases the weight of Palearctic in the data, and conversely, when two species labeled "Palearctic" are clumped into one species, it reduces the weight of "Palearctic" in the data. Most of what is happening now in research are the former - the splitting of species.

最后，动物会主动寻找适宜栖息地，回避不适宜的栖息地。全球变化已经改变了全球的气候，改变了生物群系的边界，改变了生物群落的生物生产力，也导致动物向原生境外扩散，影响了动物地理分布（National Research Council，1995；Mooney and Hobbs，2000）。此外，人类活动有意或无意引入的外来物种（Meyerson and Mooney，2007；Zhang and Jiang，2016），那些成功定殖的外来种，导致原生物种灭绝（Clavero and Garcı´a-Berthou，2005），改变了一个动物区系。这些因素增加了动物地理分布的动态性。人类活动、外来物种和全球变化对生物分布格局的影响，以及这些影响如何进一步影响生物地理值得深入探讨。

Finally, animals actively seek suitable habitats and avoid unsuitable habitats Global change has altered global climate, altered the boundaries of biomes, altered the biological productivity of biomes, led to the animals move out their original ranges, and affected the geographical distribution of animals (National Research Council, 1995, Mooney and Hobbs, 2000). In addition, alien species are introduced intentionally or unintentionally by human activities (Meyerson and Mooney 2007, Zhang and Jiang, 2016), and those successfully colonized alien species, may cause the extinction of native species (Clavero and Garcı´a-Berthou, 2005), alters fauna compositions. These factors increase the dynamics of the geographical distribution of animals. The impacts of human activities, alien species, and global change on distribution patterns and how these impacts further affect biogeography deserve further investigation.

第3章 保护

古老的自然力量，如板块构造、巨型火山和陨石曾经改变过地球的生物区系。人类脱胎于动物界之后，原始社会的人类狩猎者、采集者的群体小，史前人类对生态环境的破坏是有限的。早期农耕社会的刀耕火种、游牧的生产方式对环境的影响有限。近代人口增加，人类猎杀了相当大比例的大型动物，开始驯养家畜家禽，导致了家养动物野生近缘的灭绝(Jiang et al., 2017)。此外，大规模的森林砍伐和湿地开垦，现代农牧场、农业矿产开发和工业生产，造成了野生动物栖息地的破坏，改变了地表。随着人口的增长，人力生产力的发展，人类居住地和人类社会基础设施的建设，城市的扩张、污染物的排放、人类生态足迹的扩大，日益影响地球生物圈，危及其他物种的生存。现在人类活动把地球推入了一个新的地质时代——人类世（Anthropocene）。在地球45亿年的历史中，第一次一个单一物种越来越多地决定着地球的未来和物种的命运(Dirzo et al., 2014; Lewis and Mashin, 2018)。

3 Conservation

Ancient forces of nature, like plate tectonics, giant volcanoes and meteorites once altered Earth's biota. After human beings emerged from the animal kingdom, human density was low and the group sizes of hunters and gatherers in primitive human society were small, thus prehistoric humans' impact on the ecological environment was limited. Slash-and-burn and nomadic production modes in early agricultural society had relatively little impact on the environment. With the increase of the human population, humans hunted a large proportion of large animals and domesticated domestic animals, leading to the extinction of wild relatives of domestic animals (Jiang et al., 2017). In addition, large-scale deforestation and wetland reclamation, farming and ranching in agriculture, mining, industrial production, and expansion of human settlements have caused the destruction of wildlife habitats and changed the landscape. With the growth of population, the development of human productivity, the construction of human residents and social infrastructure, the expansion of cities, the discharge of pollutants and the ever-expansion of human ecological footprint, the Earth's biosphere is increasingly affected and the survival of other species is in peril. Now human activity has pushed the Earth into a new geological epoch, the Anthropocene. For the first time in Earth's 4.5-billion-year history, a single species, *Homo sapiens*, is increasingly determining the future of Earth and the fate of other species (Dirzo et al., 2014; Lewis and Mashin, 2018).

中国有着悠久的人类文明史，生态系统开发的历史早，许多物种种群数量下降，分布区萎缩，面临不同程度的灭绝风险。IUCN濒危物种红色名录是物种灭绝风险的测度，IUCN定期更新其濒危物种红色名录，预警全球物种的生存危机。2020年，依据IUCN受威胁物种红色名录标准和IUCN区域濒危物种红色名录标准，我们对中国哺乳动物（包括20世纪局部灭绝与野外灭绝的种类）的濒危等级进行了评估。IUCN濒危物种红色名录与国家濒危物种红色名录都是物种灭绝风险的测度，前者是全球性评估，后者则是国别研究，两者的研究空间尺度不同。IUCN濒危物种红色名录预警了全球物种的濒危状况，为全球生物多样性研究提供了大数据；国别红色名录确定了各国物种受威胁状况，填补了前者的知识空缺，两份红色名录互为补充。

China has a long history of human civilization and ecosystem exploitation. Wild species of potential value to humans or that conflict with humans in space and resources are facing different degrees of extinction risk due to declining population and shrinking distribution. The IUCN's Red List of Threatened Species measures the risk of extinction. The IUCN regularly updates its Red List of Threatened Species to alert the living status of species around the world to an approaching species extinction crisis. In 2020, we assessed the endangered status of mammals in China, including those that were regionally and extinct in the wild in the 20th century, based on the IUCN Red List Criteria for Threatened Species and the IUCN Regional Red List Criteria for Threatened Species. Both the IUCN Red List and the National Red List are measures of species extinction risk. The spatial scales of the two studies are different: the former is a global assessment, while the latter is a country-level study. The IUCN Red List of Threatened Species alerts the endangered status of global species, providing big data for global biodiversity research. The two types of red lists complement each other: the national red list fills in the knowledge gap of the IUCN Red List by identifying the threatened status of species that only has a partial range in the country.

《中国生物多样性红色名录哺乳动物卷（2021版）》评估依据IUCN受威胁物种红色名录标准和IUCN区域濒危物种红色名录标准（附录5），对中国哺乳动物的濒危等级进行评定。共评定了除智人之外的中国哺乳动物13目56科249属700种。部分种在本书成书之前依据新信息确定了濒危等级。

The status of mammals in the 2021 edition of the Mammal Volume of the Red List of Biodiversity in China was assessed according to the IUCN Red List Criteria for Threatened Species and the IUCN Regional Red List Criteria for Endangered Species (Appendix 5). A total of 700 species of mammals (*Homo sapiens* excluded) belonging to 249 genera, 56 families and 13 orders were evaluated. New information was used to determine the categories of some species before the book was finalized.

· 3.1 濒危状况

IUCN 红色名录中极危（CR）、濒危（EN）、易危（VU）三个等级的物种被称为受威胁物种。在2021年的评估中，受威胁中国哺乳动物共计181种，占已评定的种类总数的26%，其中极危（CR）54种（占已评定种类的8%），濒危（EN）62种（9%）。此外，属于近危（NT）等级的哺乳动物139种，占所有哺乳动物的20%；属于"数据缺乏"（DD）的105种；占已评定哺乳动物的15%。中国有10种哺乳动物野生灭绝或区域灭绝，其中，4种哺乳动物属于"野生灭绝"，6种属于"区域灭绝"（图3.1）。

· 3.1 Status

Species classified as Critically Endangered (CR), Endangered (EN) and Vulnerable (VU) on the IUCN Red List are collectively called Threatened Species. A total of 181 mammal species are threatened in China, accounting for 26% of the total number of assessed species, including 54 Critically Endangered species (8%), and 62 Endangered species (9%). In addition, 139 species of mammals are classified as Near Threatened (NT), accounting for 20% of all mammals. There were 105 species "data deficient" (DD), accounting for 15% of all mammals. Ten mammal species in China are either extinct in the wild or regionally, of which four are classified as " Extinct in Wild(EW)" and six as "Regional Extinct(RE)" (Figure 3.1).

中国特有哺乳动物计182种，约占中国哺乳动物总数的25%，这些特有种具有巨大的种质资源价值和遗传多样性价值。一个国家的特有种在一个国家的生存状况即是其全球生存状况。本次评估结果显示17.6%的中国哺乳动物特有种属于受威胁物种。其中，5.5%的中国特有种为"极危"种，5.5%为"濒危"种，6.6%为"易危"种(表3.1)。

There are 182 species of endemic mammals in China, accounting for about 25% of the total number of mammals in China. These endemic species have great values as germplasm resources and genetic diversity. The survival status of a country's endemic species in one country is its global survival status. The results of this assessment show that 17.6% of China's endemic mammals are classified as threatened species. Among them, 5.5% of endemic species in China are "Critically Endangered", 5.5% are "Endangered" and 7.6% are "Vulnerable" (Table 3.1).

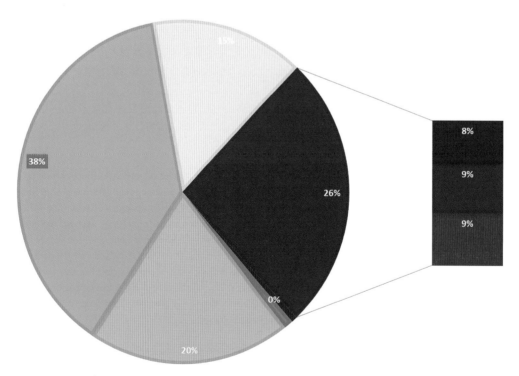

■野生灭绝（EW） ■区域灭绝(RW) ■近危（NT） ■无危（LC） ■数据缺乏(DD) ■极危(CR) ■濒危(EN) ■易危(VU)

图3.1 2021版中国哺乳动物红色名录不同等级的比例

Figure 3.1 Proportions of different categories of mammals in the Red List in China 2021

表3.1 《中国哺乳动物多样性（2023版）》收录的哺乳动物的红色名录等级

Table 3.1 Red List status of the mammals in *China'Mammal Diversity* 2023

红色名录等级 Red List Category	所有种 All Species	所有种占比 % of All Species	特有种 Endemic Species	特有种占比 % of Endemic
野生灭绝 (EW)	4	0.5%	0	0.0%
区域灭绝 (RW)	6	0.8%	0	0.0%
极危 (CR)	52	7.0%	10	5.5%
濒危 (EN)	58	7.8%	10	5.5%
易危 (VU)	64	8.7%	12	6.6%
近危 (NT)	142	19.2%	40	22.0%
无危 (LC)	257	34.8%	37	20.3%
数据缺乏 (DD)	94	12.7%	39	21.4%
未评估 （NE)	63	8.5%	34	18.7%
总计 (SUM)*	739	100%	182	100.0%

* 不包括智人 *Homo sapiens* not included

　　我们在2021版《中国生物多样性红色名录哺乳动物卷》的哺乳动物编目基础上，进一步厘定了中国哺乳动物多样性，原《中国生物多样性红色名录哺乳动物卷（2021版）》中有24哺乳动物种的分类地位或分布区存在异议，不再列入中国哺乳动物编目（尚需进一步研究的哺乳动物见附录6）。本书的新哺乳动物编目——《中国哺乳动物多样性（2022版）》增加了63个新种，其受威胁等级尚未评定或缺乏数据进行濒危等级评定(表3.1)。

　　We further defined the diversity of Mammals in China based on the mammal inventory of China's Red List of Biodiversity (2021). Classification status or distribution ranges of 24 mammal species in the original volume of Mammals, China's Red List of Biodiversity (2021 Edition) are in dispute. Thus, these species are no longer included in the Chinese Mammal Inventory 2022 (see Appendix 6 for those mammal species which require further study). The book's new mammal inventory, Mammalian Diversity in China 2022, adds 63 new species whose status has not yet been assessed (Table 3.1).

　　对中国哺乳动物红色名录已评定种类的分析发现中国哺乳动物各目中濒危物种比例是非均匀分布的(表3.2)。翼手目Chiroptera无极度濒危的种类。劳亚食虫目Eulipotyphla仅一种为极度濒危种，绝大数种没有灭绝风险。啮齿目Rodentia是中国种数最多的哺乳动物目，已经评定了220种的生存状况。啮齿目受威胁物种是中国哺乳动物中最少的，只有河狸*Castor fiber*一种被评为极度濒危(CR)。目前河狸仅分布在新疆阿尔泰山南麓的布尔根河、大小清河流域，种群数量约为700只，在中国的分布区是其边缘分布区。

　　An analysis of the species assessed in the *Red List of China's Biodiversity-Mammals* found that the proportions of endangered species in each order of China's mammals are unevenly distributed (Table 3.2). CHIROPTERA has no Critically Endangered species. Only one species of EULIPOTYPHLA is Critically Endangered, and most species are not at risk of extinction. RODENTIA is the order with the largest number of species in China and 220 species in the order have been assessed. However, Rodents are the least threatened species among mammals in China, only one species, the beaver (*Castor fiber*), is rated as Critically Endangered. The range of beavers in China is its marginal range. At present, the beaver is only distributed in the Boergen River and Major and Minor Qinghe River drainages at the southern foot of Altai Mountains in Xinjiang, with a population of about 700.

　　中国兔形目代表性濒危物种有中国特有种伊犁鼠兔*Ochotona iliensis*与海南兔*Lepus hainanus*。伊犁鼠兔1983年由李维东与马勇发现并命名，仅在中国天山山脉发现。自2000年以来，伊犁鼠兔的栖息地减少至原本分布区面积的17.05%，种群成熟个体数少于2500只，已经处于濒危状态。

表3.2 中国哺乳动物目中各IUCN受威胁物种等级种数
Table 3.2 Number of species in each IUCN Category of Threatened Species in the mammal order in China

目 Order	野生 灭绝 EW	区域 灭绝 RE	极危 CR	濒危 EN	易危 VU	近危 NT	无危 LC	数据 缺乏 DD	未评定 NE	总计 SUM
劳亚食虫目 EULIPOTYPHLA	0	0	1	1	7	29	32	14	16	100
攀鼩目 SCANDENTIA	0	0	0	0	0	0	1	0	0	1
翼手目 CHIROPTERA	0	0	0	3	15	48	39	26	11	142
灵长目 PRIMATES	0	0	15	8	4	1	1	0	0	29
鳞甲目 PHOLIDOTA	0	0	2	0	0	0	0	0	0	2
食肉目 CARNIVORA	0	0	11	17	15	13	2	6	2	66
鲸偶蹄目 CETARTIODACTYLA	3	3	17	22	12	10	17	19	0	103
海牛目 SIRENIA	0	0	1	0	0	0	0	0	0	1
长鼻目 PROBOSCIDEA	0	0	1	0	0	0	0	0	0	1
奇蹄目 PERISSODACTYLA	1	3	0	0	1	1	0	0	0	6
啮齿目 RODENTIA	0	0	2	3	8	34	143	26	31	247
兔形目 LAGOMORPHA	0	0	0	4	2	6	22	3	3	41
总计 Sum*	4	6	51	58	64	142	257	94	63	739

* 不包括智人 *Homo sapiens* not included

Representative endangered species of Lagomorphs in China are Yili Pika *Ochotona iliensis* and *Lepus hainanus*, both endemic to China. Discovered by Li Weidong and Ma Yong in 1983, the Yili Pika is found only in the Tianshan Mountains of China. Since 2000, the habitat of Yili Pika has been reduced to 17.05% of the original area. With less than 2,500 mature individuals; the population is endangered.

海南兔*Lepus hainanus*是中国海南省特有种，分布于海南岛东部丘陵台地灌草丛中，是海南热带低地陆地生态系统的典型物种，是该类生境生态系统健康状况的指示物种。海南兔繁殖力强，每窝平均产仔3～5只。然而，根据江海生团队的考察结果，由于大量的灌丛草地被开垦为耕地、建设用地，海南兔栖息地面积持续缩减，同时海南兔栖息地出现严重斑块化、破碎化的趋势。加之过度捕猎，海南兔种群数量锐减，其生存现状为极度濒危。

Hainan Rabbit (*Lepus hainanus*) is an endemic species mainly distributed in the semi-arid shrubs of hilly terraces in the eastern part of Hainan Island. This rabbit is a typical species of the tropical lowland ecosystem in Hainan and may serve as an indicator of ecosystem health in this type of habitat. Hainan Rabbits have high fecundity with litter sizes of 3-5. However, Jiang Haisheng's research team has found that due to the clear-cutting of shrub grasslands for developing farmland and construction land, Hainan rabbits' habitat is continuously being fragmented and shrinking. Additionally, its population has decreased sharply due to excessive hunting and its survival status is now Critically Endangered.

中国鳞甲目动物穿山甲*Manis pentadactyla*和马来穿山甲*M. javanica*列为极度濒危。虽然自1989年以来，穿山甲受到法律的保护。然而，在巨大市场需求和利润的刺激下，野外穿山甲被大量捕捉，种群数量急剧下降。《濒危野生动植物种国际贸易公约》第17次缔约方大会通过决议，于2017年将全球所有穿山甲列入该公约附录I，禁止穿山甲国际商业贸易活动。基于其种群与分布地现状，中国分布的两种穿山甲的濒危等级全部为极危(CR)。

Pangolins in China: Chinese Pangolin (*Manis pentadactyla*), and Malay Pangolin (*M. Javanica*) are all listed as Critically Endangered. Although pangolins have been protected by law since 1989, driven by huge market demand and profits, pangolins in the wild have been hunted and their populations have declined sharply. The 17th Conference of the Parties of the Convention on International Trade in Endangered Species of Wild Fauna and Flora (CITES) adopted a resolution to list all pangolins in the world in the Appendix I of CITES in 2017, banning international commercial trade in pangolins. Based on the status of their populations and distribution, the three pangolin species in China are all Critically Endangered (CR).

• 3.2 濒危状况

有些物种，尽管在全球还没有灭绝，但是在一个国家或者一个地区已经灭绝，称为区域灭绝，或者在野外生境中已经灭绝，称为野外灭绝。2020年中国生物多样性红色名录评定结果中有10种哺乳动物属于"野外灭绝"或"局部灭绝"等级（表3.3）。其中，属于野外灭绝等级(EW)的有野马(*Equus ferus*)、高鼻羚羊(*Saiga tatarica*)、驯鹿(*Rangifer tarandus*)、大额牛(*Bos frontalis*)等。这些"野外灭绝"或"局部灭绝"现象发生在20世纪60年代以前。Wilson et al.（2005, 2009~2017）、Groves（2011）、美国哺乳动物学会数据库（2022）均收录近期灭绝的种，故本书按照惯例收录了近期在中国灭绝的种。

• 3.2 Analysis of extinct species

Some species, though not yet extinct globally, are known to be regional extinct in a country, or extinct in the wild in their habitats. In 2020, ten mammal species were classified as "Regional Extinct" or "Extinct in Wild" in the Red List of Biodiversity in China (Table 3.3). Among them, *Equus ferus, Saiga tatarica, Rangifer tarandus*and *Bos frontalis*, are EW. These "Regional Extinct "or" Extinctions in Wild " mainly occurred before the 1960s. Wilson et al. (2005, 2009-2017), Groves (2011), and the American Mammalogist Society Database (2022) all included recently extinct species, so this book follows the convention of including recently extinct species.

表3.3 中国"区域灭绝"或"野外灭绝"的哺乳动物

Table 3.3 "Regional Extinct" or "Extinctions in Wild " mammals in China

种名 Chinese Name	学名 Scientific Name	红色名录等级 Red List Category
双角犀	*Dicerorhinus sumatrensis*	区域灭绝 RE ver. 3.1
爪哇犀	*Rhinoceros sondaicus*	区域灭绝 RE ver. 3.1
大独角犀	*Rhinoceros unicornis*	区域灭绝 RE ver. 3.1
豚鹿	*Axis porcinus*	区域灭绝 RE ver. 3.1
野水牛	*Bubalus arnee*	区域灭绝 RE ver. 3.1
爪哇野牛	*Bos javanicus*	区域灭绝 RE ver. 3.1
野马	*Equus ferus*	野外灭绝 EW ver. 3.1
驯鹿	*Rangifer tarandus*	野外灭绝 EW ver. 3.1
大额牛	*Bos frontalis*	野外灭绝 EW ver. 3.1
高鼻羚羊	*Saiga tatarica*	野外灭绝 EW ver. 3.1

犀牛在中国曾分布在云南南部与西南部、藏南地区（王应祥，2003；Chooudhury，2003；潘清华等，2007；蒋志刚等，2017；IUCN，2021）。20世纪上半叶云南南部与西南部的犀牛已经灭绝，20世纪藏南仍有独角犀分布，21世纪初，独角犀仍季节性游荡进入藏南(Chooudhury，2003)。豚鹿历史上分布在云南省临沧市耿马、沧源地区沿南汀河流域。Ding et al.（2021）通过样线调查、访问调查，在保护区布设红外相机80台拍摄了超过10000个相机日，但均未发现豚鹿踪迹。随着20世纪60、70年代的经济开发，豚鹿丧失了栖息地，加上盗猎活动，因此而灭绝。野水牛的消失与湿地开垦有关；藏南地区是否还有野水牛，尚无数据，但在20世纪藏南河流湿地常见野水牛(Choudhury，2003)。

我们在2015至2016年的野牛专项考察中，在爪哇野牛的原栖息地没有发现爪哇野牛的痕迹，爪哇野牛分布区已经退出中国。驯鹿长期以来被人们所驯养。中国现有约700头驯鹿，但这些驯鹿基本处于人工驯养状态。2017至2019年，内蒙古根河从芬兰引入了179头驯鹿。野生驯鹿在中国灭绝。本次评估将驯鹿的濒危等级定为"野外灭绝"。在中国，大额牛目前仅分布在云南高黎贡山区独龙河流域，故称之为"独龙牛"。独龙牛已经被当地牧民驯化，夏季被放养在高海拔山区，任其自由采食，冬季牧民能召唤独龙牛集群返回低海拔地区的营地。在高黎贡山区，被驯养的独龙牛数量仍不到1000只。在野外已经没有野生大额牛，故列入附录6。高鼻羚羊历史上曾分布在新疆北部、伊犁河谷、哈密盆地，20世纪60年代在中国灭绝。20世纪80年代，中国重引入了高鼻羚羊，但是由于疾病、遗传多样性等原因，目前重引入的高鼻羚羊种群仍是一个极小种群（Jiang et al.，2020）。

Rhinos were once distributed in south and southwest Yunnan and south Tibet in China (Wang, 2003, Choudhury, 2003; Pan et al., 2007; Jiang, 2017; IUCN, 2021). Those rhinos were extinct in southern and southwestern Yunnan in the first half of the 20th century. In history, there were resident One-horned Rhino populations in Zangnan in early 20 century, and some One-horned Rhinos still wandered into the region seasonally in early 21st century (Choudhury, 2003). Hog Deer was mainly distributed in Gengma and Cangyuan areas along the Nanting River in Lincang City, Yunnan Province. Ding et al. (2021) set out 80 infrared cameras in the area and photographed more than 10,000 camera days. They also conducted a line translocation survey and interview survey, but no trace of Hog Deer was found. With economic development in the 1960s and 1970s, loss of habitat and poaching, the Hog Deer became extinct (Ding et al., 2018). The disappearance of wild buffaloes is related to wetland reclamation. However, there is no data on whether there are wild buffaloes in Zangnan though they were frequently found in riverside wetland in the region in 20th century (Choudhury, 2003), the Southern Tibet. During our investigation of Gaur and Banteng in Yunnan in 2015-2016, we found no trace of Banteng in its original habitat, and the distribution area of gaur has retreated from China. Reindeer have long been domesticated and reindeer are extinct in wild in the country. There are about 700 reindeer in China, but most of them are in captivity. From 2017 to 2019, 179 reindeer were introduced from Finland to Genhe River, Inner Mongolia. In the assessment, reindeer were classified as "Extinct in the Wild." In China, *Bos frantalis* is currently only distributed in the Dulong River basin of Gaoligong Mountains in Yunnan Province, so they are called "Dulong Cattle". The Dulong Cattle have been domesticated by local herders and are free ranging in the summer at high altitudes. In winter, the herders summon the Dulong herd back to their camps at lower altitudes. There are still fewer than 1,000 domesticated Dulong Cattle in the Gaoligong Mountains. There are no *Bos frantalis* in the wild; it is assessed as "Extinct in Wild". The Saiga antelope was once found in Ili River Valley and Hami Basin of northern Xinjiang, and disappeared in China in the 1960s. In the mid-1980s, the Saiga was reintroduced to China. However, but due to diseases, low genetic diversity, and other reasons, the reintroduced Saiga population is still hanging on as the only Saiga population in Wuwei Endangered Species Breeding Center fluctuated dramatically (Jiang et al., 2020).

· 3.3 受威胁种及受胁因素

2021版中国哺乳动物受威胁物种在科级、属级分类单元之间的分布为非随机分布，各个哺乳动物科的受威胁比例相差很大。其中，儒艮科Dugongidae（1种，该科种数，下同）、懒猴科Lorisidae（2种）、獴科Herpestidae（2种）、长臂猿科Hylobatidae（8种）、熊科Ursidae（3种）、大熊猫科Ailuropodidae（1种）、小熊猫科Ailuridae（1种）、林狸科Prionodontidae（1种）、露脊鲸科Eschrichtiidae（1种）、白鱀豚科Balaenidae（1种）、鼠海豚科Lipotidae（2种）、骆驼科Camelidae（1种）、鼷鹿科Tragulidae（1种）、河狸科Castoridae（1种）、象科Elephantidae（1种）、麝科Moschidae（6种）的所有种类全部为受威胁种，受威胁比例为100%。猫科Felidae（13种，12种受威胁）、猴科Cercopithecidae（19种，17种受威胁）、灵猫科Viverridae（10种，7种受威胁）的受威胁比例达60%以上。牛科Bovidae（33种，19种受威胁）、鹿科Cervidae（22种，12种受威胁）、鼬科Mustelidae（20种，14种受威胁）、假吸血蝠科Megadermatidae（2种，1种受威胁）、狐蝠科Pteropodidae（11种，6种受威胁）、海狮科Otariidae（2种，1种受威胁）、睡鼠科Gliridae（2种，1种受威胁）的受威胁种比例均在50%以上。而跳鼠科Dipodidae（18种）、刺山鼠科Platacanthomyidae（4种）、树鼩科Tupaiidae（1种）、鞘尾蝠科Emballonuridae（2种）、犬吻蝠科Molossidae（3种）、猪科Suidae（4种）、灰鲸科Eschrictidae（1种）、喙鲸科Ziphiidae（6种）和跳鼠科Dipodidae（18种）中没有受威胁种。

总体来看，2021版已经评定的中国哺乳动物受威胁物种（极危+濒危+易危）为181种，2015版中国生物多样性红色名录中为178种；2021版的受威胁物种比例下限值为26%，最佳估值为38%，最高值为41%；2015版下限值为27%，最佳估值为30%，最高值为41%[2]。

· 3.3 Threatened species and threats

The distribution of threatened mammal species in China is non-randomly distributed among families and genera, and the proportion of threatened mammal species in different families varies greatly. All the species in Dugongidae (number of species in the family 1 species), Herpestidae (2 species), Hylobatidae (8 species), Ursidae (3 species), Ailuropodidae (1 species), Ailuridae (1 species), Prionodontidae (1 species), Eschrichtiidae (1 species), Balaenidae (1 species), Lipotidae (2 species), Camelidae (1 species), Tragulidae (1 species), Castorid

(1 species), Elephantidae (1 species), and Moschidae (6 species) are threatened. More than 60% of Felidae (13 species, 12 threatened), Cercopithecidae (19 species, 17 threatened), and Viverridae (10 species, 7 threatened) are threatened. The proportion of threatened species in Bovidae (34 species, 20 threatened), Cervidae (22 species, 12 threatened), Mustelidae (20 species, 14 threatened), Megadermatidae (2 species, 1 threatened), Pteropodidae (11 species, 6 threatened), Otariidae (2 species, 1 threatened) and Gliridae (2 species, 1 threatened) are all over 50%. There is no species in Dipodidae (18 species), Platacanthomyidae (4 species), Tupaiidae (1 species), Emballonuridae (2 species), and Molossidae (3 species), Suidae (4 species), Eschrichtidae (1 species), Ziphiidae (6 species), and Dipodidae (18 species) are threatened.

Overall, 181 mammal species are assessed as threatened (Critically Endangered + Endangered + Vulnerable) in China, and 178 species were listed in the 2015 Edition of China's Red List of Biodiversity. In 2020 Edition, the lower limit of threatened species is 26%, the best estimate is 38%, and the highest value is 41%. For the 2015 edition, the lower limit was 27%, the best estimate was 30%, and the highest was 41%.

导致物种濒危的主要原因是生境丧失和退化(表3.4)。全新世以来，人类活动改变了土地覆盖，使野生哺乳动物生境被转变为农田、人工林地、城镇居民用地，并被人类定居点、路网管线分割，导致野生哺乳动物生境破碎化，甚至生境丧失，野生哺乳动物种群数量减少甚至消失。"生境丧失"名列受威胁哺乳动物的致危因子之首，是276种已经评定的中国受威胁哺乳动物的威胁因子，占所有受威胁哺乳动物的所有致危因子的25%。"人类过度利用"和"人类干扰"名列中国受威胁哺乳动物致危因子的第二位，分别占所有受威胁哺乳动物的所有致危因子的19%。"未知"因素占所有受威胁哺乳动物的所有致危因子的14%。"自然灾害"和"种群波动"分别占所有中国受威胁哺乳动物的所有致危因子的7%与6%。在所有受威胁哺乳动物的所有致危因子中，"环境污染""人为毒杀""疾病""意外死亡"和"火灾"占所有致危因子的比例均在4%以下。说明目前中国哺乳动物的濒危主要是人类活动造成的，而物种本身的进化原因对物种的影响基本可以忽略不计。

The main threats to species are habitat loss and degradation (Table 3.4). Since the Holocene, human activities have changed the land cover of Earth. The habitat of wild mammals has been transformed into farmland and artificial forests, urban residential area, and fragmented by human settlements and road networks. Consequently, the habitat of wild mammals has been fragmented and even lost, and the number of wild mammals in many areas has been reduced or even disappeared. "Habitat Loss" tops the list of threats to mammals, accounting for 25% of all threats to mammals in China, and 276 species have been assessed as threatened by "Human Overuse" and "Human Disturbance", which ranked second among all threats to mammals in China, accounting for 19% of all threats to mammals, respectively. "Unknown" factors account for 14% of all threats to mammals. "Natural Disasters" and "Population Fluctuations" account for 7% and 6%, respectively. "Environmental Pollution", " Poisoned by Humans ", "Disease", "Accidental Death" and "Fire" accounted for less than 4% of all threats. These results indicate that the mammals in China are mainly threatened by human activities, and the impact of evolution and the intrinsic factors are negligible.

中国有几种动物处于灭绝的边缘，如虎*Panthera tigris*、金钱豹*P. pardus*和白鱀豚*Lipotes vexillifer*。虎曾是中国森林生态系统的顶级捕食者。中国曾有5个虎亚种：新疆虎（里海虎）*P. t. virgate*、孟加拉虎*P. t. tigri*、印支虎*P. t.*

---

[2] 濒危物种的最低、最佳和最高估计 （IUCN，2020）。

对评价较全面 （即 80% 以上的物种已被评估） 的群体，每一组所报告的受威胁物种均以可能值范围内的最佳估计数表示，该范围的下限估计数为限。

1. 如果所有数据缺乏（DD）物种都没有受到威胁，则受威胁物种最低估计值为：

受威胁现存物种最低估计值 =（CR + EN + VU）/（总评估 - EX）。

2. 如果所有数据缺乏（DD）种受威胁程度与评估数据充足的物种相同，则受威胁物种最佳估计值：

受威胁现存物种最佳估计值 =（CR + EN + VU）/（总评估物种数 - EX - DD）。

3. 如果所有 DD 种都受到威胁，则受威胁物种上限估计：

受威胁现存物种上限 =（CR + EN + VU + DD）/（总评估 - EX）。

The lowest, best and upper estimates of threatened species (IUCN, 2020)

For populations that are more comprehensively evaluated (i.e., more than 80% of species have been assessed), the reported threatened species in each group are expressed as the best estimate within the range of possible values, limited by estimates at the lower end of the range.

1. If all data deficient (DD) species are not threatened, the lowest estimate for threatened species is:

Minimum estimate of threatened existing species = (CR + EN + VU) / (Total assessment - EX)

2. If all data deficient (DD) species are as threatened as those for which assessment data are sufficient, the best estimate for threatened species is:

The best estimate of existing threatened species = (CR + EN + VU) / (Total estimated species – Extinct (EX)-DD)

3. If all DD species are threatened, the upper limit estimate of threatened species is:

Upper limit of threatened species = (CR + EN + VU + DD) / (Total assessment - EX)

表3.4 中国哺乳动物濒危原因分析

Table 3.4 Causes of threatened mammals in China

原因 Cause	频次 Frequency	%[a]
生境丧失 Habitat loss	276	25%
人工利用 Human exploitation	209	19%
人类干扰 Interference	208	19%
未知 Unknown	154	14%
种群波动 Population fluctuation	76	7%
自然灾害 Nature catastrophe	64	6%
环境污染 Environment pollution	38	3%
气候变化 Climate change	28	3%
人为毒杀 Poisoning	23	2%
疾病 Diseases	20	2%
意外死亡 Accidental death	14	1%
未查明 Not specified	271	–

[a] 已知威胁的百分比　% of known threats

corbetti、东北虎*P. t. altaica*和华南虎*P. t. amoyensis*，其中，新疆虎已于1916年在中国灭绝。20世纪末，西藏墨脱曾有孟加拉虎的报道，最近又在墨脱发现了孟加拉虎（王渊等，2019）。近年来在西双版纳未见印支虎报道。东北虎迁徙游荡于中国、俄罗斯边境地区，其面临的威胁有栖息地破碎化、种群分布区隔离。中国建立了东北虎豹国家公园，加强了对东北虎的保护。华南虎是中国特有虎亚种，模式标本产于福建厦门。华南虎曾经广泛分布于湖南、江西、贵州、福建、广东、广西、安徽、浙江、湖北、四川、河南、陕西、山西等地。由于过度捕杀和栖息地丧失等原因，1980年后，在华南虎分布区再没有发现野生华南虎。1990至2001年间，原国家林业局曾组织在原华南虎分布区开展过多次华南虎专项调查，未发现华南虎存在的确切证据。目前我国人工饲养的华南虎有180余只，分散在全国十几家动物园等养殖单位，受近亲繁殖、管理水平和饲养条件等因素的影响，种群发展缓慢。故华南虎亚种为"野生灭绝"。虎的整体生存状况仍定为"极度濒危"。

　　Several animals in China are on the brink of extinction, such as the tiger (*Panthera tigris*), leopards (*P. pardus*) and Baji (*Lipotes vexillifer*). Tigers were once the top predator in China's forest ecosystem. There were once five tiger subspecies in China, of which the Xinjiang tiger, also called the Caspian tiger (*P. t. virgate*), became extinct in China in 1916. The Bengal tiger (*P. t. tigris*) was reported in Motuo, Tibet in the late 20th century, and a Bengal tiger has recently been found there. In recent years, P. t. corbetti has not been reported in Xishuangbanna, Yunnan. The Amur tiger (*P. t. altaica*) is wandering along the China-Russia border, facing threats such as habitat fragmentation and population isolation. China has strengthened the protection of the Amur tiger by establishing the Siberian Tiger and Leopard National Park. The South China tiger (*P. t. amoyensis*) is a subspecies of tiger endemic to China. The South China tiger was once widely distributed in Hunan, Jiangxi, Guizhou, Fujian, Guangdong, Guangxi, Anhui, Zhejiang, Hubei, Sichuan, Henan, Shaanxi, Shanxi, and other places. Due to over-hunting and habitat loss, no wild South China Tigers have been found in their former range since 1980. From 1990 to 2001, the former State Forestry Administration organized several special expeditions on the South China Tiger in its former range, but no conclusive evidence of its existence had been found. There are more than 180 South China Tigers in captivity in China, scattered in over a dozen zoos and other breeding centers in 2020. Affected by inbreeding, management level and feeding conditions, the population development is slow. Thus, the south China tiger subspecies is "Extinct in the Wild". The overall status of the tiger remains "Critically Endangered".

金钱豹*Panthera pardus*分布地广泛，地理亚种分化多，种下分类单元尚有争论。中国有华北豹*P. p. fontanieri*、印度豹*P. p. fusea*、远东豹*P. p. orientalis*和印支豹*P. p. delacouri*等金钱豹亚种分布（蒋志刚等，2021）。近年来东北豹数量有所增加，但是在中国东北地区仅发现40余只东北豹，故其濒危等级定为极度濒危（CR）。近年来，华北豹在山西、陕西、河北、河南、甘肃等地被红外相机拍摄到，在野外仍有华北豹的可生存种群；故其濒危等级定为濒危(EN)。印度豹仅分布于喜马拉雅山脉局部低海拔区域，在西藏吉隆曾被发现记录，故其濒危等级为极度濒危（CR）；而印支豹现在中国南方野外稀少，极少见到，故其濒危等级定为极度濒危（CR）。此外，目前分布在青海以及青藏高原东缘西藏昌都、四川甘孜的豹，其分类定位未定，可能是介于华北豹和印支豹之间的中间类型。金钱豹是大型猫科动物中分布最广的一种，但是其毛皮质量高，且其骨常代替虎骨入药，因此对金钱豹的过度捕猎和非法贸易严重威胁着该物种的生存。20世纪50年代，金钱豹曾被认为是害兽而被加以杀害。加上人类活动造成的栖息地退化和丧失，导致曾经广布在中国的金钱豹的种群越来越小。金钱豹列为"濒危"等级。

The distribution of leopards (*Panthra pardus*) is wide-spreading from Eastern Asia to Southern Africa; the subspecies taxon is controversial. North China Leopard (*P. p. fontanieri*), Indian Leopard (*P. p. fusea*), the Amur Leopard (*P. p. orientalis*), and the Indochinese Leopard (*P. p. delacouri*) are distributed in China. In recent years, the number of Amur Leopard has increased, but only several-dozens Amur Leopards have been found in northeast China. Therefore, it is Critically Endangered. In recent years, the North China Leopard has been photographed by infrared camera traps in Shanxi, Shaanxi, Hebei, Henan, Gansu, Ningxia and other areas, and there are still viable populations of the North China Leopard in the wild. Therefore, it is classified as Endangered. Indian Leopards are only distributed in some low altitude areas of the Himalayas, and have only been recorded in Geelong, Tibet. Therefore, it is Critically Endangered, and Indochinese leopards are rare in the wild in southern China and are only recorded once in Xishuangbanna, Yunnan; therefore, it is Critically Endangered. In addition, the taxonomy of leopards in Qinghai, Qamdo in Tibet, and Garze in Sichuan on the eastern part of the Qinghai-Tibet Plateau have not yet been determined, and may be an intermediate type between the North China Leopard and the Indochinese Leopard (Jiang et al., 2021). Leopards are the most widely distributed big cats, but their survival is threatened by overhunting and illegal trade because of the high quality of their fur and the fact that their bones were often used in traditional medicine as substitutes for tiger bones. Leopards were thought to be vermin and killed in the 1950s. Coupled with habitat degradation and loss due to human activities, the once-widespread Chinese leopard population is getting smaller and fragmented. Thus, Leopard is listed as Endangered.

雪豹的种群和栖息地主要分布在中国（Liu et al., 2016）。近年来，中国雪豹保护初见成效，野外雪豹种群开始恢复。2018年初，IUCN宣布雪豹的濒危等级从"濒危"降为"易危"。2018年9月在深圳雪豹保护国际会议上，中国代表指出基于有效种群大小，雪豹在中国仍处于濒危状态，故2021的评估中仍将中国雪豹的濒危等级定为"濒危（EN）"。

Snow leopards live mainly in China (Liu et al., 2016). In recent years, China's snow leopard conservation has achieved initial success, and the wild snow leopard population has begun to recover. In early 2018, the IUCN announced that snow leopards had been downgraded from "Endangered" to "Vulnerable." At the Shenzhen International Conference on Snow Leopard Conservation in September 2018, Chinese delegates pointed out that snow leopards are still endangered in China based on the effective population size, the 2021 Red List of China's Biodiversity still classified the Chinese snow leopards as "Endangered (EN)".

白鱀豚是中国特有的淡水鲸，曾分布于长江中下游水系与富春江水系。20世纪70年代长江白鱀豚已经数量稀少。2006年，中国科学院水生生物研究所联合欧美及日本的科研机构开展长江淡水豚类考察，未在长江中观测到任何白鱀豚。2007年以后，有白鱀豚零星的目击报告和影像资料，白鱀豚可能还有残余个体，但种群延续困难（Smith et al., 2017）。

Baiji is a freshwater whale endemic to China, once distributed in the middle and lower reaches of the Yangtze River and the Fuchun River. By the 1970s, the Baiji had become scarce in nature. In 2006, the Institute of Hydrology of the Chinese Academy of Sciences and the scientific research institutes in Europe, America and Japan jointly carried out an investigation on the Baiji, but no Baiji has been observed in the Yangtze River during the expedition. Since 2007, there have been sporadic sightings and video data of *Lipotes vexillifer*, indicating there may be some remaining individuals, but the Baiji population is difficult to sustain (Smith et al., 2017).

· 3.4 受威胁种生境与地理分布

哺乳动物能利用多种生境。森林(包括灌丛)是中国哺乳动物的主要生境。2021年中国生物多样性红色名录评定的哺乳动物中，利用"森林"作为生境的有576种，其中有受威胁物种166种。188种利用"草原(包括草甸)"作为生境，其中有受威胁物种45种。159种生活在农田、人造建筑物、人类聚居区等生境，其中多为啮齿类。生活在人工建筑之中的哺乳类已经适应人类社会，成为"伴人动物"。74种中国哺乳动物利用"洞穴"作为生境，多为翼手目种类，其中有受威胁物种11种。42种利用"湿地（包括内陆水体、沼泽、河谷、滩涂等）"作为生境，其中有受威胁物种16种。66种生活在干旱半干旱荒漠地区，利用"荒漠生境"作为生境，其中有受威胁物种11种。43种利用"岩石（高海拔流石滩、冰缘、裸岩悬崖等）生境"作为生境，其中有受威胁物种14种。生活在"湿地"生境和"岩石生境"的中国哺乳动物濒危比例高，分别为38%和33%，以江河湖泊等水体为生境的哺乳动物受威胁比例超过50%。在其他生境中，哺乳动物受威胁比例依次为"森林"（29%）、"海洋"（26%）、"草原"（24%）、"荒漠"（17%）、"人造生境"（16%）和"洞穴"（15%）。

- ### 3.4 Habitat and geographical distribution of threatened species

Mammals make use of a variety of habitats. Forests (including shrublands) are the main habitat type for mammals in China. Of the mammal species that have been assessed in 2021 China's Red List of Biodiversity, 576 species use "Forest" as their habitat, and 166 of those species are threatened. A total of 188 mammal species in China uses "Grasslands (including meadows)" as habitats, of which 45 are threatened. 159 species, most of which are rodents, live in farmland, man-made structures, and human settlements. Mammals that live in man-made structures have adapted to human society and become "animals that taken anthropogenic landscape as habitat". Seventy-four species of Chinese mammals, most of which belong to the Order Chiroptera, use caves as habitats, among which 11 species are threatened. Forty-two species use "wetlands (including inland water bodies, swamps, riversides, tidal flats, etc.)" as their habitats, of which 16 species are threatened. Sixty-six species of mammals live in arid and semi-arid desert areas in China and use "desert habitat" as habitat, among which 11 species are threatened. Forty-three species, including 14 threatened species, use "rocky habitats (high altitude pebble beach, periglacial, bare rock cliff, etc.)" as their habitats. The proportions of threatened mammals living in "wetland" and "rock habitat" in China are high (38% and 33% respectively), and the proportion of threatened mammals living in rivers and lakes is more than 50%. Proportion of threatened mammals live in "forest" is 29%, "sea" is 26%, "the prairie" is 24%, "desert" is 17%, in "man-made habitat" is 16% and "cave" is 15%.

中国地形地貌为三级地理台阶地形地貌。第一级地理台阶是青藏高原，平均海拔4000多米。第二级地理台阶平均海拔1000米到2000米，包括内蒙古高原、云贵高原和黄土高原。第三级地理台阶主要由中国东部海拔200米以下的平原组成，其间点缀着一些丘陵和低山。中国哺乳动物种类多分布在中国第二级地理台阶。

The topography and landform of China are three geographical steps. The first geographic step is the Qinghai-Tibet Plateau, with an average elevation of more than 4,000 meters. The second level, with an average elevation of 1,000 to 2,000 meters, includes the Inner Mongolian Plateau, The Yunnan-Guizhou Plateau, and the Loess Plateau. The third geographic step consists mainly of plains in eastern China below 200 meters above sea level, interspersed with hills and low mountains. Most mammal species in China are in the second geographical step.

中国哺乳动物以分布在海拔500~1000米、1000~1500米、1500~2000米的生境中种类为多，其中，受威胁物种比例随着海拔升高而增高，分布在海拔500~1000米、1000~1500米、1500~2000米生境中哺乳动物种类受威胁比例依次为栖息在该海拔区间的哺乳动物种数的24%、24%和27%（图3.2）。然而，出乎意料的是，生活在高海拔地区哺乳动物种数虽少，但是受威胁哺乳动物种类比例却很高。生活在海拔3000~3500米、3500~4000米、4000~4500米、4500~5000米以及海拔5000米以上生境中的受威胁哺乳动物种数分别占该海拔区间的哺乳动物种数的37%、44%、49%、57%和67%（图3.2）。

Most mammals in China are distributed in habitats between 500-1,000 m, 1,000-1,500 m, and 1,500-2,000 m above sea level. The proportion of threatened mammal species at altitudes between 500-1,000 m and 1,000-1,500 m, 1,500-2,000 m are 24%, 24% and 27% of the number of mammal species at that altitude, respectively (Figure 3.2). Surprisingly, however, the number of mammal species that live at high altitudes is low, but the proportion of threatened mammal species is high. The number of threatened mammal species live in habitats at 3,000-3,500 m, 3,500-4,000 m, 4,000-4,500 m, 4,500 m, 5,000m, and above 5,000m accounted for 37%, 44%, 49%, 57% and 67% of the mammal species at these altitude ranges, respectively (Figure 3.2).

- ### 3.5 中国各省区受威胁与非受威胁哺乳动物的比例

中国各省（区、直辖市）受威胁哺乳动物在空间上的分布是不均匀的（图3.3）。西南地区哺乳动物受威胁比例高于其他地区。西藏（全区受威胁种数59种）、重庆（31种）和云南（91种）的受威胁哺乳动物比例高于全国平均值。多数省区的受威胁哺乳动物种类占本省区哺乳动物数目的20%~25%。陕西（29种）、湖北（17种）、福建（22种）、河北（12种）、宁夏（7种）、香港（7种）、北京（7种）、天津（2种）、山东（1种）和澳门（0种）受威胁哺乳动物比例低于20%。

- ### 3.5 Proportion of threatened and non-threatened mammals in provinces and regions of China

The spatial distribution of threatened mammals is also uneven across China's provinces (including the municipalities directly under the Central Government, Figure 3.3). Threatened mammal species in most provinces account for 20%-25% of the mammal population in those provinces. Proportion of mammals threatened in the provinces in southwest China is higher than in other regions. The proportion of threatened mammals in Yunnan (91 species). Tibet (59 species), and Chongqing (31 species) are higher than the national average. The proportion of threatened mammals in Shaanxi (29 species), Hubei (17 species), Fujian (22 species), Hebei (12 species), Ningxia (7 species), Hong Kong (7 species), Beijing (7 species), Tianjin (2 species), Shandong (1 species) and Macao (0 species) are less than 20% of local mammalian fauna.

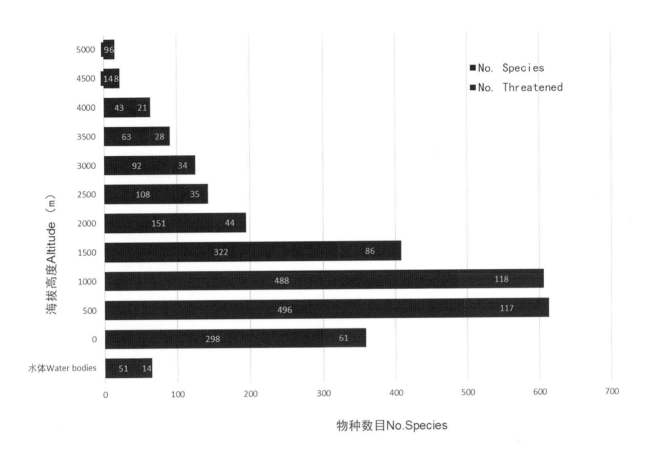

图3.2 不同海拔高度生境中的哺乳动物与受威胁的哺乳动物

Figure 3.2 Mammals and threatened mammals in habitats at different elevations

· 3.6 保护措施

联合国《生物多样性公约》于1992年5月22日通过，是1992年联合国环发大会"21世纪议程"框架下的三个称为"里约公约"的重要多边环境协议之一。1992年6月在里约热内卢联合国环境与发展大会上开放签字。1993年12月29日正式生效，目前有194个缔约方，中国在里约热内卢联合国环境与发展大会签署了该公约。生物多样性保护、生物多样性可持续利用与遗传资源惠益共享是《联合国生物多样性公约》的三大支柱。野生动物保护是生物多样性保护的重要组成部分。

· 3.6 Protection Measures

Adopted on 22 May 1992, the United Nations Convention on Biological Diversity (CBD) is one of the three important multilateral environmental agreements known as the "Rio Conventions" under the framework of Agenda 21 of the 1992 United Nations Conference on Environment and Development. Opened for signature at the United Nations Conference on Environment and Development, Rio de Janeiro, June 1992. It entered into force on December 29, 1993, and now the CBD has 194 parties. China signed the Convention at the United Nations Conference on Environment and Development in Rio de Janeiro. Biodiversity conservation, sustainable use of biodiversity and benefit-sharing of genetic resources are the three pillars of the United Nations Convention on Biological Diversity. Wildlife conservation is an important part of biodiversity conservation.

联合国《生物多样性公约》规定生物多样性保护的途径有就地保护与迁地保护。前者指在物种的生境中保育物种的种群、栖息地及其所在生态系统，后者指当物种原生境不复存在时，将濒危物种迁移到其他生境或人工生境中

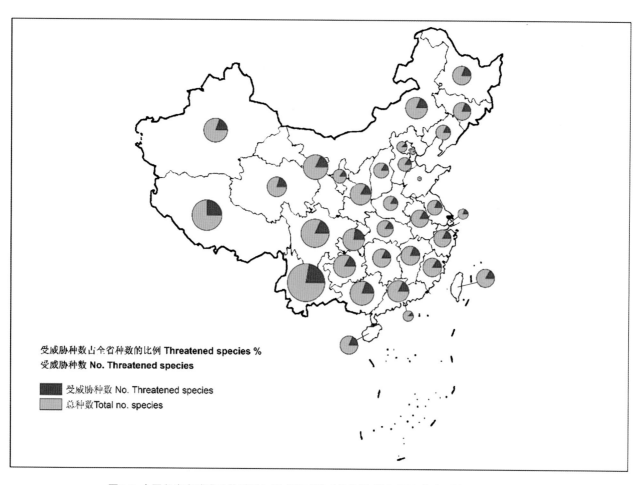

图3.3 中国各省市哺乳动物种数与受威胁哺乳动物种数（数据源自蒋志刚等，2021）

Figure 3.3 Number of mammal species and threatened mammal species across China's provinces (based on Jiang et al., 2021)

进行保育。中国积极履行生物多样性公约，保护濒危、有科学、文化和社会价值的野生动物。根据《中华人民共和国野生动物保护法》，国家野生动物主管部门主持制定了三个野生动物保护名录:《国家重点保护野生动物名录》《有重要生态、科学、社会价值的陆生野生动物名录》和《人工繁育国家重点保护陆生野生动物名录》。

According to the UN Convention on Biological Diversity (CBD), there are two ways to protect biodiversity: in situ conservation and ex situ conservation. The former refers to the conservation of species populations, habitats and ecosystems in their habitats, and the latter refers to the relocation of endangered species to other habitats or artificial habitats for conservation when the original habitat no longer exists. China has actively implemented the Convention on Biological Diversity to protect endangered wild animals of scientific, cultural, and social value. According to the Wildlife Protection Law of the People's Republic of China, the state wildlife authorities have presided over the formulation of wildlife protection lists: *The List of State Key Protected Wild Animals, The List of Terrestrial Wildlife of Important Ecological, Scientific and Social Values under State Protection and The List of Artificially bred Terrestrial Wildlife under State key Protection.*

1989年国家野生动物主管部门发布了《国家重点保护野生动物名录》。2003年和2020年分别将麝类、穿山甲所有种调升为国家一级重点保护野生动物。2017年《中华人民共和国野生动物保护法》修订后，经国务院批准，国家林业和草原局、农业农村部于2021年2月联合发布调整的《国家重点保护野生动物名录》。调整后的《名录》共列入野生动物980种和8类，其中国家一级重点保护野生动物234种和1类、国家二级重点保护野生动物746种和7类。上述物种中，686种为陆生野生动物，294种和8类为水生野生动物，其中有哺乳动物185种。此外，各省、自治区、直辖市人民政府组织制定、公布地方重点保护野生动物名录。

2000年8月1日以原国家林业局令第7号发布实施《国家保护的有益的或者有重要经济、科学研究价值的陆生野生动物名录》，简称为"三有名录"，列入其中的物种简称"三有野生动物"。2016年野生动物保护法修订后，这个名录修改为《有重要生态、科学、社会价值的陆生野生动物名录》。2023年，国家林业和草原局公布了调整的《有重要生态、科学、社会价值的陆生野生动物名录》，与2000年首次发布的原名录相比，新"三有"名录在基本保留原有种类的同时，新增了700多种野生动物，其中哺乳类91种，大幅扩大了中国野生动物保护范围。

In 1989, the State wildlife authorities issued the List of State Key Protected Wild Animals. In 2003 and 2020, all species of musk deer and pangolin were upgraded to the National First-Category Protected Wild Animals. After the Wildlife Protection Law of the People's Republic of China was amended in 2017, the National Forestry and Grassland Administration and the Ministry of Agriculture and Rural Affairs jointly issued the revised List of State Key Protected Wild Animals in February 2021. A total of 980 species and 8 categories of wildlife are included in the amended list, including 234 species and 1 category under 1st Category State Protection and 746 species and 7 categories under 2nd Category State Protection. Of those key state protected species, 686 species are terrestrial, 294 species and 8 groups are aquatic. Among them are 185 mammal species. In addition, the governments of provinces, autonomous regions, and municipalities directly under the Central Government shall formulation and promulgation local lists of wildlife under special protection.

The List of Terrestrial Wild Animal of Ecological, Economic and Scientific Values under State Protection (TWIESS) was promulgated and implemented on August 1, 2000 by Order No. 7 of the formal State Forestry Administration, which is referred to as the "Sanyou List" for short, and the species included in the list are referred to as "Sanyou" wildlife. After the Wildlife Protection Law was revised in 2016, the list was changed to List of Terrestrial Wildlife of Ecological, Scientific and Social Values under State Protection. In 2023, the National Forestry and Grassland Administration proclaimed the revised List of Terrestrial Wildlife of Important Ecological, Scientific, and Social Values. Compared with the original list first proclaimed in 2000, the new List of TWIESS retained the species of the 2000 TWIESS list while adding more than 700 species of wildlife, including 91 species of mammals, to the new List of TWIESS. Thus, the TWIESS revision significantly expands the scope of wildlife protection in China.

野生动物保护级别由主管部门组织专家评估，根据受威胁程度、稀有珍贵程度以及经济价值等多个因素而确定，其种群与栖息地受法律保护。

（1）濒危属性。IUCN受威胁物种等级标准依据种群数量、分布区大小、种群数量下降速率、分布区面积下降速率以及极小种群与分布区等参数来确定物种濒危等级。专家依据濒危等级标准评定的物种的濒危等级。濒危等级为"极危""濒危"和"易危"的物种统称为"受威胁物种"。一般地，"受威胁物种"是重点保护动物的属性，"极危""濒危"物种是国家一级重点保护野生动物的属性之一。

（2）特有属性。中国是世界上哺乳动物种最多的国家，中国特有动物是国家重点保护野生动物的属性之一，其原生种群与栖息地是中国生物多样性保护的重点。大熊猫是中国特有动物，1989年实施的《国家重点保护野生动物名录》中，大熊猫被列为国家一级重点保护野生动物。

（3）稀有属性。稀有动物指在野外罕见、数量极少，或在野外仅见于一个地点的野生动物，这些稀有动物容易灭绝。动物的稀有属性是国家重点保护野生动物的属性之一。

（4）珍贵属性。根据野生动物保护法，国家对珍贵、濒危的野生动物实行重点保护。根据野生动物及其衍生物的收藏价值、传统药用价值、战利品狩猎价值越高，其生存压力可能越大，越需要保护。因此，收藏价值、传统药用价值、战利品狩猎值高是国家重点保护野生动物的属性之一。

（5）管理属性。根据管理的需要，不仅仅将物种作为唯一的保护管理单元，可以将不同的分类阶元作为管理单元。

The level of wildlife protection is assessed by experts invited by the authorities and is determined according to several factors such as the degree of threatened, rarity, and scientific, ecological, and social values. Populations and habitats of those species are protected by law.

（1）Endangered attribute. The IUCN Red List Criteria for Threatened Species are based on population size, distribution are a rate of decline of population, rate of decline in the distribution area, and the smallest population and distribution area. The status of a species as assessed by experts according to the criteria as "Critically Endangered" "Endangered" and "Vulnerable", which are collectively referred to as "Threatened Species". Generally, "Threatened Species" shall be an attribute of wildlife under special protection, and "Critically Endangered" or "Endangered" species shall be one of the attributes of wildlife under first category protection at the state level..

（2）Uniqueness. China is the country with the largest number of mammal species in the world. Endemic animals in China are one of the attributes of wildlife under state protection. Populations and habitats of endemic species are the focus of biodiversity conservation in China, e. g. the giant panda is a unique animal in China. It has been listed in the List of State Key Protected Wild Animals in 1989.

（3）Rareness. Rare animals are wild animals that are rare in number, or found only in one place in the wild and are prone to extinction The rarity of animals is one of the attributes of national wildlife protection.

（4）Value attributes. According to the Wildlife Protection Law, the state shall give priority to the protection of rare and endangered wild animals. Depending on the values of wildlife and its derivatives for collection, traditional medicinal and trophy hunting, the greater the survival pressure due to exploitation and the greater the need for protection may be. Therefore, high collection value, traditional medicinal value and trophy hunting value are one of the attributes of national key protected wildlife.

(5) Manage attributes. According to the needs of management, not only the taxon. Species is a conservation management unit, but different taxa like a genus, a family can be taken as management units as well.

1980年，中国加入《濒危野生动植物种国际贸易公约》（CITES）。为保护野生动物和植物物种不致由于国际贸易而过度开发利用而灭绝，公约制定了3份进出口受管制的物种名单，分别为公约附录I、附录II和附录III。附录I包括所有受到和可能受到贸易的影响而有灭绝危险的物种。这些物种的国际贸易受到特别严格的管理，以防止国际贸易进一步危害其生存，并且只有在科研等特殊的情况下才能允许进行贸易。附录II包括两类情况：一类是所有那些目前虽未濒临灭绝，但如对其贸易不严加管理，以防止不利其生存的利用，就可能有灭绝危险的物种；第二类是为了使前一类指明的某些物种标本的贸易能得到有效的控制，而必须加以管理的其他物种。附录III包括任一缔约国认为属其管辖范围内，应进行管理以防止或限制开发利用而需要其他缔约国合作控制贸易的物种。这些受管制的物种标本的国际贸易，需向相关国家的公约管理机构申办允许进出口证明书，海关查验证明书后放行。制定并执行附录是《濒危野生动植物种国际贸易公约》的核心制度，为全球合作保护濒危物种发挥了十分积极的作用。

China ratified the Convention on International Trade of Species of Wild Fauna and Flora (CITES) in 1980. To protect wild animal and plant species from extinction due to over-exploitation by international trade, the Convention has established three appendixes of species whose import and export are subject to control. Appendix I includes all species that are or may be at risk of extinction as a result of international trade. International trade in these species is particularly tightly regulated to prevent international trade from further endangering their survival, trade is permitted only for scientific research. Appendix II covers two types of situations: one is for all species that are not currently endangered but are at risk of extinction if their trade is not tightly regulated to prevent exploitation that is detrimental to their survival; the second category is for other species that must be managed in order that the trade in specimens of certain species specified in the former category may be effectively controlled. Appendix III includes species within the jurisdiction of any State Party that shall be managed to prevent or limit exploitation and which require the cooperation of other State Parties to control the trade. International trade in these controlled species specimens requires import and export permits to be obtained from the convention authority of the relevant State Party. The customs will examine the certificate in exportation. The formulation and implementation of the appendix is the core of CITES, which has played a positive role in global cooperation to protect endangered species.

为了避免物种的不确定性与鉴别困难，《濒危野生动植物种国际贸易公约》（CITES）采用了在其附录中列入了比种高或者低的分类单元的做法，这些比种高或者低的分类单元通常是分类上不存在争议的，从而回避了种的分类不明确的问题。修订的《国家重点保护野生动物名录》也采取这种将一个属、一个科甚至一个目"集体列入原则"，以回避分类争议，解决执法鉴定的困难。修订的《国家重点保护野生动物名录》还参照CITES附录，对人工驯养繁殖二代以上的种群采取不同的管理方式，将这些野生动物的野生种群与人工驯养种群的保护级别分列不同的保护级别。利用"拆分列入原则"，解决物种保护的实际问题。

In order to avoid uncertainty and difficulty in species identification, CITES adopted listing higher or lower taxa in CITES appendixes, usually, these taxa of a higher or lower rank are not controversial on their taxonomy, to avoid such problem of disputing classification. The revised List of State Key Protected Wild Animals also adopts the principle of "Collective Listing" of a genus, a family or even an order to avoid disputes over taxonomy and reduce the identification difficulties in law enforcement. Many mammals are bred in captivity for commercial purposes. To solve practical problems of conservation of the artificially bred or propagated populations of those species, the principle of "Split Listing" is adapted. The revised list also considers the example of CITES appendices to exemple the wild animals and plants that have been artificially bred or propagated for more than two generations, and separates the artificial bred or propagated wild animals and plants into different protection levels.

国家对野生动物实行保护优先、规范利用、严格监管的原则。对人工繁育技术成熟稳定的国家重点保护野生动物，经科学论证，纳入国务院野生动物保护主管部门制定的《人工繁育国家重点保护陆生野生动物名录》。对列入名录的野生动物及其制品，可以凭人工繁育许可证，按照省、自治区、直辖市人民政府野生动物保护主管部门核验的年度生产数量，直接取得人工繁育国家重点保护野生动物专用标识，凭专用标识出售和利用，保证可追溯。列入人工繁育技术成熟稳定名录的国家重点保护野生动物及其制品，按照规定取得人工繁育许可证和专用标识，属于生产经营性珍贵濒危野生动物。

The state shall follow the principles of giving priority to protection, standardizing sustainable utilization and strict supervision over wildlife management. Wild animals that have matured artificial breeding technology shall, upon scientific verification, be included in the *List of Wild Animals with Matured Artificial Breeding Technology under State Key Protection*, which is formulated by the wildlife management authority. Upon applying for an artificial breeding license from the wildlife management authority of provincial governments, the listed species can be breed for commercial use. Producers should apply for annual quanta of production, and label their products with a Special Artificial Bred Wildlife Product Label before selling. The label is to ensure the traceability of wildlife products by the national wildlife management authority. In a word, the wildlife breeders in the country shall have obtained artificial breeding licenses and special labels in accordance with relevant regulations.

应注意濒危物种与保护物种的异同点。濒危物种是一个保护生物学概念，而保护物种是一个法律概念，其外延更大。濒危物种是专家依据濒危等级标准评定的物种，国家保护物种是国家法定的保护物种。濒危等级则是确定保护等级的重要因素。国家重点保护物种具有较高的经济、文化、科学、社会价值，既包括濒危物种，也包括珍贵的物种，甚至包括开发利用强度大的物种。濒危等级的评定由专家依据濒危等级标准评定，主要考虑物种的生物学属性；而保护等级由国家主管部门确定，除了生物学属性，还有社会经济属性和管理属性。野生动物保护级别则是由国家野生动物主管部门组织专家评估，根据受威胁程度、稀有珍贵程度以及经济价值等多个因素而确定为国家一级或二级重点保护野生动物，它们的种群与栖息地受法律保护（蒋志刚，2019）。

Attention should be paid to the similarities and differences between endangered species and legal protected species. Endangered species are a concept of conservation biology, whereas legal protected species are a legal concept with a larger extension. Endangered species are the species assessed by experts according to the standard of endangered species criteria, and the nationally protected species are the species legally protected by the state. Endangered category is an important attribute in determining the protection levels. The state's key protected species include endangered species also include those species with high economic, cultural, scientific, and social values, and even those that are heavily exploited and utilized. The assessment of the endangered species is made by experts according to the endangered species criteria, mainly considering the biological attributes of the species. The level of protection is determined by the wildlife authorities of the state. In addition to biological attributes, there are also socio-economic attributes and management attributes. The level of wildlife protection is assessed by experts organized by the national wildlife authorities and the species are split into the first or second category of state key protected wild animals according to the degree of threat, rarity, and economic value. Their populations and habitats are protected by law (Jiang, 2019).

《保护迁徙物种公约》（Convention on the Conservation of Migratory Species of Wild Animals，CMS）致力于保护迁徙野生动物物种及其栖息地和迁徙路线。该公约包括两份附录，经评估全部或绝大部分种群受到绝灭威胁的迁徙物种被列入该公约附录I，需要签订国际协定来加强国际合作的物种列入该公约附录II。根据2020年5月22日通过的最新决定，CMS附录I有180个物种，包括雪豹、盘羊等；CMS附录II有数千个物种，包括各种鲸类、海龟、迁徙候鸟等。中国目前不是该公约的缔约国，仅参与该公约附录物种高鼻羚羊的保护谅解备忘录工作。

The Convention on the Conservation of Migratory Species of Wild Animals (CMS) is committed to protecting migratory wildlife species, their habitats and migration routes. The CMS has two appendices, and species that have been assessed to be threatened with extinction in whole or in most of their populations are listed in the Appendix I of the Convention. Species that require international agreements to enhance international cooperation are listed in the Appendix II of the Convention. According to the latest decision adopted on 22 May 2020, there are 180 species in CMS Appendix I, including snow leopard, argali, etc. CMS Appendix II contains thousands of species, including various cetaceans, sea turtles, migratory birds, etc. China is not a party to CMS but participates only in the Memorandum of Understanding on the Conservation of the Saiga Antelope, an appendix species to the Convention.

物种就地保护途径有建立自然保护地和保护物种生境两条途径。自然保护地是通过法律或其他有效手段得到承认、专用和管理，以实现对具有相关生态系统服务和文化价值的长期自然保护之界线明确的地理空间（世界自然保护联盟www.iucn.org）。中国已经建设形成了由国家公园、自然保护区、自然公园（包括风景名胜区、森林公园、地质公园、自然文化遗产、湿地公园、沙漠公园、水产种质资源保护区、海洋公园、海洋特别保护区、自然保护小区）等自然保护地，组成了中国自然保护地体系。截至2018年，中国各类自然保护地总数1.18万处，其中国家级自然保护地3766处。各类陆域自然保护地总面积约占陆地国土面积的18%以上，已超过世界平均水平。其中，占重要地位的自然保护区数量达2729个，总面积147万平方千米，占陆地国土面积的14.8%，占所有自然保护地总面积的80%以上；风景名胜区和森林公园约占3.8%（Zhang et al., 2016）。

There are two approaches to protect species in situ: to establish protected areas and to protect species' habitat. Protected areas are defined geographic spaces recognized, dedicated, and managed by legal or other effective means to achieve the long-term conservation of nature with relevant ecosystem services and cultural values (IUCN www.iucn.org). China has formed a protected area system which is constituted of the national parks, nature reserves, the natural parks (including the scenic spots, forest parks, geological parks, natural and cultural heritage sites, wetland parks, desert parks, aquatic germplasm resource reserves, ocean parks, special marine reserve, the mini nature reserves). By 2018, China had 11,800 protected areas of more than 10 categories, of which 3,766 were state-level protected areas. The total area of protected areas accounts for more than 18% of the total land area, surpassing the world average. Among them, there are 2,729 nature reserves with a total area of 1.47 million square kilometers, accounting for 14.8% of the total land area and more than 80% of the total area of all protected areas. Scenic spots and forest parks account for about 3.8 % of the total area of all protected areas (Zhang et al., 2016).

继2021年10月联合国《生物多样性公约》第十五次缔约方大会第一阶段会议在中国云南昆明成功举办后，2022年12月19日，在加拿大蒙特尔举行的联合国《生物多样性公约》第十五次缔约方大会（CBD COP15)第二阶段会议，通过了"昆明－蒙特尔全球生物多样性框架"，提出反转生物多样性丧失曲线的宏伟目标，确立了3030目标，即2030年至少30%的陆地、内陆水域、海岸带和海洋区域得到有效保护；至少30%的生态系统得到有效恢复，使具有高度生物多样重要性的区域的丧失趋近于零。明确了今后全球生物多样性保护的目标。如何实现"昆明－蒙特尔全球生物多样性框架"设定的3030目标，中国面临着艰巨的任务。

Following the successful convening of the 15th Session of the Conference of the Parties of the United Nations Convention on Biological Diversity (COP-15 CBD) in Kunming, Yunnan, China in October 2021, on 19 December 2022, the second phase of the CBD COP15 held in Montreal, Canada, adopted the "Kunming-Montreal Global Biodiversity Framework", which proposed the ambitious goal of reversing the loss curve of biodiversity, and established the 3030 Goal. Namely, at least 30% of land, inland waters, coastal zones, and marine areas are effectively protected; At least 30% of the ecosystem has been effectively restored, consequently, making the loss of areas of high biodiversity importance close to zero. The goal of global biodiversity conservation in the future is clarified. How to achieve the 3030 Goa; set by the "Kunming-Montreal Global Biodiversity Framework", China faces an arduous task.

野生动物栖息地指野生动物野外种群生息繁衍的重要区域，国家保护野生动物及其栖息地。国家陆生野生动物主管部门已经按照2017年修订发布的《中华人民共和国野生动物保护法》着手制定野生动物重要栖息地名录，以保护野生动物重要栖息地，县级以上人民政府将制定野生动物及其栖息地相关保护规划和措施。国家实施国家公园等自然保护地建设和野生动植物保护工程。对大熊猫、亚洲象、海南长臂猿、东北虎、中华穿山甲等濒危野生动物及其栖息地继续实施抢救性保护，划定并严格保护重要栖息地，联通生态廊道，重要栖息地面积增长10%，推进野生动物及其栖息地保护。

"Habitat for wild animals" means an important area where wild animals live and reproduce. The state protects wild animals and their habitats. The state department in charge of terrestrial wildlife has formulated a list of important habitats for wildlife in accordance with the Wildlife Protection Law of the People's Republic of China, which was revised and proclaimed in 2017, to protect important habitats for wildlife. Governments at or above the county level shall formulate plans and measures for the protection of wildlife and their habitats. The state implements the construction of national parks and other protected areas and wildlife protection projects, continues to implement rescue measures for endangered wild animals such as the Giant Panda, Asian Elephant, Hainan Gibbon, Amur Tiger, and Chinese Pangolin, delimits and strictly protects important habitats, connects ecological corridors, increases the area of important habitats by 10 percent of State Key Protected Wild Animals in the near future.

· 3.7 保护成效

中国濒危物种的保护措施已见成效，重要哺乳动物的种群数量开始恢复，重要栖息地得到保护。如中国采取就地保护与迁地保护相结合方式加强对大熊猫的保护。现已建设了67处大熊猫自然保护区，覆盖了66.8%的野生大熊猫种群和53.8%的大熊猫栖息地。第四次大熊猫调查发现野生大熊猫种群数量比第三次大熊猫调查增长16.8%，达到1864只。2003年，国家组织启动了大熊猫野化放归工作，先后将11只人工繁育大熊猫放归自然。其中，9只野外放归成功。2018年，大熊猫国家公园正式挂牌。此外，2020年中国人工圈养繁殖的大熊猫已经超过600只。2015年的中国生物多样性红色名录评估中，大熊猫的濒危等级降为"易危（VU）"，2021年评估中仍将大熊猫的濒危等级定为"易危（VU）"。

藏羚*Pantholops hodgsonii*是青藏高原的代表性哺乳动物，20世纪初数以百万计的藏羚生活在高原腹地的无人区。20世纪末，由于盗猎，藏羚数量仅余6万～7万只（Schaller，1998）。20世纪末以来，中国政府在藏羚分布区先后建立了西藏羌塘，青海可可西里、三江源，新疆阿尔金山、中昆仑及西昆仑等国家级和省级自然保护区。2016年，建立了三江源国家公园，2017年，可可西里自然遗产地成功列入世界自然遗产的名录。目前青藏高原建成了中国面积最大的自然保护群。藏羚分布区各县也组建了森林公安派出所，加强了藏羚的保护和执法力度，有效地遏制了盗猎藏羚活动，加强了藏羚生境的保护。2021年，青藏高原的藏羚数量已经回升到30万余只。鉴于藏羚种群恢复状况，2015年起，中国生物多样性红色名录评估将藏羚降为近危（NT）等级，将其从受威胁物种名单中剔除。

· 3.7 Effectiveness of conservation

China's efforts to protect endangered species have brought forward positive results, as populations of important mammals have begun to recover and important habitats have been protected. For example, China has strengthened the protection of the Giant Panda by combining in-situ conservation with ex-situ conservation. There are now 67 giant panda nature reserves, covering 66.8 % of the wild giant panda population and 53.8 % of its habitat. The Fourth Giant Panda Survey reported that the number of wild giant pandas increased to 1,864 (16.8 %)

since the Third National Survey. In 2003, the State launched a campaign to reintroduce pandas into the wild, and released 11 artificially bred pandas into the wild. Among them, nine were released successfully. In addition, more than 600 giant pandas have been bred in captivity in the country in 2020. In 2018, the Giant Panda National Park was officially inaugurated. In view of the recovery of the giant panda population, the status of the Giant Panda's was downgraded to "Vulnerable" in the 2015 China's Red List of Biodiversity assessment, and it is still listed as "Vulnerable" in the 2021 assessment.

By the end of the 20th century, there were only 60,000-70,000 Tibetan Antelope survived in the heartland of the Qinghai-Tibetan Plateau due to heavy poaching of the species for this wool (Schaller, 1998). Since then, the Chinese government has set up national and provincial nature reserves in areas where Tibetan Antelope are distributed, such as Qiangtang Nature Reserve in Tibet, Hoh Xil Nature Reserve in Qinghai, The Sanjiangyuan (Source of Three Rivers) Nature Reserve in Qinghai, The Altun Mountains Nature Reserve in Xinjiang, Central Kunlun, and West Kunlun Nature Reserve. The Sanjiangyuan National Park was established in 2016. Hoh Xil Natural Heritage Site has been listed as a World Natural Heritage Site in 2017. Consequently, the largest protected natural area network in China has been established on the Qinghai-Tibet Plateau. Forest police stations have also been set up in the ranging counties of Tibetan antelope, strengthened the law enforcement of protection of Tibetan Antelope, effectively curtailing poaching and protecting the habitat of Tibetan Antelope. At present, the Tibetan Antelope population on the Qinghai-Tibetan Plateau has recovered to nearly 300,000 in 2021. In view of the population recovery, the status of the Tibetan Antelope was downgraded to the Near Threatened (NT) level and it has been removed from the threatened species list in China's Red List of Biodiversity since 2015.

普氏原羚是青藏高原地区特有的濒危有蹄类动物，是中国濒危物种保育比较成功的例子。1988年颁布的《国家重点保护野生动物名录》将其列为国家一级重点保护野生动物。2001年制定的《全国野生动植物保护和自然保护区建设工程总体规划》中将其列为全国15个亟须拯救的物种之一。2002年12月，原国家林业局编制了《全国普氏原羚保护工程总体规划》(蒋志刚等，2004)。IUCN红色名录1996年将其列为极危（CR），由于种群恢复及新种群的发现，2009年将其濒危等级由极危（CR）调整为濒危（EN）。2015年，中国生物多样性红色名录评估结果中仍将其列为极危（CR）。2020年的评估中，鉴于普氏原羚种群的恢复，其濒危等级下调为"濒危"。

麋鹿*Elaphurus davidianus*曾经一度在中国野生灭绝。20世纪初，中国最后一群麋鹿在中国北京南苑皇家猎苑毁于洪灾与战火。20世纪80年代从国外重新引入麋鹿，现在引种到24个省81个地点，已经分别建立了江苏大丰、北京南苑、湖北天鹅洲等迁地保育种群。2022年，中国麋鹿种群数量达14000多头。目前，在江苏大丰黄海海滨和洞庭湖地区分别通过人工释放和自然野化建立了野化种群。由于麋鹿的生存状况改善，在2021年中国生物多样性红色名录评估中，麋鹿的濒危等级由"野外灭绝"降为"极度濒危"。

Przewalski's Gazelle is an endangered gazelle endemic to the Qinghai-Tibet Plateau and another successful example of endangered species conservation in China. The List of State Key Protected Wild Animals, proclaimed in 1989, listed Przewalski's Gazelle as First Category State Key Protected Wild Animal Species. In the National Wildlife Protection and Nature Reserve Construction Project Master Plan in 2001, Przewalski's Gazelle was listed as one of the country's 15 species needing urgent rescue (Jiang et al., 2004). In December 2002, the State Forestry Administration compiled the National Master Plan for Conservation of Przewalski's Gazelle. The IUCN Red List listed it as Critically Endangered (CR) in 1996. Due to the recovery of the population and the discovery of new populations, its status was changed from CR to EN in the 2009. In the 2021 China's Red List of Biodiversity assessment, the status of Przewalski's Gazelle was downgraded to "Endangered" due to its population recovery.

Père David's Deer (*Elaphurus davidianus, Milu deer*) were once extinct in the wild in China. In the early 20th century, China's last herd of Père David's Deer was destroyed by floods and wars at the Nanyuan Royal Hunting Park in Beijing, China. Père David's Deer were reintroduced from England in the mid-1980s. The reintroduced Père David's Deer adapted the local environments in Dafeng of Jiangsu Province, Nanyuan of Beijing, and Tianezhou of Hubei Province and breeds successfully. Père David's Deer have now been introduced to 81 sites in 24 provinces in the country. In 2022, the population of Milu deer in China reached more than 12,000. At present, wild populations have been established in the coastal area of Yellow Sea and the wetland of Dongting Lake by artificial release and natural rewilding. In the 2021 China's Red List of Biodiversity assessment, the status of Père David's Deer was down from "Extinct in Wild" to "Critically Endangered" due to improvements in their survival status.

然而，中国人口众多，在一些地区人兽冲突问题凸显，如亚洲象*Elephas maximus*与人类的冲突。历史上，由于气候变化和人类活动的影响，亚洲象在中国分布区向南退缩。目前亚洲象的分布区从云南西双版纳和临沧市扩大到了普洱市。经过30多年的保护，尽管存在尖锐的人象冲突，如热带雨林砍伐等问题，目前中国亚洲象的种群数量从20世纪80年代100多头增长到300余头。亚洲象种群增加后，不断地寻找新的栖息地。曾有独象或小群体扩散到老挝和中国云南沧源的零星记录。20世纪90年代初，亚洲象从西双版纳扩散进入普洱市，在此定居后并不断扩展活动范围，与当地人的生活及生产发生了冲突。2020年3月一群亚洲象从云南西双版纳国家级自然保护区勐养子保护区出发，经思茅、墨江、元江、红河、石屏等地，北上500多千米，2021年6月一度进入了昆明市晋宁区境内。2021年6月另一群亚洲象进入中国科学院西双版纳植物园。人兽冲突已经成为中国野生动物保护种的一个热点问题。如何平衡经济发展与濒危物种保育问题，仍是一项重大议题。

However, China has a large human population, and in some areas, human-animal conflict intensifies as wild animal populations recover, such as the conflict between Asian Elephants (*Elephas maximus*) and humans. Historically, Asian Elephants have retreated southward in their ranges in China due to climate change and human encroachment of their habitats. Since the protection of wild Asian Elephants was implemented in the late 1980s, the population of the Asian Elephant in China is increasing from about 100 to about 300, despite the acute human-elephant conflict and the deforestation of the tropical rain forests, the wild Asian Elephants are constantly looking for new habitats. Currently, the Asian Elephant's distribution area has expanded from Xishuangbanna and Lincang to Pu 'er City in Yunnan. There were sporadic records of small living groups or lone elephants crossing the border to Laos or migrating to Cangyuan, Yunnan. In the early 1990s, Asian Elephants moved into Pu 'er City from Xishuangbanna. After settling down here, they continued to expand their range of activities, resulting in conflicts with local people's daily life and production. In March 2020, a group of Asian Elephant started from Mengyang Area of the Xishuangbanna National Nature Reserve, Yunnan, and traveled more than 500 kilometers north through Simao, Mojiang, Yuanjiang, Honghe, Shiping and other places, although their migration has been blocked by people. In June 2021, they once entered the Jining District of Kunming City. In June 2021, another group of Asian Elephants entered the Xishuangbanna Botanical Garden of the Chinese Academy of Sciences (Jiang et al., 2021). Human-animal conflict has become a hot issue in wildlife conservation in China. Balancing economic development with the conservation of endangered species remains a major issue.

中国有29种非人灵长类动物，是北半球同纬度灵长类动物多样性最高的国家。长臂猿生活在亚洲的热带雨林中。中国是世界长臂猿分布区的北缘。中国有长臂猿3属8种(长臂猿属、白眉长臂猿属和黑冠长臂猿属)：长臂长臂猿、白眉长臂猿、白眉长臂猿、高黎贡长臂猿、西方黑冠长臂猿、东方黑冠长臂猿、北方白颊长臂猿和海南长臂猿（Jiang et al., 2021）。

中国有着悠久的人类文明历史，大自然经过长期的开发，长臂猿赖以生存的完整自然植被已所剩无几。栖息地丧失和人类捕猎导致了中国长臂猿的生存危机（Li et al., 2018）。2020年6月19日，中国所有长臂猿列为国家重点保护野生动物。中国所有长臂猿均被列为《IUCN濒危物种红色名录》和《中国濒危脊椎动物红色名录》受威胁物种。中国已在所有长臂猿栖息地建立了自然保护区。

海南长臂猿是中国特有的长臂猿物种，仅在中国海南岛发现。300年前，海南长臂猿在海南岛的分布面积为2.7万平方千米。20世纪50年代，海南岛仍有2000多只长臂猿。20世纪70年代，有7～9只海南长臂猿生活在海南岛霸王岭地区仅存的热带雨林栖息地（Chivers, 2013）。

Bryant et al.（2016）报道在过去的30年里海南长臂猿的遗传多样性减少了约30%。野外标本和博物馆馆藏历史标本存在显著遗传差异。海南长臂猿在历史上经历过多次遗传瓶颈。至19世纪末，海南长臂猿有效种群开始下降（历史 $Ne = 1162.96$ ）。目前海南长臂猿有效种群数量非常小（当前 $Ne = 2.16$ ）。由于其生活史对策适应，如妊娠期和繁殖间隔较大多数长臂猿物种短，繁殖季节与植物果实开花季节重合。因此，数量极少的海南长臂猿可以存活到今天。

20世纪80年代建立自然保护区后，海南长臂猿种群开始恢复，2003年由香港嘉道理农场暨植物园、海南省林业局野生动植物保护中心以及霸王岭林业局组织的专项调查发现13只海南长臂猿，分布于不足21平方千米的原始林中。2007年底，海南长臂猿数量增长了17～20只，生活在2个家庭群中（Fellowes et al., 2008）。2013年发现海南长臂猿增长到22只。2015年新发现了2个长臂猿家庭。2020年8月调查发现了5群33只（Liu et al., 2022）。2022年12月，海南长臂猿增长到38只，生活在6个家庭群中(图3.4)。

作为世界上最濒危的灵长类动物，海南长臂猿踏上了一条惊人恢复之路，其种群数量不断增长，但海南长臂猿种群没有充分实现其繁殖潜力（Liu et al., 2022）。海南长臂猿仍面临一系列风险：干旱等气候灾害可能影响海南长臂猿的食物和栖息地；人畜共患病可能影响海南长臂猿种群健康；由于海南长臂猿种群数量过少，Allee效应可能会影响个体繁殖、种群生存；种群随机性和近交系等因素可能导致新生代性别比失衡。未来海南长臂猿仍将面临种群波动甚至灭绝的风险，应尽快制定风险防范计划。

China has 29 non-human primates, which marks the country as a country with the highest primate diversity in the Northern Hemisphere at the same latitude. Gibbons are found in Asian tropical rainforests. China is the northern margin of the world gibbon range. There are eight species of gibbons in 3 genera (*Hylobate, Hoolock,* and *Nomascus*) in China: *Hylobates lar, Hoolock hoolock, H. leuconedys,* Gaoligong Gibbon *H. tianxing,* Western Black-crested Gibbon *Nomascus concolor,* Eastern Black-crested Gibbon *N. nasutus,* Northern White-cheeked Gibbon *N. leucogenys* and Hainan Gibbon *N. hainanus* (Jiang et al., 2022).

China has a long history of human civilization, and natural ecosystems have been over-exploited. There is little intact natural vegetation left for the gibbons. Habitat loss and human hunting caused a survival crisis for gibbons in China (Li et al., 2018). On 19 June 2020, all gibbons in China were listed as one of the amended State's Key Protected Animals in China. All gibbons in China are listed as threatened species on the IUCN Red List and Critically Endangered on the Red List of China's Biodiversity. China has established nature reserves in all gibbon habitats.

The Hainan gibbon is a gibbon species endemic to China and is found only on Hainan Island, China. Three hundred years ago, Hainan gibbons were distributed in an area of 27,000 km^2 on the Hainan Island. In the 1950s, there were still more than 2,000 Hainan gibbons across the island. In the 1970s, only 7-9 Hainan gibbons lived in the remaining tropical rainforest habitat of the Bawangling Area of Hainan Island (Chivers, 2013).

Bryant et al. (2016) reported that Hainan gibbon genetic diversity had been reduced by about 30% in the past 30 years. There are significant differences between current and historical museum samples. Hainan gibbons have experienced genetic bottlenecks many times in the history. By the end of the 19th century, the effective population of Hainan gibbon had begun to decline (historical effective population size, Ne = 1162.96); The effective population of Hainan gibbon is very small (current Ne = 2.16). Hainan gibbons have acquired some life history traits to cope with their dramatically reduced natural habitat. The gestation period and breeding interval of the Hainan Gibbon are shorter than most gibbon species, and the breeding season coincides with the flowering season of the fruits of the source plants. As a result, the Hainan gibbon, with a very small population, can survive today.

After establishing nature reserves in the 1980s, the Hainan gibbon population began to recover. In 2003, a special survey organized by the KFBG (Hong Kong), the Wildlife Conservation Center of Hainan Forestry Bureau, and the Bawangling Forest Bureau found 13

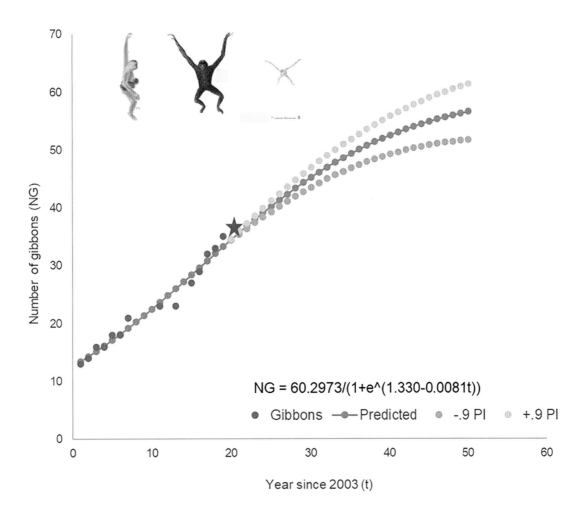

$$NG = 60.2973/(1+e^{(1.330-0.0081t)})$$

图3.4 海南长臂猿种群增长模拟曲线

Figure 3.5 The simulated curve of the Hainan Gibbon population growth

gibbons in an area of less than 21 km^2 in the primary rain-forests. By the end of 2007, the number of Hainan gibbons had increased to 17 to 20 individuals living in two family groups (Fellowes et al., 2008). In 2013, the number of Hainan gibbons increased to 22. Two new gibbon families were discovered in 2015. Thirty-three individuals were found in five family groups in August 2020 (Liu et al., 2022). In December 2022, the Hainan gibbon population grew to 38 individuals in six family groups (Figure 3.4).

As the world's most endangered primate, the Hainan Gibbon is on a remarkable path to recovery. The population of Hainan gibbons is growing, but it did not realize its full reproductive potential (Liu et al., 2022). In the future research period, the Hainan gibbon will still face population fluctuation or even extinction, so risk prevention plans should be made as soon as possible to avoid the following risks: Climate disasters, such as drought, may affect the food and habitats of Hainan gibbons. Zoonosis may also affect the survival of Hainan gibbons. Allee Effect may affect individual breeding, population survival, and the sex ratio of the newborns unbalanced because of population stochasticity.

中国有丰富的哺乳动物多样性。鉴于中国哺乳动物区系的独特性、重要性和多样性，以及如此多的哺乳动物濒临灭绝的事实。自从《中华人民共和国野生动物保护法》实施以来，通过就地保护与迁地保护两条途径，实施野外灭绝物种的重引入，一些中国濒危哺乳动物如大熊猫与藏羚的生存状况有所好转。然而，中国地形地貌的复杂性和区域经济发展不平衡性，中国哺乳动物受威胁程度高于世界平均水平。如何拯救这些濒危物种是中国生物多样性保护的一项艰巨任务。

China has a rich diversity of mammals. Given the uniqueness, importance, and diversity of China's mammalian fauna, but also the fact that so many mammals are endangered. Since the implementation of the Wildlife Protection Law of the People's Republic of China, the survival of some of China's endangered mammals, such as the Père David's Deer, Giant Panda and the Tibetan Antelope, has improved through the reintroduction of wild species through both on-site and off-site conservation. However, due to the complexity of China's landform and uneven regional economic development levels, plus the proportion of threatened mammals in China is higher than the world average. How to save these endangered species is a difficult task for the biodiversity conservation in the country.

目 Order	科 Family	属 Genus	中文名 Chinese name	学名 Scientific name	命名人 Species authority	英文名 English name	页码 page
劳亚食虫目 Eulipotyphla							
	猬科 Erinaceidae						
		毛猬属 Hylomys					
			毛猬	Hylomys suillus	Müller, 1840	Short-tailed Gymnure	0086
		新毛猬属 Neohylomys					
			海南新毛猬	Neohylomys hainanensis	Shaw & Wang, 1959	Hainan Gymnure	0088
		中国鼩猬属 Neotetracus					
			中国鼩猬	Neotetracus sinensis	Trouessart, 1909	Shrew Gymnure	0090
		刺猬属 Erinaceus					
			东北刺猬	Erinaceus amurensis	Schrenk, 1859	Amur Hedgehog	0092
		大耳猬属 Hemiechinus					
			大耳猬	Hemiechinus auritus	(Gmelin, 1770)	Long-eared Hedgehog	0094
		林猬属 Mesechinus					
			达乌尔猬	Mesechinus dauuricus	(Sundevall, 1842)	Daurian Hedgehog	0096
			侯氏猬	Mesechinus hughi	(Thomas, 1908)	Hugh's Hedgehog	0098
			小齿猬	Mesechinus miodon	(Thomas, 1908)	Small-toothed Hedgehog	0100
			高黎贡林猬	Mesechinus wangi	He, Jiang & Ai, 2018	Wang's Hedgehog	0102
	鼹科 Talpidae						
		鼩鼹属 Uropsilus					
			等齿鼩鼹	Uropsilus aequodonenia	Liu, 2013	Equivalent Teeth Shrew Mole	0104
			峨眉鼩鼹	Uropsilus andersoni	(Thomas, 1911)	Anderson's Shrew Mole	0106
			栗背鼩鼹	Uropsilus atronates	(Thomas, 1911)	Black-backed Shrew Mole	0108
			大别山鼩鼹	Uropsilus dabieshanensis	Hu, Xu, Zhang, Liu, Liao, Yang, Sun, Shi, Ban, Li, Liu, Zhang, 2021	Dabie Mountains Shrew Mole	0110
			长吻鼩鼹	Uropsilus gracilis	(Thomas, 1911)	Gracile Shrew Mole	0112
			贡山鼩鼹	Uropsilus investigator	(Thomas, 1911)	Inquisitive Shrew Mole	0114
			雪山鼩鼹	Uropsilus nivatus	(Thomas, 1911)	Snow Mountain Shrew Mole	0116
			少齿鼩鼹	Uropsilus soricipes	Milne-Edwards, 1871	Chinese Shrew Mole	0118
		高山鼹属 Alpiscaptulus					
			墨脱鼹	Alpiscaptulus medogensis	Jiang & Chen, 2021	Modong Mole	0120

目 Order	科 Family	属 Genus	中文名 Chinese name	学名 Scientific name	命名人 Species authority	英文名 English name	页码 page
		针尾鼹属 *Scaptonyx*					
			针尾鼹	*Scaptonyx fusicaudus*	Milne-Edwards, 1872	Long-tailed Mole	0122
		甘肃鼹属 *Scapanulus*					
			甘肃鼹	*Scapanulus oweni*	Thomas, 1912	Gansu Mole	0124
		东方鼹属 *Euroscaptor*					
			巨鼹	*Euroscaptor grandis*	Miller, 1940	Greater Chinese Mole	0126
			库氏长吻鼹	*Euroscaptor kuznetsovi*	Zemlemerova, 2016	Kouznetsov's Mole	0128
			长吻鼹	*Euroscaptor longirostris*	Milne-Edwards, 1870	Long-nosed Mole	0130
			奥氏长吻鼹	*Euroscaptor orlovi*	Zemlemerova, 2016	Orlov's Mole	0132
		缺齿鼹属 *Mogera*					
			海南缺齿鼹	*Mogera hainana*	Thomas, 1907	Hainan Mole	0134
			台湾缺齿鼹	*Mogera insularis*	(Swinhoe, 1863)	Taiwan Mole	0136
			鹿野氏缺齿鼹	*Mogera kanoana*	Kawada, Shinohara, Kobayashi, Harada, Oda & Lin, 2007	Kano's Mole	0138
			华南缺齿鼹	*Mogera latouchei*	Thomas, 1907	La Thou's Mole	0140
			大缺齿鼹	*Mogera robusta*	Nehring, 1891	Large Mole	0142
			钓鱼岛鼹	*Mogera uchidai*	Abe, Shiraishi & Arai, 1991	Uchida's Mole	0144
		白尾鼹属 *Parascaptor*					
			白尾鼹	*Parascaptor leucura*	(Blyth, 1850)	White-tailed Mole	0146
		麝鼹属 *Scaptochirus*					
			麝鼹	*Scaptochirus moschatus*	Milne-Edwards, 1867	Short-faced Mole	0148
	鼩鼱科 Soricidae						
		鼩鼱属 *Sorex*					
			天山鼩鼱	*Sorex asper*	Thomas, 1914	Tien Shan Shrew	0150
			小纹背鼩鼱	*Sorex bedfordiae*	Thomas, 1911	Lesser Stripe-backed Shrew	0152
			中鼩鼱	*Sorex caecutiens*	Laxmann, 1788	Laxmann's Shrew	0154
			甘肃鼩鼱	*Sorex cansulus*	Thomas, 1912	Gansu Shrew	0156
			纹背鼩鼱	*Sorex cylindricauda*	Milne-Edwards, 1872	Stripe-backed Shrew	0158
			栗齿鼩鼱	*Sorex daphaenodon*	Thomas, 1907	Large-toothed Siberian Shrew	0160
			云南鼩鼱	*Sorex excelsus*	G. M. Allen, 1923	Highland Shrew	0162
			细鼩鼱	*Sorex gracillimus*	Thomas, 1907	Slender Shrew	0164
			远东鼩鼱	*Sorex isodon*	Turov, 1924	Ever-toothed Shrew	0166
			姬鼩鼱	*Sorex minutissimus*	Zimmermann, 1780	Eurasian Least Shrew	0168
			小鼩鼱	*Sorex minutus*	Linnaeus, 1766	Eurasian Pygmy Shrew	0170
			大鼩鼱	*Sorex mirabilis*	Ognev, 1937	Ussuri Shrew	0172
			克什米尔鼩鼱	*Sorex planiceps*	Miller, 1911	Kashmir Shrew	0174

目 Order	科 Family	属 Genus	中文名 Chinese name	学名 Scientific name	命名人 Species authority	英文名 English name	页码 page
			扁颅鼩鼱	*Sorex roboratus*	Hollister, 1913	Flat-skulled Shrew	0176
			陕西鼩鼱	*Sorex sinalis*	Thomas, 1912	Chinese Shrew	0178
			藏鼩鼱	*Sorex thibetanus*	Kastschenko, 1905	Tibetan Shrew	0180
			苔原鼩鼱	*Sorex tundrensis*	Merriam, 1900	Tundra Shrew	0182
			长爪鼩鼱	*Sorex unguiculatus*	Dobson, 1890	Long-clawed Shrew	0184
		黑齿鼩鼱属 *Blarinella*					
			川鼩	*Blarinella quadraticauda*	(Milne-Edwards, 1872)	Asiatic Short-tailed Shrew	0186
			狭颅黑齿鼩鼱	*Blarinella wardi*	Thomas, 1915	Burmese Short-tailed Shrew	0188
		异黑齿鼩鼱属 *Parablarinella*					
			淡灰豹鼩	*Parablarinella griselda*	Thomas, 1912	Indochinese Short-tailed Shrew	0190
			安徽黑齿鼩鼱	*Parablarinella latimaxillata*	Chen & Jiang, 2023	Anhui Short-tailed Shrew	0192
		长尾鼩鼱属 *Soriculus*					
			墨脱大爪鼩鼱	*Soriculus medogensis*	Chen & Jiang, 2023	Soriculus medogensis	0194
			倭大爪鼩鼱	*Soriculus minor*	Dobson, 1890	Lesser large-clawed shrew	0196
			大爪长尾鼩鼱	*Soriculus nigrescens*	(Gray, 1842)	Himalayan Shrew	0198
			雪山大爪鼩鼱	*Soriculus nivatus*	Chen & Jiang, 2023	Snow Mountain large-clawed shrew	0200
		须弥鼩鼱属 *Episoriculus*					
			米什米长尾鼩鼱	*Episoriculus baileyi*	Thomas, 1914	Mishimi Brown-toothed Shrew	0202
			小长尾鼩鼱	*Episoriculus caudatus*	(Horsfield, 1851)	Hodgson's Brown-toothed Shrew	0204
			台湾长尾鼩鼱	*Episoriculus fumidus*	(Thomas, 1913)	Taiwan Brown-toothed Shrew	0206
			大长尾鼩鼱	*Episoriculus leucops*	(Horsfield, 1855)	Long-tailed Brown-toothed Shrew	0208
			缅甸长尾鼩鼱	*Episoriculus macrurus*	(Blanford, 1888)	Arboreal Brown-toothed Shrew	0210
			灰腹长尾鼩鼱	*Episoriculus sacratus*	Thomas, 1911	Grey-bellied Shrew	0212
			云南长尾鼩鼱	*Episoriculus umbrinus*	Allen, 1923	Hidden Brown-toothed Shrew	0214
		缺齿鼩属 *Chodsigoa*					
			高氏缺齿鼩	*Chodsigoa caovansunga*	Lunde, Musser & Son, 2003	Van Sung's Shrew	0216
			大别山缺齿鼩	*Chodsigoa dabieshanensis*	Chen, Hu, Pei, Yang, Yong, Xu, Qu, Onditi & Zhang, 2022	Dabieshan Long-tailed Shrew	0218
			烟黑缺齿鼩	*Chodsigoa furva*	Anthony, 1941	Dusky Long-tailed Shrew	0220
			霍氏缺齿鼩	*Chodsigoa hoffmanni*	Chen, He, Huang & Jiang, 2017	Hoffmann's Long-tailed Shrew	0222
			川西缺齿鼩鼱	*Chodsigoa hypsibia*	(de Winton, 1899)	De Winton's Shrew	0224
			云南缺齿鼩鼱	*Chodsigoa parca*	G. M. Allen, 1923	Lowe's Shrew	0226
			滇北缺齿鼩	*Chodsigoa parva*	G. M. Allen, 1923	Pygmy Brown-toothed Shrew	0228

目 Order	科 Family	属 Genus	中文名 Chinese name	学名 Scientific name	命名人 Species authority	英文名 English name	页码 page
			大缺齿鼩鼱	*Chodsigoa salenskii*	(Kastschenko, 1907)	Salenski's Shrew	0230
			斯氏缺齿鼩鼱	*Chodsigoa smithii*	Thomas, 1911	Smith's Shrew	0232
			细尾缺齿鼩鼱	*Chodsigoa sodalis*	Thomas, 1913	Lesser Taiwanese Shrew	0234
		水鼩鼱属 *Neomys*					
			水鼩鼱	*Neomys fodiens*	(Pennant, 1771)	Eurasian Water Shrew	0236
		短尾鼩属 *Anourosorex*					
			阿萨姆短尾鼩	*Anourosorex assamensis*	Anderson, 1875	Assma Mole Shrew	0238
			大短尾鼩	*Anourosorex schmidi*	Petter,1963	Giant Mole Shrew	0240
			微尾鼩	*Anourosorex squamipes*	Milne-Edwards, 1872	Mole-shrew	0242
			台湾短尾鼩	*Anourosorex yamashinai*	Kuroda, 1935	Taiwanese Mole Shrew	0244
		水麝鼩属 *Chimarrogale*					
			喜马拉雅水鼩	*Chimarrogale himalayica*	(Gray, 1842)	Himalayan Water Shrew	0246
			利安德水鼩	*Chimarrogale leander*	Thomas, 1902	Leander Water Shrew	0248
			斯氏水鼩	*Chimarrogale styani*	de Winton, 1899	Chinese Water Shrew	0250
		鼩属 *Nectogale*					
			蹼足鼩	*Nectogale elegans*	Milne-Edwards, 1870	Elegant Water Shrew	0252
			锡金蹼足鼩	*Nectogale sikhimensis*	(De Winton & Styan, 1899)	Sikkim Water Shrew	0254
		臭鼩属 *Suncus*					
			小臭鼩	*Suncus etruscus*	(Savi, 1822)	Etruscan Shrew	0256
			臭鼩	*Suncus murinus*	Linnaeus, 1766	House Shrew	0258
		麝鼩属 *Crocidura*					
			安徽麝鼩	*Crocidura anhuiensis*	Zhang, Zhang & Li, 2020	Anhui Shrew	0260
			灰麝鼩	*Crocidura attenuata*	Milne-Edwards, 1872	Grey Shrew	0262
			东阳江麝鼩	*Crocidura donyangjiangensis*	Liu Y, Chen SD, Liu SY, 2020	Dongyangjiang Shrew	0264
			白尾梢麝鼩	*Crocidura dracula*	Blyth, 1855	Dracula's Shrew	0266
			印支小麝鼩	*Crocidura indochinensis*	Robinson & Kloss, 1922	Indochinese Shrew	0268
			大麝鼩	*Crocidura lasiura*	Dobson, 1890	Ussuri Shrew	0270
			华南中麝鼩	*Crocidura rapax*	G. M. Allen, 1923	Chinese White-toothed Shrew	0272
			山东小麝鼩	*Crocidura shantungensis*	Miller, 1901	Shantung White-toothed Shrew	0274
			西伯利亚麝鼩	*Crocidura sibirica*	Dukelsky, 1930	Siberian Shrew	0276
			小麝鼩	*Crocidura suaveolens*	(Pallas, 1811)	Lesser Shrew	0278
			台湾灰麝鼩	*Crocidura tanakae*	Kuroda, 1938	Taiwanese Gray Shrew	0280
			西南中麝鼩	*Crocidura vorax*	G. M. Allen, 1923	Voracious Shrew	0282
			五指山小麝鼩	*Crocidura wuchihensis*	Wang, 1966	Hainan Island Shrew	0284
攀鼩目 Scandentia							

目 Order	科 Family	属 Genus	中文名 Chinese name	学名 Scientific name	命名人 Species authority	英文名 English name	页码 page
	树鼩科 Tupaiidae						
		树鼩属 *Tupaia*					
			北树鼩	*Tupaia belangeri*	(Wagner, 1841)	Northern Tree Shrew	0286
翼手目 Chiroptera							
	狐蝠科 Pteropodidae						
		果蝠属 *Rousettus*					
			抱尾果蝠	*Rousettus amplexicaudatus*	(É. Geoffroy Saint-Hilaire, 1810)	Geoffroy's Rousette	0288
			棕果蝠	*Rousettus leschenaultii*	(Desmarest, 1820)	Leschenault's Rousette	0290
		狐蝠属 *Pteropus*					
			琉球狐蝠	*Pteropus dasymallus*	(Temminck, 1825)	Ryukyu Flying Fox	0292
			印度大狐蝠	*Pteropus giganteus*	(Brünnich, 1782)	Indian Flying Fox	0294
		犬蝠属 *Cynopterus*					
			短耳犬蝠	*Cynopterus brachyotis*	(Müller, 1838)	Lesser Dog-faced Fruit Bat	0296
			犬蝠	*Cynopterus sphinx*	(Vahl, 1797)	Greater Shortnosed Fruit Bat	0298
		球果蝠属 *Sphaerias*					
			球果蝠	*Sphaerias blanfordi*	(Thomas, 1891)	Blandford's Fruit Bat	0300
		长舌果蝠属 *Eonycteris*					
			长舌果蝠	*Eonycteris spelaea*	(Dobson, 1871)	Dawn Bat	0302
		小长舌果蝠属 *Macroglossus*					
			安氏长舌果蝠	*Macroglossus sobrinus*	K. Andersen, 1911	Hill Long-tongued Fruit Bat	0304
		无尾果蝠属 *Megaerops*					
			无尾果蝠	*Megaerops ecaudatus*	(Temminck, 1837)	Temminck's Tailless Fruit Bat	0306
			泰国无尾果蝠	*Megaerops niphanae*	Yenbutra & Felten, 1983	Ratanaworabhan's Fruit Bat	0308
	鞘尾蝠科 Emballonuridae						
		墓蝠属 *Taphozous*					
			黑髯墓蝠	*Taphozous melanopogon*	Temminck, 1841	Black-bearded Tomb Bat	0310
			大墓蝠	*Taphozous theobaldi*	Dobson, 1872	Theobold's Bat	0312
	假吸血蝠科 Meadermatidae						
		假吸血蝠属 *Megaderma*					
			印度假吸血蝠	*Megaderma lyra*	É. Geoffroy, 1810	Greater False Vampire	0314
			马来假吸血蝠	*Megaderma spasma*	(Linnaeus, 1758)	Lesser False Vampire	0316
	菊头蝠科 Rhinolophidae						

目 Order	科 Family	属 Genus	中文名 Chinese name	学名 Scientific name	命名人 Species authority	英文名 English name	页码 page
		菊头蝠属 *Rhinolophus*					
			中菊头蝠	*Rhinolophus affinis*	Horsfield, 1823	Intermediate Horseshoe Bat	0318
			马铁菊头蝠	*Rhinolophus ferrumequinum*	(Schreber, 1774)	Greater Horseshoe Bat	0320
			台湾菊头蝠	*Rhinolophus formosae*	Sanborn, 1939	Taiwan Woolly Horseshoe Bat	0322
			短翼菊头蝠	*Rhinolophus lepidus*	Blyth, 1844	Blyth's Horseshoe Bat	0324
			大菊头蝠	*Rhinolophus luctus*	Temminck, 1834	Great Woolly Horsehoe Bat	0326
			大耳菊头蝠	*Rhinolophus macrotis*	Blyth, 1844	Big-eared Horseshoe Bat	0328
			马来菊头蝠	*Rhinolophus malayanus*	Bonhote, 1903	Malayan Horseshoe Bat	0330
			马氏菊头蝠	*Rhinolophus marshalli*	Thonglongya, 1973	Marshall's Horseshoe Bat	0332
			单角菊头蝠	*Rhinolophus monoceros*	K. Andersen, 1905	Taiwan Least Horseshoe Bat	0334
			丽江菊头蝠	*Rhinolophus osgoodi*	Sanborn, 1939	Osgood's Horseshoe Bat	0336
			皮氏菊头蝠	*Rhinolophus pearsonii*	Horsfield, 1851	Pearson's Horseshoe Bat	0338
			小菊头蝠	*Rhinolophus pusillus*	Temminck, 1834	Least Horseshoe Bat	0340
			贵州菊头蝠	*Rhinolophus rex*	G. M. Allen, 1923	King Horseshoe Bat	0342
			施氏菊头蝠	*Rhinolophus schnitzleri*	Wu & Thong, 2011	Schnitzler's Horseshoe Bat	0344
			清迈菊头蝠	*Rhinolophus siamensis*	Glydenstolpe, 1917	Thai Horseshoe Bat	0346
			中华菊头蝠	*Rhinolophus sinicus*	K. Andersen, 1905	Chinese Horseshoe Bat	0348
			小褐菊头蝠	*Rhinolophus stheno*	K. Andersen, 1905	Lesser Brown Horseshoe Bat	0350
			葛氏菊头蝠	*Rhinolophus subbadius*	Blyth, 1844	Geoffroy's Horseshoe Bat	0352
			托氏菊头蝠	*Rhinolophus thomasi*	K. Andersen, 1905	Thomas's Horseshoe Bat	0354
			锲鞍菊头蝠	*Rhinolophus xinanzhongguoensis*	Zou, Guillén-Servent, Lim, Eger, Wang & Jiang, 2009	Xinan Horseshoe Bat	0356
			云南菊头蝠	*Rhinolophus yunanensis*	Dobson, 1872	Dobson's Horseshoe Bat	0358
	蹄蝠科 Hipposideridae						
		蹄蝠属 *Hipposideros*					
			大蹄蝠	*Hipposideros armiger*	(Hodgson, 1835)	Great Leaf-nosed Bat	0360
			灰小蹄蝠	*Hipposideros cineraceus*	Blyth, 1853	Least Leaf-nosed Bat	0362
			大耳小蹄蝠	*Hipposideros fulvus*	Gray, 1838	Fulvus Leaf-nosed Bat	0364
			中蹄蝠	*Hipposideros larvatus*	(Horsfield, 1823)	Horsfield's Leaf-nosed Bat	0366
			莱氏蹄蝠	*Hipposideros lylei*	Thomas, 1913	Shield-faced Leaf-nosed Bat	0368
			小蹄蝠	*Hipposideros pomona*	K. Andersen, 1918	Andersen's Leaf-nosed Bat	0370
			普氏蹄蝠	*Hipposideros pratti*	Thomas, 1891	Pratt's Leaf-nosed Bat	0372
		三叶蹄蝠属 *Aselliscus*					
			三叶蹄蝠	*Aselliscus stoliczkanus*	(Dobson, 1871)	Stoliczka's Asian Trident Bat	0374
		无尾蹄蝠属 *Coelops*					
			无尾蹄蝠	*Coelops frithii*	Blyth, 1848	Tail-less Leaf-nosed Bat	0376

目 Order	科 Family	属 Genus	中文名 Chinese name	学名 Scientific name	命名人 Species authority	英文名 English name	页码 page
	犬吻蝠科 Molossidae						
		犬吻蝠属 Tadarida					
			宽耳犬吻蝠	Tadarida insignis	Blyth, 1862	East Asian Free-tailed Bat	0378
			华北犬吻蝠	Tadarida latouchei	Thomas, 1920	La Touche's Free-tailed Bat	0380
		小犬吻蝠属 Chaerephon					
			小犬吻蝠	Chaerephon plicatus	(Buchanan, 1800)	Wrinkle-lipped Bat	0382
	蝙蝠科 Vespertilionidae						
		盘足蝠属 Eudiscopus					
			盘足蝠	Eudiscopus denticulus	(Osgood, 1932)	Disk-footed Bat	0384
		鼠耳蝠属 Myotis					
			西南鼠耳蝠	Myotis altarium	Thomas, 1911	Szechwan Mouse-eared Bat	0386
			缺齿鼠耳蝠	Myotis annectans	(Dobson, 1871)	Hairy-faced Mouse-eared Bat	0388
			栗鼠耳蝠	Myotis badius	Tiunov, Kruskop & Feng, 2011	Bay Mouse-eared Bat	0390
			狭耳鼠耳蝠	Myotis blythii	(Tomes, 1857)	Lesser Mouse-eared Bat	0392
			远东鼠耳蝠	Myotis bombinus	Thomas, 1906	Far Eastern Mouse-eared Bat	0394
			布氏鼠耳蝠	Myotis brandtii	(Eversmann, 1845)	Brandt's Mouse-eared Bat	0396
			中华鼠耳蝠	Myotis chinensis	(Tomes, 1857)	Large Mouse-eared Bat	0398
			沼泽鼠耳蝠	Myotis dasycneme	(Boie, 1825)	Pond Mouse-eared Bat	0400
			大卫鼠耳蝠	Myotis davidii	Peters, 1869	David's Mouse-eared Bat	0402
			毛腿鼠耳蝠	Myotis fimbriatus	(Peters, 1871)	Fringed Long-footed Mouse-eared Bat	0404
			金黄鼠耳蝠	Myotis formosus	(Hodgson, 1835)	Hodgson's Bat	0406
			长尾鼠耳蝠	Myotis frater	G. M. Allen, 1923	Fraternal Mouse-eared Bat	0408
			小巨足鼠耳蝠	Myotis hasseltii	(Temminck, 1840)	Lesser Large-footed Mouse-eared Bat	0410
			霍氏鼠耳蝠	Myotis horsfieldii	Temminck, 1840	Horsfield's Mouse-eared Bat	0412
			印支鼠耳蝠	Myotis indochinensis	Son, Görföl, Francis, Motokawa, Estók, Endo, Thong, Dang, Oshida and Csorba, 2013	Indochina Mouse-eared Bat	0414
			伊氏鼠耳蝠	Myotis ikonnikovi	Ognev, 1912	Ikonnikov's Mouse-eared Bat	0416
			华南水鼠耳蝠	Myotis laniger	Peters, 1871	Chinese Water Mouse-eared Bat	0418
			长指鼠耳蝠	Myotis longipes	(Dobson, 1873)	Kashmir Cave Mouse-eared Bat	0420
			大趾鼠耳蝠	Myotis macrodactylus	(Temminck, 1840)	Big-footed Mouse-eared Bat	0422
			山地鼠耳蝠	Myotis montivagus	(Dobson, 1874)	Burmese Whiskered Mouse-eared Bat	0424
			喜山鼠耳蝠	Myotis muricola	(Gray, 1846)	Nepalese Whiskered Mouse-eared Bat	0426

目 Order	科 Family	属 Genus	中文名 Chinese name	学名 Scientific name	命名人 Species authority	英文名 English name	页码 page
			尼泊尔鼠耳蝠	*Myotis nipalensis*	(Dobson, 1871)	Nepal Mouse-eared Bat	0428
			北京鼠耳蝠	*Myotis pequinius*	Thomas, 1908	Peking Mouse-eared Bat	0430
			东亚水鼠耳蝠	*Myotis petax*	Hollister, 1912	Eastern Daubenton's Mouse-eared Bat	0432
			大足鼠耳蝠	*Myotis pilosus*	(Peters, 1869)	Rickett's Big-footed Mouse-eared Bat	0434
			渡濑氏鼠耳蝠	*Myotis rufoniger*	(Tomats, 1858)	Reddish-black Myotis	0436
			高颅鼠耳蝠	*Myotis siligorensis*	(Horsfield, 1855)	Himalayan Whiskered Mouse-eared Bat	0438
			台湾鼠耳蝠	*Myotis taiwanensis*	Linde, 1908	Taiwanese Mouse-eared Bat	0440
		宽吻鼠耳属 *Submyotodon*					
			宽吻鼠耳蝠	*Submyotodon latirostris*	Kishida, 1932	Taiwan Broad-muzzled Mouse-eared Bat	0442
		伏翼属 *Pipistrellus*					
			东亚伏翼	*Pipistrellus abramus*	(Temminck, 1838)	Japanese Pipistrelle	0444
			阿拉善伏翼	*Pipistrellus alashanicus*	(Bobrinski, 1926)	Alashanian Pipistrelle	0446
			锡兰伏翼	*Pipistrellus ceylonicus*	(Kelaart, 1852)	Kelaart's Pipistrelle	0448
			印度伏翼	*Pipistrellus coromandra*	(Gray, 1838)	Coromandel Pipistrelle	0450
			爪哇伏翼	*Pipistrellus javanicus*	(Gray, 1838)	Javan Pipistrelle	0452
			棒茎伏翼	*Pipistrellus paterculus*	Thomas, 1915	Mount Popa Pipistrelle	0454
			普通伏翼	*Pipistrellus pipistrellus*	Schreber, 1774	Common Pipistrelle	0456
			小伏翼	*Pipistrellus tenuis*	(Temminck, 1840)	Least Pipistrelle	0458
		金背伏翼属 *Arielulus*					
			大黑伏翼	*Arielulus circumdatus*	(Temminck, 1840)	Bronze Sprite	0460
		高级伏翼属 *Hypsugo*					
			茶褐伏翼	*Hypsugo affinis*	Dobson, 1871	Chocolate Pipistrelle	0462
			卡氏伏翼	*Hypsugo cadornae*	Thomas, 1916	Cadorna's pipistrelle	0464
			人灰伏翼	*Hypsugo mordax*	(Peters, 1866)	Pungent Pipistrelle	0466
			灰伏翼	*Hypsugo pulveratus*	Peters, 1871	Chinese Pipistrelle	0468
			萨氏伏翼	*Hypsugo savii*	Bonaparte, 1837	Savi's Pipistrelle	0470
		南蝠属 *Ia*					
			南蝠	*Ia io*	Thomas, 1902	Great Evening Bat	0472
		蝙蝠属 *Vespertilio*					
			双色蝙蝠	*Vespertilio murinus*	Linnaeus, 1758	Particoloured Bat	0474
			东方蝙蝠	*Vespertilio sinensis*	(Peters, 1880)	Asian Particolored Bat	0476
		棕蝠属 *Eptesicus*					
			戈壁北棕蝠	*Eptesicus gobiensis*	Bobrinski, 1926	Gobi Big Brown Bat	0478
			北棕蝠	*Eptesicus nilssonii*	(Keyserling & Blasius, 1839)	Northern Serotine	0480
			东方棕蝠	*Eptesicus pachyomus*	Tomes, 1857	Oriental Serotine	0482

目 Order	科 Family	属 Genus	中文名 Chinese name	学名 Scientific name	命名人 Species authority	英文名 English name	页码 page
			肥耳棕蝠	*Eptesicus pachyotis*	(Dobson, 1871)	Thick-eared Bat	0484
		山蝠属 *Nyctalus*					
			大山蝠	*Nyctalus aviator*	(Thomas, 1911)	Birdlike Noctule	0486
			褐山蝠	*Nyctalus noctula*	(Schreber, 1774)	Noctule	0488
			中华山蝠	*Nyctalus plancyi*	Gerbe, 1880	Chinese Noctule	0490
		扁颅蝠属 *Tylonycteris*					
			华南扁颅蝠	*Tylonycteris fulvida*	Temminck, 1840	Lesser Bamboo Bat	0492
			小扁颅蝠	*Tylonycteris pygmaeus*	Feng, Li & Wang, 2008	Minimum Bamboo Bat	0494
			托京褐扁颅蝠	*Tylonycteris tonkinensis*	Tu, Csorba, Ruedi & Hassanin, 2017	Tonkin Greater Bamboo Bat	0496
		宽耳蝠属 *Barbastella*					
			北京宽耳蝠	*Barbastella beijingensis*	Zhang, Han, Jones, Lin, Zhang, Zhu, Huang & Zhang, 2007	Beijing Barbastelle	0498
			亚洲宽耳蝠	*Barbastella leucomelas*	(Cretzschmar, 1826)	Eastern Barbastelle	0500
		斑蝠属 *Scotomanes*					
			斑蝠	*Scotomanes ornatus*	(Blyth, 1851)	Harlequin Bat	0502
		黄蝠属 *Scotophilus*					
			大黄蝠	*Scotophilus heathi*	(Horsfield, 1831)	Greater Asiatic Yellow House Bat	0504
			小黄蝠	*Scotophilus kuhlii*	Leach, 1821	Lesser Asiatic Yellow House Bat	0506
		金颈蝠属 *Thainycteris*					
			环颈蝠	*Thainycteris aureocollaris*	Kock & Storch, 1996	Collared Sprite	0508
			黄颈蝠	*Thainycteris torquatus*	Csorba & Lee, 1999	Necklace Sprite	0510
		大耳蝠属 *Plecotus*					
			大耳蝠	*Plecotus auritus*	Linnaeus, 1758	Brown Big-eared Bat	0512
			灰大耳蝠	*Plecotus austriacus*	(Fischer, 1829)	Gray Big-eared Bat	0514
			奥氏长耳蝠	*Plecotus ognevi*	(Kishida, 1927)	Ognev's Long-eared Bat	0516
			台湾大耳蝠	*Plecotus taivanus*	Yoshiyuki, 1991	Taiwan Long-eared Bat	0518
		金芒蝠属 *Harpiola*					
			金芒管鼻蝠	*Harpiola isodon*	Kuo, Fang, Csorba & Lee, 2006	Taiwan Tube-Nosed Bat	0520
		管鼻蝠属 *Murina*					
			金管鼻蝠	*Murina aurata*	Milne-Edwards, 1872	Little Tube-nosed Bat	0522
			黄胸管鼻蝠	*Murina bicolor*	Kuo, Fang, Csorba & Lee, 2009	Yellow-chested Tube-nosed Bat	0524
			金毛管鼻蝠	*Murina chrysochaetes*	Eger & Lim, 2011	Golden-haired Tube-nosed Bat	0526
			圆耳管鼻蝠	*Murina cyclotis*	Dobson, 1872	Round-eared Tube-nosed Bat	0528

目 Order	科 Family	属 Genus	中文名 Chinese name	学名 Scientific name	命名人 Species authority	英文名 English name	页码 page
			艾氏管鼻蝠	*Murina eleryi*	Furey, Thong, Bates & Csorba, 2009	Elery's Tube-nosed Bat	0530
			梵净山管鼻蝠	*Murina fanjingshanensis*	He, Xiao & Zhou, 2016	Fanjingshan Tube-nosed bats	0532
			菲氏管鼻蝠	*Murina feae*	Thomas, 1891	Murina feae	0534
			暗色管鼻蝠	*Murina fusca*	Sowerby, 1922	Dusky Tube-nosed Bat	0536
			姬管鼻蝠	*Murina gracilis*	Kuo, Fang, Csorba & Lee, 2009	Taiwanese Little Tube-nosed Bat	0538
			哈氏管鼻蝠	*Murina harrisoni*	Csorba & Bates, 2005	Harrison's Tube-nosed Bat	0540
			东北管鼻蝠	*Murina hilgendorfi*	Peters, 1880	Greater Tube-nosed Bat	0542
			中管鼻蝠	*Murina huttoni*	(Peters, 1872)	Hutton's Tube-nosed Bat	0544
			锦矗管鼻蝠	*Murina jinchui*	W.-H. Yu, Csorba and Y. Wu, 2020	Jinchu Tube-noseed Bat	0546
			白腹管鼻蝠	*Murina leucogaster*	Milne-Edwards, 1872	Rufous Tube-nosed Bat	0548
			荔波管鼻蝠	*Murina liboensis*	X.Zeng, J.Chen, H.-Q. Deng, N.Xiao & J.Zhou, 2018	Libo Tube-nosed Bat	0550
			罗蕾莱管鼻蝠	*Murina lorelieae*	Eger & Lim, 2011	Lorelie's Tube-nosed Bat	0552
			台湾管鼻蝠	*Murina puta*	Kishida, 1924	Taiwanese Tube-nosed Bat	0554
			隐姬管鼻蝠	*Murina recondita*	Kuo, Fang, Csorba & Lee, 2009	Faint-golden Little Tube-nosed Bat	0556
			榕江管鼻蝠	*Murina rongjiangensis*	J.Chen, T.Liu, H.-Q.Deng, N.Xiao & J.Zhou, 2017	Rongjiang Tube-nosed Bat	0558
			水甫管鼻蝠	*Murina shuipuensis*	Eger & Lim, 2011	Shuipu's Tube-nosed Bat	0560
			乌苏里管鼻蝠	*Murina ussuriensis*	Ognev, 1913	Ussuri Tube-nosed Bat	0562
		毛翼蝠属 *Harpiocephalus*					
			毛翼蝠	*Harpiocephalus harpia*	(Temminck, 1840)	Lesser Hairy-Winged Bat	0564
		彩蝠属 *Kerivoula*					
			暗褐彩蝠	*Kerivoula furva*	Kuo, Soisook, Ho & Rossiter, 2017	Leaf-roosting Bat	0566
			彩蝠	*Kerivoula picta*	(Pallas, 1767)	Painted Woolly Bat	0568
	长翼蝠科 Miniopteridae						
		长翼蝠属 *Miniopterus*					
			亚洲长翼蝠	*Miniopterus fuliginosus*	Hodgson, 1835	Eastern Long-fingered Bat	0570
			几内亚长翼蝠	*Miniopterus magnater*	Sanborn, 1931	Western Long-fingered Bat	0572
			南长翼蝠	*Miniopterus pusillus*	Dobson, 1876	Small Long-fingered Bat	0574
灵长目 Primates							
	懒猴科 Lorisidae						
		懒猴属 *Nycticebus*					
			蜂猴	*Nycticebus bengalensis*	(Lacépède, 1800)	Bengal Slow Loris	0578
			倭蜂猴	*Nycticebus pygmaeus*	Bonhote, 1907	Pygmy Slow Loris	0580
	猴科 Cercopithecidae						

目 Order	科 Family	属 Genus	中文名 Chinese name	学名 Scientific name	命名人 Species authority	英文名 English name	页码 page
		猕猴属 Macaca					
			短尾猴	Macaca arctoides	(I. Geoffroy, 1831)	Stump-tailed Macaque	0282
			熊猴	Macaca assamensis	M'Clelland, 1840	Assam Macaque	0584
			台湾猴	Macaca cyclopis	Swinhoe, 1863	Taiwan Macaque	0586
			北豚尾猴	Macaca leonina	Blyth, 1863	Northern Pig-tailed Macaque	0588
			白颊猕猴	Macaca leucogenys	Li, Zhao & Fan, 2015	White-cheeked Macaque	0590
			猕猴	Macaca mulatta	(Zimmermann, 1780)	Rhesus Monkey	0592
			藏南猕猴	Macaca munzala	Madhusudan & Mishra, 2005	Southern Tibet Macaque	0594
			藏酋猴	Macaca thibetana	(Milne-Edwards, 1870)	Tibetan Macaque	0596
		长尾叶猴属 Semnopithecus					
			长尾叶猴	Semnopithecus schistaceus	Hodgson, 1840	Nepal Gray Langur	0598
		乌叶猴属 Trachypithecus					
			印支灰叶猴	Trachypithecus crepusculus	(Elliot, 1909)	Indochinese Gray Langur	0600
			黑叶猴	Trachypithecus francoisi	(Pousargues, 1898)	Francois's Langur	0602
			菲氏叶猴	Trachypithecus phayrei	(Elliot, 1909)	Phayre's Leaf-monkey	0604
			戴帽叶猴	Trachypithecus pileatus	(Blyth, 1843)	Capped Langur	0606
			白头叶猴	Trachypithecus leucocephalus	Tan, 1955	White-headed Langur	0608
			萧氏叶猴	Trachypithecus shortridgei	Wroughton, 1915	Shortridge's Langur	0610
		仰鼻猴属 Rhinopithecus					
			滇金丝猴	Rhinopithecus bieti	Milne-Edwards, 1897	Black Snub-nosed Monkey	0612
			黔金丝猴	Rhinopithecus brelichi	(Thomas, 1903)	Grey Snub-nosed Monkey	0614
			川金丝猴	Rhinopithecus roxellana	(Milne-Edwards, 1870)	Golden Snub-nosed Monkey	0616
			缅甸金丝猴	Rhinopithecus strykeri	Geissmann, Ngwe Lwin, Saw Soe Aung, Thet Naing Aung, Zin Myo Aung, Tony Htin Hla, Grindley & Momberg, 2011	Stryker's Snub-nosed Monkey	0618
	长臂猿科 Hylobatidae						
		白眉长臂猿属 Hoolock					
			西白眉长臂猿	Hoolock hoolock	(Harlan, 1834)	Western Hoolock Gibbon	0620
			东白眉长臂猿	Hoolock leuconedys	Groves, 1967	Eastern Hoolock Gibbon	0622
			高黎贡白眉长臂猿	Hoolock tianxing	Fan, He, Chen, Ortiz, Zhang, Zhao, Lio, Zhang, Kimock, Wang, Groves, Turvey, Roos, Helgen & Jiang, 2017	Tianxing Hoolock Gibbon	0624
		长臂猿属 Hylobates					

目 Order	科 Family	属 Genus	中文名 Chinese name	学名 Scientific name	命名人 Species authority	英文名 English name	页码 page
			白掌长臂猿	*Hylobates lar*	(Linneaus, 1771)	Lar Gibbon	0626
		冠长臂猿属 *Nomascus*					
			西黑冠长臂猿	*Nomascus concolor*	(Harlan, 1834)	Black Crested Gibbon	0628
			东黑冠长臂猿	*Nomascus nasutus*	Kunkel d'Herculais, 1884	Cao-vit Crested Gibbon	0630
			海南长臂猿	*Nomascus hainanus*	Thomas, 1892	Hainan Gibbon	0632
			北白颊长臂猿	*Nomascus leucogenys*	Ogilby, 1840	Northern White-cheeked Gibbon	0634
	人 科 Hominidae						
		人属 *Homo*					
			智人	*Homo sapiens*	Linnaeus, 1758	Human	0636
鳞甲目 Pholidota							
	鲮鲤科 Manidae						
		鲮鲤属 *Manis*					
			马来穿山甲	*Manis javanica*	Desmarest, 1822	Sunda Pangolin	0638
			穿山甲	*Manis pentadactyla*	Linnaeus, 1758	Chinese Pangolin	0640
食肉目 Carnivora							
	犬科 Canidae						
		犬属 *Canis*					
			亚洲胡狼	*Canis aureus*	Linnaeus, 1758	Golden Jackal	0642
			狼	*Canis lupus*	Linnaeus, 1758	Gray Wolf	0644
		狐属 *Vulpes*					
			孟加拉狐	*Vulpes bengalensis*	(Shaw, 1800)	Bengal Fox	0646
			沙狐	*Vulpes corsac*	(Linnaeus, 1768)	Corsac Fox	0648
			藏狐	*Vulpes ferrilata*	Hodgson, 1842	Tibet Fox	0650
			赤狐	*Vulpes vulpes*	(Linnaeus, 1758)	Red Fox	0652
		貉属 *Nyctereutes*					
			貉	*Nyctereutes procyonoides*	(Gary, 1834)	Raccoon Dog	0654
		豺属 *Cuon*					
			豺	*Cuon alpinus*	(Pallas, 1811)	Dhole	0656
	熊科 Ursidae						
		懒熊属 *Melursus*					
			懒熊	*Melursus ursinus*	(Shaw, 1791)	Sloth Bear	0658
		熊属 *Ursus*					
			棕熊	*Ursus arctos*	Linnaeus, 1758	Brown Bear	0660
			黑熊	*Ursus thibetanus*	G. [Baron] Cuvier, 1823	Asiatic Black Bear	0662
		马来熊属 *Helarctos*					
			马来熊	*Helarctos malayanus*	(Raffles, 1821)	Sun Bear	0664

目 Order	科 Family	属 Genus	中文名 Chinese name	学名 Scientific name	命名人 Species authority	英文名 English name	页码 page
	大熊猫科 Ailuropodidae						
		熊猫属 *Ailuropoda*					
			大熊猫	*Ailuropoda melanoleuca*	(David, 1869)	Giant Panda	0666
	小熊猫科 Ailuridae						
		小熊猫属 *Ailurus*					
			小熊猫	*Ailurus fulgens*	F. G. Cuvier, 1825	Red Panda	0668
			中华小熊猫	*Ailurus styani*	Thomas, 1902	Chinese Red Panda	0670
	海狮科 Otariidae						
		海狗属 *Callorhinus*					
			北海狗	*Callorhinus ursinus*	(Linnaeus, 1758)	Northern Fur Seal	0672
		北海狮属 *Eumetopias*					
			北海狮	*Eumetopias jubatus*	(Schreber, 1776)	Steller Sea Lion	0674
	鼬科 Mustelidae						
		貂属 *Martes*					
			黄喉貂	*Martes flavigula*	(Boddaert, 1785)	Yellow-throated Marten	0676
			石貂	*Martes foina*	(Erxleben, 1777)	Stone Marten	0678
			紫貂	*Martes zibellina*	(Linnaeus, 1758)	Sable	0680
		狼獾属 *Gulo*					
			貂熊	*Gulo gulo*	(Linnaeus, 1758)	Wolverine	0682
		鼬属 *Mustela*					
			缺齿伶鼬	*Mustela aistoodonnivalis*	(Wu and Kao, 1991)	Sichuan Weasal	0684
			香鼬	*Mustela altaica*	Pallas, 1811	Altai Weasel	0686
			白鼬	*Mustela erminea*	Linnaeus, 1758	Stoat	0688
			艾鼬	*Mustela eversmanii*	Lesson, 1827	Steppe Polecat	0690
			黄腹鼬	*Mustela kathiah*	Hodgson, 1835	Yellow-bellied Weasel	0692
			伶鼬	*Mustela nivalis*	Linnaeus, 1766	Least Weasel	0694
			黄鼬	*Mustela sibirica*	Pallas, 1773	Siberian Weasel	0696
			纹鼬	*Mustela strigidorsa*	Gray, 1853	Stripe-backed Weasel	0698
		虎鼬属 *Vormela*					
			虎鼬	*Vormela peregusna*	(Güldenstädt, 1770)	European Marbled Polecat	0700
		鼬獾属 *Melogale*					
			越南鼬獾	*Melogale cucphuongensis*	Nadler, Streicher, Stefen, Schwierz & Roos, 2011	Vietnam Ferret-badger	0702
			鼬獾	*Melogale moschata*	(Gray, 1831)	Small-toothed Ferret-badger	0704
			缅甸鼬獾	*Melogale personata*	I. Geoffroy Saint-Hilaire, 1831	Large-toothed Ferret Badger	0706
		獾属 *Meles*					
			亚洲狗獾	*Meles leucurus*	(Hodgson, 1847)	Asian Badger	0708

目 Order	科 Family	属 Genus	中文名 Chinese name	学名 Scientific name	命名人 Species authority	英文名 English name	页码 page
		猪獾属 *Arctonyx*					
			猪獾	*Arctonyx collaris*	F. G. Cuvier, 1825	Hog Badger	0710
		水獭属 *Lutra*					
			水獭	*Lutra lutra*	(Linnaeus, 1758)	Eurasian Otter	0712
		江獭属 *Lutrogale*					
			江獭	*Lutrogale perspicillata*	(I. Geoffroy Saint-Hilaire, 1826)	Smooth-coated Otter	0714
		小爪水獭属 *Aonyx*					
			小爪水獭	*Aonyx cinereus*	(Illiger, 1815)	Asian Small-clawed Otter	0716
	海豹科 Phocidae						
		海豹属 *Phoca*					
			斑海豹	*Phoca largha*	(Pallas, 1811)	Spotted Seal	0718
		环斑海豹属 *Pusa*					
			环斑小头海豹	*Pusa hispida*	(Schreber, 1775)	Ringed Seal	0720
		髯海豹属 *Erignathus*					
			髯海豹	*Erignathus barbatus*	(Erxleben, 1777)	Bearded Seal	0722
	灵猫科 Viverridae						
		灵猫属 *Viverra*					
			大斑灵猫	*Viverra megaspila*	Blyth, 1862	Large-spotted Civet	0724
			大灵猫	*Viverra zibetha*	Linnaeus, 1758	Large Indian Civet	0726
		小灵猫属 *Viverricula*					
			小灵猫	*Viverricula indica*	(É. Geoffroy Saint-Hilaire, 1803)	Small Indian Civet	0728
		椰子猫属 *Paradoxurus*					
			椰子猫	*Paradoxurus hermaphroditus*	(Pallas, 1777)	Common Palm Civet	0730
		花面狸属 *Paguma*					
			果子狸	*Paguma larvata*	(C.E.H. Smith, 1827)	Masked Palm Civet	0732
		熊狸属 *Arctictis*					
			熊狸	*Arctictis binturong*	(Raffles, 1821)	Binturong	0734
		小齿狸属 *Arctogalidia*					
			小齿狸	*Arctogalidia trivirgata*	(Gray, 1832)	Small-toothed Palm Civet	0736
		缟灵猫属 *Chrotogale*					
			缟灵猫	*Chrotogale owstoni*	Thomas, 1912	Owston's Civet	0738
	林狸科 Prionodontidae						
		林狸属 *Prionodon*					
			斑林狸	*Prionodon pardicolor*	Hodgson, 1842	Spotted Linsang	0740
	獴科 Herpestidae						

目 Order	科 Family	属 Genus	中文名 Chinese name	学名 Scientific name	命名人 Species authority	英文名 English name	页码 page
		獴属 Herpestes					
			灰獴	Herpestes edwardsii	(É. Geoffroy Saint-Hilaire, 1818)	Indian Grey Mongoose	0742
			爪哇獴	Herpestes javanicus	(Hodgson, 1836)	Small Asian Mongoose	0744
			食蟹獴	Herpestes urva	(Hodgson, 1836)	Crab-eating Mongoose	0746
	猫科 Felidae						
		猫属 Felis					
			荒漠猫	Felis bieti	Milne-Edwards, 1892	Chinese Mountain Cat	0748
			丛林猫	Felis chaus	Schreber, 1777	Jungle Cat	0750
			亚非野猫	Felis lybica	Forster, 1780	Africa-asian Wild Cat	0752
		豹猫属 Prionailurus					
			渔猫	Prionailurus viverrinus	(Bennett, 1833)	Fishing cat	0754
			豹猫	Prionailurus bengalensis	(Kerr, 1792)	Leopard Cat	0756
		兔狲属 Otocolobus					
			兔狲	Otocolobus manul	(Pallas, 1776)	Pallas's Cat	0758
		猞猁属 Lynx					
			猞猁	Lynx lynx	(Linnaeus, 1758)	Eurasian Lynx	0760
		云猫属 Pardofelis					
			云猫	Pardofelis marmorata	(Martin, 1837)	Marbled Cat	0762
		金猫属 Catopuma					
			金猫	Catopuma temminckii	(Vigors & Horsfield, 1827)	Asiatic Golden Cat	0764
		云豹属 Neofelis					
			云豹	Neofelis nebulosa	(Griffith, 1821)	Clouded Leopard	0766
		豹属 Panthera					
			豹	Panthera pardus	(Linnaeus, 1758)	Leopard	0768
			虎	Panthera tigris	(Linnaeus, 1758)	Tiger	0770
			雪豹	Panthera uncia	(Schreber, 1775)	Snow Leopard	0772
奇蹄目 Perissodactyla							
	犀科 Rhinocerotidae						
		双角犀属 Dicerorhinus					
			双角犀	Dicerorhinus sumatrensis	(G. Fischer, 1814)	Sumatran Rhinoceros	0774
		独角犀属 Rhinoceros					
			爪哇犀	Rhinoceros sondaicus	Desmarest, 1822	Javan Rhinoceros	0776
			大独角犀	Rhinoceros unicornis	Linnaeus, 1758	Indian Rhinoceros	0778
	马科 Equidae						
		马属 Equus					
			野马	Equus ferus	Boddaert, 1785	Przewalski's Horse	0780

目 Order	科 Family	属 Genus	中文名 Chinese name	学名 Scientific name	命名人 Species authority	英文名 English name	页码 page
			蒙古野驴	*Equus hemionus*	Pallas, 1775	Asiatic Wild Ass	0782
			藏野驴	*Equus kiang*	Moorcroft, 1841	Kiang	0784
鲸偶蹄目 Cetartiodactyla							
	露脊鲸科 Balaenidae						
		露脊鲸属 *Eubalaena*					
			北太平洋 露脊鲸	*Eubalaena japonica*	(Lacépède, 1818)	North Pacific Right Whale	0786
	灰鲸科 Eschrichtiidae						
		灰鲸属 *Eschrichtius*					
			灰鲸	*Eschrichtius robustus*	(Lilljeborg, 1861)	Gray Whale	0788
	须鲸科 Balaenopteridae						
		须鲸属 *Balaenoptera*					
			小须鲸	*Balaenoptera acutorostrata*	Lacépède, 1804	Common Minke Whale	0790
			塞鲸	*Balaenoptera borealis*	Lesson, 1828	Sei Whale	0792
			布氏鲸	*Balaenoptera edeni*	Anderson, 1879	Bryde's Whale	0794
			蓝鲸	*Balaenoptera musculus*	(Linnaeus, 1758)	Blue Whale	0796
			大村鲸	*Balaenoptera omurai*	Wada, Oishi & Yamada, 2003	Omura's Whale	0798
			长须鲸	*Balaenoptera physalus*	(Linnaeus, 1758)	Fin Whale	0800
		大翅鲸属 *Megaptera*					
			大翅鲸	*Megaptera novaeangliae*	(Borowski, 1781)	Humpback Whale	0802
	白鱀豚科 Lipotidae						
		白鱀豚属 *Lipotes*					
			白鱀豚	*Lipotes vexillifer*	Miller, 1918	Baiji	0804
	抹香鲸科 Physeteridae						
		抹香鲸属 *Physeter*					
			抹香鲸	*Physeter macrocephalus*	Linnaeus, 1758	Sperm Whale	0806
		小抹香鲸属 *Kogia*					
			小抹香鲸	*Kogia breviceps*	(Blainville, 1838)	Pygmy Sperm Whale	0808
			侏抹香鲸	*Kogia sima*	(Owen, 1866)	Dwarf Sperm Whale	0810
	喙鲸科 Ziphiidae						
		喙鲸属 *Ziphius*					
			鹅喙鲸	*Ziphius cavirostris*	G. Cuvier, 1823	Cuvier's Beaked Whale	0812
		中喙鲸属 *Mesoplodon*					
			柏氏中喙鲸	*Mesoplodon densirostris*	(Blainville, 1817)	Blainville's Beaked Whale	0814

目 Order	科 Family	属 Genus	中文名 Chinese name	学名 Scientific name	命名人 Species authority	英文名 English name	页码 page
			银杏齿中喙鲸	*Mesoplodon ginkgodens*	Nishiwaki & Kamiya, 1958	Ginkgo-toothed Beaked Whale	0816
		贝氏喙鲸属 *Berardius*					
			贝氏喙鲸	*Berardius bairdii*	Stejneger, 1883	Baird's Beaked Whale	0818
		印太喙鲸属 *Indopacetus*					
			朗氏喙鲸	*Indopacetus pacificus*	(Longman, 1926)	Longman's Beaked Whale	0820
	鼠海豚科 Phocoenidae						
		江豚属 *Neophocaena*					
			东亚江豚	*Neophocaena sunameri*	Pilleri & Gihr, 1975	East Asian Finless Porpoise	0822
			长江江豚	*Neophocaena asiaeorientalis*	(Pilleri & Gihr, 1972)	Yangtze Finless Porpoise	0824
			印太江豚	*Neophocaena phocaenoides*	(G. Cuvier, 1829)	Indo-Pacific Finless Porpoise	0826
	恒河豚科 Platanistidae						
		恒河豚属 *Platanista*					
			恒河豚	*Platanista gangetica*	(Lebeck, 1801)	South Asian River Dolphin	0828
	海豚科 Delphinidae						
		糙齿海豚属 *Steno*					
			糙齿海豚	*Steno bredanensis*	G. Cuvier in Lesson, 1828	Rough-toothed Dolphin	0830
		白海豚属 *Sousa*					
			中华白海豚	*Sousa chinensis*	(Osbeck, 1765)	Chinese White Dolphin	0832
		原海豚属 *Stenella*					
			热带点斑原海豚	*Stenella attenuata*	(Gray, 1846)	Pantropical Spotted Dolphin	0834
			条纹原海豚	*Stenella coeruleoalba*	(Meyen, 1833)	Striped Dolphin	0836
			飞旋原海豚	*Stenella longirostris*	(Gray, 1828)	Spinner Dolphin	0838
		真海豚属 *Delphinus*					
			真海豚	*Delphinus delphis*	Linnaeus, 1758	Common dolphin	0840
		瓶鼻海豚属 *Tursiops*					
			印太瓶鼻海豚	*Tursiops aduncus*	(Ehrenberg, 1833)	Indo-pacific Bottlenose Dolphin	0842
			瓶鼻海豚	*Tursiops truncatus*	(Montagu, 1821)	Common Bottlenose Dolphin	0844
		弗氏海豚属 *Lagenodelphis*					
			弗氏海豚	*Lagenodelphis hosei*	Fraser, 1956	Fraser's Dolphin	0846
		灰海豚属 *Grampus*					
			里氏海豚	*Grampus griseus*	(G. Cuvier, 1812)	Risso's Dolphin	0848
		斑纹海豚属 *Lagenorhynchus*					

目 Order	科 Family	属 Genus	中文名 Chinese name	学名 Scientific name	命名人 Species authority	英文名 English name	页码 page
			太平洋斑纹海豚	*Lagenorhynchus obliquidens*	Gill, 1865	Pacific White-sided Dolphin	0850
		瓜头鲸属 *Peponocephala*					
			瓜头鲸	*Peponocephala electra*	(Gray, 1846)	Melon-headed Whale	0852
		虎鲸属 *Orcinus*					
			虎鲸	*Orcinus orca*	(Linnaeus, 1758)	Killer Whale	0854
		伪虎鲸属 *Pseudorca*					
			伪虎鲸	*Pseudorca crassidens*	(Owen, 1846)	False Killer Whale	0856
		小虎鲸属 *Feresa*					
			小虎鲸	*Feresa attenuata*	Gray, 1874	Pygmy Killer Whale	0858
		领航鲸属 *Globicephala*					
			短肢领航鲸	*Globicephala macrorhynchus*	Gray, 1846	Short-finned Pilot Whale	0860
	猪科 Suidae						
		猪属 *Sus*					
			野猪	*Sus scrofa*	Linnaeus, 1758	Wild Boar	0862
	骆驼科 Camelidae						
		骆驼属 *Camelus*					
			野骆驼	*Camelus ferus*	Przewalski, 1878	Bactrian Camel	0864
	鼷鹿科 Tragulidae						
		鼷鹿属 *Tragulus*					
			威氏小鼷鹿	*Tragulus williamsoni*	Kloss, 1916	Mouse-deer	0866
	麝科 Moschidae						
		麝属 *Moschus*					
			安徽麝	*Moschus anhuiensis*	Wang, Hu & Yan, 1982	Anhui Musk Deer	0868
			林麝	*Moschus berezovskii*	Flerov, 1929	Forest Musk Deer	0870
			马麝	*Moschus chrysogaster*	(Hodgson, 1839)	Alpine Musk Deer	0872
			黑麝	*Moschus fuscus*	Li, 1981	Black Musk Deer	0874
			喜马拉雅麝	*Moschus leucogaster*	Hodgson, 1839	Himalayan Musk Deer	0876
			原麝	*Moschus moschiferus*	Linnaeus, 1758	Siberian Musk Deer	0878
	鹿科 Cervidae						
		獐属 *Hydropotes*					
			獐	*Hydropotes inermis*	Swinhoe, 1870	Chinese Water Deer	0880
		毛冠鹿属 *Elaphodus*					
			毛冠鹿	*Elaphodus cephalophus*	Milne-Edwards, 1872	Tufted Deer	0882
		麂属 *Muntiacus*					
			黑麂	*Muntiacus crinifrons*	(Sclater, 1885)	Black Muntjac	0884
			林麂	*Muntiacus feae*	(Thomas & Doria, 1889)	Fea's Muntjac	0886

目 Order	科 Family	属 Genus	中文名 Chinese name	学名 Scientific name	命名人 Species authority	英文名 English name	页码 page
			贡山麂	*Muntiacus gongshanensis*	Ma in Ma, Wang & Shi, 1990	Gongshan Muntjac	0888
			海南麂	*Muntiacus nigripes*	G. M. Allen, 1930	Hainan Muntjac	0890
			叶麂	*Muntiacus putaoensis*	Amato, Egan & Rabinowitz, 1999	Leaf Muntjac	0892
			小麂	*Muntiacus reevesi*	(Ogilby, 1839)	Reeves' Muntjac	0894
			北赤麂	*Muntiacus vaginalis*	(Boaert, 1785)	Northern Red Muntjac	0896
		豚鹿属 *Axis*					
			豚鹿	*Axis porcinus*	(Zimmermann, 1780)	Hog Deer	0898
		水鹿属 *Rusa*					
			水鹿	*Rusa unicolor*	(Kerr, 1792)	Southeast Asian Sambar	0900
		鹿属 *Cervus*					
			梅花鹿	*Cervus nippon*	Swinhoe, 1864	Sika Deer	0902
			西藏马鹿	*Cervus wallichii*	Pocock, 1942	Tibet Shou	0904
			马鹿	*Cervus canadensis*	Erxleben, 1777	Manchurian Wapiti	0906
			塔里木马鹿	*Cervus hanglu*	Wagner, 1844	Yarkand stag	0908
		泽鹿属 *Rucervus*					
			坡鹿	*Rucervus eldii*	(M'Clelland, 1842)	Eastern Eld's Deer	0910
		白唇鹿属 *Przewalskium*					
			白唇鹿	*Przewalskium albirostris*	Przewalski, 1883	White-lipped Deer	0912
		麋鹿属 *Elaphurus*					
			麋鹿	*Elaphurus davidianus*	Milne-Edwards, 1866	Père David's Deer	0914
		狍属 *Capreolus*					
			东方狍	*Capreolus pygargus*	Pallas, 1771	Oriental Roe Deer	0916
		驼鹿属 *Alces*					
			驼鹿	*Alces alces*	(Linnaeus, 1758)	Moose	0918
		驯鹿属 *Rangifer*					
			驯鹿	*Rangifer tarandus*	(Linnaeus, 1758)	Reindeer	0920
	牛科 Bovidae						
		牛属 *Bos*					
			印度野牛	*Bos gaurus*	C. H. Smith, 1827	Gaur	0922
			爪哇野牛	*Bos javanicus*	d'Alton, 1823	Banteng	0924
			野牦牛	*Bos mutus*	(Przewalski, 1883)	Wild Yak	0926
		水牛属 *Bubalus*					
			野水牛	*Bubalus arnee*	(Kerr, 1792)	Asian Buffalo	0928
		原羚属 *Procapra*					
			蒙原羚	*Procapra gutturosa*	(Pallas, 1777)	Mongolian Gazelle	0930
			藏原羚	*Procapra picticaudata*	Hodgson, 1846	Tibetan Gazelle	0932
			普氏原羚	*Procapra przewalskii*	(Büchner, 1891)	Przewalski's Gazelle	0934
		羚羊属 *Gazella*					

目 Order	科 Family	属 Genus	中文名 Chinese name	学名 Scientific name	命名人 Species authority	英文名 English name	页码 page
			鹅喉羚	*Gazella subgutturosa*	(Güldenstädt, 1780)	Yarkand Goitered Gazelle	0936
		藏羚属 *Pantholops*					
			藏羚	*Pantholops hodgsonii*	(Abel, 1826)	Chiru	0938
		高鼻羚羊属 *Saiga*					
			高鼻羚羊	*Saiga tatarica*	(Linnaeus, 1766)	Mongolian Saiga	0940
		羚牛属 *Budorcas*					
			秦岭羚牛	*Budorcas bedfordi*	Thomas, 1911	Golden Takin	0942
			四川羚牛	*Budorcas tibetanus*	Milne-Edwards, 1874	Sichuan Takin	0944
			不丹羚牛	*Budorcas whitei*	Lydekker, 1907	Bhutan Takin	0946
			贡山羚牛	*Budorcas taxicolor*	Hodgson, 1850	Gongshan Takin	0948
		斑羚属 *Naemorhedus*					
			赤斑羚	*Naemorhedus baileyi*	Pocock, 1914	Red Goral	0950
			长尾斑羚	*Naemorhedus caudatus*	(Milne-Edwards, 1867)	Long-tailed Goral	0952
			缅甸斑羚	*Naemorhedus evansi*	Lydekker, 1906	Burmese Goral	0954
			喜马拉雅斑羚	*Naemorhedus goral*	(Hardwicke, 1825)	Himalayan Goral	0956
			中华斑羚	*Naemorhedus griseus*	Milne-Edwards, 1871	Chinese Goral	0958
		塔尔羊属 *Hemitragus*					
			塔尔羊	*Hemitragus jemlahicus*	C. H. Smith, 1826	Himalayan Tahr	0960
		山羊属 *Capra*					
			北山羊	*Capra sibirica*	(Pallas, 1776)	Siberian Ibex	0962
		岩羊属 *Pseudois*					
			岩羊	*Pseudois nayaur*	(Hodgson, 1833)	Bharal	0964
		羊属 *Ovis*					
			阿尔泰盘羊	*Ovis ammon*	(Linnaeus, 1758)	Argali	0966
			哈萨克盘羊	*Ovis collium*	Linnaeus, 1758	Kazkhstan argali	0968
			戈壁盘羊	*Ovis darwini*	Przewalski, 1883	Gobi Argali	0970
			西藏盘羊	*Ovis hodgsoni*	Blyth, 1841	Tibetan Argali	0972
			华北盘羊	*Ovis jubata*	Peters, 1876	Northern Chinese Argali	0974
			天山盘羊	*Ovis karelini*	Severtzov, 1873	Tianshan Argali	0976
			帕米尔盘羊	*Ovis polii*	Blyth, 1841	Pamir Argali	0978
		鬣羚属 *Capricornis*					
			中华鬣羚	*Capricornis milneedwardsii*	(Bechstein, 1799)	Chinese Serow	0980
			红鬣羚	*Capricornis rubidus*	Blyth, 1863	Red Serow	0982
			台湾鬣羚	*Capricornis swinhoei*	Gray, 1862	Taiwan Serow	0984
			喜马拉雅鬣羚	*Capricornis thar*	(Bechstein, 1799)	Himalayan Serow	0986
长鼻目 Proboscidea							
	象科 Elephantidae						

目 Order	科 Family	属 Genus	中文名 Chinese name	学名 Scientific name	命名人 Species authority	英文名 English name	页码 page
		象属 Elephas					
			亚洲象	Elephas maximus	Linnaeus, 1758	Asian Elephant	0988
海牛目 Sirenia							
	儒艮科 Dugongidae						
		儒艮属 Dugong					
			儒艮	Dugong dugon	(Müller, 1776)	Dugong	0990
啮齿目 Rodentia							
	松鼠科 Sciuridae						
		松鼠属 Sciurus					
			松鼠	Sciurus vulgaris	Linnaeus, 1758	Eurasian Red Squirrel	0994
		丽松鼠属 Callosciurus					
			金背松鼠	Callosciurus caniceps	(Gray, 1842)	Gray-bellied Squirrel	0996
			赤腹松鼠	Callosciurus erythraeus	(Pallas, 1779)	Pallas's Squirrel	0998
			印支松鼠	Callosciurus inornatus	(Gray, 1867)	Inornate Squirrel	1000
			黄足松鼠	Callosciurus phayrei	(Blyth, 1856)	Phayre's Squirrel	1002
			蓝腹松鼠	Callosciurus pygerythrus	(I. Geoffroy Saint Hilaire, 1832)	Hoary-bellied Squirrel	1004
			五纹松鼠	Callosciurus quinquestriatus	(Anderson, 1871)	Anderson's Squirrel	1006
		花松鼠属 Tamiops					
			明纹花松鼠	Tamiops macclellandi	(Horsfield, 1840)	Himalayan Striped Squirrel	1008
			倭花鼠	Tamiops maritimus	(Bonhote, 1900)	Maritime Striped Squirrel	1010
			隐纹花松鼠	Tamiops swinhoei	Milne-Edwards, 1874	Swinhoe's Striped Squirrel	1012
		长吻松鼠属 Dremomys					
			橙喉长吻松鼠	Dremomys gularis	Osgood, 1932	Red-throated Squirrel	1014
			橙腹长吻松鼠	Dremomys lokriah	(Hodgson, 1836)	Orange-bellied Himalayan Squirrel	1016
			珀氏长吻松鼠	Dremomys pernyi	(Milne-Edwards, 1867)	Perny's Long-nosed Squirrel	1018
			红腿长吻松鼠	Dremomys pyrrhomerus	(Thomas, 1895)	Red-hipped Squirrel	1020
			红颊长吻松鼠	Dremomys rufigenis	(Blanford, 1878)	Asian Red-cheeked Squirrel	1022
		巨松鼠属 Ratufa					
			巨松鼠	Ratufa bicolor	(Sparrman, 1778)	Black Giant Squirrel	1024
		条纹松鼠属 Menetes					
			条纹松鼠	Menetes berdmorei	(Blyth, 1849)	Indochinese Ground Squirrel	1026
		岩松鼠属 Sciurotamias					
			岩松鼠	Sciurotamias davidianus	(Milne-Edwards, 1867)	Pére David's Rock Squirrel	1028
		侧纹岩松鼠属 Rupestes					

目 Order	科 Family	属 Genus	中文名 Chinese name	学名 Scientific name	命名人 Species authority	英文名 English name	页码 page
			侧纹岩松鼠	*Rupestes forresti*	(Thomas, 1922)	Forrest's Rock Squirrel	1030
		花鼠属 *Eutamias*					
			北花松鼠	*Eutamias sibiricus*	(Laxmann, 1769)	Siberian Chipmunk	1032
		黄鼠属 *Spermophilus*					
			阿拉善黄鼠	*Spermophilus alashanicus*	Büchner, 1888	Alashan Ground Squirrel	1034
			短尾黄鼠	*Spermophilus brevicauda*	Brandt, 1843	Brandt's Ground Squirrel	1036
			达乌尔黄鼠	*Spermophilus dauricus*	Brandt, 1843	Daurian Ground Squirrel	1038
			赤颊黄鼠	*Spermophilus erythrogenys*	Brandt, 1841	Red-Cheeked Ground Squirrel	1040
			淡尾黄鼠	*Spermophilus pallidicauda*	(Satunin, 1903)	Pallid Ground Squirrel	1042
			长尾黄鼠	*Spermophilus parryii*	Pallas, 1778	Arctic Ground Squirrel	1044
			天山黄鼠	*Spermophilus relictus*	(Kashkarov, 1923)	Tien Shan Ground Squirrel	1046
		旱獭属 *Marmota*					
			灰旱獭	*Marmota baibacina*	Kastschenko, 1899	Gray Marmot	1048
			长尾旱獭	*Marmota caudata*	(Geoffroy, 1844)	Long-tailed Marmot	1050
			喜马拉雅旱獭	*Marmota himalayana*	(Hodgson, 1841)	Himalayan Marmot	1052
			西伯利亚旱獭	*Marmota sibirica*	(Radde, 1862)	Mongolian Marmot	1054
	鼯鼠科 Pteromyidae						
		比氏鼯鼠属 *Biswamoyopterus*					
			比氏鼯鼠	*Biswamoyopterus biswasi*	Saha, 1981	Namdapha Flying Squirrel	1056
			高黎贡 比氏鼯鼠	*Biswamoyopterus gaoligongensis*	Li, Li, Jackson, Li, Jiang, Zhao, Song & Jiang, 2019	Gaoligong Flying Squirrel	1058
		毛耳飞鼠属 *Belomys*					
			毛耳飞鼠	*Belomys pearsonii*	(Gray, 1842)	Hairy-footed Flying Squirrel	1060
		大耳飞鼠属 *Priapomys*					
			李氏小飞鼠	*Priapomys leonardi*	(Thomas, 1921)	Himalayan Large-eared Flying Squirrel	1062
		复齿鼯鼠属 *Trogopterus*					
			复齿鼯鼠	*Trogopterus xanthipes*	(Milne-Edwards, 1867)	Complex-toothed Flying Squirrel	1064
		鼯鼠属 *Petaurista*					
			栗背大鼯鼠	*Petaurista albiventer*	Gray, 1834	White-bellied Giant Flying Squirrel	1066
			红白鼯鼠	*Petaurista alborufus*	(Milne-Edwards, 1870)	Red and White Giant Flying Squirrel	1068
			灰头小鼯鼠	*Petaurista elegans*	(Müller, 1840)	Spotted Giant Flying Squirrel	1070
			海南大鼯鼠	*Petaurista hainana*	Allen, 1925	Hainan Flying Squirrel	1072
			栗褐大鼯鼠	*Petaurista magnificus*	(Hodgson, 1836)	Hodgson's Giant Flying Squirrel	1074

目 Order	科 Family	属 Genus	中文名 Chinese name	学名 Scientific name	命名人 Species authority	英文名 English name	页码 page
			麦丘卡大鼯鼠	*Petaurista mechukaensis*	Choudhury, 2007	Mechuka Giant Flying Squirrel	1076
			米什米大鼯鼠	*Petaurista mishmiensis*	Choudhury, 2009	Mishmi Giant Flying Squirrel	1078
			印支小鼯鼠	*Petaurista marica*	Thomas, 1912	Spotted Flying Squirrel	1080
			不丹大鼯鼠	*Petaurista nobilis*	(Gray, 1842)	Bhutan giant flying squirrel	1082
			红背大鼯鼠	*Petaurista petaurista*	(Pallas, 1766)	Common Giant Flying Squirrel	1084
			霜背大鼯鼠	*Petaurista philippensis*	(Elliot, 1839)	Large Brown Flying Squirrel	1086
			米博大鼯鼠	*Petaurista siangensis*	Choudhury, 2013	Mebo Giant Flying Squirrel	1088
			橙色小鼯鼠	*Petaurista sybilla*	(Thomas & Wroughton, 1916)	Chindwin Flying Squirrel	1090
			灰鼯鼠	*Petaurista xanthotis*	(Milne-Edwards, 1872)	Chinese Giant Flying Squirrel	1092
		沟牙鼯鼠属 *Aeretes*					
			沟牙鼯鼠	*Aeretes melanopterus*	(Milne-Edwards, 1867)	Northern Chinese Flying Squirrel	1094
		绒毛鼯鼠属 *Eupetaurus*					
			绒毛鼯鼠	*Eupetaurus cinereus*	Thomas, 1888	Woolly Flying Squirrel	1096
			雪山羊绒鼯鼠	*Eupetaurus nivamons*	Q. Li, Jiang, Jackson & Helgen, 2021	Yunnan Woolly Flying Squirrel	1098
			西藏羊绒鼯鼠	*Eupetaurus tibetensis*	Jackson, Helgen, Q. Li & Jiang, 2021	Tibetan Woolly Flying Squirrel	1100
		飞鼠属 *Pteromys*					
			小飞鼠	*Pteromys volans*	(Linnaeus, 1758)	Russian Flying Squirrel	1102
		箭尾飞鼠属 *Hylopetes*					
			黑白飞鼠	*Hylopetes alboniger*	(Hodgson, 1836)	Particolored Flying Squirrel	1104
			海南小飞鼠	*Hylopetes phayrei*	(Blyth, 1859)	Indochinese Flying Squirrel	1106
	河狸科 Castoridae						
		河狸属 *Castor*					
			河狸	*Castor fiber*	Linnaeus, 1758	Eurasian Beaver	1108
	仓鼠科 Cricetidae						
		原仓鼠属 *Cricetus*					
			原仓鼠	*Cricetus cricetus*	(Linnaeus, 1758)	Black-bellied Hamster	1110
		甘肃仓鼠属 *Cansumys*					
			甘肃仓鼠	*Cansumys canus*	G.M. Allen, 1928	Gansu Hamster	1112
		仓鼠属 *Cricetulus*					
			高山仓鼠	*Cricetulus alticola*	Thomas, 1917	Ladak Hamster	1114
			黑线仓鼠	*Cricetulus barabensis*	Pallas, 1773	Striped Dwarf Hamster	1116
			藏仓鼠	*Cricetulus kamensis*	(Satunin, 1903)	Kam Dwarf Hamster	1118
			长尾仓鼠	*Cricetulus longicaudatus*	(Milne-Edwards, 1867)	Long-tailed Dwarf Hamster	1120
			灰仓鼠	*Cricetulus migratorius*	(Pallas, 1773)	Gray Dwarf Hamster	1122

目 Order	科 Family	属 Genus	中文名 Chinese name	学名 Scientific name	命名人 Species authority	英文名 English name	页码 page
			索氏仓鼠	*Cricetulus sokolovi*	Orlov and Malygin, 1988	Sokolov's Dwarf Hamster	1124
		大仓鼠属 *Tscherskia*					
			大仓鼠	*Tscherskia triton*	(de Winton, 1899)	Greater Long-tailed Hamster	1126
		短尾仓鼠属 *Allocricetulus*					
			无斑短尾仓鼠	*Allocricetulus curtatus*	(G.M. Allen, 1925)	Mongolian Hamster	1128
			短尾仓鼠	*Allocricetulus eversmanni*	(Brandt, 1859)	Eversmann's Hamster	1130
		毛足鼠属 *Phodopus*					
			坎氏毛足鼠	*Phodopus campbelli*	(Thomas, 1905)	Campbell's Desert Hamster	1132
			小毛足鼠	*Phodopus roborovskii*	(Satunin, 1903)	Roborovski's Desert Hamster	1134
		鼹形田鼠属 *Ellobius*					
			鼹形田鼠	*Ellobius talpinus*	Blasius, 1884	Northern Mole Vole	1136
		旅鼠属 *Myopus*					
			林旅鼠	*Myopus schisticolor*	(Lilljeborg, 1844)	Wood Lemming	1138
		䶄属 *Myodes*					
			灰棕背䶄	*Myodes centralis*	Miller, 1906	Tien Shan Red-backed Vole	1140
			红背䶄	*Myodes rutilus*	(Pallas, 1779)	Northern Red-backed Vole	1142
		东亚䶄属 *Craseomys*					
			棕背䶄	*Craseomys rufocanus*	(Sundevall, 1846)	Grey Red-backed Vole	1144
		绒鼠属 *Eothenomys*					
			克钦绒鼠	*Eothenomys cachinus*	(Thomas, 1921)	Kachin Chinese Vole	1146
			中华绒鼠	*Eothenomys chinensis*	(Thomas, 1891)	Sichuan Red-backed Vole	1148
			西南绒鼠	*Eothenomys custos*	(Thomas, 1912)	Southwestern Chinese Red-backed Vole	1150
			滇绒鼠	*Eothenomys eleusis*	Thomas, 1911	Yunnan Red-backed Vole	1152
			康定绒鼠	*Eothenomys hintoni*	Osgood, 1932	Kangting Chinese Vole	1154
			福建绒鼠	*Eothenomys colurnu*	(Thomas, 1911)	Fujian Chinese Vole	1156
			黑腹绒鼠	*Eothenomys melanogaster*	(Milne-Edwards, 1871)	Pére David's Chinese Vole	1158
			大绒鼠	*Eothenomys miletus*	(Thomas, 1914)	Large Chinese Vole	1160
			昭通绒鼠	*Eothenomys olitor*	(Thomas, 1911)	Black-eared Red-backed Vole	1162
			玉龙绒鼠	*Eothenomys proditor*	Hinton, 1923	Yulongxuen Red-backed Vole	1164
			石棉绒鼠	*Eothenomys shimianensis*	Shaoying, 2019	Shimian Red-backed Vole	1166
			螺髻山绒鼠	*Eothenomys bialuojishanensis*	Shaoying, 2019	LuoJishan Chinese vole	1168
			美姑绒鼠	*Eothenomys meiguensis*	Shaoying, 2019	Meigu Chinese vole	1170
			金阳绒鼠	*Eothenomys jinyangensis*	Shaoying, 2019	Jinyang chinese vole	1172
			川西绒鼠	*Eothenomys tarquinius*	Thomas, 1912	Western Sichuan Red-backed Vole	1174

目 Order	科 Family	属 Genus	中文名 Chinese name	学名 Scientific name	命名人 Species authority	英文名 English name	页码 page
			丽江绒鼠	*Eothenomys fidelis*	Hinton, 1923	Lijiang Black Vole	1176
			德钦绒鼠	*Eothenomys wardi*	(Thomas, 1912)	Ward's Red-backed Vole	1178
		绒䶄属 *Caryomys*					
			洮州绒鼠	*Caryomys eva*	(Thomas, 1911)	Eva's Red-backed Vole	1180
			苛岚绒鼠	*Caryomys inez*	(Thomas, 1908)	Inez's Red-backed Vole	1182
		高山䶄属 *Alticola*					
			白尾高山䶄	*Alticola albicauda*	(True, 1894)	White -tailed Mountain Vole	1184
			银色高山䶄	*Alticola argentatus*	(Severtzov, 1879)	Silver Mountain Vole	1186
			戈壁阿尔泰高山䶄	*Alticola barakshin*	Bannikov, 1947	Gobi Altai Mountain Vole	1188
			大耳高山䶄	*Alticola macrotis*	(Radde, 1862)	Large-eared Vole	1190
			蒙古高山䶄	*Alticola semicanus*	(G. M. Allen, 1924)	Mongolian Silver Vole	1192
			斯氏高山䶄	*Alticola stoliczkanus*	(Blanford, 1875)	Stoliczka's Mountain Vole	1194
			扁颅高山䶄	*Alticola strelzowi*	(Kastschenko, 1899)	Flat-headed Vole	1196
		兔尾鼠属 *Lagurus*					
			草原兔尾鼠	*Lagurus lagurus*	Pallas, 1773	Steppe Lemming	1198
		水䶄属 *Arvicola*					
			水䶄	*Arvicola amphibius*	(Linnaeus, 1758)	European Water Vole	1200
		松田鼠属 *Neodon*					
			克氏松田鼠	*Neodon clarkei*	(Hinton, 1923)	Clarke's Vole	1202
			云南松田鼠	*Neodon forresti*	Hinton, 1923	Forrest's Mountain Vole	1204
			青海松田鼠	*Neodon fuscus*	Büchner, 1889	Plateau Pine Vole	1206
			高原松田鼠	*Neodon irene*	(Thomas, 1911)	Irene's Mountain Vole	1208
			白尾松田鼠	*Neodon leucurus*	Blyth, 1863	Blyth's Mountain Vole	1210
			林芝松田鼠	*Neodon linzhiensis*	Liu, Sun, Liu, Wang, Guo & Murphy, 2012	Linzhi Mountain Vole	1212
			墨脱松田鼠	*Neodon medogensis*	Liu, Jin, W., Liu, Murphy, Lu, Hao, Liao, Sun, Tang, Chen & Fu, 2016	Motuo Mountain Vole	1214
			聂拉木松田鼠	*Neodon nyalamensis*	Liu, Jin, W., Liu, Murphy, Lu, Hao, Liao, Sun, Tang, Chen & Fu, 2016	Niemula Mountain Vole	1216
			锡金松田鼠	*Neodon sikimensis*	(Horsfield, 1841)	Sikkim Mountain Vole	1218
		东方田鼠属 *Alexandromys*					
			东方田鼠	*Alexandromys fortis*	(Büchner, 1889)	Reed Vole	1220
			台湾田鼠	*Alexandromys kikuchii*	(Kuroda, 1920)	Taiwan Vole	1222
			柴达木根田鼠	*Alexandromys limnophilus*	(Büchner, 1889)	Lacustrine Vole	1224
			莫氏田鼠	*Alexandromys maximowiczii*	(Schrenk, 1859)	Maximowicz's Vole	1226
			蒙古田鼠	*Alexandromys mongolicus*	(Radde, 1861)	Mongolian Vole	1228

目 Order	科 Family	属 Genus	中文名 Chinese name	学名 Scientific name	命名人 Species authority	英文名 English name	页码 page
			根田鼠	*Alexandromys oeconomus*	(Pallas, 1776)	Root Vole	1230
		田鼠属 *Microtus*					
			黑田鼠	*Microtus agrestis*	(Linnaeus, 1761)	Field Vole	1232
			伊犁田鼠	*Microtus ilaeus*	Thomas, 1912	Kazakhstan Vole	1234
			帕米尔田鼠	*Microtus juldaschi*	(Severtzov, 1879)	Juniper Mountain Vole	1236
			社田鼠	*Microtus socialis*	(Pallas, 1773)	Social Vole	1238
		川西田鼠属 *Volemys*					
			四川田鼠	*Volemys millicens*	(Thomas, 1911)	Sichuan Vole	1240
			川西田鼠	*Volemys musseri*	(Lawrence, 1982)	Marie's Vole	1242
		毛足田鼠属 *Lasiopodomys*					
			布氏田鼠	*Lasiopodomys brandtii*	(Radde, 1861)	Brandt's Vole	1244
			狭颅田鼠	*Microtus gregalis*	(Pallas, 1779)	Narrow-headed Vole	1246
			棕色田鼠	*Lasiopodomys mandarinus*	(Milne-Edwards, 1871)	Mandarin Vole	1248
		东方兔尾鼠属 *Eolagurus*					
			黄兔尾鼠	*Eolagurus luteus*	(Eversmann, 1840)	Yellow Steppe Lemming	1250
			蒙古兔尾鼠	*Eolagurus przewalskii*	(Büchner, 1889)	Przewalski's Steppe Lemming	1252
		沟牙田鼠属 *Proedromys*					
			沟牙田鼠	*Proedromys bedfordi*	Thomas, 1911	Duke of Bedford's Vole	1254
			凉山沟牙田鼠	*Proedromys liangshanensis*	Liu, Sun, Zeng & Zhao, 2007	Liangshan Vole	1256
	鼠科 Muridae						
		长尾攀鼠属 *Vandeleuria*					
			长尾攀鼠	*Vandeleuria oleracea*	(Bennett, 1832)	Asiatic Long-tailed Climbing Mouse	1258
		狨鼠属 *Hapalomys*					
			小狨鼠	*Hapalomys delacouri*	Thomas, 1927	Lesser Marmoset Rat	1260
			长尾绒鼠	*Hapalomys longicaudatus*	(Blyth, 1856)	Long-tailed Marmoset Rat	1262
		笔尾树鼠属 *Chiropodomys*					
			费氏树鼠	*Chiromyscus chiropus*	(Thomas, 1891)	Indochinese Chiromyscus	1264
			笔尾树鼠	*Chiropodomys gliroides*	Blyth, 1856	Indomalayan Pencil-tailed Tree Mouse	1266
			南洋鼠	*Chiromyscus langbianis*	Robinson & Kloss, 1922	Indochinese Arboreal Niviventer	1268
		攀鼠属 *Vernaya*					
			滇攀鼠	*Vernaya fulva*	(G. M. Allen, 1927)	Vernay's Climbing Mouse	1270
		巢鼠属 *Micromys*					
			红耳巢鼠	*Micromys erythrotis*	Blyth, 1855	Red-eared Harvest Mouse	1272
			巢鼠	*Micromys minutus*	(Pallas, 1771)	Eurasian Harvest Mouse	1274

目 Order	科 Family	属 Genus	中文名 Chinese name	学名 Scientific name	命名人 Species authority	英文名 English name	页码 page
		姬鼠属 *Apodemus*					
			黑线姬鼠	*Apodemus agrarius*	(Pallas, 1771)	Striped Field Mouse	1276
			高山姬鼠	*Apodemus chevrieri*	(Milne-Edwards, 1868)	Chevrier's Field Mouse	1278
			中华姬鼠	*Apodemus draco*	(Barrett-Hamilton, 1900)	South China Field Mouse	1280
			澜沧江姬鼠	*Apodemus ilex*	Thomas,1922	Lancangjiang Field Mouse	1282
			大耳姬鼠	*Apodemus latronum*	Thomas, 1911	Large-eared Field Mouse	1284
			黑姬鼠	*Apodemus nigrus*	Deyan Ge, Anderson Feijó & Qisen Yang, 2019	Black Filed Mouse	1286
			喜马拉雅姬鼠	*Apodemus pallipes*	(Barrett-Hamilton, 1900)	Himalayan Field Mouse	1288
			大林姬鼠	*Apodemus peninsulae*	(Thomas, 1907)	Korean Field Mouse	1290
			台湾姬鼠	*Apodemus semotus*	Thomas, 1908	Taiwan Field Mouse	1292
			乌拉尔姬鼠	*Apodemus uralensis*	(Pallas, 1811)	Herb Field Mouse	1294
		硕鼠属 *Hadromys*					
			云南壮鼠	*Hadromys yunnanensis*	Yang & Wang, 1987	Yunnan Hadromys	1296
		大齿鼠属 *Dacnomys*					
			大齿鼠	*Dacnomys millardi*	Thomas, 1916	Millard's Rat	1298
		鼠属 *Rattus*					
			黑缘齿鼠	*Rattus andamanensis*	(Blyth, 1860)	Indochinese Forest Rat	1300
			缅鼠	*Rattus exulans*	(Peale, 1848)	Polynesian Rat	1302
			黄毛鼠	*Rattus losea*	(Swinhoe, 1871)	Losea Rat	1304
			大足鼠	*Rattus nitidus*	(Hodgson, 1845)	Himalayan Field Rat	1306
			褐家鼠	*Rattus norvegicus*	(Berkenhout, 1769)	Brown Rat	1308
			拟家鼠	*Rattus pyctoris*	(Hodgson 1845)	Himalayan Rat	1310
			黑家鼠	*Rattus rattus*	(Linnaeus, 1758)	Black Rat	1312
			黄胸鼠	*Rattus tanezumi*	Temminck, 1844	Oriental House Rat	1314
		白腹鼠属 *Niviventer*					
			安氏白腹鼠	*Niviventer andersoni*	(Thomas, 1911)	Anderson's Niviventer	1316
			梵鼠	*Niviventer brahma*	(Thomas, 1911)	Brahma White-bellied Rat	1318
			北社鼠	*Niviventer confucianus*	(Milne-Edwards, 1871)	Confucian Niviventer	1320
			台湾白腹鼠	*Niviventer coninga*	(Swinhoe, 1864)	Spiny Taiwan Niviventer	1322
			褐尾鼠	*Niviventer cremoriventer*	(Miller, 1900)	Sundaic Arboreal Niviventer	1324
			台湾社鼠	*Niviventer culturatus*	(Thomas, 1917)	Soft-furred Taiwan Niviventer	1326
			灰腹鼠	*Niviventer eha*	(Wroughton, 1916)	Little Himalayan Rat	1328
			川西白腹鼠	*Niviventer excelsior*	(Thomas, 1911)	Sichuan Niviventer	1330
			冯氏白腹鼠	*Niviventer fengi*	Ge, Feijó & Yang, 2020	Feng's White-bellied Mouse	1332

目 Order	科 Family	属 Genus	中文名 Chinese name	学名 Scientific name	命名人 Species authority	英文名 English name	页码 page
			针毛鼠	*Niviventer fulvescens*	(Gray, 1847)	Chestnut White-bellied Rat	1334
			剑纹小社鼠	*Niviventer gladiusmaculus*	Ge, Lu, Xia, Du, Wen, Cheng, Abramov & Yang, 2018	Least Niviventer	1336
			拟刺毛鼠	*Niviventer huang*	Bonhote, 1905	Eastern Spiny-haired Rat	1338
			海南白腹鼠	*Niviventer lotipes*	G. M. Allen, 1926	Hainan Niviventer	1340
			白腹鼠	*Niviventer niviventer*	(Hodgson, 1836)	Himalayan White-bellied Rat	1342
			湄公针毛鼠	*Niviventer mekongis*	Robinson & Kloss, 1922	Mekong White-bellied Rat	1344
			片马社鼠	*Niviventer pianmaensis*	Li et Yang, 2009	Pianma Niviventer	1346
			山东社鼠	*Niviventer sacer*	(Thomas, 1908)	Sacer Niviventer	1348
			缅甸山鼠	*Niviventer tenaster*	Thomas, 1916	Indochinese Mountain Niviventer	1350
		王鼠属 *Maxomys*					
			红毛王鼠	*Maxomys surifer*	(Miller, 1900)	Indomalayan Maxomys	1352
		长尾巨鼠属 *Berylmys*					
			大泡灰鼠	*Berylmys berdmorei*	(Blyth, 1851)	Berdmore's Berylmy	1354
			青毛巨鼠	*Berylmys bowersi*	(Anderson, 1879)	Bower's White-toothed Rat	1356
			小泡灰鼠	*Berylmys manipulus*	(Thomas, 1916)	Manipur White-toothed Rat	1358
		小泡巨鼠属 *Leopoldamys*					
			白腹巨鼠	*Leopoldamys edwardsi*	(Thomas, 1882)	Edward's Rat	1360
			耐氏大鼠	*Leopoldamys neilli*	(J.T. Marshall Jr., 1976)	Neill's Long-tailed Giant Rat	1362
		小鼠属 *Mus*					
			印度小鼠	*Mus booduga*	(Gray, 1837)	Little Indian Field Mouse	1364
			卡氏小鼠	*Mus caroli*	Bonhote, 1902	Ryukyu Mouse	1366
			仔鹿小鼠	*Mus cervicolor*	Hodgson, 1845	Fawn-colored Mouse	1368
			丛林小鼠	*Mus cookii*	Ryley, 1914	Cook's Mouse	1370
			小家鼠	*Mus musculus*	Linnaeus, 1758	House Mouse	1372
			锡金小鼠	*Mus pahari*	Thomas, 1916	Gairdner's Shrewmouse	1374
		东京鼠属 *Tonkinomys*					
			道氏东京鼠	*Tonkinomys davovantien*	Musser, Lunde & Son, 2006	Daovantien's Limestone Rat	1376
		板齿鼠属 *Bandicota*					
			小板齿鼠	*Bandicota bengalensis*	(Gray, 1835)	Lesser Bandicoot Rat	1378
			板齿鼠	*Bandicota indica*	(Bechstein, 1800)	Greater Bandicoot Rat	1380
		地鼠属 *Nesokia*					
			印度地鼠	*Nesokia indica*	(Gray, 1830)	Short-tailed Bandicoot Rat	1382
		短耳沙鼠属 *Brachiones*					
			短耳沙鼠	*Brachiones przewalskii*	(Büchner, 1889)	Przewalski's Jird	1384
		沙鼠属 *Meriones*					

目 Order	科 Family	属 Genus	中文名 Chinese name	学名 Scientific name	命名人 Species authority	英文名 English name	页码 page
			郑氏沙鼠	*Meriones chengi*	Wang, 1964	Cheng's gerbil	1386
			红尾沙鼠	*Meriones libycus*	Lichtenstein, 1823	Libyan Jird	1388
			子午沙鼠	*Meriones meridianus*	(Pallas, 1773)	Mid-day Gerbil	1390
			柽柳沙鼠	*Meriones tamariscinus*	Pallas, 1773	Tamarisk Gerbil	1392
			长爪沙鼠	*Meriones unguiculatus*	(Milne-Edwards, 1867)	Mongolian Gerbil	1394
		大沙鼠属 *Rhombomys*					
			大沙鼠	*Rhombomys opimus*	(Lichtenstein, 1823)	Great Gerbil	1396
	刺山鼠科 Platacanthomyidae						
		猪尾鼠属 *Typhlomys*					
			沙巴猪尾鼠	*Typhlomys chapensis*	Milne-Edwards, 1877	Sort-furred Tree Mouse	1398
			武夷山猪尾鼠	*Typhlomys cinereus*	Milne-Edwards, 1877	Wuyishan Tree Mouse	1400
			大猪尾鼠	*Typhlomys daloushanensis*	Wang & Li, 1996	Daloushan Pygmy Dormouse	1402
			黄山猪尾鼠	*Typhlomys huangshanensis*	(Hu & Zhang, 2021)	Huangshan Tree Mouse	1404
			小猪尾鼠	*Typhlomys nanus*	Cheng, He, Chen, Zhang, Wan, Li, Zhang & Jiang, 2017	Dwarf Tree Mouse	1406
	鼹型鼠科 Spalacidae						
		小竹鼠属 *Cannomys*					
			小竹鼠	*Cannomys badius*	(Hodgson, 1841)	Lesser Bamboo Rat	1408
		竹鼠属 *Rhizomys*					
			银星竹鼠	*Rhizomys pruinosus*	Blyth, 1851	Hoary Bamboo Rat	1410
			中华竹鼠	*Rhizomys sinensis*	Gray, 1831	Chinese Bamboo Rat	1412
			大竹鼠	*Rhizomys sumatrensis*	(Raffles, 1821)	Indomalayan Bamboo Rat	1414
		凸颅鼢鼠属 *Eospalax*					
			高原鼢鼠	*Eospalax baileyi*	(Thomas, 1911)	Plateau Zokar	1416
			甘肃鼢鼠	*Eospalax cansus*	(Lyon, 1907)	Gansu Zokor	1418
			中华鼢鼠	*Eospalax fontanierii*	(Milne-Edwards, 1867)	Chinese Zokor	1420
			木里鼢鼠	*Eospalax muliensis*	Zhang, Chen & Shi, 2022	Muli zokor	1422
			罗氏鼢鼠	*Eospalax rothschildi*	(Thomas, 1911)	Rothschild's Zokor	1424
			秦岭鼢鼠	*Eospalax rufescens*	J. Allen,1909	Qinling Mountian Zokor	1426
			斯氏鼢鼠	*Eospalax smithii*	(Thomas, 1911)	Smith's Zokor	1428
		平颅鼢鼠属 *Myospalax*					
			草原鼢鼠	*Myospalax aspalax*	(Pallas, 1776)	False Zoko	1430
			阿尔泰鼢鼠	*Myospalax myospalax*	(Laxmann, 1773)	Siberian Zokor	1432
			东北鼢鼠	*Myospalax psilurus*	(Milne-Edwards, 1874)	Transbaikal Zokor	1434
	睡鼠科 Gliridae						

目 Order	科 Family	属 Genus	中文名 Chinese name	学名 Scientific name	命名人 Species authority	英文名 English name	页码 page
		林睡鼠属 *Dryomys*					
			林睡鼠	*Dryomys nitedula*	(Pallas, 1778)	Forest Dormouse	1436
		毛尾睡鼠属 *Chaetocauda*					
			四川毛尾睡鼠	*Chaetocauda sichuanensis*	Wang, 1985	Sichuan Dormouse	1438
	蹶鼠科 Sminthidae						
		蹶鼠属 *Sicista*					
			长尾蹶鼠	*Sicista caudata*	Thomas, 1907	Long-tailed Birch Mouse	1440
			中国蹶鼠	*Sicista concolor*	(Büchner, 1892)	Chinese Birch Mouse	1442
			灰蹶鼠	*Sicista pseudonapaea*	Strautman, 1949	Grey Birch Mouse	1444
			草原蹶鼠	*Sicista subtilis*	(Pallas, 1773)	Southern Birch Mouse	1446
			天山蹶鼠	*Sicista tianshanica*	(Salensky, 1903)	Tien Shan Birch Mouse	1448
	林跳鼠科 Zopodidae						
		林跳鼠属 *Eozapus*					
			四川林跳鼠	*Eozapus setchuanus*	(Pousargues, 1896)	Chinese Jumping Mouse	1450
	跳鼠科 Dipodidae						
		五趾心颅跳鼠属 *Cardiocranius*					
			五趾心颅跳鼠	*Cardiocranius paradoxus*	Satunin, 1903	Five-toed Pygmy Jerboa	1452
		三趾心颅跳鼠属 *Salpingotus*					
			肥尾心颅跳鼠	*Salpingotus crassicauda*	Vinogradov, 1924	Thick-tailed Pygmy Jerboa	1454
			三趾心颅跳鼠	*Salpingotus kozlovi*	Vinogradov, 1922	Kozlov's Pygmy Jerboa	1456
		长耳跳鼠属 *Euchoreutes*					
			长耳跳鼠	*Euchoreutes naso*	Sclater, 1891	Long-eared Jerboa	1458
		五趾跳鼠属 *Allactaga*					
			巴里坤跳鼠	*Allactaga balikunica*	Hsia & Fang, 1964	Balikun Jerboa	1460
			巨泡五趾跳鼠	*Allactaga bullata*	Allen, 1925	Gobi Jerboa	1462
			小五趾跳鼠	*Allactaga elater*	(H. Lichtenstein, 1825)	Small Five-toed Jerboa	1464
			大五趾跳鼠	*Allactaga major*	(Kerr, 1792)	Great Jerboa	1466
			五趾跳鼠	*Allactaga sibirica*	(Forster, 1778)	Siberian Jerboa	1468
		肥尾跳鼠属 *Pygeretmus*					
			小地兔	*Pygeretmus pumilio*	(Kerr, 1792)	Dwarf Fat-tailed Jerboa	1470
		奇美跳鼠属 *Chimaerodipus*					
			奇美跳鼠	*Chimaerodipus auritus*	Shenbrot, Bannikova, Giraudoux, Quere, Raoul & Lebedev, 2018	Ningxia Three-toed Jerboa	1472
		三趾跳鼠属 *Dipus*					

目 Order	科 Family	属 Genus	中文名 Chinese name	学名 Scientific name	命名人 Species authority	英文名 English name	页码 page
			塔里木跳鼠	*Dipus deasyi*	Barrett-Hamilton, 1900	Qaidam Three-toed Jerboa	1474
			三趾跳鼠	*Dipus sagitta*	(Pallas, 1773)	Hairy-footed Jerboa	1476
		羽尾跳鼠属 *Stylodipus*					
			内蒙羽尾跳鼠	*Stylodipus andrewsi*	Allen, 1925	Andrews's Three-toed Jerboa	1478
			准噶尔 羽尾跳鼠	*Stylodipus sungorus*	Sokolov et Shenbrot, 1987	Dzungaria Three-toed Jerboa	1480
			羽尾跳鼠	*Stylodipus telum*	(Lichtenstein, 1823)	Thick-tailed Three-toed Jerboa	1482
	豪猪科 Hystricidae						
		帚尾豪猪属 *Atherurus*					
			帚尾豪猪	*Atherurus macrourus*	(Linnaeus, 1758)	Asiatic Brush-tailed Porcupine	1484
		豪猪属 *Hystrix*					
			马来豪猪	*Hystrix brachyura*	Lichtenstein, 1823	Malayan Porcupine	1486
兔形目 Lagomorpha							
	鼠兔科 Ochotonidae						
		鼠兔属 *Ochotona*					
			高山鼠兔	*Ochotona alpina*	(Pallas, 1773)	Alpine Pika	1488
			贺兰山鼠兔	*Ochotona argentata*	(Howell, 1928)	Silver Pika	1490
			间颅鼠兔	*Ochotona cansus*	Lyon, 1907	Gansu Pika	1492
			中国鼠兔	*Ochotona chinensis*	Thomas, 1911	Chinese Pika	1494
			长白山鼠兔	*Ochotona coreana*	Allen et Andrews, 1913	Changbaishan Pika	1496
			高原鼠兔	*Ochotona curzoniae*	(Hodgson, 1858)	Plateau Pika	1498
			达乌尔鼠兔	*Ochotona dauurica*	(Pallas, 1776)	Daurian Pika	1500
			红耳鼠兔	*Ochotona erythrotis*	(Büchner, 1890)	Chinese Red Pika	1502
			扁颅鼠兔	*Ochotona flatcalvariam*	Liu, Jin, Liao, Sun, Liu, 2017	Flat-cranium Pika	1504
			灰颈鼠兔	*Ochotona forresti*	Thomas, 1923	Forrest's Pika	1506
			川西鼠兔	*Ochotona gloveri*	Thomas, 1922	Glover's Pika	1508
			喜马拉雅鼠兔	*Ochotona himalayana*	Feng, 1973	Himalayan pika	1510
			黄龙鼠兔	*Ochotona huanglongensis*	Liu, Jin, Liao, Sun, Liu, 2017	Huanglong Pika	1512
			伊犁鼠兔	*Ochotona iliensis*	Li & Ma, 1986	Ili Pika	1514
			柯氏鼠兔	*Ochotona koslowi*	(Büchner, 1894)	Kozlov's Pika	1516
			拉达克鼠兔	*Ochotona ladacensis*	(Günther, 1875)	Ladak Pika	1518
			大耳鼠兔	*Ochotona macrotis*	(Günther, 1875)	Large-eared Pika	1520
			满洲里鼠兔	*Ochotona mantchurica*	Thomas, 1909	Manchurian Pika	1522
			奴布拉鼠兔	*Ochotona nubrica*	Thomas, 1922	Nubra Pika	1524
			蒙古鼠兔	*Ochotona pallasi*	(Gray, 1867)	Pallas's Pika	1526
			草原鼠兔	*Ochotona pusilla*	(Pallas, 1769)	Steppe Pika	1528

目 Order	科 Family	属 Genus	中文名 Chinese name	学名 Scientific name	命名人 Species authority	英文名 English name	页码 page
			邛崃鼠兔	*Ochotona qionglaiensis*	Liu, Jin, Liao, Sun, 2017	Qionglai Pika	1530
			灰鼠兔	*Ochotona roylei*	(Ogilby, 1839)	Royle's Pika	1532
			红鼠兔	*Ochotona rutila*	Severtzov, 1873	Turkestan Red Pika	1534
			峨眉鼠兔	*Ochotona sacraria*	(Thomas, 1923)	Ermei Pika	1536
			锡金鼠兔	*Ochotona sikimaria*	(Thomas, 1922)	Sikim Pika	1538
			秦岭鼠兔	*Ochotona syrinx*	Thomas, 1911	Tsing-Ling Pika	1540
			藏鼠兔	*Ochotona thibetana*	(Milne-Edwards, 1871)	Moupin Pika	1542
			狭颅鼠兔	*Ochotona thomasi*	Argyropulo, 1948	Thomas's Pika	1544
	兔科 Leporidae						
		粗毛兔属 *Caprolagus*					
			粗毛兔	*Caprolagus hispidus*	(Pearson, 1839)	Hispid Hare	1546
		兔属 *Lepus*					
			云南兔	*Lepus comus*	Allen, 1927	Yunnan Hare	1548
			高丽兔	*Lepus coreanus*	Thomas, 1892	Korean Hare	1550
			海南兔	*Lepus hainanus*	Swinhoe, 1870	Hainan Hare	1552
			东北兔	*Lepus mandshuricus*	Radde, 1861	Manchurian Hare	1554
			尼泊尔黑兔	*Lepus nigricollis*	F. Cuvier, 1823	Indian Hare	1556
			灰尾兔	*Lepus oiostolus*	Hodgson, 1840	Woolly Hare	1558
			华南兔	*Lepus sinensis*	Gray, 1832	Chinese Hare	1560
			中亚兔	*Lepus tibetanus*	Waterhouse, 1841	Desert Hare	1562
			雪兔	*Lepus timidus*	Linnaeus, 1758	Mountain Hare	1564
			蒙古兔	*Lepus tolai*	Pallas, 1778	Tolai Hare	1566
			塔里木兔	*Lepus yarkandensis*	Günther, 1875	Yankand Lepus	1568

1 / 毛猬

Hylomys suillus Müller, 1840

· Short-tailed Gymnure

劳亚食虫目 Eulipotyphla / 猬科 Erinaceidae / 毛猬属 *Hylomys*

科建立者及其文献 / Family Authority
G. Fischer, 1814

属建立者及其文献 / Genus Authority
Müller, 1840

亚种 / Subspecies
缅甸亚种 *Hylomys suillus peguensis* Blyth, 1859
云南西部和南部
Yunnan(western part-Yingjiang, and southern part- Xishuangbanna)

滇南亚种 *Hylomys suillus microtinus* Thomas, 1925
云南南部（江城、绿春、金平和河口）
Yunnan (Southern parts-Jiangcheng, Lüchun, Jinping and Hekou)

模式标本产地 / Type Locality
印度尼西亚
Indonesia, "Java en het andere van Sumatra"

▲ 其他名称 / Other Name(s)

其他中文名 / Other Chinese Name(s)
中国毛猬、小毛猬

其他英文名 / Other English Name(s)
Lesser Moonrat, Short-tailed Moonrat

▲ 形态及生境 / Morphology and Habitat

形态特征 / Morphological Characteristics
齿式：3.1.4.3/3.1.4.3=44。外形似鼠类，但吻更长，外耳钝圆形。头体长 90~156 mm。尾长 10~30 mm。后足 20~28 mm。体重 45~80 g。背部污黄至深棕色，背腹明显异色，腹部毛发基部浅灰色，端部白色。尾长明显短于鼩猬。
Dental formula: 3.1.4.3/3.1.4.3=44. The external morphology resembles that of a rodent, but with a longer snout and rounded ears. Head-body length 90-156 mm, tail length 10-30 mm, hindfoot length 20-28 mm, weight 45-80 g. The dorsal pelage is dirty yellow to brown in color. The ventral pelage is bicolor, the basal is grey and the tip is white. Tail is much shorter than that of a *Neoetracus sinensis*.

生境 / Habitat
亚热带热带湿润低地森林、亚热带热带湿润山地森林、灌丛
Subtropical tropical moist lowland forest, subtropical tropical moist montane forest, shrubland

▲ 地理分布 / Geographic Distribution

国内分布 / Domestic Distribution
云南 Yunnan

全球分布 / World Distribution
文莱、柬埔寨、中国、印度尼西亚、老挝、马来西亚、缅甸、泰国、
越南
Brunei, Cambodia, China, Indonesia, Laos, Malaysia, Myanmar, Thailand, Vietnam

生物地理界 / Biogeographic Realm
印度马来界 Indomalaya

WWF 生物群系 / WWF Biome
热带和亚热带湿润阔叶林 Tropical & Subtropical Moist Broadleaf Forests

动物地理分布型 / Zoogeographic Distribution Type
Wa

分布标注 / Distribution Note
非特有种 Non-Endemic

▲ 濒危状况 / Threatened Status

中国生物多样性红色名录等级 / CB RL Category (2021)
无危 LC

IUCN 红色名录 / IUCN Red List (2021)
无危 LC

威胁因子 / Threats
无 None

▲ 法律保护地位 / Legal Protection Status

国家重点保护野生动物等级 / Category of National Key Protected Wild Animals (2021)
未列入 Not listed

"三有"名录 / TWIESSV (2023)
列入 Listed

CITES 附录等级 / CITES Appendix (2023)
未列入 Not listed

迁徙物种公约附录 / CMS Appendix (2020)
未列入 Not listed

保护行动 / Conservation Action
尚无保护行动 No conservation action so far

▲ 参考文献 / References

Jiang et al. (蒋志刚等), 2021; Burgin et al., 2020; IUCN, 2020; Tong and Lu (仝磊和路纪琪), 2010; Smith et al., 2009; Pan et al. (潘清华等), 2007; Wang (王应祥), 2003; Zhang (张荣祖), 1997

2 / 海南新毛猬

Neohylomys hainanensis Shaw & Wang, 1959

· Hainan Gymnure

▲ 分类地位 / Taxonomy

劳亚食虫目 Eulipotyphla / 猬科 Erinaceidae / 新毛猬属 *Neohylomys*

科建立者及其文献 / Family Authority
G. Fischer, 1814

属建立者及其文献 / Genus Authority
Shaw and Wong, 1959

亚种 / Subspecies
无 None

模式标本产地 / Type Locality
中国
China, "Pai-sa Hsian, Hainan Island" (Baisha Xian, an administrative unit at 19°3'N, 109°6'E)

李玉春 / 供图

▲ 其他名称 / Other Name(s)

其他中文名 / Other Chinese Name(s)
无 None

其他英文名 / Other English Name(s)
无 None

▲ 形态及生境 / Morphology and Habitat

形态特征 / Morphological Characteristics

齿式：3.1.4.3/3.1.3.3=42。体型中等的食虫类。头体长 120~147 mm。后足长 24~29 mm。尾长 36~43 mm。外耳短，钝圆形。背腹毛色明显异色，背部深棕黄色，背脊有清晰黑色纵向条纹。腹部毛发基部为浅灰色，端部浅黄色。

Dental formula: 3.1.4.3/3.1.3.3=42. A medium-sized insectivore. Head-body length 120-147 mm, hindfoot length 24-29 mm, tail length 36-43 mm. Ears are short and blunt rounded. Dorsal pelage dark brown with a black dorsal longitudinal stripe. Ventral pelage is bicolor, and the basal is pale gray and the tip is pale yellow.

生境 / Habitat

海拔 250 m 的次生林以及橡胶林；海拔 600~1000 m 的山地森林
Secondary forest and rubber forest of 250 m a.s.l; mountian forest of 600-1000 m a.s.l

▲ 地理分布 / Geographic Distribution

国内分布 / Domestic Distribution
海南 Hainan

全球分布 / World Distribution
中国、越南
China, Vietnam

生物地理界 / Biogeographic Realm
印度马来界 Indomalaya

WWF 生物群系 / WWF Biome
热带和亚热带湿润阔叶林 Tropical & Subtropical Moist Broadleaf Forests

动物地理分布型 / Zoogeographic Distribution Type
J

分布标注 / Distribution Note
非特有种 Non-Endemic

▲ 濒危状况 / Threatened Status

中国生物多样性红色名录等级 / CB RL Category (2021)
濒危 EN

IUCN 红色名录 / IUCN Red List (2021)
濒危 EN

威胁因子 / Threats
森林砍伐、耕种 Logging, farming

▲ 法律保护地位 / Legal Protection Status

国家重点保护野生动物等级 / Category of National Key Protected Wild Animals (2021)
未列入 Not listed

"三有"名录 / TWIESSV (2023)
列入 Listed

CITES 附录等级 / CITES Appendix (2023)
未列入 Not listed

迁徙物种公约附录 / CMS Appendix (2020)
未列入 Not listed

保护行动 / Conservation Action
尚无保护行动 No conservation action so far

▲ 参考文献 / References

Jiang et al. (蒋志刚等), 2021; Burgin et al., 2020; IUCN, 2020; Liu et al. (刘少英等), 2020; Abramov et al., 2018; Wilson and Mittermeier, 2018; IUCN, 2015; Smith et al., 2009; Pan et al. (潘清华等), 2007; Wilson and Reeder, 2005; Wang (王应祥), 2003; Zhang (张荣祖), 1997; Stone, 1995; Xia (夏武平), 1988; Shou et al. (寿振黄等), 1966; Shou and Wang (寿振黄和汪松), 1959

3 / 中国鼩猬

Neotetracus sinensis Trouessart, 1909

· Shrew Gymnure

▲ 分类地位 / Taxonomy

劳亚食虫目 Eulipotyphla / 猬科 Erinaceidae / 中国鼩猬属 *Neotetracus*

科建立者及其文献 / Family Authority
G. Fischer, 1814

属建立者及其文献 / Genus Authority
Trouessart, 1909

亚种 / Subspecies
指名亚种 *Neotetracus sinensis sinensis* (Trouessart, 1909)
四川西部、云南东北部和贵州西北部
Sichuan (western part), Yunnan (northeastern part) and Guizhou (northwestern part)

滇西亚种 *Neotetracus sinensis cuttingi* (Anthony, 1941)
云南西部（泸水）
Yunnan (western part-Lushui)

模式标本产地 / Type Locality
中国
"Ta-tsien-lou, province of Se-tchouen (China Occidental) at an altitude of 2454 meters" (Kangding, Sichuan Sheng, 30°7'N, 102°2'E)

何锴 / 供图

▲ 其他名称 / Other Name(s)

其他中文名 / Other Chinese Name(s)
鼩猬

其他英文名 / Other English Name(s)
无 None

同物异名 / Synonym(s)
Hylomys sinensis (Trouessart, 1909)

▲ 形态及生境 / Morphology and Habitat

形态特征 / Morphological Characteristics
齿式：3.1.3.3/3.1.3.3=40。外形似毛猬，但尾更长。头体长 91~125 mm，尾长 56~78 mm。后足长 21~36 mm。耳长 17~19 mm。背部毛色污黄色，腹部毛色深灰。吻部较短，眼睛大。尾长约为头体长的一半。上犬齿小，缺少第一上前白齿。

Dental formula: 3.1.3.3/3.1.3.3=40. The external morphology is similar to that of Hylomys suillus with a longer tail. Head-body length 91-125 mm, tail length 56-78 mm, hindfoot length 21-36 mm, ear length 17-19 mm. Dorsal pelage dirty yellow, ventral dark grey. Snout is short and eyes are large. Tail is about half of the head-body length. Upper canine is small, and the first upper premolar (P1) is missing.

生境 / Habitat
温带阔叶和混交林
Temperate broadleaf and mixed forests

▲ 地理分布 / Geographic Distribution

国内分布 / Domestic Distribution
四川、贵州、云南、广东
Sichuan, Guizhou, Yunnan, Guangdong

全球分布 / World Distribution
中国、缅甸、越南
China, Myanmar, Vietnam

生物地理界 / Biogeographic Realm
印度马来界 Indomalaya

WWF 生物群系 / WWF Biome
温带阔叶和混交林
Temperate Broadleaf & Mixed Forests

动物地理分布型 / Zoogeographic Distribution Type
Sd

分布标注 / Distribution Note
非特有种 Non-Endemic

▲ 濒危状况 / Threatened Status

中国生物多样性红色名录等级 / CB RL Category (2021)
无危 LC

IUCN 红色名录 / IUCN Red List (2021)
无危 LC

威胁因子 / Threats
无 None

▲ 法律保护地位 / Legal Protection Status

国家重点保护野生动物等级 / Category of National Key Protected Wild Animals (2021)
未列入 Not listed

"三有" 名录 / TWIESSV (2023)
列入 Listed

CITES 附录等级 / CITES Appendix (2023)
未列入 Not listed

迁徙物种公约附录 / CMS Appendix (2020)
未列入 Not listed

保护行动 / Conservation Action
尚无保护行动 No conservation action so far

▲ 参考文献 / References

Jiang et al. (蒋志刚等), 2021; Burgin et al., 2020; IUCN, 2020; Liu et al. (刘少英等), 2020; Wilson and Mittermeier, 2018; Deng et al. (邓可等), 2013; Gong et al. (龚晓俊等), 2013; Sun et al. (孙治宇等), 2013; Tu et al., 2012; Smith et al., 2009; Pan et al. (潘清华等), 2007; Wilson and Reeder, 2005; Wang (王应祥), 2003; Zhang (张荣祖), 1997

4 / 东北刺猬

Erinaceus amurensis Schrenk, 1859

· Amur Hedgehog

▲ 分类地位 / Taxonomy

劳亚食虫目 Eulipotyphla / 猬科 Erinaceidae / 刺猬属 *Erinaceus*

科建立者及其文献 / Family Authority
G. Fischer, 1814

属建立者及其文献 / Genus Authority
Linnaeus, 1758

亚种 / Subspecies
无共识 No concensus

模式标本产地 / Type Locality
俄罗斯
Russia, E Siberia, "In der Ne der Stadt Aigun, im mandschurischen Dorfe Gulssoja am Amur"

▲ 其他名称 / Other Name(s)

其他中文名 / Other Chinese Name(s)
普通刺猬、猬鼠

其他英文名 / Other English Name(s)
无 None

同物异名 / Synonym(s)
无 None

▲ 形态及生境 / Morphology and Habitat

形态特征 / Morphological Characteristics

齿式：3.1.3.3/2.1.2.3=36。中国体型最大的刺猬。体重 800~1200 g。头体长 158~287 mm。尾长 17~42 mm。后足长 34~54 mm。耳长 16~26 mm。耳长接近于其周围棘刺的长度。吻部与中国其他属的刺猬相比较长。脸部的毛色棕灰色至污黄色，背腹毛色灰白色至污黄色。从上额至臀部覆盖有棘刺，在受惊或者遇到天敌时，身体蜷曲成刺球状。头部棘刺中央有一块头皮裸露。棘刺有两种颜色：部分棘刺为纯白色，其余大部分棘刺中段颜色偏黑，两端颜色偏浅，但个体之间差异很大。前后足均具 5 趾。

Dental formula: 3.1.3.3/2.1.2.3=36. The largest hedgehog in China. Bodyweight 800-1,200 g, head-body length 158-287 mm, tail length 17-42 mm, hindfoot length 34-54 mm, ear length 16-26 mm. Ear is similar to the surrounding spines in length. Snout is longer than other hedgehogs in China. Face is unicolor, whitish-grey to dusty yellow. Pelage, except for the non-spiny area on the face, legs, and underparts have dense, smooth spines. The body can curl into a spiny ball when frightened or when confronted by a predator. A narrow spineless area on the scalp is present. Spines have two colorations. Most spines have a dark ring in the middle and both basal and tip are white. The other spines are wholly white. The coloration is highly variable between individuals. All feet have five toes.

生境 / Habitat

亚热带湿润低地森林、亚热带湿润高地森林、灌丛、草地、泰加林、耕地、城市、乡村种植园
Subtropical moist lowland forest, subtropical moist highland forest, shrubland, grassland, taiga, arable land, urban area rural garden

▲ 地理分布 / Geographic Distribution

国内分布 / Domestic Distribution
黑龙江、吉林、辽宁、内蒙古、北京、河北、河南、陕西、甘肃、山西、山东、上海、江苏、浙江、安徽、江西、湖北、湖南、福建、广东、贵州、重庆
Heilongjiang, Jilin, Liaoning, Inner Mongolia, Beijing, Hebei, Henan, Shaanxi, Gansu, Shanxi, Shandong, Shanghai, Jiangsu, Zhejiang, Anhui, Jiangxi, Hubei, Hunan, Fujian, Guangdong, Guizhou, Chongqing

全球分布 / World Distribution
中国、朝鲜、韩国、俄罗斯
China, Democratic People's Republic of Korea, Republic of Korea, Russia

生物地理界 / Biogeographic Realm
古北界 Palearctic

WWF 生物群系 / WWF Biome
温带阔叶和混交林 Temperate Broadleaf & Mixed Forests

动物地理分布型 / Zoogeographic Distribution Type
Ma

分布标注 / Distribution Note
非特有种 Non-Endemic

▲ 濒危状况 / Threatened Status

中国生物多样性红色名录等级 / CB RL Category (2021)
无危 LC

IUCN 红色名录 / IUCN Red List (2021)
无危 LC

威胁因子 / Threats
无 None

▲ 法律保护地位 / Legal Protection Status

国家重点保护野生动物等级 / Category of National Key Protected Wild Animals (2021)
未列入 Not listed

"三有"名录 / TWIESSV (2023)
列入 Listed

CITES 附录等级 / CITES Appendix (2023)
未列入 Not listed

迁徙物种公约附录 / CMS Appendix (2020)
未列入 Not listed

保护行动 / Conservation Action
尚无保护行动 No conservation action so far

▲ 参考文献 / References

Jiang et al. (蒋志刚等), 2021; Burgin et al., 2020; IUCN, 2020; Liu et al. (刘少英等), 2020; Wilson and Mittermeier, 2018; Smith et al., 2009; Pan et al. (潘清华等), 2007; Wilson and Reeder, 2005; Wang (王应祥), 2003; Corbet, 1988

5 / 大耳猬

Hemiechinus auritus (Gmelin, 1770)

· Long-eared Hedgehog

▲ 分类地位 / Taxonomy

劳亚食虫目 Eulipotyphla / 猬科 Erinaceidae / 大耳猬属 *Hemiechinus*

科建立者及其文献 / Family Authority
G. Fischer, 1814

属建立者及其文献 / Genus Authority
Fitzinger, 1866

亚种 / Subspecies
南疆亚种 *Hemiechinus auritus albulus* (Stolicka, 1872)
新疆
Xinjiang

模式标本产地 / Type Locality
俄罗斯
Russia, "in regione Astrachanensi", (Astrakhan, 46°1'N, 48°3'E)

邢睿 / 供图

▲ 其他名称 / Other Name(s)

其他中文名 / Other Chinese Name(s)
无 None

其他英文名 / Other English Name(s)
无 None

同物异名 / Synonym(s)
无 None

▲ 形态及生境 / Morphology and Habitat

形态特征 / Morphological Characteristics
齿式：3.1.3.3/2.1.2.3=36。头体长 170~230 mm。尾长 18~28 mm。耳长
31~40 mm，明显超过耳周围棘刺的长度。脸颊较短，毛色污黄，腹部毛色
雪白色。头顶正中无裸露头皮，亦不形成向后延伸的深沟。背部布满棘刺，
从额头缘延伸到臀部，受惊或者遇到天敌时，身体可以蜷曲成刺球状。棘
刺靠根部 三分之二为白色，其上为黑色圆环，刺尖为白色。前后足均 5 趾。
Dental formula: 3.1.3.3/2.1.2.3=36. Head-body length 170-230 mm, tail length 18-28 mm,
ear length 31-40 mm in length and are obviously longer than the surrounding spines. The
snout is short, and the face is dusty-yellow in color. Ventral pelage is snow white. Spines
extend from forehead to hip. The body can curl into a spiny ball when frightened or when
confronted by a predator. Absence of a spineless area on the scalp. The spines are white at
2/3 the basal part, followed by a black ring, tip white. All feet have five toes.

生境 / Habitat
草地、灌丛、耕地、温带沙漠
Grassland, shrubland, arable land, temperate desert

▲ 地理分布 / Geographic Distribution

国内分布 / Domestic Distribution
内蒙古、新疆、陕西、宁夏、甘肃、青海、四川
Inner Mongolia, Xinjiang, Shaanxi, Ningxia, Gansu, Qinghai, Sichuan

全球分布 / World Distribution
阿富汗、中国、塞浦路斯、埃及、伊朗、伊拉克、以色列、吉尔吉斯斯坦、黎巴嫩、利比亚、蒙古国、巴基斯坦、叙利亚、塔吉克斯坦、土耳其、土库曼斯坦、乌克兰、乌兹别克斯坦
Afghanistan, China, Cyprus, Egypt, Iran, Iraq, Israel, Kyrgyzstan, Lebanon, Libya, Mongolia, Pakistan, Syria, Tajikistan, Turkey, Turkmenistan, Ukraine, Uzbekistan

生物地理界 / Biogeographic Realm
古北界 Palearctic

WWF 生物群系 / WWF Biome
沙漠和干燥的灌木地 Deserts & Xeric Shrublands

动物地理分布型 / Zoogeographic Distribution Type
D

分布标注 / Distribution Note
非特有种 Non-Endemic

▲ 濒危状况 / Threatened Status

中国生物多样性红色名录等级 / CB RL Category (2021)
无危 LC

IUCN 红色名录 / IUCN Red List (2021)
无危 LC

威胁因子 / Threats
无 None

▲ 法律保护地位 / Legal Protection Status

国家重点保护野生动物等级 / Category of National Key Protected Wild Animals (2021)
未列入 Not listed

"三有"名录 / TWIESSV (2023)
列入 Listed

CITES 附录等级 / CITES Appendix (2023)
未列入 Not listed

迁徙物种公约附录 / CMS Appendix (2020)
未列入 Not listed

保护行动 / Conservation Action
尚无保护行动 No conservation action so far

▲ 参考文献 / References

Jiang et al. (蒋志刚等), 2021; Burgin et al., 2020; IUCN, 2020; Liu et al. (刘少英等), 2020; Wilson and Mittermeier, 2018; Smith et al., 2009; Aptimi·Apdukader (阿布力米提·阿都卡迪尔), 2002; Hou (侯兰新), 2000; Wang (王香亭), 1991; Corbet, 1988; Pan et al. (潘清华等), 2007; Wang (王应祥), 2003; Zhang (张荣祖), 1997

6 / 达乌尔猬

Mesechinus dauuricus (Sundevall, 1842)

- Daurian Hedgehog

▲ 分类地位 / Taxonomy

劳亚食虫目 Eulipotyphla / 猬科 Erinaceidae / 林猬属 *Mesechinus*

科建立者及其文献 / Family Authority
G. Fischer, 1814

属建立者及其文献 / Genus Authority
Ognev, 1951

亚种 / Subspecies
无 None

模式标本产地 / Type Locality
俄罗斯
Russia, Transbaikalia, "Dauuria" = Dauryia (49°7'N, 116°5'E)

▲ 其他名称 / Other Name(s)

其他中文名 / Other Chinese Name(s)
达呼尔刺猬、达斡尔猬

其他英文名 / Other English Name(s)
无 None

同物异名 / Synonym(s)
无 None

▲ 形态及生境 / Morphology and Habitat

形态特征 / Morphological Characteristics

齿式：3.1.3.3/2.1.2.3=36。中国体型较大的刺猬之一。头体长 175~261 mm。尾长 17~30 mm。后足 18~41 mm。耳长 22~34 mm。吻部较长且颜色较深。脸部和腹部毛色浅棕色或白色。耳长与周围棘刺的长度相当。头顶正中没有裸露头皮形成的深沟。额头至臀部背部布满棘刺，受惊或者遇到天敌时，身体蜷曲成刺球状。棘刺长度 21~23 mm，基部三分之二为白色，向上为深色圆环，深色圆环与刺尖之间为白色圆环，刺尖为黑色。

Dental formula: 3.1.3.3/2.1.2.3=36. A large-sized hedgehog in China. Head-body length 175-261 mm, tail length 17-30 mm, hindfoot length 18-41 mm, ear length 22-34 mm. The snout is long and dark in color. The face and ventral pelage pale brown or white in color. Ear similar to surrounding spines in length. Absence of a spineless area on the scalp. Spines extend from forehead to hip. When frightened or encounter a predator, the body can be curled into a spiny ball. Spine length 21-23 mm, white for two-thirds of length, followed by a black ring, narrow light ring, and black tip.

生境 / Habitat
草地 Grassland

▲ 地理分布 / Geographic Distribution

国内分布 / Domestic Distribution
黑龙江、吉林、辽宁、内蒙古、河北、山西、陕西、宁夏
Heilongjiang, Jilin, Liaoning, Inner Mongolia, Hebei, Shanxi, Shaanxi, Ningxia

全球分布 / World Distribution
中国、蒙古国、俄罗斯
China, Mongolia, Russia

生物地理界 / Biogeographic Realm
古北界 Palearctic

WWF 生物群系 / WWF Biome
温带草原、热带稀树草原和灌木地
Temperate Grasslands, Savannas & Shrublands

动物地理分布型 / Zoogeographic Distribution Type
Dn

分布标注 / Distribution Note
非特有种 Non-Endemic

▲ 濒危状况 / Threatened Status

中国生物多样性红色名录等级 / CB RL Category (2021)
无危 LC

IUCN 红色名录 / IUCN Red List (2021)
无危 LC

威胁因子 / Threats
无 None

▲ 法律保护地位 / Legal Protection Status

国家重点保护野生动物等级 / Category of National Key Protected Wild Animals (2021)
未列入 Not listed

"三有" 名录 / TWIESSV (2023)
列入 Listed

CITES 附录等级 / CITES Appendix (2023)
未列入 Not listed

迁徙物种公约附录 / CMS Appendix (2020)
未列入 Not listed

保护行动 / Conservation Action
尚无保护行动 No conservation action so far

▲ 参考文献 / References

Jiang et al. (蒋志刚等), 2021; Burgin et al., 2020; IUCN, 2020; Liu et al. (刘少英等), 2020; Wilson and Mittermeier, 2018; Smith et al., 2009; Zhang and Yu (张杰和于洪伟), 2005; Corbet, 1988; Pan et al. (潘清华等), 2007; Wilson and Reeder, 2005; Wang (王应祥), 2003; Zhang (张荣祖), 1997; Xia (夏武平), 1988

7 / 侯氏猬

Mesechinus hughi (Thomas, 1908)

· Hugh's Hedgehog

▲ 分类地位 / Taxonomy

劳亚食虫目 Eulipotyphla / 猬科 Erinaceidae / 林猬属 *Mesechinus*

科建立者及其文献 / Family Authority
G. Fischer, 1814

属建立者及其文献 / Genus Authority
Ognev, 1951

亚种 / Subspecies
无 None

模式标本产地 / Type Locality
中国
China, Shaanxi Prov., "Paochi, Shen-si" Baoji

▲ 其他名称 / Other Name(s)

其他中文名 / Other Chinese Name(s)
秦巴刺猬、秦岭刺猬

其他英文名 / Other English Name(s)
无 None

同物异名 / Synonym(s)
无 None

▲ 形态及生境 / Morphology and Habitat

形态特征 / Morphological Characteristics

齿式：3.1.3.3/2.1.2.3 = 36。体重 112~750 g。头体长 148~232 mm。尾长 12~24 mm。后足长 30~47 mm。耳长 16~33 mm。耳长接近于其周围棘刺长度。体表被毛深棕色或浅棕色。全身被棘刺，额头正中没有裸露的头皮。几乎没有纯白色棘刺，绝大多数棘刺的色素沉积模式相同：棘刺尖端黑色或深褐色，朝根部方向依次为 1~3 mm 宽的白色圆环，3~5 mm 宽的黑色或深褐色圆环，以及 50% 以上的白色区域。

Dental formula: 3.1.3.3/2.1.2.3 = 36. Bodyweight 112-750 g, head-body length 148-232 mm, tail length 12-24 mm, hindfoot length 30-47 mm, ear length 16-33 mm. Ear length similar to surrounding spines. Pelage dark brown to light brown. Spines extend from forehead to hip. Absence of a spineless area on the scalp. There are few holly white spines, and the pigmentation pattern of most spines is black or dark brown spines at the tip, followed by white rings 1-3 mm wide toward the root, black or dark brown rings 3-5 mm wide, and white for more than 50% of the spines' length.

生境 / Habitat

落叶阔叶林 Deciduous broad-leaved forest

▲ 地理分布 / Geographic Distribution

国内分布 / Domestic Distribution
陕西、山西、甘肃、湖北、四川、重庆、安徽、浙江
Shaanxi, Shanxi, Gansu, Hubei, Sichuan, Chongqing, Anhui, Zhejiang

全球分布 / World Distribution
中国 China

生物地理界 / Biogeographic Realm
古北界 Palearctic

WWF 生物群系 / WWF Biome
温带阔叶和混交林
Temperate Broadleaf & Mixed Forests

动物地理分布型 / Zoogeographic Distribution Type
Bc

分布标注 / Distribution Note
特有种 Endemic

▲ 濒危状况 / Threatened Status

中国生物多样性红色名录等级 / CB RL Category (2021)
近危 NT

IUCN 红色名录 / IUCN Red List (2021)
无危 LC

威胁因子 / Threats
狩猎及采集陆生植物 Hunting, gathering terrestrial plants

▲ 法律保护地位 / Legal Protection Status

国家重点保护野生动物等级 / Category of National Key Protected Wild Animals (2021)
未列入 Not listed

"三有"名录 / TWIESSV (2023)
列入 Listed

CITES 附录等级 / CITES Appendix (2023)
未列入 Not listed

迁徙物种公约附录 / CMS Appendix (2020)
未列入 Not listed

保护行动 / Conservation Action
尚无保护行动 No conservation action so far

▲ 参考文献 / References

Jiang et al. (蒋志刚等), 2021; Burgin et al., 2020; Chen et al. (陈正忠等), 2020; IUCN, 2020; Liu et al. (刘少英等), 2020; Wilson and Mittermeier, 2018; Smith et al., 2009; Qin et al. (秦岭等), 2007; Pan et al. (潘清华等), 2007; Wilson and Reeder, 2005; Wang (王应祥), 2003; Zhang (张荣祖), 1997

8 / 小齿猬

Mesechinus miodon (Thomas, 1908)

· Small-toothed Hedgehog

劳亚食虫目 Eulipotyphla / 猬科 Erinaceidae / 林猬属 *Mesechinus*

科建立者及其文献 / Family Authority
G. Fischer, 1814

属建立者及其文献 / Genus Authority
Ognev, 1951

亚种 / Subspecies
无 None

模式标本产地 / Type Locality
中国
Ordos desert, Shaanxi

▲ 其他名称 / Other Name(s)

其他中文名 / Other Chinese Name(s)
无 None

其他英文名 / Other English Name(s)
无 None

同物异名 / Synonym(s)
无 None

▲ 形态及生境 / Morphology and Habitat

形态特征 / Morphological Characteristics
齿式：3.1.3.3/2.1.2.3=36。体重 230~750 g。头体长 120~220 mm。尾长 25~43 mm。后足 35~40 mm。耳长 24~34.5 mm。外形与达乌尔猬相似。耳长与周围的棘刺相似。头顶中部无裸露的头皮。分类地位有待进一步厘定。

Dental formula: 3.1.3.3/2.1.2.3 = 36. Head-body length 120-220 mm, tail length 25-43 mm, hindfoot length 35-40 mm, ear length 24-34.5 mm. External morphology analog to that of a Daurian hedgehog. Ear length similar to the surrounding spines. Absence of a spineless area on the scalp. The taxonomic state needs to be further revised.

生境 / Habitat
荒漠、草地、耕地
Desert, grassland, farmland

▲ 地理分布 / Geographic Distribution

国内分布 / Domestic Distribution
陕西、宁夏 Shaanxi, Ningxia

全球分布 / World Distribution
中国 China

生物地理界 / Biogeographic Realm
古北界 Palearctic

WWF 生物群系 / WWF Biome
沙漠和干燥的灌木地
Deserts & Xeric Shrublands

动物地理分布型 / Zoogeographic Distribution Type
Ga

分布标注 / Distribution Note
特有种 Endemic

▲ 濒危状况 / Threatened Status

中国生物多样性红色名录等级 / CB RL Category (2021)
数据缺乏 DD

IUCN 红色名录 / IUCN Red List (2021)
未评定 NE

威胁因子 / Threats
未知 Unknown

▲ 法律保护地位 / Legal Protection Status

国家重点保护野生动物等级 / Category of National Key Protected Wild Animals (2021)
未列入 Not listed

"三有"名录 / TWIESSV (2023)
列入 Listed

CITES 附录等级 / CITES Appendix (2023)
未列入 Not listed

迁徙物种公约附录 / CMS Appendix (2020)
未列入 Not listed

保护行动 / Conservation Action
尚无保护行动 No conservation action so far

▲ 参考文献 / References

Jiang et al. (蒋志刚等), 2021; Wilson and Mittermeier, 2018; Pan et al. (潘清华等), 2007; Pan et al. (潘清华等), 2007; Wang (王应祥), 2003

9 | 高黎贡林猬

Mesechinus wangi He, Jiang & Ai, 2018

• Wang's Hedgehog

▲ 分类地位 / Taxonomy

劳亚食虫目 Eulipotyphla / 猬科 Erinaceidae / 林猬属 *Mesechinus*

科建立者及其文献 / Family Authority
Fischer, 1814

属建立者及其文献 / Genus Authority
Ognev, 1951

亚种 / Subspecies
无 None

模式标本产地 / Type Locality
中国
Gaoligongshan National Nature Reserve (24°500N, 98°450E), Baoshan, Yunnan, China, on 1 September 2010 at an altitude of 2 215 m a. s. l.

何锴 / 供图

何锴 / 供图

▲ 其他名称 / Other Name(s)

其他中文名 / Other Chinese Name(s)
王氏林猬

其他英文名 / Other English Name(s)
Wang's Forest Hedgehog

同物异名 / Synonym(s)
无 None

▲ 形态及生境 / Morphology and Habitat

形态特征 / Morphological Characteristics

齿式：3.1.3.4/2.1.2.3 = 38。有第四上白齿。头体长 177~240 mm。尾长 14~18 mm。后足 45~48 mm。耳长 28~32 mm，耳长与耳周围棘刺长度相当。浑身被毛，毛色为浅棕色，背腹毛色相近，脸部毛色均一。背部长满棘刺，头顶中部无裸露的头皮。受惊或者遇到天敌时，身体蜷曲成刺球状。棘刺长 22~25 mm，绝大部分棘刺双色，靠近根部的三分之二 为白色，靠近尖端的三分之一 为黑色。后足大拇趾发育良好。
Dental formula: 3.1.3.4/2.1.2.3 = 38. The 4th upper molar is present. Head-body length 177-240 mm, tail length 14-18 mm, hindfoot length 45-48 mm, ear length 28-32 mm. Ear length similar to the surrounding spines. The whole body is covered with light brown hairs and the hairs on the face are unicolor. Spines extend from forehead to hip. When frightened or encounter a predator, the body is curled into a spiny ball. Absence of a spineless area on the scalp. The spines are 22-25 mm long, and most spines are bicolor. White for 2/3 of the length and the rest 1/3 black. The thumb toe of the hind feet well-developed.

生境 / Habitat

落叶阔叶林 Deciduous broad-leaved forest

▲ 地理分布 / Geographic Distribution

国内分布 / Domestic Distribution
云南 Yunnan

全球分布 / World Distribution
中国 China

生物地理界 / Biogeographic Realm
印度马来界 Indomalaya

WWF 生物群系 / WWF Biome
热带和亚热带湿润阔叶林
Tropical & Subtropical Moist Broadleaf Forests

动物地理分布型 / Zoogeographic Distribution Type
Hc

分布标注 / Distribution Note
特有种 Endemic

▲ 濒危状况 / Threatened Status

中国生物多样性红色名录等级 / CB RL Category (2021)
数据缺乏 DD

IUCN 红色名录 / IUCN Red List (2021)
未评定 NE

威胁因子 / Threats
未知 Unknown

▲ 法律保护地位 / Legal Protection Status

国家重点保护野生动物等级 / Category of National Key Protected Wild Animals (2021)
未列入 Not listed

"三有"名录 / TWIESSV (2023)
列入 Listed

CITES 附录等级 / CITES Appendix (2023)
未列入 Not listed

迁徙物种公约附录 / CMS Appendix (2020)
未列入 Not listed

保护行动 / Conservation Action
尚无保护行动 No conservation action so far

▲ 参考文献 / References

Jiang et al. (蒋志刚等), 2021; Liu et al. (刘少英等), 2020; Wilson and Mittermeier, 2018; Ai et al., 2018

10 / 等齿鼩鼹

Uropsilus aequodonenia Liu, 2013

· Equivalent Teeth Shrew Mole

▲ 分类地位 / Taxonomy

劳亚食虫目 Eulipotyphla / 鼹科 Talpidae / 鼩鼹属 *Uropsilus*

科建立者及其文献 / Family Authority
G. Fischer, 1814

属建立者及其文献 / Genus Authority
Milne-Edwards, 1871

亚种 / Subspecies
无 None

模式标本产地 / Type Locality
中国
Luoji shan, Gelang county, Sichuan

▲ 其他名称 / Other Name(s)

其他中文名 / Other Chinese Name(s)
无 None

其他英文名 / Other English Name(s)
无 None

同物异名 / Synonym(s)
无 None

▲ 形态及生境 / Morphology and Habitat

形态特征 / Morphological Characteristics

齿式：2.1.3.3/2.1.3.3=36。上、下齿列牙齿总数相等，3枚上前白齿自前向后依次增大。头体长 72~82 mm。后足 14~16 mm。尾长 67~73 mm。鼩型鼹亚科鼩鼹属的物种外形十分相似：外形似鼩鼱，但吻部更长，且吻部皮肤裸露，仅覆盖有稀疏的短毛，外耳未退化。尾细长，细小鳞片组成环状纹路。前爪较小，未发生特化。

Dentation:2.1.3.3/2.1.3.3=36. Numbers of teeth on the upper and lower jaws are equal. The upper premolars successively increase in size from P1 to P4. Head-body length 72-82 mm, hindfoot length 14-16 mm, tail length 67-73 mm. All species in this extant genus of *Uropsilinae* have similar external morphology, assembling that of a shrew, but with a longer snout. The snout is almost naked only covered with sparse hairs. External ears appear. Forefeet small, unspecialized for burrowing. Tail is long and slender and covered with rings of small scales.

生境 / Habitat

针叶林、落叶阔叶林、灌丛
Coniferous forest, deciduous broad-leaved forest, shrubland

▲ 地理分布 / Geographic Distribution

国内分布 / Domestic Distribution
四川 Sichuan

全球分布 / World Distribution
中国 China

生物地理界 / Biogeographic Realm
古北界 Palearctic

WWF 生物群系 / WWF Biome
温带阔叶和混交林
Temperate Broadleaf & Mixed Forests

动物地理分布型 / Zoogeographic Distribution Type
Id

分布标注 / Distribution Note
特有种 Endemic

▲ 濒危状况 / Threatened Status

中国生物多样性红色名录等级 / CB RL Category (2021)
近危 NT

IUCN 红色名录 / IUCN Red List (2021)
未评定 NE

威胁因子 / Threats
未知 Unknown

▲ 法律保护地位 / Legal Protection Status

国家重点保护野生动物等级 / Category of National Key Protected Wild Animals (2021)
未列入 Not listed

"三有"名录 / TWIESSV (2023)
未列入 Not listed

CITES 附录等级 / CITES Appendix (2023)
未列入 Not listed

迁徙物种公约附录 / CMS Appendix (2020)
未列入 Not listed

保护行动 / Conservation Action
尚无保护行动 No conservation action so far

▲ 参考文献 / References

Jiang et al. (蒋志刚等), 2021; Wilson and Mittermeier, 2018; Liu et al. (刘洋等), 2013

11 / 峨眉鼩鼹

Uropsilus andersoni (Thomas, 1911)

• Anderson's Shrew Mole

▲ 分类地位 / Taxonomy

劳亚食虫目 Eulipotyphla / 鼹科 Talpidae / 鼩鼹属 *Uropsilus*

科建立者及其文献 / Family Authority
G. Fischer, 1814

属建立者及其文献 / Genus Authority
Milne-Edwards, 1871

亚种 / Subspecies
无 None

模式标本产地 / Type Locality
中国
China, Sichuan, "Omi-san" Emei-Shan

▲ 其他名称 / Other Name(s)

其他中文名 / Other Chinese Name(s)
无 None

其他英文名 / Other English Name(s)
无 None

同物异名 / Synonym(s)
无 None

▲ 形态及生境 / Morphology and Habitat

形态特征 / Morphological Characteristics
齿式：2.1.4.3/2.1.4.3=38。第二上前白齿明显大于犬齿，是第一上前白齿的 2 倍大小，第三上前白齿严重退化。体重 7.4~11.2 g。头体长 65~83 mm。后足 14~17.5 mm。尾长 59~72 mm。背部深咖啡色，腹部深灰色。尾巴背腹颜色均一。

Dental formula: 2.1.4.3/2.1.4.3=38. Upper P2 larger than the upper canine and is about twice as large as upper P1. Upper P3 highly reduced. Bodyweight 7.4-11.2 g, head-body length 65-83 mm, hindfoot length 14-17.5 mm, tail length 59-72 mm. Dorsal pelage dark brown in color, ventral dark gray. Tail unicolor.

生境 / Habitat
针阔混交林、针叶林
Coniferous and broad-leaved mixed forest, coniferous forest

▲ 地理分布 / Geographic Distribution

国内分布 / Domestic Distribution
四川 Sichuan

全球分布 / World Distribution
中国 China

生物地理界 / Biogeographic Realm
古北界 Palearctic

WWF 生物群系 / WWF Biome
温带阔叶和混交林
Temperate Broadleaf & Mixed Forests

动物地理分布型 / Zoogeographic Distribution Type
Hc

分布标注 / Distribution Note
特有种 Endemic

▲ 濒危状况 / Threatened Status

中国生物多样性红色名录等级 / CB RL Category (2021)
易危 VU

IUCN 红色名录 / IUCN Red List (2021)
数据缺乏 DD

威胁因子 / Threats
未知 Unknown

▲ 法律保护地位 / Legal Protection Status

国家重点保护野生动物等级 / Category of National Key Protected Wild Animals (2021)
未列入 Not listed

"三有"名录 / TWIESSV (2023)
未列入 Not listed

CITES 附录等级 / CITES Appendix (2023)
未列入 Not listed

迁徙物种公约附录 / CMS Appendix (2020)
未列入 Not listed

保护行动 / Conservation Action
尚无保护行动 No conservation action so far

▲ 参考文献 / References

Jiang et al. (蒋志刚等), 2021; Burgin et al., 2020; IUCN, 2020; Liu et al. (刘少英等), 2020; Wilson and Mittermeier, 2018; Zhang et al. (张君等), 2010; Zhu et al. (朱红艳等), 2010; Smith et al., 2009; Liang et al. (梁艺于等), 2009; Li et al. (李艳红等), 2007; Pan et al. (潘清华等), 2007; Wilson and Reeder, 2005; Wang (王应祥), 2003; Hoffman, 1987

12 / 栗背鼩鼹

Uropsilus atronates (Thomas, 1911)

• Black-backed Shrew Mole

劳亚食虫目 Eulipotyphla / 鼹科 Talpidae / 鼩鼹属 *Uropsilus*

科建立者及其文献 / Family Authority
G. Fischer, 1814

属建立者及其文献 / Genus Authority
Milne-Edwards, 1871

亚种 / Subspecies
无 None

模式标本产地 / Type Locality
中国
China, Yunnan, Mucheng (Muchang Village, Zhenkang County)

▲ 其他名称 / Other Name(s)

其他中文名 / Other Chinese Name(s)
无 None

其他英文名 / Other English Name(s)
无 None

同物异名 / Synonym(s)
无 None

▲ 形态及生境 / Morphology and Habitat

形态特征 / Morphological Characteristics

齿齿式：2.1.4.3/1.1.4.3=38。头体长 61~71 mm。尾长 55~72 mm。后足 11.5~14 mm。体重 6~8.8 g。背腹毛色差异明显，存在一条明显的分界线。背毛呈较为鲜亮的栗红色，腹部青灰色。

Dental formula: 2.1.4.3/1.1.4.3=38. Head-body length 61-71 mm, tail length 55-72 mm, hindfoot 11.5-14 mm, bodyweight 6-8.8 g. Dorsal and ventral pelage obviously bicolor. Dorsal pelage is chestnut red and the venter is slate gray.

生境 / Habitat
亚热带湿润山地森林
Subtropical moist montane forest

▲ 地理分布 / Geographic Distribution

国内分布 / Domestic Distribution
云南 Yunnan

全球分布 / World Distribution
中国 China

生物地理界 / Biogeographic Realm
印度马来界 Indomalaya

WWF 生物群系 / WWF Biome
热带和亚热带湿润阔叶林
Tropical & Subtropical Moist Broadleaf Forests

动物地理分布型 / Zoogeographic Distribution Type
Sh

分布标注 / Distribution Note
特有种 Endemic

▲ 濒危状况 / Threatened Status

中国生物多样性红色名录等级 / CB RL Category (2021)
未评定 NE

IUCN 红色名录 / IUCN Red List (2021)
未评定 NE（作为长吻鼩鼱 *U. gracilis* 的亚种）

威胁因子 / Threats
无 None

▲ 法律保护地位 / Legal Protection Status

国家重点保护野生动物等级 / Category of National Key Protected Wild Animals (2021)
未列入 Not listed

"三有"名录 / TWIESSV (2023)
未列入 Not listed

CITES 附录等级 / CITES Appendix (2023)
未列入 Not listed

迁徙物种公约附录 / CMS Appendix (2020)
未列入 Not listed

保护行动 / Conservation Action
尚无保护行动 No conservation action so far

▲ 参考文献 / References

Tu et al., 2015

13 / 大别山鼩鼹

Uropsilus dabieshanensis
Hu, Xu, Zhang, Liu, Liao, Yang, Sun,
Shi, Ban, Li, Liu, Zhang, 2021

· Dabie Mountains Shrew Mole

劳亚食虫目 Eulipotyphla / 鼹科 Talpidae / 鼩鼹属 *Uropsilus*

科建立者及其文献 / Family Authority
G. Fischer, 1814

属建立者及其文献 / Genus Authority
Milne-Edwards, 1871

亚种 / Subspecies
无 None

模式标本产地 / Type Locality
中国
Dabieshan, Anhui, China

▲ 其他名称 / Other Name(s)

其他中文名 / Other Chinese Name(s)
无 None

其他英文名 / Other English Name(s)
无 None

同物异名 / Synonym(s)
无 None

▲ 形态及生境 / Morphology and Habitat

形态特征 / Morphological Characteristics
齿式：2.1.4.3/1.1.4.3=38。体重 6.2~8.9 g。头体长 118.6~136.4 mm。耳长 8.1~8.9 mm。后足 12.8~12.9 mm。尾长 52.4~54.1 mm。背部毛色深褐色，腹部毛石板灰色。尾双色，背面黑色，腹面浅咖啡色。
Dental formula: 2.1.4.3/1.1.4.3=38. Bodyweight 6.2-8.9 g. Head and body length 118.6-136.4 mm, ear length 8.1-8.9 mm, hindfoot length 12.8-12.9 mm, tail length 52.4-54.1 mm. Dorsal pelage dark brown and ventral slate gray. Tail bicolor, upperpart black and underpart light coffee color.

生境 / Habitat
落叶、常绿阔叶混交林
Deciduous, evergreen broad-leaved mixed forest

▲ 地理分布 / Geographic Distribution

国内分布 / Domestic Distribution
安徽 Anhui

全球分布 / World Distribution
中国 China

生物地理界 / Biogeographic Realm
古北界 Palearctic

WWF 生物群系 / WWF Biome
热带和亚热带湿润阔叶林
Tropical & Subtropical Moist Broadleaf Forests

动物地理分布型 / Zoogeographic Distribution Type
Si

分布标注 / Distribution Note
特有种 Endemic

▲ 濒危状况 / Threatened Status

中国生物多样性红色名录等级 / CB RL Category (2021)
未评定 NE

IUCN 红色名录 / IUCN Red List (2021)
未评定 NE

威胁因子 / Threats
未知 Unknown

▲ 法律保护地位 / Legal Protection Status

国家重点保护野生动物等级 / Category of National Key Protected Wild Animals (2021)
未列入 Not listed

"三有" 名录 / TWIESSV (2023)
未列入 Not listed

CITES 附录等级 / CITES Appendix (2023)
未列入 Not listed

迁徙物种公约附录 / CMS Appendix (2020)
未列入 Not listed

保护行动 / Conservation Action
尚无保护行动 No conservation action so far

▲ 参考文献 / References

Hu et al., 2021

14 / 长吻鼩鼹

Uropsilus gracilis (Thomas, 1911)

· Gracile Shrew Mole

劳亚食虫目 Eulipotyphla / 鼹科 Talpidae / 鼩鼹属 *Uropsilus*

科建立者及其文献 / Family Authority
G. Fischer, 1814

属建立者及其文献 / Genus Authority
Milne-Edwards, 1871

亚种 / Subspecies
无 None

模式标本产地 / Type Locality
中国
China, Chongqing, near Nan-chwan (Nanchuan), Mt. Chin-fu-san (Jingfu Shan)

何锴 / 供图

▲ 其他名称 / Other Name(s)

其他中文名 / Other Chinese Name(s)
无 None

其他英文名 / Other English Name(s)
无 None

同物异名 / Synonym(s)
无 None

▲ 形态及生境 / Morphology and Habitat

形态特征 / Morphological Characteristics
齿式：2.1.4.3/1.1.4.3=38。头体长 65~85 mm。后足长 13~18 mm。尾长 55~78 mm。有明显的外耳郭。背毛棕褐色至棕黄色，腹毛颜色较浅，为棕灰色，背腹毛异色不明显。前后足细小，尾细长。
Dental formula: 2.1.4.3/1.1.4.3=38. Head-body length 65-85 mm, hindfoot length 13-18 mm, tail length 55-78 mm. External ears appear. Dorsal and ventral pelage not obvious bicolor. Dorsal pelage dark brown to yellowish-brown, ventral pelage brownish gray. Forefeet and hindfeet slender. Tail is long.

生境 / Habitat
阔叶林、竹林
Broad-leaved forest, bamboo forest

▲ 地理分布 / Geographic Distribution

国内分布 / Domestic Distribution
湖北、陕西、四川、重庆、贵州、云南
Hubei, Shaanxi, Sichuan, Chongqing, Guizhou, Yunnan

全球分布 / World Distribution
中国 China

生物地理界 / Biogeographic Realm
古北界、印度马来界 Palearctic, Indomalaya

WWF 生物群系 / WWF Biome
热带和亚热带湿润阔叶林，温带阔叶和混交林
Tropical & Subtropical Moist Broadleaf Forests; Temperate Broadleaf & Mixed Forests

动物地理分布型 / Zoogeographic Distribution Type
Sh

分布标注 / Distribution Note
特有种 Endemic

▲ 濒危状况 / Threatened Status

中国生物多样性红色名录等级 / CB RL Category (2021)
无危 LC

IUCN 红色名录 / IUCN Red List (2021)
无危 LC

威胁因子 / Threats
无 None

▲ 法律保护地位 / Legal Protection Status

国家重点保护野生动物等级 / Category of National Key Protected Wild Animals (2021)
未列入 Not listed

"三有" 名录 / TWIESSV (2023)
未列入 Not listed

CITES 附录等级 / CITES Appendix (2023)
未列入 Not listed

迁徙物种公约附录 / CMS Appendix (2020)
未列入 Not listed

保护行动 / Conservation Action
尚无保护行动 No conservation action so far

▲ 参考文献 / References

Jiang et al. (蒋志刚等), 2021; Burgin et al., 2020; IUCN, 2020; Wilson and Mittermeier, 2018; Smith et al., 2009; Pan et al. (潘清华等), 2007; Wilson and Reeder, 2005; Wang (王应祥), 2003; Zhang (张荣祖), 1997; Hoffmann, 1984

15 / 贡山鼩鼹

Uropsilus investigator (Thomas, 1911)

· Inquisitive Shrew Mole

▲ 分类地位 / Taxonomy

劳亚食虫目 Eulipotyphla / 鼹科 Talpidae / 鼩鼹属 *Uropsilus*

科建立者及其文献 / Family Authority
G. Fischer, 1814

属建立者及其文献 / Genus Authority
Milne-Edwards, 1871

亚种 / Subspecies
无 None

模式标本产地 / Type Locality
中国
China, Yunnan, Kui-chiang-Salween divide at 3353 m

陈俊海 / 供图

▲ 其他名称 / Other Name(s)

其他中文名 / Other Chinese Name(s)
无 None

其他英文名 / Other English Name(s)
无 None

同物异名 / Synonym(s)
无 None

▲ 形态及生境 / Morphology and Habitat

形态特征 / Morphological Characteristics
齿式：2.1.4.3/1.1.4.3=38。头体长 67~83 mm。尾长 57~75 mm。后足 13~16 mm。外耳郭明显。吻部细长，皮肤裸露，上面覆盖稀疏的短毛。背腹明显异色，背部为棕褐色或棕黄色，腹部为灰色或青灰色。
Dental formula: 2.1.4.3/1.1.4.3=38. Head-body 67-83 mm, tail length 57-75 mm, hindfoot length 13-16 mm. External ears appear. The snout is long, covered with sparse hairs. Pelage obviously bicolor, dorsal pelage dark brown or brownish-yellow, ventral gray or slate gray.

生境 / Habitat
冷杉林、竹林和杜鹃灌丛
Fir forest, bamboo forest and rhododendron thicket

▲ 地理分布 / Geographic Distribution

国内分布 / Domestic Distribution
云南 Yunnan

全球分布 / World Distribution
中国、缅甸 China, Myanmar

生物地理界 / Biogeographic Realm
古北界 Palearctic

WWF 生物群系 / WWF Biome
温带针叶林、热带和亚热带湿润阔叶林
Temperate Conifer Forests, Tropical & Subtropical Moist Broadleaf Forests

动物地理分布型 / Zoogeographic Distribution Type
Hc

分布标注 / Distribution Note
非特有种 Non-Endemic

▲ 濒危状况 / Threatened Status

中国生物多样性红色名录等级 / CB RL Category (2021)
近危 NT

IUCN 红色名录 / IUCN Red List (2021)
数据缺乏 DD

威胁因子 / Threats
未知 Unknown

▲ 法律保护地位 / Legal Protection Status

国家重点保护野生动物等级 / Category of National Key Protected Wild Animals (2021)
未列入 Not listed

"三有"名录 / TWIESSV (2023)
未列入 Not listed

CITES 附录等级 / CITES Appendix (2023)
未列入 Not listed

迁徙物种公约附录 / CMS Appendix (2020)
未列入 Not listed

保护行动 / Conservation Action
尚无保护行动 No conservation action so far

▲ 参考文献 / References

Jiang et al. (蒋志刚等), 2021; Burgin et al., 2020; IUCN, 2020; Liu et al. (刘少英等), 2020; Wilson and Mittermeier, 2018; Smith et al., 2009; Pan et al. (潘清华等), 2007; Wilson and Reeder, 2005; Wang (王应祥), 2003; Hoffmann, 1984

16 / 雪山鼩鼹

Uropsilus nivatus (Thomas, 1911)

· Snow Mountain Shrew Mole

▲ 分类地位 / Taxonomy

劳亚食虫目 Eulipotyphla / 鼹科 Talpidae / 鼩鼹属 *Uropsilus*

科建立者及其文献 / Family Authority
G. Fischer, 1814

属建立者及其文献 / Genus Authority
Milne-Edwards, 1871

亚种 / Subspecies
无 None

模式标本产地 / Type Locality
中国
China, Yunnan, Jade Dragon Snow Mountain (=Yulong Snow Mountain)

▲ 其他名称 / Other Name(s)

其他中文名 / Other Chinese Name(s)
无 None

其他英文名 / Other English Name(s)
无 None

同物异名 / Synonym(s)
无 None

▲ 形态及生境 / Morphology and Habitat

形态特征 / Morphological Characteristics
齿式：2.1.4.3/1.1.4.3=38。头体长 56~76 mm。后足 11~16 mm。尾长 55~72 mm。体重 6.1~11.3 g。背腹毛色差异明显，背毛呈现乌金色，腹部青灰色。
Dental formula: 2.1.4.3/1.1.4.3=38. Head-body length 56-76 mm, hindfoot 11-16 mm, tail length 55-72 mm, bodyweight 6.1-11.3 g. Dorsal and ventral pelage obviously bicolor. Dorsal pelage is black gold and the venter is slate gray.

生境 / Habitat
针阔混交林、高山针叶林和山顶矮林
Coniferous and broad-leaved mixed forest, alpine coniferous forest and shrubon mountain ridge

▲ 地理分布 / Geographic Distribution

国内分布 / Domestic Distribution
云南 、四川、西藏 Yunnan, Sichuan, Tibet

全球分布 / World Distribution
中国 China

生物地理界 / Biogeographic Realm
古北界 Palearctic

WWF 生物群系 / WWF Biome
温带阔叶和混交林
Temperate Broadleaf & Mixed Forests

动物地理分布型 / Zoogeographic Distribution Type
Sh

分布标注 / Distribution Note
未知 Unknown

▲ 濒危状况 / Threatened Status

中国生物多样性红色名录等级 / CB RL Category (2021)
未评定 NE

IUCN 红色名录 / IUCN Red List (2021)
未评定 NE（作为长吻鼩鼱 *U. gracilis* 的亚种）

威胁因子 / Threats
未知 Unknown

▲ 法律保护地位 / Legal Protection Status

国家重点保护野生动物等级 / Category of National Key Protected Wild Animals (2021)
未列入 Not listed

"三有" 名录 / TWIESSV (2023)
未列入 Not listed

CITES 附录等级 / CITES Appendix (2023)
未列入 Not listed

迁徙物种公约附录 / CMS Appendix (2020)
未列入 Not listed

保护行动 / Conservation Action
尚无保护行动 No conservation action so far

▲ 参考文献 / References

Burgin et al., 2020; IUCN, 2020; Wilson and Mittermeier, 2018; Wan et al., 2013

17 / 少齿鼩鼹

Uropsilus soricipes Milne-Edwards, 1871

· Chinese Shrew Mole

▲ 分类地位 / Taxonomy

劳亚食虫目 Eulipotyphla / 鼹科 Talpidae / 鼩鼹属 *Uropsilus*

科建立者及其文献 / Family Authority
G. Fischer, 1814

属建立者及其文献 / Genus Authority
Milne-Edwards, 1871

亚种 / Subspecies
无 None

模式标本产地 / Type Locality
中国
China, Sichuan, Moupin

▲ 其他名称 / Other Name(s)

其他中文名 / Other Chinese Name(s)
无 None

其他英文名 / Other English Name(s)
无 None

同物异名 / Synonym(s)
无 None

▲ 形态及生境 / Morphology and Habitat

形态特征 / Morphological Characteristics
齿式：2.1.3.3/1.1.3.3=34。上前白齿仅3枚，不存在P3，P1~P4依次增大。头体长66~80 mm。后足14~17 mm。尾长50~69 mm。背毛深棕色，腹毛石板灰色。

Dental formula: 2.1.3.3/1.1.3.3=34. Only three upper premolars present, and P3 is missing, the sizes gradually increase from P1 to P4. Head-body length 66-80 mm, hindfoot length 14-17 mm, tail length 50-69 mm. Dorsal pelage dark brown; ventral surface slate gray.

生境 / Habitat
常绿阔叶林和箭竹林
Evergreen forest and arrow bamboo grove

▲ 地理分布 / Geographic Distribution

国内分布 / Domestic Distribution
陕西、甘肃、四川
Shaanxi, Gansu, Sichuan

全球分布 / World Distribution
中国 China

生物地理界 / Biogeographic Realm
古北界 Palearctic

WWF 生物群系 / WWF Biome
温带阔叶和混交林
Temperate Broadleaf & Mixed Forests

动物地理分布型 / Zoogeographic Distribution Type
Qh+Leg

分布标注 / Distribution Note
特有种 Endemic

▲ 濒危状况 / Threatened Status

中国生物多样性红色名录等级 / CB RL Category (2021)
无危 LC

IUCN 红色名录 / IUCN Red List (2021)
无危 LC

威胁因子 / Threats
未知 Unknown

▲ 法律保护地位 / Legal Protection Status

国家重点保护野生动物等级 / Category of National Key Protected Wild Animals (2021)
未列入 Not listed

"三有"名录 / TWIESSV (2023)
未列入 Not listed

CITES 附录等级 / CITES Appendix (2023)
未列入 Not listed

迁徙物种公约附录 / CMS Appendix (2020)
未列入 Not listed

保护行动 / Conservation Action
尚无保护行动 No conservation action so far

▲ 参考文献 / References

Jiang et al. (蒋志刚等), 2021; Burgin et al., 2020; IUCN, 2020; Liu et al. (刘少英等), 2020; Wilson and Mittermeier, 2018; Pan et al. (潘清华等), 2007; Wilson and Reeder, 2005; Wang (王应祥), 2003; Zhang (张荣祖), 1997; Hoffmann, 1984; Shi and Zhao (施白南和赵尔宓), 1980

18 / 墨脱鼹

Alpiscaptulus medogensis
Jiang & Chen, 2021

· Modong Mole

▲ 分类地位 / Taxonomy

劳亚食虫目 Eulipotyphla / 鼹科 Talpidae / 高山鼹属 *Alpiscaptulus*

科建立者及其文献 / Family Authority
G. Fischer, 1814

属建立者及其文献 / Genus Authority
Chen Z et al., 2021

亚种 / Subspecies
无 None

模式标本产地 / Type Locality
中国
Mt Namjagbarwa in the eastern Himalayas, Tibet, China

▲ 其他名称 / Other Name(s)

其他中文名 / Other Chinese Name(s)
无 None

其他英文名 / Other English Name(s)
无 None

同物异名 / Synonym(s)
无 None

▲ 形态及生境 / Morphology and Habitat

形态特征 / Morphological Characteristics

齿式：3.1.4.3/3.1.4.3 = 44。头体长 100 mm。尾长 41 mm。后足长 18 mm。2021年被发现并命名的单型属物种。与甘肃鼹（*Scapanulus oweni*）同为美洲鼹族（Scalopini）在亚洲的孑遗类群。具有适应地下生活的特征，身体圆柱状，毛短而致密，眼极小，被毛发遮盖，外耳退化，耳道被毛发遮盖，前掌很大，且向身体外侧翻转。亦具有美洲鼹族的主要特征，尾粗大，密布深褐色刚毛。以上颌第一门齿显著大于第二门齿及犬齿，区别于鼹族（Talpini）物种。背部毛色深灰色，腹部毛色略浅。鼻和脸颊有白色毛发。后足大拇趾向身体内侧弯曲，偏离其余四趾。

Dental formula: 3.1.4.3/3.1.4.3 = 44. Head-body length 100 mm, tail length 41 mm, hindfoot length 18 mm. A newly discovered species in 2021 in a monotypic genus. *Alpiscaptulus medogensis* and *Scapanulus oweni* are the only two relict taxa of the tribe Scalopini in Asia. The animal is adapted to fully fossorial life, body cylindrical, covered with short and dense fur. Eyes minute, concealed in fur; external ears absent. Forefeet enlarged and projected outwards. Several typical Scalopini characters appear, such as a long and fat tail covered with short and dense hairs. The 1st upper incisor is obviously larger than the 2nd upper incisor and the canine, which is distinct from Talpini species. Dorsal pelage dark grey, ventral pelage imperceptibly lighter. White hairs present on the nose and the chin. First toe of the hindfeet is curved outward at a slight angle against the remaining toes.

生境 / Habitat

高山灌丛、竹林
Alpine scrub, bamboo forest

▲ 地理分布 / Geographic Distribution

国内分布 / Domestic Distribution
西藏 Tibet

全球分布 / World Distribution
中国 China

生物地理界 / Biogeographic Realm
古北界 Palearctic

WWF 生物群系 / WWF Biome
山地草原和灌丛
Montane Grasslands & Shrublands

动物地理分布型 / Zoogeographic Distribution Type
Ha

分布标注 / Distribution Note
未知 Unknown

▲ 濒危状况 / Threatened Status

中国生物多样性红色名录等级 / CB RL Category (2021)
未评定 NE

IUCN 红色名录 / IUCN Red List (2021)
未评定 NE

威胁因子 / Threats
未知 Unknown

▲ 法律保护地位 / Legal Protection Status

国家重点保护野生动物等级 / Category of National Key Protected Wild Animals (2021)
未列入 Not listed

"三有" 名录 / TWIESSV (2023)
未列入 Not listed

CITES 附录等级 / CITES Appendix (2023)
未列入 Not listed

迁徙物种公约附录 / CMS Appendix (2020)
未列入 Not listed

保护行动 / Conservation Action
尚无保护行动 No conservation action so far

▲ 参考文献 / References

Chen et al., 2021

19 / 针尾鼹

Scaptonyx fusicauda Milne-Edwards, 1872

· Long-tailed Mole

▲ 分类地位 / Taxonomy

劳亚食虫目 Eulipotyphla / 鼹科 Talpidae / 针尾鼹属 *Scaptonyx*

科建立者及其文献 / Family Authority
G. Fischer, 1814

属建立者及其文献 / Genus Authority
Milne-Edwards, 1872

亚种 / Subspecies
指名亚种 *S. f. dafusicauda* Milne-Edwards, 1871
四川、陕西、重庆、贵州
Sichuan, Shaanxi, Chongqing, Guizhou

滇北亚种 *S. f. affinis* Thomas,1912
云南 Yunnan

模式标本产地 / Type Locality
中国
"Frontie du Kokonoor", vicinity of Kukunor (Lake), China

张冬莉 / 供图

▲ 其他名称 / Other Name(s)

其他中文名 / Other Chinese Name(s)
甘肃长尾鼹

其他英文名 / Other English Name(s)
无 None

同物异名 / Synonym(s)
Scaptonyx fusicaudatus (Milne-Edwards, 1872)

▲ 形态及生境 / Morphology and Habitat

形态特征 / Morphological Characteristics

齿式：3.1.4.3/2.1.4.3=42。上犬齿略长于上门齿。头体长 72~90 mm。后足长 16~17 mm。尾长 26~45 mm。体重 9.8~13.8 g。单型属。外形和体型大小介于鼩型鼹亚科和真鼹类（鼹族、美洲鼹族）之间的过渡类型。身体纺锤形，前爪略大于鼩型鼹类的前爪，向外侧翻转。背腹毛色差异不明显，背部深灰或者黑色，腹部颜色略浅于背部，背腹毛色无明显分界线。眼睛和外耳明显退化，均隐藏在毛发之下。尾较粗，但根部很细，表面覆盖有稀疏的针毛，环状纹路清晰。

Dental formula: 3.1.4.3/2.1.4.3=42. Head-body length 72-90 mm, hindfoot length 16-17 mm, tail length 26-45 mm, bodyweight 9.8-13.8 g. Monotypic genus. External morphology is in between of a shrew-like mole and a true mole (i. e., Talpini and Scalopini species). Body shape fusiform, forefeet slightly larger than that of a shrew-like mole, and projected outwards. Dorsal pelage black or dark gray, and the ventral is lighter, and there is no obvious boundary in between. Eyes minute, concealed in fur; external ears absent. Tail is thick, tapering from enlargement near the base. Tail covered with rings of small scales with long sparse hairs. Upper canine slightly larger than the incisors.

生境 / Habitat

阔叶林、针阔混交林、针叶林
Broad-leaved forest, coniferous and broad-leaved mixed forest, coniferous forest

▲ 地理分布 / Geographic Distribution

国内分布 / Domestic Distribution
陕西、四川、湖北、重庆、贵州、云南
Shaanxi, Sichuan, Hubei, Chongqing, Guizhou, Yunnan

全球分布 / World Distribution
中国、越南、缅甸
China, Vietnam, Myanmar

生物地理界 / Biogeographic Realm
印度马来界、古北界 Indomalaya, Palearctic

WWF 生物群系 / WWF Biome
热带和亚热带湿润阔叶林、温带阔叶和混交林
Tropical & Subtropical Moist Broadleaf Forests, Temperate Broadleaf & Mixed Forests

动物地理分布型 / Zoogeographic Distribution Type
Q+Hc

分布标注 / Distribution Note
非特有种 Non-Endemic

▲ 濒危状况 / Threatened Status

中国生物多样性红色名录等级 / CB RL Category (2021)
无危 LC

IUCN 红色名录 / IUCN Red List (2021)
无危 LC

威胁因子 / Threats
未知 Unknown

▲ 法律保护地位 / Legal Protection Status

国家重点保护野生动物等级 / Category of National Key Protected Wild Animals (2021)
未列入 Not listed

"三有" 名录 / TWIESSV (2023)
未列入 Not listed

CITES 附录等级 / CITES Appendix (2023)
未列入 Not listed

迁徙物种公约附录 / CMS Appendix (2020)
未列入 Not listed

保护行动 / Conservation Action
尚无保护行动 No conservation action so far

▲ 参考文献 / References

Jiang et al. (蒋志刚等), 2021; Burgin et al., 2020; IUCN, 2020; Liu et al. (刘少英等), 2020; Smith et al., 2009; Pan et al. (潘清华等), 2007; Wilson and Reeder, 2005; Wang (王应祥), 2003; Zhang (张荣祖), 1997

20 / 甘肃鼹

Scapanulus oweni Thomas, 1912

· Gansu Mole

▲ 分类地位 / Taxonomy

劳亚食虫目 Eulipotyphla / 鼹科 Talpidae / 甘肃鼹属 *Scapanulus*

科建立者及其文献 / Family Authority
G. Fischer, 1814

属建立者及其文献 / Genus Authority
Thomas, 1912

亚种 / Subspecies
无 None

模式标本产地 / Type Locality
中国
China, Kansu, "37 km S. E. of Tao-chou

何锴 / 供图

▲ 其他名称 / Other Name(s)

其他中文名 / Other Chinese Name(s)
无 None

其他英文名 / Other English Name(s)
无 None

同物异名 / Synonym(s)
无 None

▲ 形态及生境 / Morphology and Habitat

形态特征 / Morphological Characteristics

齿式：2.1.3.3/2.1.3.3=36。具有典型的美洲鼹族的特征，第一上门齿大于第二上门齿、上犬齿。颅全长 27~32 mm。头体长 100~136 mm。后足长 14~20 mm。尾长 35~45 mm。通体黑色，背腹同色，眼睛小，无外耳结构，眼睛和耳道被毛发覆盖。前掌宽厚，远大于后掌，明显向外翻转。尾粗大，密布深褐色针毛。后足大拇趾朝外侧弯曲，远离其余四趾。
Dental formula: 2.1.3.3/2.1.3.3=36. The greatest length of the skull 27-32 mm. Head-body length 100-136 mm, hindfoot length 14-20 mm, tail length 35-45 mm. Dorsal and ventral pelage black. Eyes minute, concealed in fur; external ears absent. Forefeet enlarged and projected outwards. Tail thick, densely covered with dark brown hairs. First toe of the hindfeet curved and twisted away from the other digits. 1st upper incisor is larger than 2nd upper incisor and the upper canine, which is a typical Scalopini characteristic.

生境 / Habitat
阔叶林、针叶林与灌丛
Broad-leaved forest, coniferous forest, shrubland

▲ 地理分布 / Geographic Distribution

国内分布 / Domestic Distribution
陕西、甘肃、湖北、青海、重庆、四川
Shaanxi, Gansu, Hubei, Qinghai, Chongqing, Sichuan

全球分布 / World Distribution
中国 China

生物地理界 / Biogeographic Realm
古北界 Palearctic

WWF 生物群系 / WWF Biome
温带阔叶和混交林
Temperate Broadleaf & Mixed Forests

动物地理分布型 / Zoogeographic Distribution Type
Q+P

分布标注 / Distribution Note
特有种 Endemic

▲ 濒危状况 / Threatened Status

中国生物多样性红色名录等级 / CB RL Category (2021)
近危 NT

IUCN 红色名录 / IUCN Red List (2021)
无危 LC

威胁因子 / Threats
未知 Unknown

▲ 法律保护地位 / Legal Protection Status

国家重点保护野生动物等级 / Category of National Key Protected Wild Animals (2021)
未列入 Not listed

"三有" 名录 / TWIESSV (2023)
未列入 Not listed

CITES 附录等级 / CITES Appendix (2023)
未列入 Not listed

迁徙物种公约附录 / CMS Appendix (2020)
未列入 Not listed

保护行动 / Conservation Action
尚无保护行动 No conservation action so far

▲ 参考文献 / References

Jiang et al. (蒋志刚等), 2021; Burgin et al., 2020; IUCN, 2020; Wilson and Mittermeier, 2018; Pan et al. (潘清华等), 2007; Wilson and Reeder, 2005; Wang (王应祥), 2003; Wang and Hu (王酉之和胡锦矗), 1999; Wang (汪松), 1998; Zhang (张荣祖), 1997; Stone, 1995; Northwest Institute of Plateau Biology/ Chinese Academy of Sciences, 1989

21 / 巨鼹

Euroscaptor grandis Miller, 1940

· Greater Chinese Mole

▲ 分类地位 / Taxonomy

劳亚食虫目 Eulipotyphla / 鼹科 Talpidae / 东方鼹属 *Euroscaptor*

科建立者及其文献 / Family Authority
G. Fischer, 1814

属建立者及其文献 / Genus Authority
Miller, 1940

亚种 / Subspecies
指名亚种 *E. g. grandis* Miller, 1940
四川西部
Sichuan

滇西亚种 *E. g. yunnanensis* Wang, 2002
云南西部
Yunnan

模式标本产地 / Type Locality
中国
China, Sichuan, "Mount Omei, alt.1524 m", Omei-Shan

▲ 其他名称 / Other Name(s)

其他中文名 / Other Chinese Name(s)
无 None

其他英文名 / Other English Name(s)
无 None

同物异名 / Synonym(s)
无 None

▲ 形态及生境 / Morphology and Habitat

形态特征 / Morphological Characteristics

齿式：3.1.4.3/3.1.4.3=44。鼹族东方鼹属物种。上犬齿明显长于上颚的其他牙齿，是鼹族的鉴别特征。上前白齿 4 枚。头体长约 150 mm。后足长 18 mm。尾长约 10 mm。通体为棕黄色，背腹毛发几乎同色，背部被毛有金属光泽。腹部颜色略微浅于背部。尾短棒状、覆盖着稀疏的棕色长毛。

Dental formula: 3.1.4.3/3.1.4.3=44. A Talpini species of the genus *Euroscaptor*. The upper canine is obviously longer than the other teeth on the upper jaw, which is a typical Talpini characteristic. Four upper premolars appear. Head-body length about 150 mm, hindfoot length 18 mm, tail length about 10 mm. Dorsal and ventral pelage brownish yellow, and ventral is only slightly paler. Dorsal pelage slightly metallic luster. Tail is short and rod-liked, covered with sparse brown hairs.

生境 / Habitat
森林、农田、果园
Forest, farmland, orchard

▲ 地理分布 / Geographic Distribution

国内分布 / Domestic Distribution
四川 Sichuan

全球分布 / World Distribution
中国 China

生物地理界 / Biogeographic Realm
印度马来界、古北界 Indomalaya, Palearctic

WWF 生物群系 / WWF Biome
热带和亚热带湿润阔叶林
Tropical & Subtropical Moist Broadleaf Forests

动物地理分布型 / Zoogeographic Distribution Type
Hm

分布标注 / Distribution Note
特有种 Endemic

▲ 濒危状况 / Threatened Status

中国生物多样性红色名录等级 / CB RL Category (2021)
易危 VU

IUCN 红色名录 / IUCN Red List (2021)
无危 LC

威胁因子 / Threats
公路建设 Highway construction

▲ 法律保护地位 / Legal Protection Status

国家重点保护野生动物等级 / Category of National Key Protected Wild Animals (2021)
未列入 Not listed

"三有" 名录 / TWIESSV (2023)
未列入 Not listed

CITES 附录等级 / CITES Appendix (2023)
未列入 Not listed

迁徙物种公约附录 / CMS Appendix (2020)
未列入 Not listed

保护行动 / Conservation Action
尚无保护行动 No conservation action so far

▲ 参考文献 / References

Jiang et al. (蒋志刚等), 2021; Burgin et al., 2020; IUCN, 2020; Wilson and Mittermeier, 2018; Smith et al., 2009; Pan et al. (潘清华等), 2007; Wilson and Reeder, 2005; Wang (王应祥), 2003; Miller, 1940

22 / 库氏长吻鼹

Euroscaptor kuznetsovi Zemlemerova, 2016

- Kouznetsov's Mole

▲ 分类地位 / Taxonomy

劳亚食虫目 Eulipotyphla / 鼹科 Talpidae / 东方鼹属 *Euroscaptor*

科建立者及其文献 / Family Authority
G. Fischer, 1814

属建立者及其文献 / Genus Authority
Miller, 1940

亚种 / Subspecies
无 None

模式标本产地 / Type Locality
越南
Vietnam, Vinh Phuc Province

何锴 / 供图

▲ 其他名称 / Other Name(s)

其他中文名 / Other Chinese Name(s)
无 None

其他英文名 / Other English Name(s)
无 None

同物异名 / Synonym(s)
无 None

▲ 形态及生境 / Morphology and Habitat

形态特征 / Morphological Characteristics

齿式：3.1.4.3/3.1.4.3=44。上犬齿明显长于上颚其他牙齿。头体长 132~136 mm。尾长 14~17 mm。身体颜色深棕色，部分部位棕黄色，有金属光泽。短棒状尾巴覆盖有稀疏的白色针毛。具有适应于地下生活特征，前掌宽厚，向外侧翻转。

Dental formula: 3.1.4.3/3.1.4.3=44. Forefeet enlarged and projected outwards. The upper canine is obviously longer than the other teeth on the upper jaw. Head-body length 132-136 mm, tail length is 14-17 mm. Body pelage overall brown, and partly brownish yellow, metallic luster. The tail is short rod-like and covered with sparse white hairs. It has adapted to fully fossorial life.

生境 / Habitat
森林、农田
Forest, farmland

▲ 地理分布 / Geographic Distribution

国内分布 / Domestic Distribution
云南、江西、广东、浙江
Yunnan, Jiangxi, Guangdong, Zhejiang

全球分布 / World Distribution
中国、越南 China, Vietnam

生物地理界 / Biogeographic Realm
印度马来界 Indomalaya

WWF 生物群系 / WWF Biome
热带和亚热带湿润阔叶林
Tropical & Subtropical Moist Broadleaf Forests

动物地理分布型 / Zoogeographic Distribution Type
Hm+Sg

分布标注 / Distribution Note
非特有种 Non-Endemic

▲ 濒危状况 / Threatened Status

中国生物多样性红色名录等级 / CB RL Category (2021)
未评定 NE

IUCN 红色名录 / IUCN Red List (2021)
未评定 NE

威胁因子 / Threats
未知 Unknown

▲ 法律保护地位 / Legal Protection Status

国家重点保护野生动物等级 / Category of National Key Protected Wild Animals (2021)
未列入 Not listed

"三有"名录 / TWIESSV (2023)
未列入 Not listed

CITES 附录等级 / CITES Appendix (2023)
未列入 Not listed

迁徙物种公约附录 / CMS Appendix (2020)
未列入 Not listed

保护行动 / Conservation Action
尚无保护行动 No conservation action so far

▲ 参考文献 / References

Liu et al. (刘少英等), 2020; Wilson and Mittermeier, 2018

23 / 长吻鼹

Euroscaptor longirostris
Milne-Edwards, 1870

- Long-nosed Mole

劳亚食虫目 Eulipotyphla / 鼹科 Talpidae / 东方鼹属 *Euroscaptor*

科建立者及其文献 / Family Authority
G. Fischer, 1814

属建立者及其文献 / Genus Authority
Miller, 1940

亚种 / Subspecies
无 None

模式标本产地 / Type Locality
中国
China, Sichuan, Moupin

李成 / 供图

▲ 其他名称 / Other Name(s)

其他中文名 / Other Chinese Name(s)
无 None

其他英文名 / Other English Name(s)
无 None

同物异名 / Synonym(s)
无 None

▲ 形态及生境 / Morphology and Habitat

形态特征 / Morphological Characteristics
齿式：3.1.4.3/3.1.4.3=44。上颌犬齿明显大于门齿和前白齿。头体长 90~145 mm。后足 14~23 mm。尾长 11~25 mm。体表毛发短、柔软而致密。背黑色或深灰色，腹部毛色略浅于背部。前手掌宽大，远大于后足，爪长而锋利。后足趾间有不完全发育的蹼。和大多数真鼹类一样，前爪大拇指外侧，由籽骨构成额外附生指。尾呈短棒状，覆盖有稀疏针毛。
Dental formula: 3.1.4.3/3.1.4.3=44. The upper canine is obviously longer than the upper incisors and premolars. Head-body length 90-145 mm, hindfoot 14-23 mm, tail length 11-25 mm. Body hairs short, dense and soft. Dorsal pelage is black or dark gray, and the ventral is only slightly paler. Forefeet enlarged and projected outwards, claws long and sharp. Hindfeet are partly webbed. A sesamoid is present next to the thumb, forming a prepollex. Tail is short rod-like, covered with sparse black hairs.

生境 / Habitat
山地阔叶林、针叶林
Montane broad-leaved forest, coniferous forest

▲ 地理分布 / Geographic Distribution

国内分布 / Domestic Distribution
陕西、四川
Shaanxi, Sichuan

全球分布 / World Distribution
中国 China

生物地理界 / Biogeographic Realm
印度马来界、古北界 Indomalaya, Palearctic

WWF 生物群系 / WWF Biome
热带和亚热带湿润阔叶林
Tropical & Subtropical Moist Broadleaf Forests

动物地理分布型 / Zoogeographic Distribution Type
Sd

分布标注 / Distribution Note
特有种 Endemic

▲ 濒危状况 / Threatened Status

中国生物多样性红色名录等级 / CB RL Category (2021)
无危 LC

IUCN 红色名录 / IUCN Red List (2021)
无危 LC

威胁因子 / Threats
无 None

▲ 法律保护地位 / Legal Protection Status

国家重点保护野生动物等级 / Category of National Key Protected Wild Animals (2021)
未列入 Not listed

"三有" 名录 / TWIESSV (2023)
未列入 Not listed

CITES 附录等级 / CITES Appendix (2023)
未列入 Not listed

迁徙物种公约附录 / CMS Appendix (2020)
未列入 Not listed

保护行动 / Conservation Action
尚无保护行动 No conservation action so far

▲ 参考文献 / References

Jiang et al. (蒋志刚等), 2021; Burgin et al., 2020; IUCN, 2020; Wilson and Mittermeier, 2018; Tu et al., 2012; Pan et al. (潘清华等), 2007; Qin et al. (秦岭等), Wilson and Reeder, 2005; Wang (王应祥), 2003; Zhang (张荣祖), 1997

24 / 奥氏长吻鼹

Euroscaptor orlovi Zemlemerova, 2016

· Orlov's Mole

▲ 分类地位 / Taxonomy

劳亚食虫目 Eulipotyphla / 鼹科 Talpidae / 东方鼹属 *Euroscaptor*

科建立者及其文献 / Family Authority
G. Fischer, 1814

属建立者及其文献 / Genus Authority
Miller, 1940

亚种 / Subspecies
无 None

模式标本产地 / Type Locality
越南
VietnamVinh Phuc Province

▲ 其他名称 / Other Name(s)

其他中文名 / Other Chinese Name(s)
无 None

其他英文名 / Other English Name(s)
无 None

同物异名 / Synonym(s)
无 None

▲ 形态及生境 / Morphology and Habitat

形态特征 / Morphological Characteristics

齿式：3.1.4.3/3.1.4.3=44。头体长 115~129 mm。后足 15.5 mm。尾长 15~17.5 mm。背腹深棕色。尾短棒状，长有针毛。

Dental formula: 3.1.4.3/3.1.4.3=44. Head-body length 115-129m, hindfoot length 15.5 mm, tail length 15-17.5 mm. Pelage unicolor which is dark brown. Tail is short rod-like and covered with long hairs.

生境 / Habitat

森林 Forest

▲ 地理分布 / Geographic Distribution

国内分布 / Domestic Distribution
云南 Yunnan

全球分布 / World Distribution
中国、越南
China, Vietnam

生物地理界 / Biogeographic Realm
印度马来界 Indomalaya

WWF 生物群系 / WWF Biome
热带和亚热带湿润阔叶林
Tropical & Subtropical Moist Broadleaf Forests

动物地理分布型 / Zoogeographic Distribution Type
Hm

分布标注 / Distribution Note
非特有种 Non-Endemic

▲ 濒危状况 / Threatened Status

中国生物多样性红色名录等级 / CB RL Category (2021)
未评定 NE

IUCN 红色名录 / IUCN Red List (2021)
未评定 NE

威胁因子 / Threats
未知 Unknown

▲ 法律保护地位 / Legal Protection Status

国家重点保护野生动物等级 / Category of National Key Protected Wild Animals (2021)
未列入 Not listed

"三有"名录 / TWIESSV (2023)
未列入 Not listed

CITES 附录等级 / CITES Appendix (2023)
未列入 Not listed

迁徙物种公约附录 / CMS Appendix (2020)
未列入 Not listed

保护行动 / Conservation Action
尚无保护行动 No conservation action so far

▲ 参考文献 / References

Liu et al. (刘少英等), 2020; Wilson and Mittermeier, 2018

25 / 海南缺齿鼹

Mogera hainana Thomas, 1907

· Hainan Mole

▲ 分类地位 / Taxonomy

劳亚食虫目 Eulipotyphla / 鼹科 Talpidae / 缺齿鼹属 *Mogera*

科建立者及其文献 / Family Authority
G. Fischer, 1814

属建立者及其文献 / Genus Authority
Pomel, 1848

亚种 / Subspecies
无 None

模式标本产地 / Type Locality
中国
Hainan, China

杨川 / 供图

▲ 其他名称 / Other Name(s)

其他中文名 / Other Chinese Name(s)
无 None

其他英文名 / Other English Name(s)
无 None

同物异名 / Synonym(s)
无 None

▲ 形态及生境 / Morphology and Habitat

形态特征 / Morphological Characteristics

齿式：3.1.4.3/3.1.4.3=44。上颌犬齿明显长于上颌门齿和前白齿。体重约 66 g。头体长 120~134 mm。后足长 14~18 mm。尾长 9~14 mm。外形与华南缺齿鼹相似，通体棕黄色。前掌宽厚，向外侧翻转。

Dental formula: 3.1.4.3/3.1.4.3=44. The upper canine is obviously longer than upper incisors and premolars. Bodyweight 66 g, head-body length 120-134 mm, hindfoot length 14-18 mm, tail length 9-14 mm. External morphology is similar to that of a *Mogera latouchei*. Pelage dark brownish yellow. Dorsal and ventral pelage similar in color. Forefeet enlarged and projected outwards.

生境 / Habitat

阔叶林、农田、果园
Broad-leaved forest, farmland, orchard

▲ 地理分布 / Geographic Distribution

国内分布 / Domestic Distribution
海南 Hainan

全球分布 / World Distribution
中国 China

生物地理界 / Biogeographic Realm
印度马来界 Indomalaya

WWF 生物群系 / WWF Biome
热带和亚热带湿润阔叶林
Tropical & Subtropical Moist Broadleaf Forests

动物地理分布型 / Zoogeographic Distribution Type
He

分布标注 / Distribution Note
特有种 Endemic

▲ 濒危状况 / Threatened Status

中国生物多样性红色名录等级 / CB RL Category (2021)
未评定 NE

IUCN 红色名录 / IUCN Red List (2021)
未评定 NE

威胁因子 / Threats
未知 Unknown

▲ 法律保护地位 / Legal Protection Status

国家重点保护野生动物等级 / Category of National Key Protected Wild Animals (2021)
未列入 Not listed

"三有" 名录 / TWIESSV (2023)
未列入 Not listed

CITES 附录等级 / CITES Appendix (2023)
未列入 Not listed

迁徙物种公约附录 / CMS Appendix (2020)
未列入 Not listed

保护行动 / Conservation Action
尚无保护行动 No conservation action so far

▲ 参考文献 / References

Liu et al. (刘少英等), 2020

26 / 台湾缺齿鼹

Mogera insularis (Swinhoe, 1863)

· Taiwan Mole

▲ 分类地位 / Taxonomy

劳亚食虫目 Eulipotyphla / 鼹科 Talpidae / 缺齿鼹属 *Mogera*

科建立者及其文献 / Family Authority
G. Fischer, 1814

属建立者及其文献 / Genus Authority
Pomel, 1848

亚种 / Subspecies
无 None

模式标本产地 / Type Locality
中国
Tatachia, Yushan National Park, Nantou, Taiwan, China

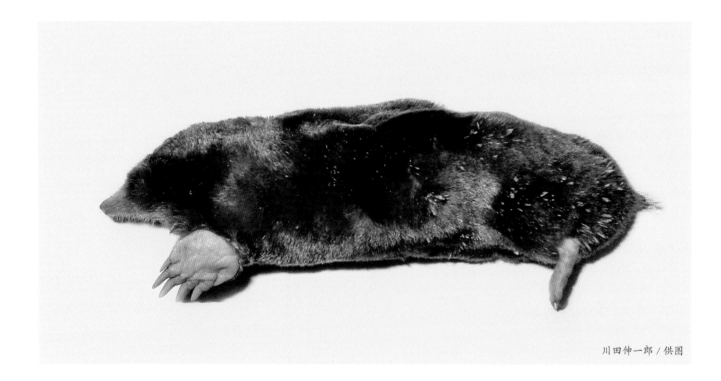

川田伸一郎 / 供图

▲ 其他名称 / Other Name(s)

其他中文名 / Other Chinese Name(s)
无 None

其他英文名 / Other English Name(s)
无 None

同物异名 / Synonym(s)
无 None

▲ 形态及生境 / Morphology and Habitat

形态特征 / Morphological Characteristics
齿式：3.1.4.3/2.1.4.3=42。上颌犬齿明显长于上颌门齿和前白齿。体重23~59 g。头体长 112~140 mm。后足 13.5~15 mm。尾长 8.5~13.5 mm。背部为棕色，背腹不明显异色，腹部毛色较背部略浅，且略泛金黄色。尾很短，长有较长的针毛。前爪大而宽，前肢短而粗壮，适合掘土，前后爪背面几乎裸露，后足趾间有半蹼。

Dental formula: 3.1.4.3/2.1.4.3=42. The upper canine is obviously longer than the upper incisors and premolars. Body weighs 23-59 g, head-body length 112-140 mm, hindfoot length 13.5-15 mm, tail length 8.5-13.5 mm. Dorsal pelage brown, ventral is paler with slightly golden color. Forefeet enlarged and projected outwards, adapting to digging. The upper part of both fore- and hindfeet are naked, and hindfeet are partly webbed.

生境 / Habitat
阔叶林、农田、果园
Broad-leaved forest, farmland, orchard

▲ 地理分布 / Geographic Distribution

国内分布 / Domestic Distribution
台湾、安徽
Taiwan, Anhui

全球分布 / World Distribution
中国 China

生物地理界 / Biogeographic Realm
印度马来界 Indomalaya

WWF 生物群系 / WWF Biome
热带和亚热带湿润阔叶林
Tropical & Subtropical Moist Broadleaf Forests

动物地理分布型 / Zoogeographic Distribution Type
J

分布标注 / Distribution Note
特有种 Endemic

▲ 濒危状况 / Threatened Status

中国生物多样性红色名录等级 / CB RL Category (2021)
无危 LC

IUCN 红色名录 / IUCN Red List (2021)
数据缺乏 DD

威胁因子 / Threats
无 None

▲ 法律保护地位 / Legal Protection Status

国家重点保护野生动物等级 / Category of National Key Protected Wild Animals (2021)
未列入 Not listed

"三有"名录 / TWIESSV (2023)
未列入 Not listed

CITES 附录等级 / CITES Appendix (2023)
未列入 Not listed

迁徙物种公约附录 / CMS Appendix (2020)
未列入 Not listed

保护行动 / Conservation Action
尚无保护行动 No conservation action so far

▲ 参考文献 / References

Jiang et al. (蒋志刚等), 2021; Burgin et al., 2020; IUCN, 2020; Wilson and Mittermeier, 2018; Kawada et al., 2007; Pan et al. (潘清华等), 2007; Wilson and Reeder, 2005; Wang (王应祥), 2003; Zhang (张荣祖), 1997

27 / 鹿野氏缺齿鼹

Mogera kanoana
Kawada, Shinohara, Kobayashi, Harada,
Oda & Lin, 2007

• Kano's Mole

▲ 分类地位 / Taxonomy

劳亚食虫目 Eulipotyphla / 鼹科 Talpidae / 缺齿鼹属 *Mogera*

科建立者及其文献 / Family Authority
G. Fischer, 1814

属建立者及其文献 / Genus Authority
Pomel, 1848

亚种 / Subspecies
无 None

模式标本产地 / Type Locality
中国
Taiwan, China

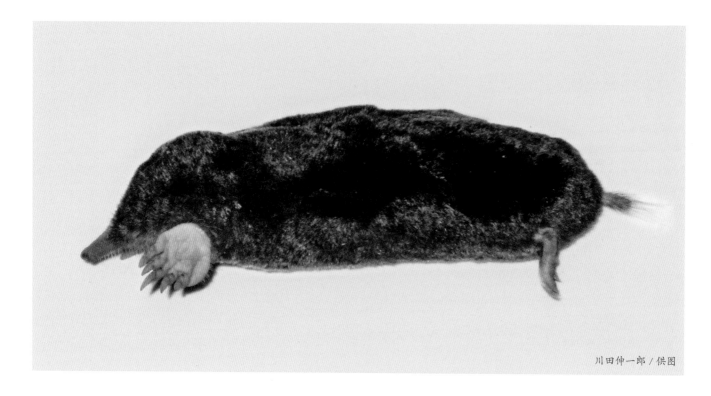

川田伸一郎 / 供图

▲ 其他名称 / Other Name(s)

其他中文名 / Other Chinese Name(s)
无 None

其他英文名 / Other English Name(s)
无 None

同物异名 / Synonym(s)
无 None

▲ 形态及生境 / Morphology and Habitat

形态特征 / Morphological Characteristics

齿式：3.1.4.3/2.1.4.3=42。上颌犬齿明显长于上颌门齿和前白齿。与东方鼹属相比，下颌少一枚门齿。体重 23~59 g。头体长 113~134 mm。后足长 13.5~15 mm。尾长 8.5~13.5 mm。背部为近乎黑色，腹部毛色略浅于背部。前爪大而宽，前肢短而粗壮，爪长而锋利，适合掘土，前后爪背面几乎裸露，后足趾间有蹼。尾巴较短，上覆稀疏黑色针毛。

Dental formula: 3.1.4.3/2.1.4.3=42. The upper canine is obviously longer than the upper incisors and premolars. The lower i2 is missing, so the number of teeth is less than that of a Euroscaptor mole. Bodyweight 23-59 g, head-body length is 113-134 mm, hindfoot length 13.5-15 mm, tail length 8.5-13.5 mm. Dorsal pelage is black and the ventral is only slightly paler. Forefeet enlarged and projected outwards, claws long and sharp, adapted to digging. Upper part of fore and hind feet naked, and the hind feet are webbed. Tail is very short, covered with sparse black hairs.

生境 / Habitat
山地阔叶林、农田
Mountainous broadleaf forest, farmland

▲ 地理分布 / Geographic Distribution

国内分布 / Domestic Distribution
台湾 Taiwan

全球分布 / World Distribution
中国 China

生物地理界 / Biogeographic Realm
印度马来界 Indomalaya

WWF 生物群系 / WWF Biome
热带和亚热带湿润阔叶林
Tropical & Subtropical Moist Broadleaf Forests

动物地理分布型 / Zoogeographic Distribution Type
He

分布标注 / Distribution Note
特有种 Endemic

▲ 濒危状况 / Threatened Status

中国生物多样性红色名录等级 / CB RL Category (2021)
数据缺乏 DD

IUCN 红色名录 / IUCN Red List (2021)
未评定 NE

威胁因子 / Threats
未知 Unknown

▲ 法律保护地位 / Legal Protection Status

国家重点保护野生动物等级 / Category of National Key Protected Wild Animals (2021)
未列入 Not listed

"三有" 名录 / TWIESSV (2023)
未列入 Not listed

CITES 附录等级 / CITES Appendix (2023)
未列入 Not listed

迁徙物种公约附录 / CMS Appendix (2020)
未列入 Not listed

保护行动 / Conservation Action
尚无保护行动 No conservation action so far

▲ 参考文献 / References

Jiang et al. (蒋志刚等), 2021; Liu et al. (刘少英等), 2020; Wilsonand Mittermeier, 2018

28 / 华南缺齿鼹

Mogera latouchei Thomas, 1907

· La Thou's Mole

▲ 分类地位 / Taxonomy

劳亚食虫目 Eulipotyphla / 鼹科 Talpidae / 缺齿鼹属 *Mogera*

科建立者及其文献 / Family Authority
G. Fischer, 1814

属建立者及其文献 / Genus Authority
Pomel, 1848

亚种 / Subspecies
无 None

模式标本产地 / Type Locality
中国
China, Taiwan

▲ 其他名称 / Other Name(s)

其他中文名 / Other Chinese Name(s)
缺齿鼹

其他英文名 / Other English Name(s)
无 None

同物异名 / Synonym(s)
无 None

▲ 形态及生境 / Morphology and Habitat

形态特征 / Morphological Characteristics
齿式：3.1.4.3/2.1.4.3=42。头体长 87~115 mm。后足 13.5~14 mm。尾长
15~20 mm。浑身毛发短而光滑，略带有金属光泽。背腹几乎同色，背
部毛色深棕色，腹部毛色略浅。前掌宽厚，向外翻转，前后足都有鳞状
黑斑。尾短棒状，上覆盖有稀疏的黑色针毛。
Dental formula: 3.1.4.3/2.1.4.3=42. Head-body length 87-115 mm, tail length 15-20
mm, hindfoot length 13.5-14 mm. Body fur short and dense and is a metallic luster.
Dorsal and ventral pelage is nearly unicolor, the dorsal pelage is dark brown and the
dorsal is slightly lighter. Forefeet enlarged and projected outwards, adapting to digging.
The upper part of both fore- and hind feet are covered with black scales. Tail is short
rod-like, covered with long black hairs.

生境 / Habitat
阔叶林、农田、果园
Broad-leaved forest, farmland, orchard

▲ 地理分布 / Geographic Distribution

国内分布 / Domestic Distribution
江苏、浙江、安徽、江西、湖南、四川、贵州、广西、广东、福建
Jiangsu, Zhejiang, Anhui, Jiangxi, Hunan, Sichuan, Guizhou, Guangxi, Guangdong,
Fujian

全球分布 / World Distribution
中国、越南 China, Vietnam

生物地理界 / Biogeographic Realm
古北界、印度马来界 Palearctic, Indomalaya

WWF 生物群系 / WWF Biome
热带和亚热带湿润阔叶林
Tropical & Subtropical Moist Broadleaf Forests

动物地理分布型 / Zoogeographic Distribution Type
Sc

分布标注 / Distribution Note
非特有种 Non-Endemic

▲ 濒危状况 / Threatened Status

中国生物多样性红色名录等级 / CB RL Category (2021)
未评定 NE

IUCN 红色名录 / IUCN Red List (2021)
未评定 NE

威胁因子 / Threats
未知 Unknown

▲ 法律保护地位 / Legal Protection Status

国家重点保护野生动物等级 / Category of National Key Protected Wild Animals (2021)
未列入 Not listed

"三有"名录 / TWIESSV (2023)
未列入 Not listed

CITES 附录等级 / CITES Appendix (2023)
未列入 Not listed

迁徙物种公约附录 / CMS Appendix (2020)
未列入 Not listed

保护行动 / Conservation Action
尚无保护行动 No conservation action so far

▲ 参考文献 / References

Liu et al. (刘少英等), 2020; Tu et al. (涂飞云等), 2014; Su et al. (粟海军等), 2013; Pan et al. (潘清华等), 2007

29 / 大缺齿鼹

Mogera robusta Nehring, 1891

• Large Mole

▲ 分类地位 / Taxonomy

劳亚食虫目 Eulipotyphla / 鼹科 Talpidae / 缺齿鼹属 *Mogera*

科建立者及其文献 / Family Authority
G. Fischer, 1814

属建立者及其文献 / Genus Authority
Pomel, 1848

亚种 / Subspecies
无 None

模式标本产地 / Type Locality
俄罗斯
Vladivostok, Eastern Siberia

▲ 其他名称 / Other Name(s)

其他中文名 / Other Chinese Name(s)
缺齿鼹

其他英文名 / Other English Name(s)
无 None

同物异名 / Synonym(s)
无 None

▲ 形态及生境 / Morphology and Habitat

形态特征 / Morphological Characteristics

齿式：3.1.4.3/2.1.4.3=42。头体长 147~165 mm。后足长 18~21 mm。尾长 16~20.5 mm。体重 95.9~127.3 g。中国体型最大的鼹科动物。头骨致密，吻部较宽。前爪很大，向外侧翻转。体表毛发短而致密，具金属光泽。背面毛色棕灰色，腹部毛色略浅，部分区域橙黄色。尾巴覆盖棕色针毛。
Dental formula: 3.1.4.3/2.1.4.3=42. Head-body length 147-165 mm, hindfoot 18-21 mm, tail length 16-20.5 mm. Bodyweight 95.9-127.3 g. The largest species of Talpidae in China. Skull is stout, and the rostrum is broad. Forefeet enlarged and projected outwards, adapting to digging. Body fur is short and dense and is a metallic luster. Dorsal pelage brownish grey, ventral is paler and partly yellow-orange. Tail is covered with long brown hairs.

生境 / Habitat
森林、草地、农田
Forest, grassland, farmland

0142

▲ 地理分布 / Geographic Distribution

国内分布 / Domestic Distribution
黑龙江、吉林、辽宁
Heilongjiang, Jilin, Liaoning

全球分布 / World Distribution
中国、朝鲜、韩国、俄罗斯
China, Democratic People's Republic of Korea, Republic of Korea, Russia

生物地理界 / Biogeographic Realm
古北界 Palearctic

WWF 生物群系 / WWF Biome
温带阔叶和混交林
Temperate Broadleaf & Mixed Forests

动物地理分布型 / Zoogeographic Distribution Type
Kb

分布标注 / Distribution Note
非特有种 Non-Endemic

▲ 濒危状况 / Threatened Status

中国生物多样性红色名录等级 / CB RL Category (2021)
无危 LC

IUCN 红色名录 / IUCN Red List (2021)
无危 LC

威胁因子 / Threats
无 None

▲ 法律保护地位 / Legal Protection Status

国家重点保护野生动物等级 / Category of National Key Protected Wild Animals (2021)
未列入 Not listed

"三有" 名录 / TWIESSV (2023)
未列入 Not listed

CITES 附录等级 / CITES Appendix (2023)
未列入 Not listed

迁徙物种公约附录 / CMS Appendix (2020)
未列入 Not listed

保护行动 / Conservation Action
尚无保护行动 No conservation action so far

▲ 参考文献 / References

Jiang et al. (蒋志刚等), 2021; Burgin et al., 2020; IUCN, 2020; Wilson and Mittermeier, 2018; Smith et al., 2009; Pan et al. (潘清华等), 2007; Wang (王应祥), 2003; Zhang (张荣祖), 1997; Abe, 1995; Xia (夏武平), 1988; Shou (寿正黄), 1962; Institute of Zoology/Chinese Academy of Sciences, 1958

30 / 钓鱼岛鼹

Mogera uchidai
Abe, Shiraishi & Arai, 1991

• Uchida's mole

▲ 分类地位 / Taxonomy

劳亚食虫目 Eulipotyphla / 鼹科 Talpidae / 缺齿鼹属 *Mogera*

科建立者及其文献 / Family Authority
G. Fischer, 1814

属建立者及其文献 / Genus Authority
Pomel, 1848

亚种 / Subspecies
无 None

模式标本产地 / Type Locality
中国
Diaoyu Island

▲ 其他名称 / Other Name(s)

其他中文名 / Other Chinese Name(s)
岛鼹

其他英文名 / Other English Name(s)
无 None

同物异名 / Synonym(s)
无 None

▲ 形态及生境 / Morphology and Habitat

形态特征 / Morphological Characteristics
齿式：3.1.3.3/2.1.3.3=38。1991 年被发现命名。仅正模标本一号标本。齿式不同于缺齿鼹其他物种。头体长 130 mm。尾长 12 mm。后足长 16 mm。体重 43 g。吻部短而宽。背部毛发深灰棕色，腹部毛发颜色略浅。喉部和脸颊棕色。尾及后足覆盖有浅褐色长毛。

Dental formula: 3.1.3.3/2.1.3.3=38. Discovered and named in 1991 based on a single holotype specimen. Its dental formular is different from other species of Mogera. Head-body length 130 mm, tail 12 mm, hindfoot length 16 mm, bodyweight 43 g. Dorsal pelage is dark grayish brown and venter is slightly paler. Throat and chest are brown. Tail and upper part of hindfoot covered with drab long hairs.

生境 / Habitat
草地 Grassland

▲ 地理分布 / Geographic Distribution

国内分布 / Domestic Distribution
钓鱼岛 Diaoyu Island

全球分布 / World Distribution
中国 China

生物地理界 / Biogeographic Realm
印度马来界 Indomalaya

WWF 生物群系 / WWF Biome
热带和亚热带湿润阔叶林
Tropical & Subtropical Moist Broadleaf Forests

动物地理分布型 / Zoogeographic Distribution Type
J

分布标注 / Distribution Note
特有种 Endemic

▲ 濒危状况 / Threatened Status

中国生物多样性红色名录等级 / CB RL Category (2021)
数据缺乏 DD

IUCN 红色名录 / IUCN Red List (2021)
易危 VU

威胁因子 / Threats
外来入侵物种 Alien Species Invasion

▲ 法律保护地位 / Legal Protection Status

国家重点保护野生动物等级 / Category of National Key Protected Wild Animals (2021)
未列入 Not listed

"三有" 名录 / TWIESSV (2023)
未列入 Not listed

CITES 附录等级 / CITES Appendix (2023)
未列入 Not listed

迁徙物种公约附录 / CMS Appendix (2020)
未列入 Not listed

保护行动 / Conservation Action
尚无保护行动 No conservation action so far

▲ 参考文献 / References

Jiang et al. (蒋志刚等), 2021; Burgin et al., 2020; IUCN, 2020; Wilson and Mittermeier, 2018; Pan et al. (潘清华等), 2007; Wang (王应祥), 2003

31 / 白尾鼹

Parascaptor leucura (Blyth, 1850)

· White-tailed Mole

▲ 分类地位 / Taxonomy

劳亚食虫目 Eulipotyphla / 鼹科 Talpidae / 白尾鼹属 *Parascaptor*

科建立者及其文献 / Family Authority
G. Fischer, 1814

属建立者及其文献 / Genus Authority
Gill, 1875

亚种 / Subspecies
无 None

模式标本产地 / Type Locality
印度
India, Assam, Khasi Hills, Cherrapunji

何锴 / 供图

▲ 其他名称 / Other Name(s)

其他中文名 / Other Chinese Name(s)
无 None

其他英文名 / Other English Name(s)
无 None

同物异名 / Synonym(s)
无 None

▲ 形态及生境 / Morphology and Habitat

形态特征 / Morphological Characteristics

齿式：3.1.3.3/3.1.4.3=42。上犬齿明显长于上门齿和上前白齿，上前白齿3枚，少于缺齿鼹属上前白齿数（4枚）。头体长100~115 mm。后足长15~16 mm。尾长5~15 mm。单型属。通体被毛黑色或深灰色，背腹近乎同色。吻部短，皮肤裸露仅覆盖稀疏的短毛。前爪大而宽厚，明显外翻。尾纺锤形，表皮裸露，被覆稀疏白色刚毛。

Dental formula: 3.1.3.3/3.1.4.3=42. Upper canine is larger than the upper incisors and upper premolars. One less pair of upper premolar than that of a Mogera species. Head-body length 100-115 mm, hindfoot length 15-16 mm, tail 5-15 mm. A monotypic genus. Pelage is black or dark gray, and the dorsal and ventral pelage is similar in color. Snout is short, naked, and covered with sparse short hairs. Forefeet enlarged and projected outwards, adapting to digging. Tail is short rod-like and covered with white hairs.

生境 / Habitat

亚热带湿润山地森林、灌丛、草地
Subtropical moist montane forest, shrubland, grassland

▲ 地理分布 / Geographic Distribution

国内分布 / Domestic Distribution
四川、云南 Sichuan, Yunnan

全球分布 / World Distribution
孟加拉国、中国、印度、缅甸
Bengladesh, China, India, Myanmar

生物地理界 / Biogeographic Realm
古北界、印度马来界 Palearctic, Indomalaya

WWF 生物群系 / WWF Biome
热带和亚热带湿润阔叶林
Tropical & Subtropical Moist Broadleaf Forests

动物地理分布型 / Zoogeographic Distribution Type
Wb

分布标注 / Distribution Note
非特有种 Non-Endemic

▲ 濒危状况 / Threatened Status

中国生物多样性红色名录等级 / CB RL Category (2021)
近危 NT

IUCN 红色名录 / IUCN Red List (2021)
无危 LC

威胁因子 / Threats
森林砍伐、耕种、农业林业污染
Logging, farming, agricultural forestry or effluents

▲ 法律保护地位 / Legal Protection Status

国家重点保护野生动物等级 / Category of National Key Protected Wild Animals (2021)
未列入 Not listed

"三有" 名录 / TWIESSV (2023)
未列入 Not listed

CITES 附录等级 / CITES Appendix (2023)
未列入 Not listed

迁徙物种公约附录 / CMS Appendix (2020)
未列入 Not listed

保护行动 / Conservation Action
尚无保护行动 No conservation action so far

▲ 参考文献 / References

Jiang et al. (蒋志刚等), 2021; Burgin et al., 2020; IUCN, 2020; He (何锴), 2013; Duan et al. (段海生等), 2011; Smith et al., 2009; Qin et al. (秦岭等), 2007; Pan et al. (潘清华等), 2007; Hutterer, 2005; Wilson and Reeder, 2005; Wang (王应祥), 2003; Zhang (张荣祖), 1997

32 / 麝鼹

Scaptochirus moschatus
Milne-Edwards, 1867

• Short-faced Mole

▲ 分类地位 / Taxonomy

劳亚食虫目 Eulipotyphla / 鼹科 Talpidae / 麝鼹属 *Scaptochirus*

科建立者及其文献 / Family Authority
G. Fischer, 1814

属建立者及其文献 / Genus Authority
Milne-Edwards, 1867

亚种 / Subspecies
指名亚种 *S. m. moschatus* Milne-Edwards, 1867
北京、内蒙古、河北、山东、江苏、河南
Beijing, Inner Mongolia, Hebei, Shandong, Jiangsu, Henan

西北亚种 *S. m. gilliesi* Thomas, 1910
宁夏、甘肃、陕西、湖北、山西
Ningxia, Gansu, Shaanxi, Hubei, Shanxi

东北亚种 *S. m. grandidens* (Stroganov, 1941)
内蒙古、黑龙江、辽宁
Inner Mongolia, Heilongjiang, Liaoning

模式标本产地 / Type Locality
中国
"En Mongolie"; Swanhwafu(=Zhangjiakou, Xuanhua), 161 km NW of Peking, China

乔轶伦 / 供图

▲ 其他名称 / Other Name(s)

其他中文名 / Other Chinese Name(s)
无 None

其他英文名 / Other English Name(s)
无 None

同物异名 / Synonym(s)
无 None

▲ 形态及生境 / Morphology and Habitat

形态特征 / Morphological Characteristics
齿式：3.1.3.3/3.1.3.3=40。上前白齿、下前白齿均 3 枚。头体长 100~150 mm。尾长 14~23 mm。后足 15~19 mm。头骨致密，吻部短粗。体表被毛短而致密光滑，有金属光泽。背部毛色棕灰色，吻部和前肢毛色略浅，腹部锈黄色。前掌宽大，向外翻转。尾巴纤细，略短于后足，覆盖稀疏的针毛，尾尖有一簇毛。

Dental formula: 3.1.3.3/3.1.3.3=40. Three upper and lower premolars present. Head-body length 100-150 mm, hindfoot length 15-19 mm, tail length 14-23 mm. Skull is large and stout, rostrum is short and broad. Body fur short, dense, and smooth, with a metallic luster. Dorsal pelage dark brown, paler around moue and on forearms. Venter rust yellow. Forefeet enlarged and projected outwards, adapting to digging. Tail is slender, slighter shorter than hindfeet, and sparsely covered with hairs and a small tuft at the tip.

生境 / Habitat
森林、草原、荒漠、牧场、农田、果园
Forest, grassland, desert, meadow, farmland, orchard

▲ 地理分布 / Geographic Distribution

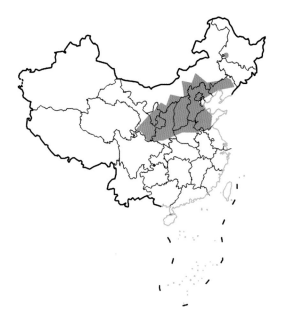

国内分布 / Domestic Distribution

黑龙江、辽宁、内蒙古、北京、甘肃、河北、河南、江苏、宁夏、山西、陕西、山东

Heilongjiang, Liaoning, Inner Mongolia, Beijing, Gansu, Hebei, Henan, Jiangsu, Ningxia, Shanxi, Shaanxi, Shandong

全球分布 / World Distribution

中国 China

生物地理界 / Biogeographic Realm

古北界 Palearctic

WWF 生物群系 / WWF Biome

温带阔叶和混交林、温带草原、稀树草原和灌丛

Temperate Broadleaf & Mixed Forests, Temperate Grasslands, Savannas and Shrublands

动物地理分布型 / Zoogeographic Distribution Type

Ba

分布标注 / Distribution Note

特有种 Endemic

▲ 濒危状况 / Threatened Status

中国生物多样性红色名录等级 / CB RL Category (2021)

近危 NT

IUCN 红色名录 / IUCN Red List (2021)

无危 LC

威胁因子 / Threats

未知 Unknown

▲ 法律保护地位 / Legal Protection Status

国家重点保护野生动物等级 / Category of National Key Protected Wild Animals (2021)

未列入 Not listed

"三有"名录 / TWIESSV (2023)

未列入 Not listed

CITES 附录等级 / CITES Appendix (2023)

未列入 Not listed

迁徙物种公约附录 / CMS Appendix (2020)

未列入 Not listed

保护行动 / Conservation Action

尚无保护行动 No conservation action so far

▲ 参考文献 / References

Jiang et al. (蒋志刚等), 2021; Burgin et al., 2020; IUCN, 2020; Wilson and Mittermeier, 2018; Smith et al., 2009; China Species Information Service, 2008; Pan et al. (潘清华等), 2007; Wilson and Reeder, 2005; Wang (王应祥), 2003; Zhang (张荣祖), 1997; Xia (夏武平), 1988

33 / 天山鼩鼱

Sorex asper Thomas, 1914

· Tien Shan Shrew

▲ 分类地位 / Taxonomy

劳亚食虫目 Eulipotyphla / 鼩鼱科 Soricidae / 鼩鼱属 *Sorex*

科建立者及其文献 / Family Authority
G. Fischer, 1814

属建立者及其文献 / Genus Authority
G. Fischer, 1814

亚种 / Subspecies
无 None

模式标本产地 / Type Locality
中国
"Thian-shan (Tien-shan), Tekes Valley". Note on type specimen tag says "Jigalong" (Dzhergalan?, see Hoffmann, 1987:119); Narynko'skii r-n., Alma-Ata Obl., Kazakhstan

▲ 其他名称 / Other Name(s)

其他中文名 / Other Chinese Name(s)
无 None

其他英文名 / Other English Name(s)
无 None

同物异名 / Synonym(s)
无 None

▲ 形态及生境 / Morphology and Habitat

形态特征 / Morphological Characteristics

齿式：3.1.3.3/2.0.1.3=32。头体长 65~77 mm。尾长 34~45 mm。后足长 11~13 mm。体重 6~10.3 g。形态与苔原鼩鼱 *S. tundrensis* 相似。背腹异色，背部深棕色，腹部灰白色至浅灰色。尾双色。鼩鼱亚科物种除短尾鼩属、亚洲水鼩属及蹼足鼩属，牙齿或多或少存在色素沉积。鼩鼱属物种牙齿均有红褐色色素沉积，均 5 枚上单尖齿。

Dental formula: 3.1.3.3/2.0.1.3=32. Head-body length 65-77 mm, tail length 34-45 mm, hindfoot length 11-13 mm, weight 6-10.3 g. Morphologically analog a tundra shrew (*Sorex tundrensis*). Pelage is bicolored. Dorsal pelage light brown to dark brown, and dorsum is grayish-white to slate gray. Tail is bicolored. Teeth of all Soricinae species are more or less pigmented except for species in *Anourosorex*, *Chimarrogale* and *Nectogale*. All teeth are reddish-brown pigmented as in all *Sorex* species. It has five upper unicuspids as in all *Sorex* species.

生境 / Habitat
高山草甸、灌丛、泰加林
Alpine meadow, shrubland, taiga

▲ 地理分布 / Geographic Distribution

国内分布 / Domestic Distribution
新疆 Xinjiang

全球分布 / World Distribution
中国、哈萨克斯坦、吉尔吉斯斯坦
China, Kazakhstan, Kyrgyzstan

生物地理界 / Biogeographic Realm
古北界 Palearctic

WWF 生物群系 / WWF Biome
温带针叶树森林
Temperate Conifer Forests

动物地理分布型 / Zoogeographic Distribution Type
Dp

分布标注 / Distribution Note
非特有种 Non-Endemic

▲ 濒危状况 / Threatened Status

中国生物多样性红色名录等级 / CB RL Category (2021)
近危 NT

IUCN 红色名录 / IUCN Red List (2021)
无危 LC

威胁因子 / Threats
未知 Unknown

▲ 法律保护地位 / Legal Protection Status

国家重点保护野生动物等级 / Category of National Key Protected Wild Animals (2021)
未列入 Not listed

"三有"名录 / TWIESSV (2023)
未列入 Not listed

CITES 附录等级 / CITES Appendix (2023)
未列入 Not listed

迁徙物种公约附录 / CMS Appendix (2020)
未列入 Not listed

保护行动 / Conservation Action
尚无保护行动 No conservation action so far

▲ 参考文献 / References

Jiang et al. (蒋志刚等), 2021; Burgin et al., 2020; IUCN, 2020; Liu et al. (刘少英等), 2020; Wilson and Mittermeier, 2018; Smith et al., 2009; Fumagalli et al., 1999; Pan et al. (潘清华等), 2007; Wilson and Reeder, 2005; Wang (王应祥), 2003

34 / 小纹背鼩鼱

Sorex bedfordiae Thomas, 1911

· Lesser Stripe-backed Shrew

▲ 分类地位 / Taxonomy

劳亚食虫目 Eulipotyphla / 鼩鼱科 Soricidae / 鼩鼱属 *Sorex*

科建立者及其文献 / Family Authority
G. Fischer, 1814

属建立者及其文献 / Genus Authority
G. Fischer, 1814

亚种 / Subspecies
指名亚种 *S. b. bedfordiae* (Allen, 1923)
四川和陕西
Sichuan and Shaanxi

甘肃亚种 *S. b. wardi* (Thomas, 1911)
甘肃和青海
Gansu and Qinghai

石南亚种 *S. b. gowphus* (G. Allen, 1923)
云南
Yunnan

模式标本产地 / Type Locality
中国
"Omi-san, Sze-chwan" (China, Sichuan, Emei Shan.)

何锴 / 供图

▲ 其他名称 / Other Name(s)

其他中文名 / Other Chinese Name(s)
无 None

其他英文名 / Other English Name(s)
无 None

同物异名 / Synonym(s)
无 None

▲ 形态及生境 / Morphology and Habitat

形态特征 / Morphological Characteristics

齿式：3.1.3.3/2.0.1.3=32。头体长 47~76 mm。尾长 47~66 mm。后足长 11~15 mm。背脊部有一道深色条纹。背部毛色棕灰色，腹部毛色较背部毛色浅，背腹毛色逐渐过渡。尾双色，尾尖皮肤裸露。

Dental formula: 3.1.3.3/2.0.1.3=32. Head length 47-76 mm. Tail length 47-66 mm. Hindfoot length 11-15 mm. Dorsal pelage is dark brown and the ventral is lighter with white hair present, but the pelage is not sharply bicolor. The tail is unicolor and the tip is naked.

生境 / Habitat

泰加林 Taiga

▲ 地理分布 / Geographic Distribution

国内分布 / Domestic Distribution
陕西、甘肃、青海、四川、云南，湖北、重庆、西藏
Shaanxi, Gansu, Qinghai, Sichuan, Yunnan, Hubei, Chongqing, Tibet

全球分布 / World Distribution
中国、缅甸 China, Myanmar

生物地理界 / Biogeographic Realm
古北界、印度马来界 Palearctic, Indomalaya

WWF 生物群系 / WWF Biome
温带针叶树森林
Temperate Conifer Forests

动物地理分布型 / Zoogeographic Distribution Type
Hc+Q+L

分布标注 / Distribution Note
非特有种 Non-Endemic

▲ 濒危状况 / Threatened Status

中国生物多样性红色名录等级 / CB RL Category (2021)
无危 LC

IUCN 红色名录 / IUCN Red List (2021)
无危 LC

威胁因子 / Threats
无 None

▲ 法律保护地位 / Legal Protection Status

国家重点保护野生动物等级 / Category of National Key Protected Wild Animals (2021)
未列入 Not listed

"三有"名录 / TWIESSV (2023)
未列入 Not listed

CITES 附录等级 / CITES Appendix (2023)
未列入 Not listed

迁徙物种公约附录 / CMS Appendix (2020)
未列入 Not listed

保护行动 / Conservation Action
尚无保护行动 No conservation action so far

▲ 参考文献 / References

Jiang et al. (蒋志刚等), 2021; Burgin et al., 2020; IUCN, 2020; Wilson and Mittermeier, 2018; Smith et al., 2009; Pan et al. (潘清华等), 2007; Wilson and Reeder, 2005; Wang (王应祥), 2003; Zhang (张荣祖), 1997

35 / 中鼩鼱

Sorex caecutiens Laxmann, 1788

• Laxmann's Shrew

▲ 分类地位 / Taxonomy

劳亚食虫目 Eulipotyphla / 鼩鼱科 Soricidae / 鼩鼱属 *Sorex*

科建立者及其文献 / Family Authority
G. Fischer, 1814

属建立者及其文献 / Genus Authority
G. Fischer, 1814

亚种 / Subspecies
指名亚种 *S. c. macropygmaeus* Miller, 1901
黑龙江和吉林
Heilongjiang and Jilin

模式标本产地 / Type Locality
俄罗斯
Russia, Buryatskaya ASSR, SW shore of Lake Baikal (Pavlinov and Rossolimo, 1987:17)

▲ 其他名称 / Other Name(s)

其他中文名 / Other Chinese Name(s)
无 None

其他英文名 / Other English Name(s)
无 None

同物异名 / Synonym(s)
无 None

▲ 形态及生境 / Morphology and Habitat

形态特征 / Morphological Characteristics
齿式：3.1.3.3/2.0.1.3=32。头体长 50~73 mm。尾长 28~45 mm。后足长 10~12 mm。体重 4~12.5 g。背部棕色，腹部通常淡黄色。尾明显异色。
Dental formula: 3.1.3.3/2.0.1.3=32. Head-body length 50-73 mm, tail length 28-45 mm, hindfoot length 10-12 mm, weight 4-12.5 g. Dorsal pelage brown, ventral pale yellow. Tail is sharply bicolored.

生境 / Habitat
泰加林、苔原、沼泽
Taiga, tundra, swamp

▲ **地理分布** / Geographic Distribution

国内分布 / Domestic Distribution
黑龙江、吉林、内蒙古、新疆
Heilongjiang, Jilin, Inner Mongolia, Xinjiang

全球分布 / World Distribution
白俄罗斯、中国、爱沙尼亚、芬兰、日本、哈萨克斯坦、朝鲜、韩国、
拉脱维亚、立陶宛、蒙古国、挪威、波兰、俄罗斯、瑞典、乌克兰
Belarus, China, Estonia, Finland, Japan, Kazakhstan, Democratic People's Republic
of Korea, Republic of Korea, Latvia, Lithuania, Mongolia, Norway, Poland, Russia,
Sweden, Ukraine

生物地理界 / Biogeographic Realm
古北界 Palearctic

WWF 生物群系 / WWF Biome
北方森林 / 针叶林
Boreal Forests/Coniferous Forests

动物地理分布型 / Zoogeographic Distribution Type
Ue

分布标注 / Distribution Note
非特有种 Non-Endemic

▲ **濒危状况** / Threatened Status

中国生物多样性红色名录等级 / CB RL Category (2021)
近危 NT

IUCN 红色名录 / IUCN Red List (2021)
无危 LC

威胁因子 / Threats
森林砍伐 Logging

▲ **法律保护地位** / Legal Protection Status

国家重点保护野生动物等级 / Category of National Key Protected Wild Animals (2021)
未列入 Not listed

"三有" 名录 / TWIESSV (2023)
未列入 Not listed

CITES 附录等级 / CITES Appendix (2023)
未列入 Not listed

迁徙物种公约附录 / CMS Appendix (2020)
未列入 Not listed

保护行动 / Conservation Action
尚无保护行动 No conservation action so far

▲ **参考文献** / References

Jiang et al. (蒋志刚等), 2021; Burgin et al., 2020; IUCN, 2020; Wilson and Mittermeier, 2018; Smith et al., 2009; Pan et al. (潘清华等), 2007;
Wilson and Reeder, 2005; Wang (王应祥), 2003; Aptimi·Apdukader (阿布力米提·阿都卡迪尔), 2002; Zhang (张荣祖), 1997

36 / 甘肃鼩鼱

Sorex cansulus Thomas, 1912

· Gansu Shrew

▲ 分类地位 / Taxonomy

劳亚食虫目 Eulipotyphla / 鼩鼱科 Soricidae / 鼩鼱属 *Sorex*

科建立者及其文献 / Family Authority
G. Fischer, 1814

属建立者及其文献 / Genus Authority
G. Fischer, 1814

亚种 / Subspecies
无 None

模式标本产地 / Type Locality
中国
China, Gansu, "46 miles (74 km) south-east of SE Taochou" (Lintan)

▲ 其他名称 / Other Name(s)

其他中文名 / Other Chinese Name(s)
无 None

其他英文名 / Other English Name(s)
无 None

同物异名 / Synonym(s)
无 None

▲ 形态及生境 / Morphology and Habitat

形态特征 / Morphological Characteristics
齿式：3.1.3.3/2.0.1.3=32。头体长 62~64 mm。尾长 38~43 mm。后足长
12 mm。小型鼩鼱，曾被认为是中鼩鼱的一个亚种，外形与后者相似。
背部毛皮灰褐色，腹部皮毛淡褐色。前后足背浅棕色。尾双色，背面深
褐色，腹面毛色稍浅。吻部与中鼩鼱相比较短且较宽。
Dental formula: 3.1.3.3/2.0.1.3=32. Head-body length 62-64 mm, tail length 38-43
mm, hindfoot length 12 mm. It was considered a subspecies of, and morphologically
similar to a Laxmann's shrew (*Sorex caecutiens*), but smaller. Dorsal pelage greyish
brown, ventrum drab or broccoli-brown. Their hands and feet are brownish white.
Tail dark brown above and lighter below. Rostrum relatively short and wide, especially
when compared with *S. caecutiens*.

生境 / Habitat
森林、灌丛、沼泽
Forest, shrubland, swamp

▲ 地理分布 / Geographic Distribution

国内分布 / Domestic Distribution
甘肃 Gansu

全球分布 / World Distribution
中国 China

生物地理界 / Biogeographic Realm
古北界 Palearctic

WWF 生物群系 / WWF Biome
山地草原和灌丛
Montane Grasslands & Shrublands

动物地理分布型 / Zoogeographic Distribution Type
Lel

分布标注 / Distribution Note
特有种 Endemic

▲ 濒危状况 / Threatened Status

中国生物多样性红色名录等级 / CB RL Category (2021)
近危 NT

IUCN 红色名录 / IUCN Red List (2021)
数据缺乏 DD

威胁因子 / Threats
未知 Unknown

▲ 法律保护地位 / Legal Protection Status

国家重点保护野生动物等级 / Category of National Key Protected Wild Animals (2021)
未列入 Not listed

"三有" 名录 / TWIESSV (2023)
未列入 Not listed

CITES 附录等级 / CITES Appendix (2023)
未列入 Not listed

迁徙物种公约附录 / CMS Appendix (2020)
未列入 Not listed

保护行动 / Conservation Action
尚无保护行动 No conservation action so far

▲ 参考文献 / References

Jiang et al. (蒋志刚等), 2021; Burgin et al., 2020; IUCN, 2020; Wilson and Mittermeier, 2018; Smith et al., 2009; Pan et al. (潘清华等), 2007; Wilson and Reeder, 2005; Wang (王应祥), 2003; Wang (王香亭), 1991; Hoffmann, 1987

37 | 纹背鼩鼱

Sorex cylindricauda
Milne-Edwards, 1872

• Stripe-backed Shrew

▲ 分类地位 / Taxonomy

劳亚食虫目 Eulipotyphla / 鼩鼱科 Soricidae / 鼩鼱属 *Sorex*

科建立者及其文献 / Family Authority
G. Fischer, 1814

属建立者及其文献 / Genus Authority
G. Fischer, 1814

亚种 / Subspecies
无 None

模式标本产地 / Type Locality
中国
China, Sichuan, Moupin (Baoxing)

▲ 其他名称 / Other Name(s)

其他中文名 / Other Chinese Name(s)
大纹背鼩鼱

其他英文名 / Other English Name(s)
无 None

同物异名 / Synonym(s)
无 None

▲ 形态及生境 / Morphology and Habitat

形态特征 / Morphological Characteristics
齿式：3.1.3.3/2.0.1.3=32。头体长 54~77 mm。尾长 55~62 mm。后足 12.5~16 mm。颅全长约 17.5 mm。外形与小纹背鼩鼱相似，但体型略大，尾长略长，背部深色条纹较小纹背鼩鼱更明显。背部毛色肉桂色，腹部深棕灰。尾长接近头体长或略短。
Dental formula: 3.1.3.3/2.0.1.3=32. Head-body length 56-67 mm, tail length 56-67 mm hind-foot 12.5-16 mm, condyloincisive length about 17.5 mm. Morphologically similar to *S. bedfordiae*, but larger with a relatively long tail, and a more distinct dorsal strip. The dorsal pelage is cinnamon-brown, and the ventral is dark grayish brown. The tail length is equal to, or slightly short than the head-body length.

生境 / Habitat
阔叶林、针叶林、针叶阔叶混交林、杜鹃林
Broad-leaved forest, coniferous forest, coniferous and broad-leaved mixed forest and alpine rhododendron

▲ 地理分布 / Geographic Distribution

国内分布 / Domestic Distribution
陕西、甘肃、四川、云南
Shaanxi, Gansu, Sichuan, Yunnan

全球分布 / World Distribution
中国 China

生物地理界 / Biogeographic Realm
古北界 Palearctic

WWF 生物群系 / WWF Biome
温带阔叶和混交林
Temperate Broadleaf & Mixed Forests

动物地理分布型 / Zoogeographic Distribution Type
Hc

分布标注 / Distribution Note
特有种 Endemic

▲ 濒危状况 / Threatened Status

中国生物多样性红色名录等级 / CB RL Category (2021)
近危 NT

IUCN 红色名录 / IUCN Red List (2021)
无危 LC

威胁因子 / Threats
未知 Unknown

▲ 法律保护地位 / Legal Protection Status

国家重点保护野生动物等级 / Category of National Key Protected Wild Animals (2021)
未列入 Not listed

"三有"名录 / TWIESSV (2023)
未列入 Not listed

CITES 附录等级 / CITES Appendix (2023)
未列入 Not listed

迁徙物种公约附录 / CMS Appendix (2020)
未列入 Not listed

保护行动 / Conservation Action
尚无保护行动 No conservation action so far

▲ 参考文献 / References

Jiang et al. (蒋志刚等), 2021; Burgin et al., 2020; IUCN, 2020; Wilson and Mittermeier, 2018; Tu et al. (涂飞云等), 2012; Smith et al., 2009; Liu et al. (刘少英等), 2005; Pan et al. (潘清华等), 2007; Wilson and Reeder, 2005; Wang (王应祥), 2003; Zhang (张荣祖), 1997

38 / 栗齿鼩鼱

Sorex daphaenodon Thomas, 1907

• Large-toothed Siberian Shrew

劳亚食虫目 Eulipotyphla / 鼩鼱科 Soricidae / 鼩鼱属 *Sorex*

科建立者及其文献 / Family Authority
G. Fischer, 1814

属建立者及其文献 / Genus Authority
G. Fischer, 1814

亚种 / Subspecies
指名亚种 *S. d. daphaenodon* Thomas, 1907
内蒙古、吉林和黑龙江
Inner Mongolia, Jilin and Heilongjiang

模式标本产地 / Type Locality
俄罗斯
Russia, Sakhalin Isl, "Darine 40 km N. W. of Korsakoff, Saghalien"

▲ 其他名称 / Other Name(s)

其他中文名 / Other Chinese Name(s)
大齿鼩鼱

其他英文名 / Other English Name(s)
无 None

同物异名 / Synonym(s)
无 None

▲ 形态及生境 / Morphology and Habitat

形态特征 / Morphological Characteristics
齿式：3.1.3.3/2.0.1.3=32。头体长 48~76 mm。尾长 25~39 mm。后足长 10~13 mm。体型比中鼩鼱 *S. caecutiens* 大。背腹明显异色，背部深棕色，腹部浅灰棕色。尾巴约为头体长的一半。牙齿色素沉积面积较大。
Dental formula: 3.1.3.3/2.0.1.3=32. Head-body length 48-76 mm, tail length 25-39 mm, hindfoot length 10-13 mm. Size larger than Laxmann's Shrew *S. caecutiens*. Dorsal and ventral bicolor. Dorsal pelage dark brown and ventral hairs lighter grayish brown. Tail about half of head and body length. Massive reddish brown pigment on the teeth.

生境 / Habitat
冲积平原的草地、湿地
Meadow in floodplain

▲ 地理分布 / Geographic Distribution

国内分布 / Domestic Distribution
黑龙江、吉林、内蒙古
Heilongjiang, Jilin, Inner Mongolia

全球分布 / World Distribution
中国、哈萨克斯坦、蒙古国、俄罗斯
China, Kazakhstan, Mongolia, Russia

生物地理界 / Biogeographic Realm
古北界 Palearctic

WWF 生物群系 / WWF Biome
北方森林 / 针叶林
Boreal Forests / Coniferous Forests

动物地理分布型 / Zoogeographic Distribution Type
M

分布标注 / Distribution Note
非特有种 Non-Endemic

▲ 濒危状况 / Threatened Status

中国生物多样性红色名录等级 / CB RL Category (2021)
近危 NT

IUCN 红色名录 / IUCN Red List (2021)
无危 LC

威胁因子 / Threats
森林砍伐、火灾 Logging, fire

▲ 法律保护地位 / Legal Protection Status

国家重点保护野生动物等级 / Category of National Key Protected Wild Animals (2021)
未列入 Not listed

"三有"名录 / TWIESSV (2023)
未列入 Not listed

CITES 附录等级 / CITES Appendix (2023)
未列入 Not listed

迁徙物种公约附录 / CMS Appendix (2020)
未列入 Not listed

保护行动 / Conservation Action
尚无保护行动 No conservation action so far

▲ 参考文献 / References

Jiang et al. (蒋志刚等), 2021; Burgin et al., 2020; IUCN, 2020; Liu et al. (刘少英等), 2020; Wilson and Mittermeier, 2018; Smith et al., 2009; Pan et al. (潘清华等), 2007; Wilson and Reeder, 2005; Wang (王应祥), 2003; Zhang (张荣祖), 1997; Wilson and Reeder, 1993; Xia (夏武平), 1988; Wang (汪松), 1959

39 / 云南鼩鼱

Sorex excelsus G. M. Allen, 1923

· Highland Shrew

▲ 分类地位 / Taxonomy

劳亚食虫目 Eulipotyphla / 鼩鼱科 Soricidae / 鼩鼱属 *Sorex*

科建立者及其文献 / Family Authority
G. Fischer, 1814

属建立者及其文献 / Genus Authority
G. Fischer, 1814

亚种 / Subspecies
无 None

模式标本产地 / Type Locality
中国
"summit of Ho-shan (Xue Shan), Pae-tai, 48 km south of Chung-tien (Zhongdian), Yunnan, China, altitude 3962 m"

▲ 其他名称 / Other Name(s)

其他中文名 / Other Chinese Name(s)
无 None

其他英文名 / Other English Name(s)
无 None

同物异名 / Synonym(s)
无 None

▲ 形态及生境 / Morphology and Habitat

形态特征 / Morphological Characteristics
齿式：3.1.3.3/2.0.1.3=32。头体长 60~68 mm，尾长 50~61 mm。后足 13~16 mm。最大颅骨长 18~20 mm。体重 5~10 g。背部毛色棕色，腹部银灰色，背腹异色明显。尾双色。云南鼩鼱与小纹背鼩鼱有很近的亲缘关系，二者的分类关系仍有待研究。
Dental formula: 3.1.3.3/2.0.1.3=32. Head-body length 60-68 mm, tail length 50-61 mm, hindfoot 13-16 mm, condyloincisive length 18-20 mm. Weight 5-10 g. Pelage is obviously bicolored, dorsal is grayish brown, and ventral is silvery gray. Tail is bicolored. *S. excelsus* has a close relationship with *S. bedfordiae* and a taxonomic revision of both species is warranted.

生境 / Habitat
亚热带、温带湿润山地森林
Subtropical and temperate moist montane forest

▲ 地理分布 / Geographic Distribution

国内分布 / Domestic Distribution
青海、四川、云南、西藏
Qinghai, Sichuan, Yunnan, Tibet

全球分布 / World Distribution
中国 China

生物地理界 / Biogeographic Realm
古北界 Palearctic

WWF 生物群系 / WWF Biome
热带和亚热带湿润阔叶林
Tropical & Subtropical Moist Broadleaf Forests

动物地理分布型 / Zoogeographic Distribution Type
Hc

分布标注 / Distribution Note
特有种 Endemic

▲ 濒危状况 / Threatened Status

中国生物多样性红色名录等级 / CB RL Category (2021)
无危 LC

IUCN 红色名录 / IUCN Red List (2021)
无危 LC

威胁因子 / Threats
未知 Unknown

▲ 法律保护地位 / Legal Protection Status

国家重点保护野生动物等级 / Category of National Key Protected Wild Animals (2021)
未列入 Not listed

"三有"名录 / TWIESSV (2023)
未列入 Not listed

CITES 附录等级 / CITES Appendix (2023)
未列入 Not listed

迁徙物种公约附录 / CMS Appendix (2020)
未列入 Not listed

保护行动 / Conservation Action
尚无保护行动 No conservation action so far

▲ 参考文献 / References

Jiang et al. (蒋志刚等), 2021; Burgin et al., 2020; IUCN, 2020; Wilson and Mittermeier, 2018; Jiang et al. (蒋志刚等), 2015; Deng et al. (邓可等), 2013; Pan et al. (潘清华等), 2007; Wilson and Reeder, 2005; Wang (王应祥), 2003; Corbet, 1988

40 / 细鼩鼱

Sorex gracillimus Thomas, 1907

• Slender Shrew

▲ 分类地位 / Taxonomy

劳亚食虫目 Eulipotyphla / 鼩鼱科 Soricidae / 鼩鼱属 *Sorex*

科建立者及其文献 / Family Authority
G. Fischer, 1814

属建立者及其文献 / Genus Authority
G. Fischer, 1814

亚种 / Subspecies
大兴安岭亚种 *S. g. hyojiconis* (Kuroda, 1939)
内蒙古、黑龙江和吉林
Inner Mongolia, Heilongjiang and Jilin

模式标本产地 / Type Locality
俄罗斯
Russia, Sakhalin Isl, "Darine 40 km N. W. of Korsakoff, Saghalien"

摄影 / 伊阁

▲ 其他名称 / Other Name(s)

其他中文名 / Other Chinese Name(s)
瘦鼩鼱

其他英文名 / Other English Name(s)
无 None

同物异名 / Synonym(s)
无 None

▲ 形态及生境 / Morphology and Habitat

形态特征 / Morphological Characteristics
齿式：3.1.3.3/2.0.1.3=32。头体长 45~66 mm。后足 10~11 mm。尾长 36~49 mm。体重 2.5~5.3 g。体型很小的鼩鼱。头部毛色较身体更深，略泛橄榄色光泽。背腹毛色明显异色，背部通常为深棕色或浅棕色，腹部通常为灰色或银灰色。尾较粗，表面毛密而长，尾端有一簇短毛，尾长通常为头体长的 75%。

Dental formula: 3.1.3.3/2.0.1.3 =32. Head-body length 45-66 mm, hindfoot 10-11 mm, tail length 36-49 mm, usually more than 75% of the head-body length. Weight 2.5-5.3 g. A small-sized shrew. Head color is darker than body and with an olive green tint. Dorsal pelage usually light brown to dark brown and ventrum is usually gray or silvery gray. Tail is thick, covered by dense hairs. A tuft of hair on the tip of the tail.

生境 / Habitat
泰加林、阔叶林、针叶阔叶混交林、竹林、湿地
Taiga, broad-leaved forest, coniferous and broad-leaved mixed forest, bamboo forest, meadow

▲ 地理分布 / Geographic Distribution

国内分布 / Domestic Distribution
内蒙古、黑龙江、吉林
Inner Mongolia, Heilongjiang, Jilin

全球分布 / World Distribution
中国、日本、朝鲜、俄罗斯
China, Japan, Democratic People's Republic of Korea, Russia

生物地理界 / Biogeographic Realm
古北界 Palearctic

WWF 生物群系 / WWF Biome
北方森林 / 针叶林
Boreal Forests/Coniferous Forests

动物地理分布型 / Zoogeographic Distribution Type
M

分布标注 / Distribution Note
非特有种 Non-Endemic

▲ 濒危状况 / Threatened Status

中国生物多样性红色名录等级 / CB RL Category (2021)
无危 LC

IUCN 红色名录 / IUCN Red List (2021)
无危 LC

威胁因子 / Threats
未知 Unknown

▲ 法律保护地位 / Legal Protection Status

国家重点保护野生动物等级 / Category of National Key Protected Wild Animals (2021)
未列入 Not listed

"三有" 名录 / TWIESSV (2023)
未列入 Not listed

CITES 附录等级 / CITES Appendix (2023)
未列入 Not listed

迁徙物种公约附录 / CMS Appendix (2020)
未列入 Not listed

保护行动 / Conservation Action
尚无保护行动 No conservation action so far

▲ 参考文献 / References

Jiang et al. (蒋志刚等), 2021; Burgin et al., 2020; IUCN, 2020; Wilson and Mittermeier, 2018; Liu et al. (刘铸等), 2018; Smith et al., 2009; Pan et al. (潘清华等), 2007; Wang (王应祥), 2003; Hutterer, 2005; Hoffmann, 1987

41 / 远东鼩鼱

Sorex isodon Turov, 1924

· Ever-toothed Shrew

▲ 分类地位 / Taxonomy

劳亚食虫目 Eulipotyphla / 鼩鼱科 Soricidae / 鼩鼱属 *Sorex*

科建立者及其文献 / Family Authority
G. Fischer, 1814

属建立者及其文献 / Genus Authority
G. Fischer, 1814

亚种 / Subspecies
黑龙江亚种 *S. i. gravesi* Goodwin, 1933
黑龙江
Heilongjiang

模式标本产地 / Type Locality
俄罗斯
Russia, Siberia, NE of Lake Baikal, Barguzinsk taiga, River Sosovka

▲ 其他名称 / Other Name(s)

其他中文名 / Other Chinese Name(s)
无 None

其他英文名 / Other English Name(s)
Taiga Shrew

同物异名 / Synonym(s)
无 None

▲ 形态及生境 / Morphology and Habitat

形态特征 / Morphological Characteristics
齿式：3.1.3.3/2.0.1.3=32。头长 54~86 mm。尾长 37~55 mm。后足长 13~15 mm。体重 6.1~16.3 g。背腹近乎同色，成体通常深棕色。
Dental formula: 3.1.3.3/2.0.1.3=32. Head-body length 54-86 mm, tail length 37-55 mm, hindfoot length 13-15 mm, weight 6.1-16.3 g. Pelage is almost unicolored, and blackish-brown in color with no difference in color between back and belly.

生境 / Habitat
温带森林 Temperate forest

▲ 地理分布 / Geographic Distribution

国内分布 / Domestic Distribution
内蒙古、黑龙江、吉林、辽宁
Inner Mongolia, Heilongjiang, Jilin, Liaoning

全球分布 / World Distribution
中国、白俄罗斯、芬兰、哈萨克斯坦、朝鲜、蒙古国、挪威、俄罗斯、瑞典
China, Belarus, Finland, Kazakhstan, Democratic People's Republic of Korea, Mongolia, Norway, Russia, Sweden

生物地理界 / Biogeographic Realm
古北界 Palearctic

WWF 生物群系 / WWF Biome
北方森林 / 针叶林
Boreal Forests/Coniferous Forests

动物地理分布型 / Zoogeographic Distribution Type
M

分布标注 / Distribution Note
非特有种 Non-Endemic

▲ 濒危状况 / Threatened Status

中国生物多样性红色名录等级 / CB RL Category (2021)
近危 NT

IUCN 红色名录 / IUCN Red List (2021)
无危 LC

威胁因子 / Threats
农业、林业污染 Agricultural or forestry effluents

▲ 法律保护地位 / Legal Protection Status

国家重点保护野生动物等级 / Category of National Key Protected Wild Animals (2021)
未列入 Not listed

"三有"名录 / TWIESSV (2023)
未列入 Not listed

CITES 附录等级 / CITES Appendix (2023)
未列入 Not listed

迁徙物种公约附录 / CMS Appendix (2020)
未列入 Not listed

保护行动 / Conservation Action
尚无保护行动 No conservation action so far

▲ 参考文献 / References

Jiang et al. (蒋志刚等), 2021; Burgin et al., 2020; IUCN, 2020; Wilson and Mittermeier, 2018; Smith et al., 2009; Pan et al. (潘清华等), 2007; Zhang and Yu, 2006; Wilson and Reeder, 2005; Wang (王应祥), 2003; Wilson and Reeder, 1993; Corbet, 1978

42 / 姬鼩鼱

Sorex minutissimus
Zimmermann, 1780

• Eurasian Least Shrew

▲ 分类地位 / Taxonomy

劳亚食虫目 Eulipotyphla / 鼩鼱科 Soricidae / 鼩鼱属 *Sorex*

科建立者及其文献 / Family Authority
G. Fischer, 1814

属建立者及其文献 / Genus Authority
G. Fischer, 1814

亚种 / Subspecies
东北亚种 *S. m. tscherskii* (Kuroda, 1939)
黑龙江
Heilongjiang

模式标本产地 / Type Locality
俄罗斯
"Yenisei"; given by Stroganov (1957:176) as "iz raiona sela Kiiskow chto na r. Kie (nyne g. Mariinsk Kemerovskoi oblasti)" [Russia, Kemerovsk. Obl., Mariinsk (Kiiskoe), bank of Kiia River (near Yenesei River)]; Restricted by Pavlinov and Rossolimo

▲ 其他名称 / Other Name(s)

其他中文名 / Other Chinese Name(s)
无 None

其他英文名 / Other English Name(s)
Least Siberian Shrew, Lesser Pygmy Shrew, Miniscule Shrew

同物异名 / Synonym(s)
无 None

▲ 形态及生境 / Morphology and Habitat

形态特征 / Morphological Characteristics

齿式：3.1.3.3/2.0.1.3=32。头长 39~55 mm。尾长 20~35 mm。后足长 7~11 mm。颅全长 12.4~14.2 mm。体重 1.4~4 g。是鼩鼱科最小的物种之一。背腹异色，背部棕灰色，腹部浅灰色。尾长通常不超过头体长的 50%，且表面覆盖有长而稀疏的针毛。吻部较短。

Dental formula: 3.1.3.3/2.0.1.3=32. Head length 39-55 mm, tail length 20-35 mm, hind foot length 7-11 mm, the greatest length of skull 12.4-14.2 mm, weight 1.4-4 g. It is one of the smallest shrew species. Pelage bicolored, dorsal pelage grayish brown, and ventrum is light gray. Tail is no longer than 50% of the head-body length and is covered with rare long hairs. Rostrum is very short blunt.

生境 / Habitat
森林、灌丛
Forest, shrubland

▲ 地理分布 / Geographic Distribution

国内分布 / Domestic Distribution
黑龙江、内蒙古、辽宁
Heilongjiang, Inner Mongolia, Liaoning

全球分布 / World Distribution
中国、爱沙尼亚、芬兰、日本、蒙古国、挪威、俄罗斯、瑞典
China, Estonia, Finland, Japan, Mongolia, Norway, Russia, Sweden

生物地理界 / Biogeographic Realm
古北界 Palearctic

WWF 生物群系 / WWF Biome
北方森林 / 针叶林
Boreal Forests/Coniferous Forests

动物地理分布型 / Zoogeographic Distribution Type
Dn

分布标注 / Distribution Note
非特有种 Non-Endemic

▲ 濒危状况 / Threatened Status

中国生物多样性红色名录等级 / CB RL Category (2021)
近危 NT

IUCN 红色名录 / IUCN Red List (2021)
无危 LC

威胁因子 / Threats
未知 Unknown

▲ 法律保护地位 / Legal Protection Status

国家重点保护野生动物等级 / Category of National Key Protected Wild Animals (2021)
未列入 Not listed

"三有"名录 / TWIESSV (2023)
未列入 Not listed

CITES 附录等级 / CITES Appendix (2023)
未列入 Not listed

迁徙物种公约附录 / CMS Appendix (2020)
未列入 Not listed

保护行动 / Conservation Action
尚无保护行动 No conservation action so far

▲ 参考文献 / References

Jiang et al. (蒋志刚等), 2021; Burgin et al., 2020; IUCN, 2020; Wilson and Mittermeier, 2018; Smith et al., 2009; Pan et al. (潘清华等), 2007; Wilson and Reeder, 2005; Wang (王应祥), 2003; Wilson and Reeder, 1993

43 / 小鼩鼱

Sorex minutus Linnaeus, 1766

· Eurasian Pygmy Shrew

▲ 分类地位 / Taxonomy

劳亚食虫目 Eulipotyphla / 鼩鼱科 Soricidae / 鼩鼱属 *Sorex*

科建立者及其文献 / Family Authority
G. Fischer, 1814

属建立者及其文献 / Genus Authority
G. Fischer, 1814

亚种 / Subspecies
指名亚种 *S. m. minutus* (Stroganov, 1957)
新疆（天山）
Xinjiang (Mt. Tianshan)

模式标本产地 / Type Locality
俄罗斯
"Yenisei"; restricted by Pavlinov and Rossolimo (1987:15) to "Krasnoyarskii kr., Krasnoyarsk." According to Ellerman and Morrison-Scott (1951:47), the type locality is Barnaul, Rus

▲ 其他名称 / Other Name(s)

其他中文名 / Other Chinese Name(s)
无 None

其他英文名 / Other English Name(s)
无 None

同物异名 / Synonym(s)
无 None

▲ 形态及生境 / Morphology and Habitat

形态特征 / Morphological Characteristics
齿式：3.1.3.3/2.0.1.3=32。头体长 39~ 60 mm。尾长 32~46 mm。后足长度 10 ~11 mm。体重 2.4~6.1 g。背腹异色，背部棕色或深棕色，腹部灰色。尾长通常为头体长的 70%~75%，近臀部明显变细。尾部覆盖有长而致密的毛发，并在尾尖形成一簇短毛。
Dental formula: 3.1.3.3/2.0.1.3=32. The head-body length 39-60 mm, tail length 32-46 mm, hindfoot length 10-11 mm, weight 2.4-6.1 g. Pelage is bicolored, dorsal brown to dark brown and ventrum is gray. Tail is usually 70%-75% of head-body length and is notably narrowed at base. Tail is covered with long hairs that lie tightly against skin and form a distinct tuft at tip.

生境 / Habitat
沼泽、草原、海地、沙丘、林地边缘、岩石区、灌木林和山地森林
Swamp, grassland, heath, sand dune, woodland edge, rocky area, shrubland and montane forest

▲ 地理分布 / Geographic Distribution

国内分布 / Domestic Distribution

新疆 Xinjiang

全球分布 / World Distribution

阿尔巴尼亚、安道尔、亚美尼亚、奥地利、阿塞拜疆、白俄罗斯、比利时、波斯尼亚和黑塞哥维那、保加利亚、中国、克罗地亚、捷克、丹麦、爱沙尼亚、芬兰、法国、格鲁吉亚、德国、希腊、匈牙利、印度、爱尔兰、意大利、拉脱维亚、列支敦士登、立陶宛、卢森堡、马其顿、摩尔多瓦、黑山、荷兰、挪威、巴基斯坦、波兰、葡萄牙、罗马尼亚、俄罗斯、塞尔维亚、斯洛伐克、斯洛文尼亚、西班牙、瑞典、瑞士、土耳其、乌克兰、英国 Albania, Andorra, Armenia, Austria, Azerbaijan, Belarus, Belgium, Bosnia and Herzegovina, Bulgaria, China, Croatia, Czech, Danmark, Estonia, Finland, France, Georgia, Germany, Greece, Hungary, India, Ireland, Italy, Latvia, Liechtenstein, Lithuania, Luxembourg, Macedonia, Moldova, Montenegro, Netherlands, Norway, Pakistan, Poland, Portugal, Romania, Russia, Serbia, Slovakia, Slovenia, Spain, Sweden, Switzerland, Turkey, Ukraine, United Kingdom

生物地理界 / Biogeographic Realm

古北界 Palearctic

WWF 生物群系 / WWF Biome

北方森林 / 针叶林 Boreal Forests/Coniferous Forests

动物地理分布型 / Zoogeographic Distribution Type

Ub

分布标注 / Distribution Note

非特有种 Non-Endemic

▲ 濒危状况 / Threatened Status

中国生物多样性红色名录等级 / CB RL Category (2021)

近危 NT

IUCN 红色名录 / IUCN Red List (2021)

无危 LC

威胁因子 / Threats

农业、林业污染 Agricultural and forestry effluents

▲ 法律保护地位 / Legal Protection Status

国家重点保护野生动物等级 / Category of National Key Protected Wild Animals (2021)

未列入 Not listed

"三有"名录 / TWIESSV (2023)

未列入 Not listed

CITES 附录等级 / CITES Appendix (2023)

未列入 Not listed

迁徙物种公约附录 / CMS Appendix (2020)

未列入 Not listed

保护行动 / Conservation Action

尚无保护行动 No conservation action so far

▲ 参考文献 / References

Jiang et al. (蒋志刚等), 2021; Burgin et al., 2020; IUCN, 2020; Wilson and Mittermeier, 2018; Smith et al., 2009; Pan et al. (潘清华等), 2007; Wilson and Reeder, 2005; Wang (王应祥), 2003; Zhang (张荣祖), 1997; Hoffmann, 1987

44 / 大鼩鼱

Sorex mirabilis Ognev, 1937

· Ussuri Shrew

▲ 分类地位 / Taxonomy

劳亚食虫目 Eulipotyphla / 鼩鼱科 Soricidae / 鼩鼱属 *Sorex*

科建立者及其文献 / Family Authority
G. Fischer, 1814

属建立者及其文献 / Genus Authority
G. Fischer, 1814

亚种 / Subspecies
指名亚种 *S. m. mirabilis* Ognev, 1937
黑龙江
Heilongjiang

模式标本产地 / Type Locality
俄罗斯
Russia, Primorskii Krai, Ussuriiskii r-n., Kamenka River (specified by Pavlinov and Rossolimo (1987)

王旭明 / 供图

▲ 其他名称 / Other Name(s)

其他中文名 / Other Chinese Name(s)
无 None

其他英文名 / Other English Name(s)
Giant Shrew

同物异名 / Synonym(s)
无 None

▲ 形态及生境 / Morphology and Habitat

形态特征 / Morphological Characteristics
齿式：3.1.3.3/2.0.1.3=32。头体长 74~97 mm。尾长 63~74 mm。后足 16~18 mm。体重 11~14.2 g。古北界鼩鼱亚科中最大的一种。身体背腹同色，成体烟灰色。尾长为头体长 85% 左右，尾双色。
Dental formula: 3.1.3.3/2.0.1.3=32. Head-body length 74-97 mm, tail length 63-74 mm, hindfoot 16-18 mm, weight 11-14.2 g. The largest red-toothed shrew in Palearctic region. Pelage is unicolor, smoky gray in adults.

生境 / Habitat
针叶阔叶混交林、温带森林地带的岩石山坡
Rocky slopes in coniferous and broad-leaved mixed Forest, temperate forest

▲ 地理分布 / Geographic Distribution

国内分布 / Domestic Distribution
黑龙江、辽宁
Heilongjiang, Liaoning

全球分布 / World Distribution
中国、朝鲜、俄罗斯
China, Democratic Republic of Korea, Russia

生物地理界 / Biogeographic Realm
古北界 Palearctic

WWF 生物群系 / WWF Biome
温带阔叶和混交林
Temperate Broadleaf & Mixed Forests

动物地理分布型 / Zoogeographic Distribution Type
Ke

分布标注 / Distribution Note
非特有种 Non-Endemic

▲ 濒危状况 / Threatened Status

中国生物多样性红色名录等级 / CB RL Category (2021)
易危 VU

IUCN 红色名录 / IUCN Red List (2021)
数据缺乏 DD

威胁因子 / Threats
未知 Unknown

▲ 法律保护地位 / Legal Protection Status

国家重点保护野生动物等级 / Category of National Key Protected Wild Animals (2021)
未列入 Not listed

"三有"名录 / TWIESSV (2023)
未列入 Not listed

CITES 附录等级 / CITES Appendix (2023)
未列入 Not listed

迁徙物种公约附录 / CMS Appendix (2020)
未列入 Not listed

保护行动 / Conservation Action
尚无保护行动 No conservation action so far

▲ 参考文献 / References

Jiang et al. (蒋志刚等), 2021; Burgin et al., 2020; IUCN, 2020; Liu et al. (刘少英等), 2020; Wilson and Mittermeier, 2018; Liu et al. (刘铸等), 2018; Smith et al., 2009; Pan et al. (潘清华等), 2007; Wang (王应祥), 2003; Zhang (张荣祖), 1997; Wilson and Reeder, 1993; Institute of Zoology/ Chinese Academy of Sciences, 1958

45 / 克什米尔鼩鼱

Sorex planiceps Miller, 191

· Kashmir Shrew

▲ 分类地位 / Taxonomy

劳亚食虫目 Eulipotyphla / 鼩鼱科 Soricidae / 鼩鼱属 *Sorex*

科建立者及其文献 / Family Authority
G. Fischer, 1814

属建立者及其文献 / Genus Authority
G. Fischer, 1814

亚种 / Subspecies
无 None

模式标本产地 / Type Locality
克什米尔地区
"Dachin, Khistwar, Kashmir (altitude, 2743 m)"

▲ 其他名称 / Other Name(s)

其他中文名 / Other Chinese Name(s)
无 None

其他英文名 / Other English Name(s)
Kashmir Pygmy Shrew

同物异名 / Synonym(s)
无 None

▲ 形态及生境 / Morphology and Habitat

形态特征 / Morphological Characteristics

齿式：3.1.3.3/2.0.1.3=32。头体长 57~74 mm。尾长 40~47 mm。后足长 11~13 mm。颅全长 16.7~17.1 mm。对该物种的研究很不充分。形态与小鼩鼱及藏鼩鼱相似，但体型更大。背侧毛皮为棕色，向腹部逐渐过渡为浅灰色。四足颜色较浅，尾双色。头骨较藏鼩鼱更为扁平，第二上单尖齿比第一、第三上单尖齿更小。

Dental formula: 3.1.3.3/2.0.1.3=32. Head-body 57-74 mm, tail 40-47 mm, hind-foot 11-13 mm, condyloincisive length 16.7-17.1 mm. Poorly studied species. Similar to *S. minutus* and *S. thibetanus* but larger. Dorsal pelage is brown, shading gradually into grayish on the ventrum. Feet are light and tail is bicolored, brown on top and light gray to white below. Skull is flatter than S. thibetanus. The second upper unicuspid is smaller than the first and the third.

生境 / Habitat
内陆岩石区域
Inland rocky area

▲ 地理分布 / Geographic Distribution

国内分布 / Domestic Distribution
西藏 Tibet

全球分布 / World Distribution
中国、印度、巴基斯坦
China, India, Pakistan

生物地理界 / Biogeographic Realm
古北界 Palearctic

WWF 生物群系 / WWF Biome
温带针叶林、温带草原、稀树草原和灌丛
Temperate Coniferous Forests, Temperate Grasslands, Savannas and Shrublands

动物地理分布型 / Zoogeographic Distribution Type
Pe

分布标注 / Distribution Note
非特有种 Non-Endemic

▲ 濒危状况 / Threatened Status

中国生物多样性红色名录等级 / CB RL Category (2021)
近危 NT

IUCN 红色名录 / IUCN Red List (2021)
无危 LC

威胁因子 / Threats
未知 Unknown

▲ 法律保护地位 / Legal Protection Status

国家重点保护野生动物等级 / Category of National Key Protected Wild Animals (2021)
未列入 Not listed

"三有"名录 / TWIESSV (2023)
未列入 Not listed

CITES 附录等级 / CITES Appendix (2023)
未列入 Not listed

迁徙物种公约附录 / CMS Appendix (2020)
未列入 Not listed

保护行动 / Conservation Action
尚无保护行动 No conservation action so far

▲ 参考文献 / References

Jiang et al. (蒋志刚等), 2021; Burgin et al., 2020; IUCN, 2020; Wilson and Mittermeier, 2018; Smith et al., 2009; Pan et al. (潘清华等), 2007; Hoffmann, 1987

46 / 扁颅鼩鼱

Sorex roboratus Hollister, 1913

· Flat-skulled Shrew

▲ 分类地位 / Taxonomy

劳亚食虫目 Eulipotyphla / 鼩鼱科 Soricidae / 鼩鼱属 *Sorex*

科建立者及其文献 / Family Authority
G. Fischer, 1814

属建立者及其文献 / Genus Authority
G. Fischer, 1814

亚种 / Subspecies
无 None

模式标本产地 / Type Locality
俄罗斯
Russia, Gorno-Altaisk A. O., "8 km S Dapuchu (Altai Mtns, Tapucha)"

▲ 其他名称 / Other Name(s)

其他中文名 / Other Chinese Name(s)
无 None

其他英文名 / Other English Name(s)
无 None

同物异名 / Synonym(s)
无 None

▲ 形态及生境 / Morphology and Habitat

形态特征 / Morphological Characteristics

齿式：3.1.3.3/2.0.1.3=32。头体长 58~86 mm。尾长 30~48 mm。后足长 12~15.5 mm。体重 7~15.5 g。背腹异色，背部浅棕至深棕色，腹部浅灰色。尾长通常为头体长的 40%~60%。头部长而尖，背侧毛皮深棕色，侧面较浅，腹侧毛皮浅灰色。尾巴毛色双色分明，尾背面深棕色，腹面颜色稍浅。

Dental formula: 3.1.3.3/2.0.1.3=32. Head-body length 58-86 mm, tail length 30-48 mm, hindfoot length 12-15.5 mm, weight 7-15.5 g. Dorsal pelage deep dark brown, sides lighter, ventral pelage buffy gray. Tail sharply bicolored, dark brown above, paler below.

生境 / Habitat

针叶阔叶混交林、温带森林地带的岩石山坡
Rocky slopes in coniferous and broad-leaved mixed forest

▲ 地理分布 / Geographic Distribution

国内分布 / Domestic Distribution
新疆、黑龙江
Xinjiang, Heilongjiang

全球分布 / World Distribution
中国、蒙古国、俄罗斯
China, Mongolia, Russia

生物地理界 / Biogeographic Realm
古北界 Palearctic

WWF 生物群系 / WWF Biome
北方森林 / 针叶林
Boreal Forests / Coniferous Forests

动物地理分布型 / Zoogeographic Distribution Type
Uc

分布标注 / Distribution Note
非特有种 Non-Endemic

▲ 濒危状况 / Threatened Status

中国生物多样性红色名录等级 / CB RL Category (2021)
易危 VU

IUCN 红色名录 / IUCN Red List (2021)
无危 LC

威胁因子 / Threats
森林砍伐、火灾、干旱 Logging, fire, drought

▲ 法律保护地位 / Legal Protection Status

国家重点保护野生动物等级 / Category of National Key Protected Wild Animals (2021)
未列入 Not listed

"三有" 名录 / TWIESSV (2023)
未列入 Not listed

CITES 附录等级 / CITES Appendix (2023)
未列入 Not listed

迁徙物种公约附录 / CMS Appendix (2020)
未列入 Not listed

保护行动 / Conservation Action
尚无保护行动 No conservation action so far

▲ 参考文献 / References

Jiang et al. (蒋志刚等), 2021; Burgin et al., 2020; IUCN, 2020; Liu et al. (刘少英等), 2020; Wilson and Mittermeier, 2018; Liu et al. (刘铸等), 2018; Smith et al., 2009

47 / 陕西鼩鼱

Sorex sinalis Thomas, 1912

· Chinese Shrew

▲ 分类地位 / Taxonomy

劳亚食虫目 Eulipotyphla / 鼩鼱科 Soricidae / 鼩鼱属 *Sorex*

科建立者及其文献 / Family Authority
G. Fischer, 1814

属建立者及其文献 / Genus Authority
G. Fischer, 1814

亚种 / Subspecies
无 None

模式标本产地 / Type Locality
中国
China, Shaanxi, "72 km S. E. of Feng-siang-fu (Feng Xian), Shen-si, 3200 m"

▲ 其他名称 / Other Name(s)

其他中文名 / Other Chinese Name(s)
无 None

其他英文名 / Other English Name(s)
Dusky Shrew

同物异名 / Synonym(s)
无 None

▲ 形态及生境 / Morphology and Habitat

形态特征 / Morphological Characteristics
齿式：3.1.3.3/2.0.1.3=32。头体长 64~85 mm。尾长 49~68 mm。后足长 13~17 mm。颅全长 20~22 mm。体型较大的鼩鼱。毛色总体呈棕灰色，背腹异色，背部偏栗色，腹部深棕色，较背部颜色更深。尾较长。头骨吻部较长，脑颅较窄。牙齿色素沉积与同属物种相比颜色较淡。
Dental formula: 3.1.3.3/2.0.1.3=32. Head-body length 64-85 mm, tail length 49-68 mm, hindfoot length 13-17 mm, the greatest length of skull 20-22 mm. A large shrew with a long tail. General color greyish brown, dorsal pelage hazel, ventral color drab-brown and darker than the dorsal. Rostrum is long and braincase is narrow. Pigmentation of teeth is light compared with congeneric species.

生境 / Habitat
针叶阔叶混交林、温带森林地带的岩石山坡
Rocky slopes in coniferous and broad-leaved mixed forest

▲ 地理分布 / Geographic Distribution

国内分布 / Domestic Distribution
陕西、四川、甘肃
Shaanxi, Sichuan, Gansu

全球分布 / World Distribution
中国 China

生物地理界 / Biogeographic Realm
古北界 Palearctic

WWF 生物群系 / WWF Biome
温带阔叶和混交林
Temperate Broadleaf & Mixed Forests

动物地理分布型 / Zoogeographic Distribution Type
Le

分布标注 / Distribution Note
特有种 Endemic

▲ 濒危状况 / Threatened Status

中国生物多样性红色名录等级 / CB RL Category (2021)
近危 NT

IUCN 红色名录 / IUCN Red List (2021)
数据缺乏 DD

威胁因子 / Threats
未知 Unknown

▲ 法律保护地位 / Legal Protection Status

国家重点保护野生动物等级 / Category of National Key Protected Wild Animals (2021)
未列入 Not listed

"三有"名录 / TWIESSV (2023)
未列入 Not listed

CITES 附录等级 / CITES Appendix (2023)
未列入 Not listed

迁徙物种公约附录 / CMS Appendix (2020)
未列入 Not listed

保护行动 / Conservation Action
尚无保护行动 No conservation action so far

▲ 参考文献 / References

Jiang et al. (蒋志刚等), 2021; Burgin et al., 2020; IUCN, 2020; Wilson and Mittermeier, 2018; Jiang et al. (姜雪松等), 2013; Liu et al. (刘洋等), 2013; Smith et al., 2009; Pan et al. (潘清华等), 2007; Qin and Meng (秦岭和孟祥明), 2007; Wilson and Reeder, 2005; Wang (王应祥), 2003

48 / 藏鼩鼱

Sorex thibetanus Kastschenko, 1905

· Tibetan Shrew

▲ 分类地位 / Taxonomy

劳亚食虫目 Eulipotyphla / 鼩鼱科 Soricidae / 鼩鼱属 *Sorex*

科建立者及其文献 / Family Authority
G. Fischer, 1814

属建立者及其文献 / Genus Authority
G. Fischer, 1814

亚种 / Subspecies
指名亚种 *S. th. thibetanus* Kastscheko, 1905
青海、甘肃
Qinghai and Gansu

玉树亚种 *S. th. kozlovi* Stroganov, 1952
青海、西藏和四川
Qinghai, Tibet and Sichuan

模式标本产地 / Type Locality
中国
"Tsaidam" (Haixi, Qinghai, China)

何锴 / 供图

▲ 其他名称 / Other Name(s)

其他中文名 / Other Chinese Name(s)
无 None

其他英文名 / Other English Name(s)
无 None

同物异名 / Synonym(s)
无 None

▲ 形态及生境 / Morphology and Habitat

形态特征 / Morphological Characteristics

齿式: 3.1.3.3/2.0.1.3=32。头体长 51~64 mm。后足长 12~13 mm。尾长 32~54 mm。与小鼩鼱外形相似但体型更大。背部毛色为暗沉棕灰色，腹部毛色浅灰色。尾部密布短毛，尾尖有一簇短毛。几乎所有牙齿的齿尖都有红褐色色素沉积。

Dental formula: 3.1.3.3/2.0.1.3=32. Head-body length 51-64 mm, hindfoot length 12-13 mm, tail length 32-54 mm. It is similar to but larger than *Sorex minutus*. Dorsal pelage grayish brown, and ventral is light gray. Tail is covered by dense hairs and with a tuft of hair at the tip. Almost all teeth have reddish-brown pigmentation deposits on the tips.

生境 / Habitat
草地、灌丛、沼泽
Grassland, shrubland, swamp

▲ 地理分布 / Geographic Distribution

国内分布 / Domestic Distribution
四川、甘肃、青海、西藏
Sichuan, Gansu, Qinghai, Tibet

全球分布 / World Distribution
中国 China

生物地理界 / Biogeographic Realm
古北界 Palearctic

WWF 生物群系 / WWF Biome
温带针叶林
Temperate Coniferous Forests

动物地理分布型 / Zoogeographic Distribution Type
Ic

分布标注 / Distribution Note
特有种 Endemic

▲ 濒危状况 / Threatened Status

中国生物多样性红色名录等级 / CB RL Category (2021)
近危 NT

IUCN 红色名录 / IUCN Red List (2021)
数据缺乏 DD

威胁因子 / Threats
未知 Unknown

▲ 法律保护地位 / Legal Protection Status

国家重点保护野生动物等级 / Category of National Key Protected Wild Animals (2021)
未列入 Not listed

"三有"名录 / TWIESSV (2023)
未列入 Not listed

CITES 附录等级 / CITES Appendix (2023)
未列入 Not listed

迁徙物种公约附录 / CMS Appendix (2020)
未列入 Not listed

保护行动 / Conservation Action
尚无保护行动 No conservation action so far

▲ 参考文献 / References

Jiang et al. (蒋志刚等), 2021; Burgin et al., 2020; IUCN, 2020; Wilson and Mittermeier, 2018; Liao et al. (廖锐等), 2015; Xie et al. (谢文华等), 2014; Smith et al., 2009; Peng et al. (彭基泰等), 2007; Pan et al. (潘清华等), 2007; Wang (王应祥), 2003

49 / 苔原鼩鼱

Sorex tundrensis Merriam, 1900

· Tundra Shrew

▲ 分类地位 / Taxonomy

劳亚食虫目 Eulipotyphla / 鼩鼱科 Soricidae / 鼩鼱属 *Sorex*

科建立者及其文献 / Family Authority
G. Fischer, 1814

属建立者及其文献 / Genus Authority
G. Fischer, 1814

亚种 / Subspecies
无 None

模式标本产地 / Type Locality
美国
USA, Alaska, St. Michaels

▲ 其他名称 / Other Name(s)

其他中文名 / Other Chinese Name(s)
无 None

其他英文名 / Other English Name(s)
Holarctic Shrew

同物异名 / Synonym(s)
无 None

▲ 形态及生境 / Morphology and Habitat

形态特征 / Morphological Characteristics
齿式：3.1.3.3/2.0.1.3=32。头体长 60~85 mm。尾长 22~36 mm。体重 5~10 g。夏天毛皮是三色的，背毛深棕色，体侧毛浅灰褐色，腹部毛浅灰色。冬季毛皮双色，背部毛深棕色，腹部毛浅灰色。

Dental formula: 3.1.3.3/2.0.1.3=32. Head-body length 60-85 mm, tail length 22-36 mm, weight 5-10 g. Summer pelage is tricolored, dark brown dorsally, pale grayish-brown on the sides, and pale gray on the belly. Winter pelage bicolored, dark brown above, pale gray ventral part.

生境 / Habitat
灌木丛、草丛、沼泽附近的干脊
Shrubs, grassy vegetation or dry ridge near marshe

▲ 地理分布 / Geographic Distribution

国内分布 / Domestic Distribution
黑龙江、内蒙古、山西
Heilongjiang, Inner Mongolia, Shanxi

全球分布 / World Distribution
加拿大、中国、蒙古国、俄罗斯、美国
Canada, China, Mongolia, Russia, United States

生物地理界 / Biogeographic Realm
古北界、新北界
Palearctic, Nearctic

WWF 生物群系 / WWF Biome
北方森林 / 针叶林
Boreal Forests/Coniferous Forests

动物地理分布型 / Zoogeographic Distribution Type
Ua

分布标注 / Distribution Note
非特有种 Non-Endemic

▲ 濒危状况 / Threatened Status

中国生物多样性红色名录等级 / CB RL Category (2021)
近危 NT

IUCN 红色名录 / IUCN Red List (2021)
无危 LC

威胁因子 / Threats
未知 Unknown

▲ 法律保护地位 / Legal Protection Status

国家重点保护野生动物等级 / Category of National Key Protected Wild Animals (2021)
未列入 Not listed

"三有" 名录 / TWIESSV (2023)
未列入 Not listed

CITES 附录等级 / CITES Appendix (2023)
未列入 Not listed

迁徙物种公约附录 / CMS Appendix (2020)
未列入 Not listed

保护行动 / Conservation Action
尚无保护行动 No conservation action so far

▲ 参考文献 / References

Jiang et al. (蒋志刚等), 2021; Burgin et al., 2020; IUCN, 2020; Liu et al. (刘少英等), 2020; Wilson and Mittermeier, 2018; Liu et al. (刘铸等), 2018; Liu et al. (刘洋等), 2010; Smith et al., 2009; Pan et al. (潘清华等), 2007; Wilson and Reeder, 2005

50 / 长爪鼩鼱

Sorex unguiculatus Dobson, 1890

· Long-clawed Shrew

劳亚食虫目 Eulipotyphla / 鼩鼱科 Soricidae / 鼩鼱属 *Sorex*

科建立者及其文献 / Family Authority
G. Fischer, 1814

属建立者及其文献 / Genus Authority
G. Fischer, 1814

亚种 / Subspecies
无 None

模式标本产地 / Type Locality
俄罗斯
Russia, "Saghalien (Sakhalin) Island; Nikolajewsk, at the mouth of the Amur River."
Ognev (1928:204) and Ellerman and Morrison-Scott (1951:52) both restricted the
type locality to Sakhalin Isl

▲ 其他名称 / Other Name(s)

其他中文名 / Other Chinese Name(s)
无 None

其他英文名 / Other English Name(s)
无 None

同物异名 / Synonym(s)
无 None

▲ 形态及生境 / Morphology and Habitat

形态特征 / Morphological Characteristics
齿式：3.1.3.3/2.0.1.3=32。头体长 70~91 mm。尾长 41~51 mm。后足
长 12~14 mm。颅全长 19~21 mm。体重 6.9~15 g。体型较大的鼩鼱。
背侧和腹侧毛皮颜色深棕色，腹部毛色灰色或棕灰色。前掌宽大，爪
长 3~5 mm，适应于挖掘。前爪明显长于后爪。
Dental formula: 3.1.3.3/2.0.1.3=32. Head-body length 70-91 mm, tail length 41-51
mm, hind foot length 12-14 mm, the greatest length of skull 19-21 mm, weight 6.9-
15 g. A large-sized shrew. Dorsal pelage dark brown, and gradually changes to brown
on the sides. Ventral pelage gray or grayish brown. Forefeet broad, with claws 3-5 mm
long, much longer than claws on hind feet, adapted to digging.

生境 / Habitat
草地、泰加林、针叶阔叶混交林
Grassland, taiga, coniferous and broad-leaved mixed Forest

▲ 地理分布 / Geographic Distribution

国内分布 / Domestic Distribution
黑龙江、内蒙古、吉林
Heilongjiang, Inner Mongolia, Jilin

全球分布 / World Distribution
中国、日本、朝鲜、俄罗斯
China, Japan, Democratic People's Republic of Korea, Russia

生物地理界 / Biogeographic Realm
古北界 Palearctic

WWF 生物群系 / WWF Biome
北方森林 / 针叶林
Boreal Forests/Coniferous Forests

动物地理分布型 / Zoogeographic Distribution Type
Mg

分布标注 / Distribution Note
非特有种 Non-Endemic

▲ 濒危状况 / Threatened Status

中国生物多样性红色名录等级 / CB RL Category (2021)
近危 NT

IUCN 红色名录 / IUCN Red List (2021)
无危 LC

威胁因子 / Threats
未知 Unknown

▲ 法律保护地位 / Legal Protection Status

国家重点保护野生动物等级 / Category of National Key Protected Wild Animals (2021)
未列入 Not listed

"三有"名录 / TWIESSV (2023)
未列入 Not listed

CITES 附录等级 / CITES Appendix (2023)
未列入 Not listed

迁徙物种公约附录 / CMS Appendix (2020)
未列入 Not listed

保护行动 / Conservation Action
尚无保护行动 No conservation action so far

▲ 参考文献 / References

Jiang et al. (蒋志刚等), 2021; Burgin et al., 2020; IUCN, 2020; Liu et al. (刘少英等), 2020; Wilson and Mittermeier, 2018; Liu et al. (刘铸等), 2018; Liu et al. (刘洋等), 2013; Smith et al., 2009; Pan et al. (潘清华等), 2007; Hutterer, 2005; Wang (王应祥), 2003; Zhang (张荣祖), 1997

51 / 川鼩

Blarinella quadraticauda
(Milne-Edwards, 1872)

• Asiatic Short-tailed Shrew

▲ 分类地位 / Taxonomy

劳亚食虫目 Eulipotyphla / 鼩鼱科 Soricidae / 黑齿鼩鼱属 *Blarinella*

科建立者及其文献 / Family Authority
G. Fischer, 1814

属建立者及其文献 / Genus Authority
Thomas, 1911

亚种 / Subspecies
无共识 No consensus

模式标本产地 / Type Locality
中国
China, Sichuan

何锴 / 供图

▲ 其他名称 / Other Name(s)

其他中文名 / Other Chinese Name(s)
黑齿鼩鼱、四川短尾鼩鼱

其他英文名 / Other English Name(s)
Chinese Short-tailed Shrew, Northern
Short-tailed Shrew

同物异名 / Synonym(s)
无 None

▲ 形态及生境 / Morphology and Habitat

形态特征 / Morphological Characteristics

齿式：3.1.3.3/2.0.1.3=32。头体长 54~81 mm。尾长 31~60 mm。后足长 13~16 mm。颅全长 19.2~20.0 mm。体重 5~8 g。通体毛发短而致密，背部毛色黑色或深灰色，略具金属光泽，腹部毛色略浅，灰色或灰白色，背腹毛色无明显分界。尾长通常为头体长的一半。尾部毛色上部呈深灰色或黑色，下部为灰白色。前足较大，前爪锋利。每颗牙齿均有棕红色或黑色色素沉积，上下门齿尤为明显。上单尖齿 5 枚，第四、第五枚齿很小。上颌白齿发育良好，近四边形。

Dental formula: 3.1.3.3/2.0.1.3=32. Head-body length 54-81 mm, tail length 31-60 mm, hindfoot 13-16 mm, cranial length 19.2-20.0 mm, weight 5-8 g. Dorsal pelage black or dark gray, and the ventrum is gray or whitish gray. Tail is bicolor, upper part black or dark gray, lower part grayish white. Forefeet broad with sharp claws. All teeth have reddish-brown or black pigmentation and pigmentation heavy on the upper and lower incisors. Five upper unicuspids present, and the 4th and 5th teeth are very small. Upper molars well developed and nearly quadrilateral.

生境 / Habitat

阔叶林、针阔混交林、竹林和杜鹃林
Broadleaf forest, coniferous and broad-leaved mixed forest, bamboo forest and alpine rhododendron

▲ 地理分布 / Geographic Distribution

国内分布 / Domestic Distribution
四川、云南、陕西、湖北、贵州、重庆
Sichuan, Yunnan, Shaanxi, Hubei, Guizhou, Chongqing

全球分布 / World Distribution
中国、越南
China, Vietnam

生物地理界 / Biogeographic Realm
古北界 Palearctic

WWF 生物群系 / WWF Biome
温带阔叶和混交林，热带和亚热带湿润阔叶林
Temperate Broadleaf & Mixed Forests, Tropical & Subtropical Moist Broadleaf
Forests

动物地理分布型 / Zoogeographic Distribution Type
Hc+Le+D

分布标注 / Distribution Note
非特有种 Non-Endemic

▲ 濒危状况 / Threatened Status

中国生物多样性红色名录等级 / CB RL Category (2021)
无危 LC

IUCN 红色名录 / IUCN Red List (2021)
近危 NT

威胁因子 / Threats
无 None

▲ 法律保护地位 / Legal Protection Status

国家重点保护野生动物等级 / Category of National Key Protected Wild Animals (2021)
未列入 Not listed

"三有"名录 / TWIESSV (2023)
未列入 Not listed

CITES 附录等级 / CITES Appendix (2023)
未列入 Not listed

迁徙物种公约附录 / CMS Appendix (2020)
未列入 Not listed

保护行动 / Conservation Action
尚无保护行动 No conservation action so far

▲ 参考文献 / References

Jiang et al. (蒋志刚等), 2021; Burgin et al., 2020; IUCN, 2020; Liu et al. (刘少英等), 2020; Wilson and Mittermeier, 2018; Sun et al. (孙治宇等), 2013; Tu et al. (涂飞云等), 2012; Smith et al., 2009; Pan et al. (潘清华等), 2007; Wilson and Reeder, 2005; Wang (王应祥), 2003; Zhang (张荣祖), 1997

52 / 狭颅黑齿鼩鼱

Blarinella wardi Thomas, 1915

· Burmese Short-tailed Shrew

▲ 分类地位 / Taxonomy

劳亚食虫目 Eulipotyphla / 鼩鼱科 Soricidae / 黑齿鼩鼱属 *Blarinella*

科建立者及其文献 / Family Authority
G. Fischer, 1814

属建立者及其文献 / Genus Authority
Thomas, 1911

亚种 / Subspecies
无 None

模式标本产地 / Type Locality
缅甸
Burma (Myanmar), "Hpimaw, Upper Burma (Myanmar), about 26⁰, 98⁰5'E. 2400 m"

▲ 其他名称 / Other Name(s)

其他中文名 / Other Chinese Name(s)
滇缅短尾鼩

其他英文名 / Other English Name(s)
Southern Short-tailed Shrew, Ward's
Short-tailed Shrew

同物异名 / Synonym(s)
无 None

▲ 形态及生境 / Morphology and Habitat

形态特征 / Morphological Characteristics

齿式：3.1.3.3/2.0.1.3=32。头体长 60~69 mm。后足长 10.5~13 mm。尾长 32~43 mm。颅全长 18.5~19.9 mm。该物种形态与川鼩十分相似，但体型更小，且脑颅较窄。大多数牙齿有黑色素沉积，且上、下门齿最为明显。上单尖齿 5 枚，第三枚上单尖牙明显小于第二枚。

Dental formula: 3.1.3.3/2.0.1.3=32. Head-body length 60-69 mm, hind-foot length 10.5-13 mm, tail length 32-43 mm, the greatest length of skull 18.5-19.9 mm. This species is similar to, but smaller than *B. quadraticauda*, characterized by a narrower skull and braincase. Most teeth are heavily pigmented, which is most obvious on upper and lower incisors. It has five upper unicuspids. The third upper unicuspid is much smaller than the second.

生境 / Habitat
温带森林
Temperate forest

▲ 地理分布 / Geographic Distribution

国内分布 / Domestic Distribution
云南 Yunnan

全球分布 / World Distribution
中国、缅甸 China, Myanmar

生物地理界 / Biogeographic Realm
印度马来界 Indomalaya

WWF 生物群系 / WWF Biome
热带和亚热带湿润阔叶林
Tropical & Subtropical Moist Broadleaf Forests

动物地理分布型 / Zoogeographic Distribution Type
Sc

分布标注 / Distribution Note
非特有种 Non-Endemic

▲ 濒危状况 / Threatened Status

中国生物多样性红色名录等级 / CB RL Category (2021)
近危 NT

IUCN 红色名录 / IUCN Red List (2021)
无危 LC

威胁因子 / Threats
未知 Unknown

▲ 法律保护地位 / Legal Protection Status

国家重点保护野生动物等级 / Category of National Key Protected Wild Animals (2021)
未列入 Not listed

"三有"名录 / TWIESSV (2023)
未列入 Not listed

CITES 附录等级 / CITES Appendix (2023)
未列入 Not listed

迁徙物种公约附录 / CMS Appendix (2020)
未列入 Not listed

保护行动 / Conservation Action
尚无保护行动 No conservation action so far

▲ 参考文献 / References

Jiang et al. (蒋志刚等), 2021; Burgin et al., 2020; IUCN, 2020; Liu et al. (刘少英等), 2020; Wilson and Mittermeier, 2018; Deng et al. (邓可等), 2013; Sun et al. (孙治宇等), 2013; Tu et al. (涂飞云等), 2012; Smith et al., 2009; Pan et al. (潘清华等), 2007; Wilson and Reeder, 2005; Wang (王应祥), 2003; Hoffmann, 1987

53 | 淡灰豹鼩

Parablarinella griselda
Thomas, 1912

· Indochinese Short-tailed Shrew

劳亚食虫目 Eulipotyphla / 鼩鼱科 Soricidae / 异黑齿鼩鼱属 *Parablarinella*

科建立者及其文献 / Family Authority
G. Fischer, 1814

属建立者及其文献 / Genus Authority
Thomas, 1911

亚种 / Subspecies
无共识 No consensus

模式标本产地 / Type Locality
中国
China, Gansu, 68 km SE Taochou, 3048 m

▲ 其他名称 / Other Name(s)

其他中文名 / Other Chinese Name(s)
淡灰黑齿鼩鼱

其他英文名 / Other English Name(s)
无 None

同物异名 / Synonym(s)
无 None

▲ 形态及生境 / Morphology and Habitat

形态特征 / Morphological Characteristics
齿式：3.1.3.3/2.0.1.3=32。头体长 52~79 mm。后足长 8~14 mm。尾长 31~42 mm。颅全长 19~21 mm。体重 8 g。体型中等大小的鼩鼱。背部深灰近黑色，腹部颜色较背部略浅。外耳郭被毛发遮盖。前掌宽厚，前爪较长。牙齿有红褐色或黑色色素沉积，且上下门齿最明显。
Dental formula: 3.1.3.3/2.0.1.3=32. Head-body length 52-79 mm, greatest skull length 19-21 mm, hind food length 8-14 mm, tail length 31-42 mm. Weight 8 g. A medium-sized shrew. Dorsal pelage is dark gray near black, and ventral is slightly paler. Ears are under the fur and not visible. Forehands broad and claws are long. All teeth are heavily pigmented, which is most obvious on upper and lower incisors.

生境 / Habitat
森林 Forest

▲ 地理分布 / Geographic Distribution

国内分布 / Domestic Distribution
四川、陕西、甘肃、宁夏
Sichuan, Shaanxi, Gansu, Ningxia

全球分布 / World Distribution
中国、越南 China, Vietnam

生物地理界 / Biogeographic Realm
古北界 Palearctic

WWF 生物群系 / WWF Biome
温带阔叶混交林
Temperate Broadleaf & Mixed Forests

动物地理分布型 / Zoogeographic Distribution Type
Le

分布标注 / Distribution Note
非特有种 Non-Endemic

▲ 濒危状况 / Threatened Status

中国生物多样性红色名录等级 / CB RL Category (2021)
无危 LC

IUCN 红色名录 / IUCN Red List (2021)
未评定 NE

威胁因子 / Threats
未知 Unknown

▲ 法律保护地位 / Legal Protection Status

国家重点保护野生动物等级 / Category of National Key Protected Wild Animals (2021)
未列入 Not listed

"三有"名录 / TWIESSV (2023)
未列入 Not listed

CITES 附录等级 / CITES Appendix (2023)
未列入 Not listed

迁徙物种公约附录 / CMS Appendix (2020)
未列入 Not listed

保护行动 / Conservation Action
尚无保护行动 No conservation action so far

▲ 参考文献 / References

Pu et al. (普缨婷等), 2021; Bannikova et al., 2019; He et al., 2018

0191

54 / 安徽黑齿鼩鼱

Parablarinella latimaxillata
Chen & Jiang, 2023

· Anhui Short-tailed Shrew

▲ 分类地位 / Taxonomy

劳亚食虫目 Eulipotyphla / 鼩鼱科 Soricidae / 异黑齿鼩鼱属 *Parablarinella*

科建立者及其文献 / Family Authority
G. Fischer, 1814

属建立者及其文献 / Genus Authority
Anna A. Bannikova, et al. 2019

亚种 / Subspecies
无 None

模式标本产地 / Type Locality
中国（安徽）
China, Anhui, Dabie Mountains, Yaoluoping Natural Reserve. Altitude: 1100~1700 m

▲ 其他名称 / Other Name(s)

其他中文名 / Other Chinese Name(s)
无 None

其他英文名 / Other English Name(s)
无 None

同物异名 / Synonym(s)
无 None

▲ 形态及生境 / Morphology and Habitat

形态特征 / Morphological Characteristics

齿式：3.1.3.3/1.1.1.3=32。头体长 67~76 mm。尾长 32~39 mm。后足长 11~12 mm。尾巴约为头体长的一半。毛发深棕色，背腹同色。前爪和后足背部毛色较淡灰黑齿鼩鼱及黑齿鼩鼱属物种略浅。头骨粗大，头骨的吻部在前颌骨区域变窄，P4~M3 齿列唇侧略呈弧形弯曲，区别于淡灰黑齿鼩鼱和黑齿鼩鼱属物种。与淡灰黑齿鼩鼱相比，安徽黑齿鼩鼱的上齿列及 P4~M3 齿列长度更大，M2~M2 和颧板较宽。大部分牙齿有近黑色的色素沉积，与淡灰黑齿鼩鼱和黑齿鼩鼱属的物种相似。P4 舌侧呈一定弧度，牙齿咬合面三角形，区别于黑齿鼩鼱属物种。

Dental formula: 3.1.3.3/ 1.1.1.3=32. Head-body length 67-76 mm. Tail length 32-39 mm. Hind foot length 11-12 mm. The tail is about half the head and body length. The hair is dark brown, the back and abdomen are the same color. The backs of hands and feet are much paler than those of *P. griselda* and *Blarinella* spp. The skull is relatively stout and broad, the rostrum abruptly narrows in the premaxillary region, and the outline of the tooth row of P4-M3 is curved or rounded, which distinguishes it from *P. griselda* and *Blarinella* species. It has longer UTR and P4-M3, and broader M2-M2 and zygomatic plate than *P. griselda*. The teeth are heavily pigmented, similar to *P. griselda* and *Blarinella* spp. The lingual margin of P4 is curved, and the tooth is triangular in occlusal view, distinguishing it from *Blarinella* species.

生境 / Habitat

海拔 1100~1700 米的落叶阔叶林
Deciduous broad-leaved forest at 1100~1700 m above sea level

▲ 地理分布 / Geographic Distribution

国内分布 / Domestic Distribution
安徽 Anhui

全球分布 / World Distribution
中国 China

生物地理界 / Biogeographic Realm
东方界 Oriental

WWF 生物群系 / WWF Biome
温带阔叶林 Temperate Broadleaf Forests

动物地理分布型 / Zoogeographic Distribution Type
Sn

分布标注 / Distribution Note
特有种 Endemic

▲ 濒危状况 / Threatened Status

中国生物多样性红色名录等级 / CB RL Category (2021)
未评定 NE

IUCN 红色名录 / IUCN Red List (2021)
未评定 NE

威胁因子 / Threats
未知 Unknown

▲ 法律保护地位 / Legal Protection Status

国家重点保护野生动物等级 / Category of National Key Protected Wild Animals (2021)
未列入 Not listed

"三有" 名录 / TWIESSV (2023)
列入 Listed

CITES 附录等级 / CITES Appendix (2023)
未列入 Not listed

迁徙物种公约附录 / CMS Appendix (2020)
未列入 Not listed

保护行动 / Conservation Action
尚无保护行动 No conservation action so far

▲ 参考文献 / References

Chen et al., 2023

55 | 墨脱大爪鼩鼱

Soriculus medogensis Chen & Jiang, 2023

· Soriculus medogensis

▲ 分类地位 / Taxonomy

劳亚食虫目 Eulipotyphla / 鼩鼱科 Soricidae / 长尾鼩鼱属 *Soriculus*

科建立者及其文献 / Family Authority
G. Fischer, 1814

属建立者及其文献 / Genus Authority
Blyth, 1854

亚种 / Subspecies
无 None

模式标本产地 / Type Locality
中国（西藏）
China, Medog, southeastern Tibet, Damu Town

▲ 其他名称 / Other Name(s)

其他中文名 / Other Chinese Name(s)
无 None

其他英文名 / Other English Name(s)
无 None

同物异名 / Synonym(s)
无 None

▲ 形态及生境 / Morphology and Habitat

形态特征 / Morphological Characteristics

齿式：33/2.1/0.2/1.3/3=30。体重 12.0~14.1 g。头体长 83~85 mm。尾长 43~54 mm。后肢长 14~16 mm。耳长 9~12 mm。外形特征及体型大小与雪山大爪鼩鼱相似，但毛色更暗。背部毛发致密而柔软，冬毛长约 8mm。尾背面和腹面颜色差异不明显。前爪大，与其他长尾鼩属物种相似。头骨粗壮、骨质致密、棱角分明。脑颅顶部低平，不向上隆起。枕骨略向后延伸。肩胛脊和矢状脊发育良好。基底枕骨和基底蝶骨不完全融合，在中部收窄。下颌骨比该属的其他物种更加发达。冠状突呈铲状。上门齿两个齿尖：前尖短，匕首状，向前延伸；后尖低于 U1。四颗上单尖齿；U2 略大于 U1，U3 约为 U2 高度的一半；U4 高度退化。M1 与 M2 大小相似，M3 退化。下颌门齿（I1）长，仅有一个齿尖，齿尖略向前延伸并略向上弯曲。下颌单尖齿（U1）紧靠 P4。M1 比 M2 略大，M3 退化。牙齿只在齿尖有较浅的铁锈色色素沉积。

Dental formula: 3/2.1/0.2/1.3/3=30. External morphology and size similar to *S. nivatus*, but the pelage is slightly darker. The dorsal pelage is dense and soft, about 8 mm long in winter. The tail is not sharply bicolored. The foreclaws are enlarged, similar to those of other *Soriculus* species. The skull is robust and bony. Braincase is low and relatively flattened, and the posterior of the skull is angular rather than rounded. The lambdoidal and sagittal crest are well developed and pronounced. The basioccipital and basisphenoid are fused and underdeveloped, contracted into a line in the central region. The mandible is more developed than in other species in the genus. The coronoid process is spatula shaped, rising straight upwards. The first upper incisor is bifid; the principal (anterior) cusp is short, dagger like, and extends anteriorly; the posterior cusp is lower than U1. Four upper unicuspid teeth are present; U2 is slightly larger than U1, U3 is about half the height of U2; U4 is minute. M1 and M2 are similar in size, whereas M3 is reduced. The lower incisor (I1) is long with only a low cusp. The lower unicuspid (U1) and P4 are crowded. M1 slightly larger than M2, and M3 is reduced. The tips of the teeth are pigmented only lightly with orange.

生境 / Habitat

海拔 2100~2830m 的山地针阔叶混交林或针叶林带和亚高山暗针叶林带
Mountain coniferous broad-leaved mixed forest or coniferous forest belt and subalpine dark coniferous forest belt at elevations from 2100~2830 m

▲ 地理分布 / Geographic Distribution

国内分布 / Domestic Distribution
西藏 Tibet

全球分布 / World Distribution
中国 China

生物地理界 / Biogeographic Realm
东方界 Oriental

WWF 生物群系 / WWF Biome
东喜马拉雅山脉亚高山针叶林
Eastern Himalayan Subalpine Conifer Forests

动物地理分布型 / Zoogeographic Distribution Type
P 或 I

分布标注 / Distribution Note
特有种 Endemic

▲ 濒危状况 / Threatened Status

中国生物多样性红色名录等级 / CB RL Category (2021)
未评定 NE

IUCN 红色名录 / IUCN Red List (2021)
未评定 NE

威胁因子 / Threats
未知 Unknown

▲ 法律保护地位 / Legal Protection Status

国家重点保护野生动物等级 / Category of National Key Protected Wild Animals (2021)
未列入 Not listed

"三有"名录 / TWIESSV (2023)
列入 Listed

CITES 附录等级 / CITES Appendix (2023)
未列入 Not listed

迁徙物种公约附录 / CMS Appendix (2020)
未列入 Not listed

保护行动 / Conservation Action
尚无保护行动 No conservation action so far

▲ 参考文献 / References

Chen et al., 2023

56 / 倭大爪鼩鼱

Soriculus minor Dobson, 1890

• Lesser large-clawed shrew

▲ 分类地位 / Taxonomy

劳亚食虫目 Eulipotyphla / 鼩鼱科 Soricidae / 长尾鼩鼱属 *Soriculus*

科建立者及其文献 / Family Authority
G. Fischer, 1814

属建立者及其文献 / Genus Authority
Blyth, 1854

亚种 / Subspecies
无 None

模式标本产地 / Type Locality
印度
Manipur, Assam

▲ 其他名称 / Other Name(s)

其他中文名 / Other Chinese Name(s)
无 None

其他英文名 / Other English Name(s)
无 None

同物异名 / Synonym(s)
Soriculus radulus Tomas, 1922
Type locality: Mishmi Hills, Tibet

▲ 形态及生境 / Morphology and Habitat

形态特征 / Morphological Characteristics

齿式：3.1.2.3 / 2.0.1.3 = 30。体重 7.7~12.1 g。头体长 62~77 mm。尾长 32~43 mm。后足长 11~14 mm。背部暗褐色至黑色，腹部颜色稍浅。尾背面深棕色，腹面浅灰色。牙齿的色素沉积很浅，只存在于齿尖。上单尖齿 4 枚，U1 最大，U2 略大于 U3，U4 很小。

Dental formula: 3.1.2.3 / 2.0.1.3 = 30. Weight 7.7-12.1 g. The head length is 62-77 mm. The tail length is 32-43 mm. The length of hindfoot is 11-14 mm. The back is dark brown to black, and the abdomen is slightly lighter. The tail is two-colored, but not obvious. The top is dark brown and the bottom is light gray. The pigmentation of the teeth is lighter and only exists at the tip of the tooth. Four upper unicuspids (U1–U4) are present; U1 is largest in size, U2 is slightly larger than U3 and U4 is minute.

生境 / Habitat

阔叶林、针阔混交林、针叶林
Broad-leaved forest, coniferous and broad-leaved mixed forest, coniferous forest

国内分布 / Domestic Distribution
西藏、云南
Tibet, Yunnan

全球分布 / World Distribution
中国、不丹、印度、缅甸
China, Bhutan, India, Myanmar

生物地理界 / Biogeographic Realm
东方界 Oriental

WWF 生物群系 / WWF Biome
亚热带湿润阔叶林、针叶林、混交林
Subtropical Moist Broadleaf Forests, Coniferous Forests, Mixed Forests

动物地理分布型 / Zoogeographic Distribution Type
Ha

分布标注 / Distribution Note
非特有种 Non-Endemic

▲ 濒危状况 / Threatened Status

中国生物多样性红色名录等级 / CB RL Category (2021)
未评定 NE

IUCN 红色名录 / IUCN Red List (2021)
未评定 NE

威胁因子 / Threats
未知 Unknown

▲ 法律保护地位 / Legal Protection Status

国家重点保护野生动物等级 / Category of National Key Protected Wild Animals (2021)
未列入 Not listed

"三有"名录 / TWIESSV (2023)
列入 Listed

CITES 附录等级 / CITES Appendix (2023)
未列入 Not listed

迁徙物种公约附录 / CMS Appendix (2020)
未列入 Not listed

保护行动 / Conservation Action
尚无保护行动 No conservation action so far

▲ 参考文献 / References

Chen et al., 2023; Burgin and He, 2018; Jenkins, 2013

57 / 大爪长尾鼩鼱

Soriculus nigrescens (Gray, 1842)

· Himalayan Shrew

▲ 分类地位 / Taxonomy

劳亚食虫目 Eulipotyphla / 鼩鼱科 Soricidae / 长尾鼩鼱属 *Soriculus*

科建立者及其文献 / Family Authority
G. Fischer, 1814

属建立者及其文献 / Genus Authority
Blyth, 1854

亚种 / Subspecies
无共识 No consensus

模式标本产地 / Type Locality
印度
"India", West Bengal, Darjeeling

▲ 其他名称 / Other Name(s)

其他中文名 / Other Chinese Name(s)
无 None

其他英文名 / Other English Name(s)
无 None

同物异名 / Synonym(s)
无 None

▲ 形态及生境 / Morphology and Habitat

形态特征 / Morphological Characteristics
齿式：3.1.2.3/2.0.1.3=30。头体长 70~94 mm。后足长 12~17 mm。尾长 32~50 mm。体重 7~9 g。背部毛色暗灰色或深棕色，腹部毛色浅灰色。尾长通常为头体长的一半。前爪与其他同体型大小的鼩鼱相比大而粗壮，长有较为锋利的指甲。头骨致密，脑颅较为扁平。牙齿色素沉积颜色很浅且仅在齿尖分布，门齿无明显色素沉积。该属物种均上单尖齿 4 枚，第四枚上单尖齿高度退化。

Dental formula: 3.1.2.3/2.0.1.3=30. Head-body length 70-94 mm, hindfoot length 12-17 mm, tail length 32-50 mm, weight 7-9 g. Dorsal pelage dust color, and ventral is slight paler. Tail is short and only half of the head-body length. Forefeet and claws enlarged, and larger than hindfeet, adopted for digging. Skull is bony and the braincase is low. Pigmentation of teeth is very light and if exists only presents on the tips of teeth. Four upper unicuspids present and the 4th upper unicuspid is heavily reduced as in all *Episoriculus* species.

生境 / Habitat
落叶针叶混交林、针叶杜鹃林、树线以上的高山区
Mixed deciduous-coniferous forest, conifer-rhododendron forest, alpine zone (above timberline)

▲ 地理分布 / Geographic Distribution

国内分布 / Domestic Distribution
西藏 Tibet

全球分布 / World Distribution
中国、不丹、印度、尼泊尔、缅甸
China, Bhutan, India, Nepal, Myanmar

生物地理界 / Biogeographic Realm
古北界 Palearctic

WWF 生物群系 / WWF Biome
热带和亚热带湿润阔叶林，山地草原和灌丛
Tropical & Subtropical Moist Broadleaf Forests, Montane Grasslands and Shrublands

动物地理分布型 / Zoogeographic Distribution Type
Ha

分布标注 / Distribution Note
非特有种 Non-Endemic

▲ 濒危状况 / Threatened Status

中国生物多样性红色名录等级 / CB RL Category (2021)
近危 NT

IUCN 红色名录 / IUCN Red List (2021)
无危 LC

威胁因子 / Threats
耕种 Farming

▲ 法律保护地位 / Legal Protection Status

国家重点保护野生动物等级 / Category of National Key Protected Wild Animals (2021)
未列入 Not listed

"三有" 名录 / TWIESSV (2023)
未列入 Not listed

CITES 附录等级 / CITES Appendix (2023)
未列入 Not listed

迁徙物种公约附录 / CMS Appendix (2020)
未列入 Not listed

保护行动 / Conservation Action
尚无保护行动 No conservation action so far

▲ 参考文献 / References

Jiang et al. (蒋志刚等), 2021; Burgin et al., 2020; IUCN, 2020; Liu et al. (刘少英等), 2020; Wilson and Mittermeier, 2018; Liao et al. (廖锐等), 2015; Smith et al., 2009; Pan et al. (潘清华等), 2007; Wilson and Reeder, 2005; Wang (王应祥), 2003; Zhang (张荣祖), 1997; Chen et al., 2023

58 / 雪山大爪鼩鼱

Soriculus nivatus Chen & Jiang, 2023

· Snow Mountain large-clawed shrew

▲ 分类地位 / Taxonomy

劳亚食虫目 Eulipotyphla / 鼩鼱科 Soricidae / 长尾鼩鼱属 *Soriculus*

科建立者及其文献 / Family Authority
G. Fischer, 1814

属建立者及其文献 / Genus Authority
Blyth, 1854

亚种 / Subspecies
无 None

模式标本产地 / Type Locality
中国（西藏）
K81 of Motuo Highway (29.742° N, 95.683° E, 3619 m a.s.l.), Damu Town, Medog, Southeast Tibet, China

普昌智／供图

▲ 其他名称 / Other Name(s)

其他中文名 / Other Chinese Name(s)
无 None

其他英文名 / Other English Name(s)
无 None

同物异名 / Synonym(s)
无 None

▲ 形态及生境 / Morphology and Habitat

形态特征 / Morphological Characteristics

齿式：3.1.2.3/2.0.1.3=30。体重 9.5~15.3 g。头体长 70~90 mm。尾长 46~58 mm。后足长 13~17 mm。耳长 7~12 mm。背腹为深棕色。背部毛发夏季长约 5 mm，冬季长约 8 mm。尾背腹颜色差异较小，背面深棕色，腹面颜色略浅。前后足背覆盖浅棕色至黑色的毛发（Chen et al., 2023）。头骨坚固且骨质致密。与其他长尾鼩属的物种相比，雪山大爪鼩鼱的吻部和腭部相对较窄，眶间距较宽，边缘略呈锥形。脑颅顶部呈半球型，但不向上隆起。枕骨中央区域向后延伸。人字脊明显，矢状脊不明显。基枕骨和基蝶骨在中部融合并收窄。下颌骨冠状突粗壮，近铲形，与平行枝垂直。髁突发育良好，与冠状突形成大约 45° 的角度。角突细长且直。第一上门齿齿尖两枚。前尖长度中等，镰刀状，齿尖垂直向下；后尖小且高度低于 U1。上单尖齿 4 枚（U1‒U4）；U2 略大于 U1，U3 的高度大约是 U2 的一半高，U4 非常小。P4 后缘深凹。M1 和 M2 大小相似，M3 退化。下门齿（I1）较长，齿尖略向前延伸，并向上呈 45 度弯曲。下单尖齿（U1）较小。M3 退化。大部分个体牙齿的齿尖有较淡的棕色色素沉积。

Dental formula: 3.1.2.3 / 2.0.1.3 = 30. Weight is 9.5-15.3 g. Head-body length 70-90 mm. Tail length 46-58 mm. Hindfoot length 13-17 mm. Ear length 7-12 mm. The dorsal belly is dark brown. The length of back hair is about 5 mm in summer and 8 mm in winter. The tail is an inconspicuous two-color, the top is dark brown, and the bottom is light color. The backs of hands and feet are covered with light brown to black hair. The skull is robust and bony. Relative to other species of *Soriculus*, the rostrum and maxillary of *S. nivatus* are relatively narrower; the interorbital region is broad with slightly tapering edges. Braincase is domed but low, and the posterior of the skull is rounded. The lambdoidal crest is modestly developed, and the sagittal crest is indistinct. The basioccipital and basisphenoid are fused and narrowed markedly in the middle region, forming a thread-like structure. The coronoid process is robust with a spatulate tip, rising straight upwards. The condyloid process is heavy, forming an angle at roughly 45° with the coronoid process. The angular process is long, straight, and very thin. The first upper incisor bifid; the principal (anterior) cusp moderately long, falciform, and the apex pointing straight downwards; the posterior cusp is small and lower than the following U1. Four upper unicuspids (U1–U4) are present; U2 is largest in size, U1 is slightly smaller, U3 is about half as high as U2, and U4 is minute. P4 is deeply excavated in the posterior borders. M1 and M2 are similar in size, whereas M3 is reduced. The low incisor (I1) is long, the tooth tip extends slightly forward and bends slightly upward. The lower unicuspid (U1) is small. M3 is reduced. The tips of the cusps of all teeth are pigmented with orange.

生境 / Habitat

阔叶林、针阔混交林、针叶林
Broad-leaved forest, coniferous and broad-leaved mixed forest, coniferous forest

▲ 地理分布 / Geographic Distribution

国内分布 / Domestic Distribution
西藏 Tibet

全球分布 / World Distribution
中国 China

生物地理界 / Biogeographic Realm
东方界 Oriental

WWF 生物群系 / WWF Biome
亚热带湿润阔叶林、针叶林、混交林
Subtropical Moist Broadleaf Forests, Coniferous Forest, Mixed Forests

动物地理分布型 / Zoogeographic Distribution Type
Ha

分布标注 / Distribution Note
特有种 Endemic

▲ 濒危状况 / Threatened Status

中国生物多样性红色名录等级 / CB RL Category (2021)
未评定 NE

IUCN 红色名录 / IUCN Red List (2021)
未评定 NE

威胁因子 / Threats
未知 Unknown

▲ 法律保护地位 / Legal Protection Status

国家重点保护野生动物等级 / Category of National Key Protected Wild Animals (2021)
未列入 Not listed

"三有"名录 / TWIESSV (2023)
列入 Listed

CITES 附录等级 / CITES Appendix (2023)
未列入 Not listed

迁徙物种公约附录 / CMS Appendix (2020)
未列入 Not listed

保护行动 / Conservation Action
尚无保护行动 No conservation action so far

▲ 参考文献 / References

Chen et al., 2023

59 | 米什米长尾鼩鼱

Episoriculus baileyi Thomas, 1914

• Mishmi Brown-toothed Shrew

▲ 分类地位 / Taxonomy

劳亚食虫目 Eulipotyphla / 鼩鼱科 Soricidae / 须弥鼩鼱属 *Episoriculus*

科建立者及其文献 / Family Authority
G. Fischer, 1814

属建立者及其文献 / Genus Authority
Ellerman & Morrison-Scott, 1966

亚种 / Subspecies
无 None

模式标本产地 / Type Locality
中国
Tsu River, Mishmi Hills at 2286 m (=Mt Qilinggong, Tibet)

▲ 其他名称 / Other Name(s)

其他中文名 / Other Chinese Name(s)
无 None

其他英文名 / Other English Name(s)
无 None

同物异名 / Synonym(s)
无 None

▲ 形态及生境 / Morphology and Habitat

形态特征 / Morphological Characteristics

齿式：3.1.2.3/2.0.1.3=30。头体长 63~81 mm。尾长 60~76 mm。后足长 13~14.5 mm。颅全长 19~20.6 mm。该属中体型较大的物种。尚未得到充分研究。毛色背部深棕色或深青灰色，腹部浅棕色。尾长接近头体长，尾覆盖有短毛。

Dental formula: 3.1.2.3/2.0.1.3=30. Head-body length 63-81 mm, tail length 60-76 mm, hind-foot length 13-14.5 mm, condyloincisive length 19-20.6 mm. A large Episoriculus species and understudied species. Dorsal pelage dark brown or dark slate-grey washed with blackish, and ventral is light brown. Tail is shorter than or similar to the head-body length, and covered with short hairs along its entire length.

生境 / Habitat
针叶林、落叶阔叶林
Coniferous forest, deciduous broad-leaved forest

▲ 地理分布 / Geographic Distribution

国内分布 / Domestic Distribution
西藏 Tibet

全球分布 / World Distribution
中国、尼泊尔、缅甸、越南
China, Nepal, Myanmar, Vietnam

生物地理界 / Biogeographic Realm
印度马来界 Indomalaya

WWF 生物群系 / WWF Biome
热带和亚热带湿润阔叶林
Tropical & Subtropical Moist Broadleaf Forests

动物地理分布型 / Zoogeographic Distribution Type
Pe

分布标注 / Distribution Note
非特有种 Non-Endemic

▲ 濒危状况 / Threatened Status

中国生物多样性红色名录等级 / CB RL Category (2021)
数据缺乏 DD

IUCN 红色名录 / IUCN Red List (2021)
未评定 NE

威胁因子 / Threats
未知 Unknown

▲ 法律保护地位 / Legal Protection Status

国家重点保护野生动物等级 / Category of National Key Protected Wild Animals (2021)
未列入 Not listed

"三有"名录 / TWIESSV (2023)
未列入 Not listed

CITES 附录等级 / CITES Appendix (2023)
未列入 Not listed

迁徙物种公约附录 / CMS Appendix (2020)
未列入 Not listed

保护行动 / Conservation Action
尚无保护行动 No conservation action so far

▲ 参考文献 / References

Jiang et al. (蒋志刚等), 2021; Wilson and Mittermeier, 2018; Corbet and Hill, 1986

60 / 小长尾鼩鼱

Episoriculus caudatus
(Horsfield, 1851)

· Hodgson's Brown-toothed Shrew

劳亚食虫目 Eulipotyphla / 鼩鼱科 Soricidae / 须弥鼩鼱属 *Episoriculus*

科建立者及其文献 / Family Authority
G. Fischer, 1814

属建立者及其文献 / Genus Authority
Ellerman & Morrison-Scott, 1966

亚种 / Subspecies
无共识 No consensus

模式标本产地 / Type Locality
尼泊尔
Nepal

▲ 其他名称 / Other Name(s)

其他中文名 / Other Chinese Name(s)
无 None

其他英文名 / Other English Name(s)
无 None

同物异名 / Synonym(s)
无 None

▲ 形态及生境 / Morphology and Habitat

形态特征 / Morphological Characteristics
齿式：3.1.2.3/2.0.1.3=30。头体长 52~70 mm。尾长 46~65 mm。后足 10~16 mm。颅全长 17~19 mm。为该属中体型中等大小的物种。脑颅拱形隆起。对该物种的研究十分欠缺。

Dental formula: 3.1.2.3/2.0.1.3=30. Head-body 52-70 mm, tail 46-65 mm, hind-foot 10-16 mm, condyloincisive length 17-19 mm. This is a middle-sized Episoriculus. Braincase is well-domed and high. It is understudied species.

生境 / Habitat
常绿阔叶林、针叶林、杜鹃林
Evergreen broad-leaved forest, rhododendron forest and coniferous forest

▲ 地理分布 / Geographic Distribution

国内分布 / Domestic Distribution
西藏 Tibet

全球分布 / World Distribution
中国、印度、尼泊尔
China, India, Nepal

生物地理界 / Biogeographic Realm
印度马来界 Indomalaya

WWF 生物群系 / WWF Biome
温带阔叶和混交林
Temperate Broadleaf & Mixed Forests

动物地理分布型 / Zoogeographic Distribution Type
Hm

分布标注 / Distribution Note
非特有种 Non-Endemic

▲ 濒危状况 / Threatened Status

中国生物多样性红色名录等级 / CB RL Category (2021)
无危 LC

IUCN 红色名录 / IUCN Red List (2021)
无危 LC

威胁因子 / Threats
无 None

▲ 法律保护地位 / Legal Protection Status

国家重点保护野生动物等级 / Category of National Key Protected Wild Animals (2021)
未列入 Not listed

"三有" 名录 / TWIESSV (2023)
未列入 Not listed

CITES 附录等级 / CITES Appendix (2023)
未列入 Not listed

迁徙物种公约附录 / CMS Appendix (2020)
未列入 Not listed

保护行动 / Conservation Action
尚无保护行动 No conservation action so far

▲ 参考文献 / References

Jiang et al. (蒋志刚等), 2021; Burgin et al., 2020; IUCN, 2020; Liu et al. (刘少英等), 2020; Wilson and Mittermeier, 2018; Deng et al. (邓可等), 2013; Smith et al., 2009; Pan et al. (潘清华等), 2007; Hutterer, 2005; Wilson and Reeder, 2005; Zhang (张荣祖), 1997

61 / 台湾长尾鼩鼱

Episoriculus fumidus (Thomas, 1913)

• Taiwan Brown-toothed Shrew

▲ 分类地位 / Taxonomy

劳亚食虫目 Eulipotyphla / 鼩鼱科 Soricidae / 须弥鼩鼱属 *Episoriculus*

科建立者及其文献 / Family Authority
G. Fischer, 1814

属建立者及其文献 / Genus Authority
Ellerman & Morrison-Scott, 1966

亚种 / Subspecies
无 None

模式标本产地 / Type Locality
中国
Taiwan, Chiai Hsien, "Mt. Arisan (Alishan); Central Taiwan, China 2438 m"

何锴 / 供图

▲ 其他名称 / Other Name(s)

其他中文名 / Other Chinese Name(s)
台湾烟尖鼠

其他英文名 / Other English Name(s)
无 None

同物异名 / Synonym(s)
无 None

▲ 形态及生境 / Morphology and Habitat

形态特征 / Morphological Characteristics

齿式：3.1.2.3/2.0.1.3=30。头体长 47~73 mm。尾长 37~52 mm。后足 11~15 mm。背腹异色，背部棕褐色，腹部颜色较浅，近灰白色，背腹毛色无明显的分界线。尾长小于头体长。外耳不发达，上面覆盖有浓密的毛发。门齿和单尖齿的齿尖有棕红色的色素沉积，白齿几乎没有色素沉积。有研究认为该物种与大陆物种为不同属。

Dental formula: 3.1.2.3/2.0.1.3=30. Head-body length 47-73 mm, tail length 37-52 mm, hindfoot length 11-15 mm. Dorsal pelage smoky blackish, and the ventral lighter. Tail is shorter than the head-body length. Ears are under the fur. Incisors and unicuspids have pigmentation on tips, and molars are not pigmented. It might represent a distinct genus from the mainland species.

生境 / Habitat
阔叶林、针叶林和箭竹林
Broad-leaved forest, coniferous forest and arrow bamboogrove

国内分布 / Domestic Distribution
台湾 Taiwan

全球分布 / World Distribution
中国 China

生物地理界 / Biogeographic Realm
印度马来界 Indomalaya

WWF 生物群系 / WWF Biome
热带和亚热带湿润阔叶林、山地草地与灌丛
Tropical & Subtropical Moist Broadleaf Forests, Montane Grasslands and Shrublands

动物地理分布型 / Zoogeographic Distribution Type
J

分布标注 / Distribution Note
特有种 Endemic

▲ 濒危状况 / Threatened Status

中国生物多样性红色名录等级 / CB RL Category (2021)
无危 LC

IUCN 红色名录 / IUCN Red List (2021)
无危 LC

威胁因子 / Threats
未知 Unknown

▲ 法律保护地位 / Legal Protection Status

国家重点保护野生动物等级 / Category of National Key Protected Wild Animals (2021)
未列入 Not listed

"三有"名录 / TWIESSV (2023)
未列入 Not listed

CITES 附录等级 / CITES Appendix (2023)
未列入 Not listed

迁徙物种公约附录 / CMS Appendix (2020)
未列入 Not listed

保护行动 / Conservation Action
尚无保护行动 No conservation action so far

▲ 参考文献 / References

Jiang et al. (蒋志刚等), 2021; Burgin et al., 2020; IUCN, 2020; Wilson and Mittermeier, 2018; Smith et al., 2009; Pan et al. (潘清华等), 2007; Wilson and Reeder, 2005; Motokawa et al., 1998; Motokawa et al., 1997; Yu, 1994; Yu, 1993; Zhang (张荣祖), 1997; Jameson and Jones, 1977

62 / 大长尾鼩鼱

Episoriculus leucops (Horsfield, 1855)

• Long-tailed Brown-toothed Shrew

▲ 分类地位 / Taxonomy

劳亚食虫目 Eulipotyphla / 鼩鼱科 Soricidae / 须弥鼩鼱属 *Episoriculus*

科建立者及其文献 / Family Authority
G. Fischer, 1814

属建立者及其文献 / Genus Authority
Ellerman & Morrison-Scott, 1966

亚种 / Subspecies
无共识 No consensus

模式标本产地 / Type Locality
尼泊尔
Nepal

何锴 / 供图

▲ 其他名称 / Other Name(s)

其他中文名 / Other Chinese Name(s)
印度长尾鼩

其他英文名 / Other English Name(s)
无 None

同物异名 / Synonym(s)
无 None

▲ 形态及生境 / Morphology and Habitat

形态特征 / Morphological Characteristics
齿式：3.1.2.3/2.0.1.3=30。头体长 71~82 mm。尾长 63~76 mm。后足长 15~16 mm。背腹颜色相近，通体深灰色或深棕色。尾长接近于头体长。外耳郭明显。脑颅隆起。

Dental formula: 3.1.2.3/2.0.1.3=30. The head-body length is 71-82 mm, tail length 63-76 mm, hindfoot length 15-16 mm. The whole body is covered with uniform dark brown hairs. Tail length is similar to the head-body length. Braincase is dome-shaped and high.

生境 / Habitat
针叶林、高山杜鹃林、阔叶林、竹林
Coniferous forest, alpine rhododendron forest, broad-leaved forest, bamboo forest

▲ 地理分布 / Geographic Distribution

国内分布 / Domestic Distribution
西藏、云南 Tibet, Yunnan

全球分布 / World Distribution
中国、印度、尼泊尔
China, India, Nepal

生物地理界 / Biogeographic Realm
古北界 Palearctic

WWF 生物群系 / WWF Biome
热带和亚热带湿润阔叶林、温带阔叶和混交林
Tropical & Subtropical Moist Broadleaf Forests, Temperate Broadleaf & Mixed Forests

动物地理分布型 / Zoogeographic Distribution Type
Hm

分布标注 / Distribution Note
非特有种 Non-Endemic

▲ 濒危状况 / Threatened Status

中国生物多样性红色名录等级 / CB RL Category (2021)
无危 LC

IUCN 红色名录 / IUCN Red List (2021)
无危 LC

威胁因子 / Threats
未知 Unknown

▲ 法律保护地位 / Legal Protection Status

国家重点保护野生动物等级 / Category of National Key Protected Wild Animals (2021)
未列入 Not listed

"三有" 名录 / TWIESSV (2023)
未列入 Not listed

CITES 附录等级 / CITES Appendix (2023)
未列入 Not listed

迁徙物种公约附录 / CMS Appendix (2020)
未列入 Not listed

保护行动 / Conservation Action
尚无保护行动 No conservation action so far

▲ 参考文献 / References

Jiang et al. (蒋志刚等), 2021; Burgin et al., 2020; IUCN, 2020; Wilson and Mittermeier, 2018; Smith et al., 2009; Pan et al. (潘清华等), 2007; Hutterer, 2005; Wilson and Reeder, 2005; Chakraborty et al., 2004; Zhang (张荣祖), 1997

63 / 缅甸长尾鼩鼱

Episoriculus macrurus (Blanford, 1888)

- Arboreal Brown-toothed Shrew

劳亚食虫目 Eulipotyphla / 鼩鼱科 Soricidae / 须弥鼩鼱属 *Episoriculus*

科建立者及其文献 / Family Authority
G. Fischer, 1814

属建立者及其文献 / Genus Authority
Ellerman & Morrison-Scott, 1966

亚种 / Subspecies
川西亚种 *S. m. irene* Thomas, 1912
四川西部（宝兴、石棉、康定、汶川和盐津），云南西北部（贡山、泸水、碧江、德钦和维西）
Sichuan (western parts-Baoxing, Shimian, Kangding, Wenchuan and Yan jin) and Yunnan (northwestern parts-Gongshan, Lushui, Bijiang, Deqin and Weixi)

模式标本产地 / Type Locality
印度
India, Darjeeling

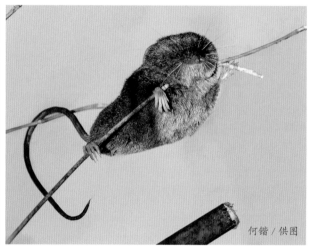

何锴 / 供图

何锴 / 供图

▲ 其他名称 / Other Name(s)

其他中文名 / Other Chinese Name(s)
褐腹长尾鼩鼱、小长尾鼩鼱

其他英文名 / Other English Name(s)
无 None

同物异名 / Synonym(s)
无 None

▲ 形态及生境 / Morphology and Habitat

形态特征 / Morphological Characteristics
齿式：3.1.2.3/2.0.1.3=30。头体长 47~73 mm。尾长 76~101 mm。后足长 14~18 mm。颅全长 17~19 mm。体型较小的鼩鼱，尾长通常为头体长的 150%。身体纤细，四肢细长。背腹异色，背部灰黄色，腹部污白。头骨脑颅隆起，吻部短于同属的其他物种。上单尖齿近四边形。
Dental formula: 3.1.2.3/2.0.1.3=30. Head-body 47-73 mm, tail 76-101 mm, hindfoot 14-18 mm, condyloincisive length 17-19 mm. A small-sized shrew with a long tail which is usually 150% of its head-body length. Body and limbs are slender. Dorsal pelage yellowish-grey, and the ventrum is dirty white. Braincase is well-dome shaped. Rostrum is shorter than the other *Episoriculus* species. Upper unicuspids are quadrate to wider than long.

生境 / Habitat
阔叶林、高山杜鹃林
Broad-leaved forest, alpine rhododendron forest

▲ 地理分布 / Geographic Distribution

国内分布 / Domestic Distribution
四川、西藏、云南
Sichuan, Tibet, Yunnan

全球分布 / World Distribution
中国、印度、缅甸、越南、尼泊尔
China, India, Myanmar, Vietnam, Nepal

生物地理界 / Biogeographic Realm
古北界、印度马来界 Palearctic, Indomalaya

WWF 生物群系 / WWF Biome
热带和亚热带湿润阔叶林、温带阔叶和混交林
Tropical & Subtropical Moist Broadleaf Forests, Temperate Broadleaf & Mixed Forests

动物地理分布型 / Zoogeographic Distribution Type
Hm

分布标注 / Distribution Note
非特有种 Non-Endemic

▲ 濒危状况 / Threatened Status

中国生物多样性红色名录等级 / CB RL Category (2021)
无危 LC

IUCN 红色名录 / IUCN Red List (2021)
无危 LC

威胁因子 / Threats
未知 Unknown

▲ 法律保护地位 / Legal Protection Status

国家重点保护野生动物等级 / Category of National Key Protected Wild Animals (2021)
未列入 Not listed

"三有"名录 / TWIESSV (2023)
未列入 Not listed

CITES 附录等级 / CITES Appendix (2023)
未列入 Not listed

迁徙物种公约附录 / CMS Appendix (2020)
未列入 Not listed

保护行动 / Conservation Action
尚无保护行动 No conservation action so far

▲ 参考文献 / References

Jiang et al. (蒋志刚等), 2021; Burgin et al., 2020; IUCN, 2020; Wilson and Mittermeier, 2018; Deng et al. (邓可等), 2013; Tu et al. (涂飞云等), 2012; Smith et al., 2009; Pan et al. (潘清华等), 2007; Wilson and Reeder, 2005

64 / 灰腹长尾鼩鼱

Episoriculus sacratus Thomas, 1911

• Grey-bellied Shrew

▲ 分类地位 / Taxonomy

劳亚食虫目 Eulipotyphla / 鼩鼱科 Soricidae / 须弥鼩鼱属 *Episoriculus*

科建立者及其文献 / Family Authority
G. Fischer, 1814

属建立者及其文献 / Genus Authority
Ellerman & Morrison-Scott, 1966

亚种 / Subspecies
无 None

模式标本产地 / Type Locality
中国
China, Sichuan, Omi-san. Altitude: 1800 m

陈广岳 / 供图

▲ 其他名称 / Other Name(s)

其他中文名 / Other Chinese Name(s)
无 None

其他英文名 / Other English Name(s)
无 None

同物异名 / Synonym(s)
无 None

▲ 形态及生境 / Morphology and Habitat

形态特征 / Morphological Characteristics

齿式：3.1.2.3/2.0.1.3=30。头体长 58~74 mm。尾长 48~69 mm。后足长 13~16 mm。背部毛色存在季节变化，可能为浅灰色或深棕色，腹部毛色为烟灰色。尾双色，尾长与头体长接近。

Dental formula: 3.1.2.3/2.0.1.3=30. Head-body length 58-74 mm. Tail length 48-69 mm. Hind foot length 13-16 mm. Body color changes seasonally. Dorsal pelage slate grey to dark brown, and ventral pelage smoky grey. Tail is bicolored and similar to the head-body in length.

生境 / Habitat

常绿阔叶林、落叶阔叶林、针叶林、杜鹃林
Evergreen broad-leaved forest, deciduous broad-leaved forest, coniferous forest, azalea forest

▲ 地理分布 / Geographic Distribution

国内分布 / Domestic Distribution
四川 Sichuan

全球分布 / World Distribution
中国 China

生物地理界 / Biogeographic Realm
古北界 Palearctic

WWF 生物群系 / WWF Biome
温带阔叶和混交林
Temperate Broadleaf & Mixed Forests

动物地理分布型 / Zoogeographic Distribution Type
Hc

分布标注 / Distribution Note
特有种 Endemic

▲ 濒危状况 / Threatened Status

中国生物多样性红色名录等级 / CB RL Category (2021)
数据缺乏 DD

IUCN 红色名录 / IUCN Red List (2021)
未评定 NE

威胁因子 / Threats
未知 Unknown

▲ 法律保护地位 / Legal Protection Status

国家重点保护野生动物等级 / Category of National Key Protected Wild Animals (2021)
未列入 Not listed

"三有" 名录 / TWIESSV (2023)
未列入 Not listed

CITES 附录等级 / CITES Appendix (2023)
未列入 Not listed

迁徙物种公约附录 / CMS Appendix (2020)
未列入 Not listed

保护行动 / Conservation Action
已发现的种群位于保护区内
Known populations are in nature reserves

▲ 参考文献 / References

Jiang et al. (蒋志刚等), 2021; Liu et al. (刘少英等), 2020; Wilson and Mittermeier, 2018; Smith et al., 2009; Pan et al. (潘清华等), 2007; Wang (王应祥), 2003

65 / 云南长尾鼩鼱

Episoriculus umbrinus Allen, 1923

· Hidden Brown-toothed Shrew

▲ 分类地位 / Taxonomy

劳亚食虫目 Eulipotyphla / 鼩鼱科 Soricidae / 须弥鼩鼱属 *Episoriculus*

科建立者及其文献 / Family Authority
G. Fischer, 1814

属建立者及其文献 / Genus Authority
Ellerman & Morrison-Scott, 1966

亚种 / Subspecies
无 None

模式标本产地 / Type Locality
中国
Yunnan, China

▲ 其他名称 / Other Name(s)

其他中文名 / Other Chinese Name(s)
无 None

其他英文名 / Other English Name(s)
无 None

同物异名 / Synonym(s)
无 None

▲ 形态及生境 / Morphology and Habitat

形态特征 / Morphological Characteristics

齿式：3.1.2.3/2.0.1.3=30。头体长 47~74 mm。尾长 42~58 mm。后足长 10~14 mm。背腹毛色之间没有明显分界线，背部深棕色，腹部毛色逐渐过渡到灰白色。头骨脑颅隆起，但吻部较短。上门齿较短，但向前突出，超出上颌骨的前缘。牙齿有明显的红褐色色素沉积，但主要着色于牙齿的尖端。

Dental formula: 3.1.2.3/2.0.1.3=30. Head-body 44-74 mm, tail 42-58 mm, hind-foot 10-14 mm. Dorsal pelage is seal brown and ventral color only slightly paler than the dorsal, or brussels brown, tinged with gray. Tail length is similar to the head-body. The skull is well-domed and high. Rostrum is short. The apex of the upper incisor is short and projects anteriorly. Tip of teeth pigmented with light brown or red color.

生境 / Habitat

常绿阔叶林、落叶阔叶林、针叶林、杜鹃灌丛、高山草甸
Evergreen broad-leaved forest, deciduous broad-leaved forest, coniferous forest, rhododendron thicket, alpine meadow

▲ 地理分布 / Geographic Distribution

国内分布 / Domestic Distribution
云南、西藏、贵州
Yunnan, Tibet, Guizhou

全球分布 / World Distribution
中国、印度、缅甸、越南
China, India, Myanmar, Vietnam

生物地理界 / Biogeographic Realm
古北界、印度马来界
Palearctic, Indomalaya

WWF 生物群系 / WWF Biome
热带和亚热带湿润阔叶林、温带阔叶和混交林、温带针叶树森林
Temperate Conifer Forests Temperate Broadleaf & Mixed Forests, Tropical &
Subtropical Moist Broadleaf Forests, Temperate Conifer Forests

动物地理分布型 / Zoogeographic Distribution Type
Sb

分布标注 / Distribution Note
特有种 Endemic

▲ 濒危状况 / Threatened Status

中国生物多样性红色名录等级 / CB RL Category (2021)
未评定 NE

IUCN 红色名录 / IUCN Red List (2021)
未评定 NE

威胁因子 / Threats
未知 Unknown

▲ 法律保护地位 / Legal Protection Status

国家重点保护野生动物等级 / Category of National Key Protected Wild Animals (2021)
未列入 Not listed

"三有"名录 / TWIESSV (2023)
未列入 Not listed

CITES 附录等级 / CITES Appendix (2023)
未列入 Not listed

迁徙物种公约附录 / CMS Appendix (2020)
未列入 Not listed

保护行动 / Conservation Action
尚无保护行动 No conservation action so far

▲ 参考文献 / References

Burgin et al., 2020; Wilson and Mittermeier, 2018

66 / 高氏缺齿鼩

Chodsigoa caovansunga
Lunde, Musser & Son, 2003

• Van Sung's Shrew

▲ 分类地位 / Taxonomy

劳亚食虫目 Eulipotyphla / 鼩鼱科 Soricidae / 缺齿鼩属 *Chodsigoa*

科建立者及其文献 / Family Authority
G. Fischer, 1814

属建立者及其文献 / Genus Authority
Kastchenko, 1907

亚种 / Subspecies
无 None

模式标本产地 / Type Locality
越南
Vietnam, Ha Giang province, Vi Xuyen distr., Cao Bo Co mmune, Mt. Tay Con Linh II, 1500 m (22°5'27"N, 104°9'49"E)

▲ 其他名称 / Other Name(s)

其他中文名 / Other Chinese Name(s)
无 None

其他英文名 / Other English Name(s)
无 None

同物异名 / Synonym(s)
无 None

▲ 形态及生境 / Morphology and Habitat

形态特征 / Morphological Characteristics
齿式：3.1.1.3/2.0.1.3=28。头体长 58~74 mm。尾长 51~83 mm。后足长 14~16 mm。背部毛色灰褐色，腹部淡灰色。尾长接近头体长。尾明显双色，尾尖无毛。 头骨脑颅隆起，吻部较短较窄。上单尖齿 3 枚，为该属的属级分类特征。大部分牙齿仅齿尖存在浅褐色色素沉积。
Dental formula: 3.1.1.3/2.0.1.3=28. Head-body 58-74 mm, tail 51-83 mm, hind-foot 14-16 mm. Dorsal pelage is dark brownish grey, ventral paler. Tail is longer than or equal to the head-body. Tail is distinctly bicolored, and the tip is naked. Braincase is well-domed. Rostrum is short, and narrow. It has three upper unicuspids which is the characteristic of the genus. Pigmentation only presents on the tips of most teeth.

生境 / Habitat
亚热带常绿阔叶林、沟谷雨林、种植园
Subtropical moist forest, ravine rainforest, plantation

▲ 地理分布 / Geographic Distribution

国内分布 / Domestic Distribution
云南、广东 Yunnan, Guangdong

全球分布 / World Distribution
中国、越南 China, Vietnam

生物地理界 / Biogeographic Realm
印度马来界 Indomalaya

WWF 生物群系 / WWF Biome
热带和亚热带湿润阔叶林
Tropical & Subtropical Moist Broadleaf Forests

动物地理分布型 / Zoogeographic Distribution Type
Sb

分布标注 / Distribution Note
非特有种 Non-Endemic

▲ 濒危状况 / Threatened Status

中国生物多样性红色名录等级 / CB RL Category (2021)
极危 CR

IUCN 红色名录 / IUCN Red List (2021)
数据缺乏 DD

威胁因子 / Threats
未知 Unknown

▲ 法律保护地位 / Legal Protection Status

国家重点保护野生动物等级 / Category of National Key Protected Wild Animals (2021)
未列入 Not listed

"三有" 名录 / TWIESSV (2023)
未列入 Not listed

CITES 附录等级 / CITES Appendix (2023)
未列入 Not listed

迁徙物种公约附录 / CMS Appendix (2020)
未列入 Not listed

保护行动 / Conservation Action
尚无保护行动 No conservation action so far

▲ 参考文献 / References

Jiang et al. (蒋志刚等), 2021; Burgin et al., 2020; IUCN, 2020; Wilson and Mittermeier, 2018; He et al. (何锴等), 2012; Smith et al., 2009; Lunde et al., 2003

67 / 大别山缺齿鼩

Chodsigoa dabieshanensis
Chen, Hu, Pei, Yang, Yong, Xu, Qu, Onditi
& Zhang, 2022

• Dabieshan Long-tailed Shrew

劳亚食虫目 Eulipotyphla / 鼩鼱科 Soricidae / 缺齿鼩属 *Chodsigoa*

科建立者及其文献 / Family Authority
G. Fischer, 1814

属建立者及其文献 / Genus Authority
Kastchenko, 1907

亚种 / Subspecies
无 None

模式标本产地 / Type Locality
中国
Foziling Natural Reserve (31.119° N, 116.245° E, 1187 m a. s. l.), at the Dabie Mountains, Huoshan County, Luan City, Anhui Province, China

▲ 其他名称 / Other Name(s)

其他中文名 / Other Chinese Name(s)
无 None

其他英文名 / Other English Name(s)
无 None

同物异名 / Synonym(s)
无 None

▲ 形态及生境 / Morphology and Habitat

形态特征 / Morphological Characteristics
齿式：3.1.1.3/2.0.1.3=28。门齿、单尖齿和第 4 前白齿的齿尖有栗色或棕红色色素沉积。头体长 67~73 mm。颅全长 18.5~19.5 mm。耳长 8~9 mm。后足长 13~14 mm。尾长 54~64 mm。头骨脑颅扁平。背部深褐色，腹部颜色略浅，背腹毛色无明显分界线。足背面浅棕色，边缘变淡。尾短于头体长，上面深褐色，下面稍淡，异色不明显，尾末端有一簇短毛。
Dental formula: 3.1.1.3/2.0.1.3=28. The tips of incisors, unicuspids, and the 4th premolars are light chestnut or reddish-brown pigmented. Head and body length 67-73 mm, condyloincisive length 18.5-19.5 mm, hindfoot length 13-14 mm, ear length 8-9 mm, tail length 54-64 mm. The braincase is markedly flattened. Dorsal pelage is dark brown, the ventral pelage is slightly paler. Dorsum of hands and feet are covered with short brown hairs, lighter at the margin. Tail shorter than the head-body length, brown above, slightly paler below, and with a small tuft of longer hairs at the tip.

生境 / Habitat
常绿阔叶林
Evergreen broadleaved forest

▲ 地理分布 / Geographic Distribution

国内分布 / Domestic Distribution
安徽 Anhui

全球分布 / World Distribution
中国 China

生物地理界 / Biogeographic Realm
古北界 Palearctic

WWF 生物群系 / WWF Biome
常绿阔叶林
Evergreen Broadleaved Forests

动物地理分布型 / Zoogeographic Distribution Type
Sn(j)

分布标注 / Distribution Note
特有种 Endemic

▲ 濒危状况 / Threatened Status

中国生物多样性红色名录等级 / CB RL Category (2021)
未评定 NE

IUCN 红色名录 / IUCN Red List (2021)
未评定 NE

威胁因子 / Threats
未知 Unknown

▲ 法律保护地位 / Legal Protection Status

国家重点保护野生动物等级 / Category of National Key Protected Wild Animals (2021)
未列入 Not listed

"三有" 名录 / TWIESSV (2023)
未列入 Not listed

CITES 附录等级 / CITES Appendix (2023)
未列入 Not listed

迁徙物种公约附录 / CMS Appendix (2020)
未列入 Not listed

保护行动 / Conservation Action
已发现的种群位于保护区内
Known populations are in nature reserves

▲ 参考文献 / References

Chen et al., 2022

68 / 烟黑缺齿鼩

Chodsigoa furva Anthony, 1941

· Dusky Long-tailed Shrew

劳亚食虫目 Eulipotyphla / 鼩鼱科 Soricidae / 缺齿鼩属 *Chodsigoa*

科建立者及其文献 / Family Authority
G. Fischer, 1814

属建立者及其文献 / Genus Authority
Kastchenko, 1907

亚种 / Subspecies
无 None

模式标本产地 / Type Locality
缅甸
Imaw Bum, North Myanmar. Altitude: 2750 m

▲ 其他名称 / Other Name(s)

其他中文名 / Other Chinese Name(s)
无 None

其他英文名 / Other English Name(s)
Dark Brown-toothed Shrew, Dusky Long-tailed Shrew

同物异名 / Synonym(s)
无 None

▲ 形态及生境 / Morphology and Habitat

形态特征 / Morphological Characteristics
齿式：3.1.1.3/2.0.1.3=28。头体长 69~75 mm。后足 16~18 mm。尾长 84~87 mm。颅全长 19~23 mm。背部毛色深灰色，腹部毛色灰白色。尾长短于头体长，尾尖裸露。脑颅隆起程度较低。上单尖齿 3 枚。门齿至第一前臼齿部分齿尖有猩红色色素沉淀。
Dental formula: 3.1.1.3/2.0.1.3=28. Head-body 69-75 mm, tail 84-87 mm, hind-foot 16-18 mm, condyloincisive length 19-23 mm. Dorsal pelage dark grey, ventral pelage whitish gray. No tuft of longer hair on the tip of the tail. Braincase is dome-shaped and low. It has three upper unicuspids. Incisors through the first molar are pigmented on tips of the teeth.

生境 / Habitat
海拔 2000 m 以上的山地
Mountians with elevations higher than 2000 m

▲ 地理分布 / Geographic Distribution

国内分布 / Domestic Distribution
云南、西藏 Yunnan, Tibet

全球分布 / World Distribution
中国、缅甸 China, Myanmar

生物地理界 / Biogeographic Realm
古北界 Palearctic

WWF 生物群系 / WWF Biome
温带针叶树森林
Temperate Conifer Forests

动物地理分布型 / Zoogeographic Distribution Type
Hc

分布标注 / Distribution Note
非特有种 Non-Endemic

▲ 濒危状况 / Threatened Status

中国生物多样性红色名录等级 / CB RL Category (2021)
数据缺乏 DD

IUCN 红色名录 / IUCN Red List (2021)
未评定 NE

威胁因子 / Threats
未知 Unknown

▲ 法律保护地位 / Legal Protection Status

国家重点保护野生动物等级 / Category of National Key Protected Wild Animals (2021)
未列入 Not listed

"三有" 名录 / TWIESSV (2023)
未列入 Not listed

CITES 附录等级 / CITES Appendix (2023)
未列入 Not listed

迁徙物种公约附录 / CMS Appendix (2020)
未列入 Not listed

保护行动 / Conservation Action
尚无保护行动 No conservation action so far

▲ 参考文献 / References

Jiang et al. (蒋志刚等), 2021; Liu et al. (刘少英等), 2020; Wilson and Mittermeier, 2018; Chen et al., 2017; Zhang et al.(张敏等), 2021

69 / 霍氏缺齿鼩

Chodsigoa hoffmanni
Chen, He, Huang & Jiang, 2017

• Hoffmann's Long-tailed Shrew

▲ 分类地位 / Taxonomy

劳亚食虫目 Eulipotyphla / 鼩鼱科 Soricidae / 缺齿鼩属 *Chodsigoa*

科建立者及其文献 / Family Authority
G. Fischer, 1814

属建立者及其文献 / Genus Authority
Kastchenko, 1907

亚种 / Subspecies
无 None

模式标本产地 / Type Locality
中国
The eastern slope of the Ailao Mountain (24.590 N, 101.508 E, 2600 m a. s. l.), Shuangbai, Chuxiong, Yunnan, China

何锴 / 供图

▲ 其他名称 / Other Name(s)

其他中文名 / Other Chinese Name(s)
无 None

其他英文名 / Other English Name(s)
无 None

同物异名 / Synonym(s)
无 None

▲ 形态及生境 / Morphology and Habitat

形态特征 / Morphological Characteristics

齿式：3.1.1.3/2.0.1.3=28。上单尖齿 3 枚。头体长 58~75 mm。尾长 74~88 mm。后足 14~17 mm。耳长 7~11 mm。背部颜色深灰，腹部颜色略浅，背腹毛色无明显分界线。尾尖有一簇短毛。后足和尾较长，可能是适应树栖生活的特征。牙齿有棕红色的色素沉积，但只有门齿、单尖齿和第四前白齿的齿尖色素沉积较为明显。

Dental formula: 3.1.1.3/2.0.1.3=28. Head-body length 58-75 mm, tail length 74-88 mm, hind foot length 14-17 mm, ear length 7-11 mm. Dorsal pelage slate grey, ventral pelage paler than the dorsum. The tail and the hindfeet are long, adapted to an arboreal lifestyle. Dorsal surface of tail pigmented brown, ventral surface paler. Tip of tail white with a tuft of slightly longer hair. Reddish brown pigmentation present on incisors, unicuspids, and the upper P4.

生境 / Habitat

常绿阔叶林 Evergreen broad-leaved forest

▲ 地理分布 / Geographic Distribution

国内分布 / Domestic Distribution
云南、贵州、四川、重庆、湖北
Yunnan, Guizhou, Sichuan, Chongqing, Hubei

全球分布 / World Distribution
中国、越南 China, Vietnam

生物地理界 / Biogeographic Realm
古北界、印度马来界 Palearctic, Indomalaya

WWF 生物群系 / WWF Biome
温带针叶树森林
Temperate Conifer Forests

动物地理分布型 / Zoogeographic Distribution Type
Sb

分布标注 / Distribution Note
非特有种 Non-Endemic

▲ 濒危状况 / Threatened Status

中国生物多样性红色名录等级 / CB RL Category (2021)
无危 LC

IUCN 红色名录 / IUCN Red List (2021)
未评定 NE

威胁因子 / Threats
未知 Unknown

▲ 法律保护地位 / Legal Protection Status

国家重点保护野生动物等级 / Category of National Key Protected Wild Animals (2021)
未列入 Not listed

"三有"名录 / TWIESSV (2023)
未列入 Not listed

CITES 附录等级 / CITES Appendix (2023)
未列入 Not listed

迁徙物种公约附录 / CMS Appendix (2020)
未列入 Not listed

保护行动 / Conservation Action
尚无保护行动 No conservation action so far

▲ 参考文献 / References

Bo (柏阳), 2021; Hiroaki et al., 2021; Jiang et al. (蒋志刚等), 2021; Let at al. (雷博宇等), 2019; Wilson and Mittermeier, 2018; Chen et al., 2017

70 | 川西缺齿鼩鼱

Chodsigoa hypsibia (de Winton, 1899)

- De Winton's Shrew

▲ 分类地位 / Taxonomy

劳亚食虫目 Eulipotyphla / 鼩鼱科 Soricidae / 缺齿鼩属 *Chodsigoa*

科建立者及其文献 / Family Authority
G. Fischer, 1814

属建立者及其文献 / Genus Authority
Kastchenko, 1907

亚种 / Subspecies
指名亚种 *C. h. hypsibia* Winton, 1899
四川、云南、西藏、青海、陕西、河南
Sichuan, Yunnan, Tibet, Qinghai, Shaanxi, Henan

华北亚种 *C. h. larvarum* Thomas 1915
河北、北京、山西
Hebei, Beijing, Shanxi

模式标本产地 / Type Locality
中国
China, Sichuan, "Yang-liu-pa"

刘洋 / 供图

▲ 其他名称 / Other Name(s)

其他中文名 / Other Chinese Name(s)
无 None

其他英文名 / Other English Name(s)
De Winton's Brown-toothed Shrew

同物异名 / Synonym(s)
无 None

▲ 形态及生境 / Morphology and Habitat

形态特征 / Morphological Characteristics
齿式：3.1.1.3/2.0.1.3=28。头体长 62~86 mm。尾长 56~73 mm。后足 13~18 mm。颅全长 19~22.6 mm。背部毛色青灰色，腹部棕灰色。尾长短于头体长，尾背腹异色不明显，尾尖有一撮短毛。头骨脑颅不隆起，颅部扁平。上单尖齿 3 枚。上门齿前缘向前突出，齿尖向下。
Dental formula: 3.1.1.3/2.0.1.3=28. Head-body 62-86 mm, tail 56-73 mm, hind-foot 13-18 mm, condyloincisive length 19-22.6 mm. Dorsal pelage is slate grey, ventral brownish grey. Tail is shorter than the head-body, and not sharply bicolored. A tuft of slightly longer hair on tip of the tail. Skull is low and braincase markedly flattened. It has three upper unicuspids. The apex of the upper incisor projects anteriorly and the tip is perpendicular to the skull.

生境 / Habitat
海拔 300~3500 m 的灌木、常绿阔叶林、针叶林以及农田
Shrubs, evergreen broad-leaved forests, coniferous forest and farmland at altitudes of 300-3500 meters

▲ 地理分布 / Geographic Distribution

国内分布 / Domestic Distribution
北京、河北、河南、四川、云南、甘肃、西藏、陕西、山西、重庆
Beijing, Hebei, Henan, Sichuan, Yunnan, Gansu, Tibet, Shaanxi, Shanxi, Chongqing

全球分布 / World Distribution
中国 China

生物地理界 / Biogeographic Realm
古北界 Palearctic

WWF 生物群系 / WWF Biome
热带和亚热带湿润阔叶林
Tropical & Subtropical Moist Broadleaf Forests

动物地理分布型 / Zoogeographic Distribution Type
Hc+D+L

分布标注 / Distribution Note
特有种 Endemic

▲ 濒危状况 / Threatened Status

中国生物多样性红色名录等级 / CB RL Category (2021)
无危 LC

IUCN 红色名录 / IUCN Red List (2021)
无危 LC

威胁因子 / Threats
未知 Unknown

▲ 法律保护地位 / Legal Protection Status

国家重点保护野生动物等级 / Category of National Key Protected Wild Animals (2021)
未列入 Not listed

"三有" 名录 / TWIESSV (2023)
未列入 Not listed

CITES 附录等级 / CITES Appendix (2023)
未列入 Not listed

迁徙物种公约附录 / CMS Appendix (2020)
未列入 Not listed

保护行动 / Conservation Action
尚无保护行动 No conservation action so far

▲ 参考文献 / References

Jiang et al. (蒋志刚等), 2021; Burgin et al., 2020; IUCN, 2020; Wilson and Mittermeier, 2018; Tu et al. (涂飞云等), 2012; Liu et al. (刘洋等), 2011; Smith et al., 2009; China Species Information Service, 2008; Pan et al. (潘清华等), 2007; Liu et al. (刘少英等), 2005; Wilson and Reeder, 2005; Wang (王应祥), 2003; Lunde et al., 2003

71 / 云南缺齿鼩鼱

Chodsigoa parca G. M. Allen, 1923

• Lowe's Shrew

▲ 分类地位 / Taxonomy

劳亚食虫目 Eulipotyphla / 鼩鼱科 Soricidae / 缺齿鼩属 *Chodsigoa*

科建立者及其文献 / Family Authority
G. Fischer, 1814

属建立者及其文献 / Genus Authority
Kastchenko, 1907

亚种 / Subspecies
指名亚种 *G. p. parca* G. Allen, 1923
云南、四川和贵州
Yunnan, Sichuan and Guizhou

模式标本产地 / Type Locality
中国
"Ho-mu-shu Pass, Western Yunnan, China, 2438 m"

▲ 其他名称 / Other Name(s)

其他中文名 / Other Chinese Name(s)
无 None

其他英文名 / Other English Name(s)
无 None

同物异名 / Synonym(s)
无 None

▲ 形态及生境 / Morphology and Habitat

形态特征 / Morphological Characteristics
齿式：3.1.1.3/2.0.1.3=28。头体长 62~77 mm，尾长 77~99 mm，后足长 15~18 mm，颅全长 20.1~20.9 mm。外形与霍氏缺齿鼩相似，但体型更大。背面毛皮石板灰色，腹部颜色略浅。尾双色，尾尖有一簇短毛。脑颅拱形隆起。

Dental formula: 3.1.1.3/2.0.1.3=28. Head-body 62-77 mm, tail 77-99 mm, hind-foot 15-18 mm, condyloincisive length 20.1-20.9 mm. It is similar to *C. hoffmanni* but larger. Dorsal pelage slate grey and ventral pelage paler. Tail is longer than the head-body. Tail bicolored, dorsal surface brown, ventral surface creamy white. A tuft of slightly longer hair is located on the tip of tail. Braincase dome-shaped.

生境 / Habitat
亚热带湿润山地森林
Subtropical moist montane forest

▲ 地理分布 / Geographic Distribution

国内分布 / Domestic Distribution
云南 Yunnan

全球分布 / World Distribution
中国、缅甸、泰国、越南
China, Myanmar, Thailand, Vietnam

生物地理界 / Biogeographic Realm
印度马来界 Indomalaya

WWF 生物群系 / WWF Biome
温带阔叶和混交林
Temperate Broadleaf & Mixed Forests

动物地理分布型 / Zoogeographic Distribution Type
Sb

分布标注 / Distribution Note
非特有种 Non-Endemic

▲ 濒危状况 / Threatened Status

中国生物多样性红色名录等级 / CB RL Category (2021)
无危 LC

IUCN 红色名录 / IUCN Red List (2021)
无危 LC

威胁因子 / Threats
未知 Unknown

▲ 法律保护地位 / Legal Protection Status

国家重点保护野生动物等级 / Category of National Key Protected Wild Animals (2021)
未列入 Not listed

"三有" 名录 / TWIESSV (2023)
未列入 Not listed

CITES 附录等级 / CITES Appendix (2023)
未列入 Not listed

迁徙物种公约附录 / CMS Appendix (2020)
未列入 Not listed

保护行动 / Conservation Action
尚无保护行动 No conservation action so far

▲ 参考文献 / References

Jiang et al. (蒋志刚等), 2021; Burgin et al., 2020; IUCN, 2020; Liu et al. (刘少英等), 2020; Smith et al., 2009; Allen, 1923

72 / 滇北缺齿鼩

Chodsigoa parva G. M. Allen, 1923

· Pygmy Brown-toothed Shrew

▲ 分类地位 / Taxonomy

劳亚食虫目 Eulipotyphla / 鼩鼱科 Soricidae / 缺齿鼩属 *Chodsigoa*

科建立者及其文献 / Family Authority
G. Fischer, 1814

属建立者及其文献 / Genus Authority
Kastchenko, 1907

亚种 / Subspecies
无 None

模式标本产地 / Type Locality
中国
China, W Yunnan, Likiang Range, Ssushancheng

▲ 其他名称 / Other Name(s)

其他中文名 / Other Chinese Name(s)
滇北长尾鼩

其他英文名 / Other English Name(s)
Pygmy Red-toothed Shrew

同物异名 / Synonym(s)
无 None

▲ 形态及生境 / Morphology and Habitat

形态特征 / Morphological Characteristics
齿式：3.1.1.3/2.0.1.3=28。头体长 47~64 mm。尾长 41~52 mm。后足 10~12 mm。颅全长 15.1~16.2 mm。与川西缺齿鼩比例相似但更小，是缺齿鼩属最小的物种。背部毛色暗灰色，腹部毛色淡灰色。尾长小于头体长，尾尖有一簇毛。头骨脑颅不隆起，颅骨扁平。

Dental formula: 3.1.1.3/2.0.1.3=28. Head-body 47-64 mm, tail 41-52 mm, hind-foot 10-12 mm, condyloincisive length 15.1-16.2 mm. It is a small-sized shrew and is the smallest species *of Chodsigoa*. Proportionally similar to *C. hypsibia*. Dorsal pelage dark grey, and ventral pelage slightly paler grey. Tail is shorter than the head-body. A small tuft of hair at the tip of the tail. Skull is small and braincase is flattened.

生境 / Habitat
山地湿润阔叶林
Moist broadleaf forest in mountain area

▲ 地理分布 / Geographic Distribution

国内分布 / Domestic Distribution
云南、四川 Yunnan, Sichuan

全球分布 / World Distribution
中国 China

生物地理界 / Biogeographic Realm
古北界 Palearctic

WWF 生物群系 / WWF Biome
温带阔叶和混交林
Temperate Broadleaf & Mixed Forests

动物地理分布型 / Zoogeographic Distribution Type
Sb

分布标注 / Distribution Note
特有种 Endemic

▲ 濒危状况 / Threatened Status

中国生物多样性红色名录等级 / CB RL Category (2021)
近危 NT

IUCN 红色名录 / IUCN Red List (2021)
数据缺乏 DD

威胁因子 / Threats
未知 Unknown

▲ 法律保护地位 / Legal Protection Status

国家重点保护野生动物等级 / Category of National Key Protected Wild Animals (2021)
未列入 Not listed

"三有" 名录 / TWIESSV (2023)
未列入 Not listed

CITES 附录等级 / CITES Appendix (2023)
未列入 Not listed

迁徙物种公约附录 / CMS Appendix (2020)
未列入 Not listed

保护行动 / Conservation Action
尚无保护行动 No conservation action so far

▲ 参考文献 / References

Jiang et al. (蒋志刚等), 2021; Burgin et al., 2020; IUCN, 2020; Liu et al. (刘少英等), 2020; Wilson and Mittermeier, 2018; Smith et al., 2009; Pan et al. (潘清华等), 2007; Wilson and Reeder, 2005; Wang and Xie, 2004

73 / 大缺齿鼩鼱

Chodsigoa salenskii (Kastschenko, 1907)

· Salenski's Shrew

▲ 分类地位 / Taxonomy

劳亚食虫目 Eulipotyphla / 鼩鼱科 Soricidae / 缺齿鼩属 *Chodsigoa*

科建立者及其文献 / Family Authority
G. Fischer, 1814

属建立者及其文献 / Genus Authority
Kastchenko, 1907

亚种 / Subspecies
无 None

模式标本产地 / Type Locality
中国
China, Sichuan, "Lun-ngan'-fu" (Long'an fu=Pingwu)

▲ 其他名称 / Other Name(s)

其他中文名 / Other Chinese Name(s)
无 None

其他英文名 / Other English Name(s)
Salenski's Brown-toothed Shrew

同物异名 / Synonym(s)
无 None

▲ 形态及生境 / Morphology and Habitat

形态特征 / Morphological Characteristics
齿式：3.1.1.3/2.0.1.3=28。头体长 78 mm。尾长 110 mm。后足长 22 mm。颅全长 25 mm。仅模式标本一号标本。形态上与斯氏缺齿鼩鼱相似，但体型更大，二者分类学关系仍有待于厘定，二者可能为同物异名。
Dental formula: 3.1.1.3/2.0.1.3=28. Head-body length 78 mm, hind foot length 22 mm, tail length 110 mm, greatest skull length 25 mm. It is only represented by a single holotype specimen. Morphologically similar to *Chodsigoa smithii*, but larger. The taxonomic relationships between *C. smithii* and *C. salenskii* need to be revised. These two might be synonymous.

生境 / Habitat
温带森林
Temperate forest

▲ 地理分布 / Geographic Distribution

国内分布 / Domestic Distribution
四川 Sichuan

全球分布 / World Distribution
中国 China

生物地理界 / Biogeographic Realm
古北界 Palearctic

WWF 生物群系 / WWF Biome
温带阔叶和混交林
Temperate Broadleaf & Mixed Forests

动物地理分布型 / Zoogeographic Distribution Type
Qd

分布标注 / Distribution Note
特有种 Endemic

▲ 濒危状况 / Threatened Status

中国生物多样性红色名录等级 / CB RL Category (2021)
数据缺乏 DD

IUCN 红色名录 / IUCN Red List (2021)
数据缺乏 DD

威胁因子 / Threats
未知 Unknown

▲ 法律保护地位 / Legal Protection Status

国家重点保护野生动物等级 / Category of National Key Protected Wild Animals (2021)
未列入 Not listed

"三有"名录 / TWIESSV (2023)
未列入 Not listed

CITES 附录等级 / CITES Appendix (2023)
未列入 Not listed

迁徙物种公约附录 / CMS Appendix (2020)
未列入 Not listed

保护行动 / Conservation Action
尚无保护行动 No conservation action so far

▲ 参考文献 / References

Jiang et al. (蒋志刚等), 2021; Burgin et al., 2020; IUCN, 2020; Liu et al. (刘少英等), 2020; Wilson and Mittermeier, 2018; Chen et al., 2017; Sun et al. (孙治宇等), 2013; Tu et al. (涂飞云等), 2012; Smith et al., 2009; Qin et al. (秦岭等), 2007; Pan et al. (潘清华等), 2007; Wilson and Reeder, 2005; Wang (王应祥), 2003

74 / 斯氏缺齿鼩鼱

Chodsigoa smithii Thomas, 1911

· Smith's Shrew

▲ 分类地位 / Taxonomy

劳亚食虫目 Eulipotyphla / 鼩鼱科 Soricidae / 缺齿鼩属 *Chodsigoa*

科建立者及其文献 / Family Authority
G. Fischer, 1814

属建立者及其文献 / Genus Authority
Kastchenko, 1907

亚种 / Subspecies
无 None

模式标本产地 / Type Locality
中国
China, Sichuan, "Ta-tsien-lu"

何鑫 / 供图

▲ 其他名称 / Other Name(s)

其他中文名 / Other Chinese Name(s)
大长尾鼩 、史密斯长尾鼩

其他英文名 / Other English Name(s)
Smith's Brown-toothed Shrew

同物异名 / Synonym(s)
无 None

▲ 形态及生境 / Morphology and Habitat

形态特征 / Morphological Characteristics

齿式：3.1.1.3/2.0.1.3=28。头体长 76~84 mm。尾长 93~110 mm。后足长 16~20 mm。颅全长 21.5~23.1 mm。背腹均为深灰色。尾背腹近乎同色，长有稀疏的短毛，尾尖无毛。头骨脑颅隆起程度较低。上颌两侧上单尖齿近乎平行。

Dental formula: 3.1.1.3/2.0.1.3=28. Head-body length 76-84 mm, tail length 93-110 mm, hind foot length 16-20 mm, cranial length 21.5-23.1 mm. Dorsal and ventral pelage is generally dark grey. Tail is longer than the head-body length and is not sharply bicolored; it is covered by short hairs and naked at the tip. Braincase low. Rostrum is narrow, and sharply narrowed in the front.

生境 / Habitat
温带森林
Temperate forest

▲ 地理分布 / Geographic Distribution

国内分布 / Domestic Distribution
四川、陕西、贵州、云南
Sichuan, Shaanxi, Guizhou, Yunnan

全球分布 / World Distribution
中国 China

生物地理界 / Biogeographic Realm
古北界 Palearctic

WWF 生物群系 / WWF Biome
温带阔叶和混交林
Temperate Broadleaf & Mixed Forests

动物地理分布型 / Zoogeographic Distribution Type
Sb

分布标注 / Distribution Note
特有种 Endemic

▲ 濒危状况 / Threatened Status

中国生物多样性红色名录等级 / CB RL Category (2021)
无危 LC

IUCN 红色名录 / IUCN Red List (2021)
近危 NT

威胁因子 / Threats
未知 Unknown

▲ 法律保护地位 / Legal Protection Status

国家重点保护野生动物等级 / Category of National Key Protected Wild Animals (2021)
未列入 Not listed

"三有"名录 / TWIESSV (2023)
未列入 Not listed

CITES 附录等级 / CITES Appendix (2023)
未列入 Not listed

迁徙物种公约附录 / CMS Appendix (2020)
未列入 Not listed

保护行动 / Conservation Action
尚无保护行动 No conservation action so far

▲ 参考文献 / References

Jiang et al. (蒋志刚等), 2021; Burgin et al., 2020; IUCN, 2020; Wilson and Mittermeier, 2018; Chen et al., 2017; Tu et al. (涂飞云等), 2012; Smith et al., 2009; Qin et al. (秦岭等), 2007; Pan et al. (潘清华等), 2007; Wilson and Reeder, 2005; Wang (王应祥), 2003

75 | 细尾缺齿鼩鼱

Chodsigoa sodalis Thomas, 1913

• Lesser Taiwanese Shrew

▲ 分类地位 / Taxonomy

劳亚食虫目 Eulipotyphla / 鼩鼱科 Soricidae / 缺齿鼩属 *Chodsigoa*

科建立者及其文献 / Family Authority
G. Fischer, 1814

属建立者及其文献 / Genus Authority
Kastchenko, 1907

亚种 / Subspecies
无 None

模式标本产地 / Type Locality
中国
China, Central Taiwan, Mt. Arizan, 2438 m

▲ 其他名称 / Other Name(s)

其他中文名 / Other Chinese Name(s)
阿里山天鹅绒尖鼠、细尾长尾鼩

其他英文名 / Other English Name(s)
Koshun Shrew

同物异名 / Synonym(s)
无 None

▲ 形态及生境 / Morphology and Habitat

形态特征 / Morphological Characteristics
齿式：3.1.1.3/2.0.1.3=28。头体长 65~71 mm。后足长 13~15 mm。尾长 64~73 mm。颅全长 18~18.4 mm。头骨骨质致密，脑颅隆起。背部深灰棕色，向腹部逐渐过渡为深灰色。尾长接近于头体长，尾暗褐色。
Dental formula: 3.1.1.3/2.0.1.3=28. Head-body 65-71 mm, tail 64-73 mm, hindfoot 13-15 mm, condyloincisive length 18-18.4 mm. Skull is robust, braincase domeshaped but low. Dorsal pelage is dark grey, ventral pelage is slightly paler. Tail is solid brown in color, tail length approximately equal to the head-body length.

生境 / Habitat
亚热带高海拔灌丛
Subtropical high altitude shrubland

▲ 地理分布 / Geographic Distribution

国内分布 / Domestic Distribution
台湾 Taiwan

全球分布 / World Distribution
中国 China

生物地理界 / Biogeographic Realm
印度马来界 Indomalaya

WWF 生物群系 / WWF Biome
热带和亚热带湿润阔叶林
Tropical & Subtropical Moist Broadleaf Forests

动物地理分布型 / Zoogeographic Distribution Type
J

分布标注 / Distribution Note
特有种 Endemic

▲ 濒危状况 / Threatened Status

中国生物多样性红色名录等级 / CB RL Category (2021)
数据缺乏 DD

IUCN 红色名录 / IUCN Red List (2021)
数据缺乏 DD

威胁因子 / Threats
未知 Unknown

▲ 法律保护地位 / Legal Protection Status

国家重点保护野生动物等级 / Category of National Key Protected Wild Animals (2021)
未列入 Not listed

"三有" 名录 / TWIESSV (2023)
未列入 Not listed

CITES 附录等级 / CITES Appendix (2023)
未列入 Not listed

迁徙物种公约附录 / CMS Appendix (2020)
未列入 Not listed

保护行动 / Conservation Action
尚无保护行动 No conservation action so far

▲ 参考文献 / References

Jiang et al. (蒋志刚等), 2021; Burgin et al., 2020; IUCN, 2020; Liu et al. (刘少英等), 2020; Wilson and Mittermeier, 2018; Smith et al., 2009; Pan et al. (潘清华等), 2007; Wilson and Reeder, 2005; Wang (王应祥), 2003; Motokawa et al., 1998; Yu et al.,1993

76 / 水鼩鼱

Neomys fodiens (Pennant, 1771)

· Eurasian Water Shrew

劳亚食虫目 Eulipotyphla / 鼩鼱科 Soricidae / 水鼩鼱属 *Neomys*

科建立者及其文献 / Family Authority
G. Fischer, 1814

属建立者及其文献 / Genus Authority
Kaup, 1829

亚种 / Subspecies
远东亚种 *N. f. orientis* Thomas, 1914
吉林和新疆
Jilin and Xinjiang

模式标本产地 / Type Locality
德国
Germany, Berlin

▲ 其他名称 / Other Name(s)

其他中文名 / Other Chinese Name(s)
无 None

其他英文名 / Other English Name(s)
Northern Water Shrew, Water Shrew

同物异名 / Synonym(s)
无 None

▲ 形态及生境 / Morphology and Habitat

形态特征 / Morphological Characteristics

齿式：3.1.2.3/2.0.1.3=30。头体长 75~103 mm。后足长 16~21 mm。尾长 58~73 mm。体重 8.5~25 g。毛发较陆生鼩鼱更短。背部和两侧暗棕色，腹部银灰色。尾长通常超过头体长的 65%，尾背面形成龙骨状凸起。尾双色，背面黑色，腹面银灰色。前后足侧面由白色的毛发构成流苏。

Dental formula: 3.1.2.3/2.0.1.3=30. Head-body length 75-103 mm, hind foot length 16-21 mm, tail length 58-73 mm, weight 8.5-25 g. Pelage is relatively short, back and sides are blackish brown or black, belly is silvery gray. Tail is usually longer than 65% of head-body length. Keel occurs along entire length of tail. Tail is bicolored, with dark color above contrasting with silvery gray of below. Swimming borders on feet are well formed with long stiff hairs.

生境 / Habitat

森林、草地 Forest, grassland

▲ 地理分布 / Geographic Distribution

国内分布 / Domestic Distribution
黑龙江、吉林、新疆 Heilongjiang, Jilin, Xinjiang

全球分布 / World Distribution
阿尔巴尼亚、奥地利、白俄罗斯、比利时、波斯尼亚和黑塞哥维那、保加利亚、中国、克罗地亚、捷克、丹麦、爱沙尼亚、芬兰、法国、德国、希腊、匈牙利、意大利、哈萨克斯坦、朝鲜、拉脱维亚、列支敦士登、立陶宛、卢森堡、马其顿、摩尔多瓦、蒙古国、黑山、荷兰、挪威、波兰、罗马尼亚、俄罗斯、塞尔维亚、斯洛伐克、斯洛文尼亚、西班牙、瑞典、瑞士、土耳其、乌克兰、英国
Albania, Austria, Belarus, Belgium, Bosnia and Herzegovina, Bulgaria, China, Croatia, Czech, Danmark, Estonia, Finland, France, Germany, Greece, Hungary, Italy, Kazakhstan, Democratic People's Republic of Korea, Latvia, Liechtenstein, Lithuania, Luxembourg, Macedonia, Moldova, Mongolia, Montenegro, Netherlands, Norway, Poland, Romania, Russia, Serbia, Slovakia, Slovenia, Spain, Sweden, Switzerland, Turkey, Ukraine, United Kingdom

生物地理界 / Biogeographic Realm
古北界 Palearctic

WWF 生物群系 / WWF Biome
北方森林 / 针叶林 Boreal Forests/Coniferous Forests

动物地理分布型 / Zoogeographic Distribution Type
Ub

分布标注 / Distribution Note
非特有种 Non-Endemic

▲ 濒危状况 / Threatened Status

中国生物多样性红色名录等级 / CB RL Category (2021)
易危 VU

IUCN 红色名录 / IUCN Red List (2021)
无危 LC

威胁因子 / Threats
耕地、农业林业污染、湿地排水 Non-timber crops, agricultural or forestry effluents, wetland drainage

▲ 法律保护地位 / Legal Protection Status

国家重点保护野生动物等级 / Category of National Key Protected Wild Animals (2021)
未列入 Not listed

"三有" 名录 / TWIESSV (2023)
未列入 Not listed

CITES 附录等级 / CITES Appendix (2023)
未列入 Not listed

迁徙物种公约附录 / CMS Appendix (2020)
未列入 Not listed

保护行动 / Conservation Action
尚无保护行动 No conservation action so far

▲ 参考文献 / References

Jiang et al. (蒋志刚等), 2021; Burgin et al., 2020; IUCN, 2020; Liu et al. (刘少英等), 2020; Wilson and Mittermeier, 2018; Smith et al., 2009; Pan et al. (潘清华等), 2007; Wilson and Reeder, 2005; Wang (王应祥), 2003; Huang et al. (黄乃伟等), 2012; Zhang (张荣祖), 1997; Wang (王东风), 1993; Pu and Yu (朴仁峰和俞曙林), 1990; Xia (夏武平), 1988; Wang (汪松), 1958

77 | 阿萨姆短尾鼩

Anourosorex assamensis Anderson, 1875

• Assma Mole Shrew

▲ 分类地位 / Taxonomy

劳亚食虫目 Eulipotyphla / 鼩鼱科 Soricidae / 短尾鼩属 *Anourosorex*

科建立者及其文献 / Family Authority
G. Fischer, 1814

属建立者及其文献 / Genus Authority
Milne-Edwards, 1872

亚种 / Subspecies
无 None

模式标本产地 / Type Locality
中国
Bombdila, Zangnan, China, 2700 m

▲ 其他名称 / Other Name(s)

其他中文名 / Other Chinese Name(s)
无 None

其他英文名 / Other English Name(s)
无 None

同物异名 / Synonym(s)
无 None

▲ 形态及生境 / Morphology and Habitat

形态特征 / Morphological Characteristics
齿式：3.0.1.3/2.0.1.3=26。头体长 85~119 mm。后足长 10~15 mm。尾长 14~20 mm。形态与微尾鼩 *Anourosorex squamipes* 相似，但体型更大。通体毛色黑色，足背和尾白色。眼很小，外耳严重退化，被毛发遮盖。上颌骨和下颌骨粗壮。仅有 2 枚上单尖齿，第一枚较长，第二枚退化明显。该属物种牙齿没有色素沉积。

Dental formula: 3.0.1.3/2.0.1.3=26. Head-body length 85-119 mm, hindfoot length 10-15 mm, tail length 14-20 mm. Morphologically similar to *Anourosorex squamipes* but larger. Pelage is uniformly black, feet and tail are white. Eyes minute. Ears are highly reduced and covered by the fur. Skull and mandible are robust. It has two upper unicuspids, the first one is long and the second is much reduced. Cusps of teeth are unpigmented as in all Anourosorex species.

生境 / Habitat
森林、农田、村庄
Forest, farmland, village

▲ 地理分布 / Geographic Distribution

国内分布 / Domestic Distribution
西藏 Tibet

全球分布 / World Distribution
中国、印度、缅甸
China, India, Myanmar

生物地理界 / Biogeographic Realm
印度马来界 Indomalaya

WWF 生物群系 / WWF Biome
热带和亚热带湿润阔叶林
Tropical & Subtropical Moist Broadleaf Forests

动物地理分布型 / Zoogeographic Distribution Type
Ha

分布标注 / Distribution Note
非特有种 Non-Endemic

▲ 濒危状况 / Threatened Status

中国生物多样性红色名录等级 / CB RL Category (2021)
未评定 NE

IUCN 红色名录 / IUCN Red List (2021)
无危 LC

威胁因子 / Threats
未知 Unknown

▲ 法律保护地位 / Legal Protection Status

国家重点保护野生动物等级 / Category of National Key Protected Wild Animals (2021)
未列入 Not listed

"三有"名录 / TWIESSV (2023)
未列入 Not listed

CITES 附录等级 / CITES Appendix (2023)
未列入 Not listed

迁徙物种公约附录 / CMS Appendix (2020)
未列入 Not listed

保护行动 / Conservation Action
尚无保护行动 No conservation action so far

▲ 参考文献 / References

Burgin et al., 2020; IUCN, 2020; Wilson and Mittermeier, 2018; Kawada et al., 2014

78 / 大短尾鼩

Anourosorex schmidi Petter, 1963

· Giant Mole Shrew

劳亚食虫目 Eulipotyphla / 鼩鼱科 Soricidae / 短尾鼩属 *Anourosorex*

科建立者及其文献 / Family Authority
G. Fischer, 1814

属建立者及其文献 / Genus Authority
Milne-Edwards, 1872

亚种 / Subspecies
无 None

模式标本产地 / Type Locality
中国
Bombdila (= Cona, Shannan, Tibet), China, 2700 m

▲ 其他名称 / Other Name(s)

其他中文名 / Other Chinese Name(s)
无 None

其他英文名 / Other English Name(s)
无 None

同物异名 / Synonym(s)
无 None

▲ 形态及生境 / Morphology and Habitat

形态特征 / Morphological Characteristics
齿式：3.0.1.3/2.0.1.3=26。缺乏研究的物种。形态与同属其他物种相似，但头骨更长。
Dental formula: 3.0.1.3/2.0.1.3=26. It was described as similar to but different from the other species of *Anourosorex* by a longer skull. Understudied species.

生境 / Habitat
亚热带湿润山地森林
Subtropical moist montane forest

▲ 地理分布 / Geographic Distribution

国内分布 / Domestic Distribution
西藏 Tibet

全球分布 / World Distribution
中国、不丹、印度
China, Bhutan, India

生物地理界 / Biogeographic Realm
印度马来界 Indomalaya

WWF 生物群系 / WWF Biome
热带和亚热带湿润阔叶林
Tropical & Subtropical Moist Broadleaf Forests

动物地理分布型 / Zoogeographic Distribution Type
Ha

分布标注 / Distribution Note
非特有种 Non-Endemic

▲ 濒危状况 / Threatened Status

中国生物多样性红色名录等级 / CB RL Category (2021)
数据缺乏 DD

IUCN 红色名录 / IUCN Red List (2021)
数据缺乏 DD

威胁因子 / Threats
未知 Unknown

▲ 法律保护地位 / Legal Protection Status

国家重点保护野生动物等级 / Category of National Key Protected Wild Animals (2021)
未列入 Not listed

"三有"名录 / TWIESSV (2023)
未列入 Not listed

CITES 附录等级 / CITES Appendix (2023)
未列入 Not listed

迁徙物种公约附录 / CMS Appendix (2020)
未列入 Not listed

保护行动 / Conservation Action
尚无保护行动 No conservation action so far

▲ 参考文献 / References

Burgin et al., 2020; IUCN, 2020; Wilson and Mittermeier, 2018

79 / 微尾鼩

Anourosorex squamipes
Milne-Edwards, 1872

· Mole-shrew

▲ 分类地位 / Taxonomy

劳亚食虫目 Eulipotyphla / 鼩鼱科 Soricidae / 短尾鼩属 *Anourosorex*

科建立者及其文献 / Family Authority
G. Fischer, 1814

属建立者及其文献 / Genus Authority
Milne-Edwards, 1872

亚种 / Subspecies
无共识 No consensus

模式标本产地 / Type Locality
中国
China, Sichuan Prov., probably Moupin (Baoxing)

何锴 / 供图

▲ 其他名称 / Other Name(s)

其他中文名 / Other Chinese Name(s)
短尾鼩、四川短尾鼩

其他英文名 / Other English Name(s)
Chinese Short-tailed Shrew, Sichuan
Burrowing Shrew

同物异名 / Synonym(s)
无 None

▲ 形态及生境 / Morphology and Habitat

形态特征 / Morphological Characteristics
齿式：3.0.1.3/2.0.1.3=26。头体长 74~110 mm。后足长 11~15 mm。尾长 8~19 mm。外形与鼹科鼩鼹类动物相似，但前肢明显更小。毛发致密松软。背部毛色黑色或深灰色，腹部毛色略浅。眼小，外耳退化。前掌较其他鼩鼱宽大，前爪长。尾短，深色。头骨致密，脑颅较低。上 P4 至 M1 为四边形。齿尖无色素沉积。有臭腺。

Dental formula: 3.0.1.3/2.0.1.3=26. Head-body 74-110 mm, tail 8-19 mm, hindfoot 11-15 mm. The external appearance assembles a shrew mole but with smaller claws. Body fur dense and lax. Dorsal pelage is black or dark grayish brown, and the ventral is slightly paler. Eyes minute. The external ears are completely reduced. Forefeet have lengthened claws. Tail is short and dark brown in color. The skull is stout, and the braincase is low. Upper P4 and M1 are quadrangle shaped. Cusps of teeth are unpigmented. It has very strong smell.

生境 / Habitat
森林、竹林、灌丛、草地、农田、城市环境
Forest, bamboo forest, thickets, grassland, farmland, urban environment

▲ 地理分布 / Geographic Distribution

国内分布 / Domestic Distribution
陕西、甘肃、湖北、重庆、四川、贵州、广东、云南、广西
Shaanxi, Gansu, Hubei, Chongqing, Sichuan, Guizhou, Guangdong, Yunnan, Guangxi

全球分布 / World Distribution
中国、老挝、缅甸、泰国、越南
China, Laos, Myanmar, Thailand, Vietnam

生物地理界 / Biogeographic Realm
古北界、印度马来界 Palearctic, Indomalaya

WWF 生物群系 / WWF Biome
热带和亚热带湿润阔叶林、温带阔叶和混交林、温带针叶树森林、山地草原和灌丛
Tropical & Subtropical Moist Broadleaf Forests, Temperate Broadleaf & Mixed Forests, Temperate Conifer Forests, Montane Grasslands & Shrublands

动物地理分布型 / Zoogeographic Distribution Type
Sd

分布标注 / Distribution Note
非特有种 Non-Endemic

▲ 濒危状况 / Threatened Status

中国生物多样性红色名录等级 / CB RL Category (2021)
无危 LC

IUCN 红色名录 / IUCN Red List (2021)
无危 LC

威胁因子 / Threats
无 None

▲ 法律保护地位 / Legal Protection Status

国家重点保护野生动物等级 / Category of National Key Protected Wild Animals (2021)
未列入 Not listed

"三有"名录 / TWIESSV (2023)
未列入 Not listed

CITES 附录等级 / CITES Appendix (2023)
未列入 Not listed

迁徙物种公约附录 / CMS Appendix (2020)
未列入 Not listed

保护行动 / Conservation Action
尚无保护行动 No conservation action so far

▲ 参考文献 / References

Jiang et al. (蒋志刚等), 2021; Burgin et al., 2020; IUCN, 2020; Liu et al. (刘少英等), 2020; Wilson and Mittermeier, 2018; Xie et al. (谢文华等), 2014; Wang et al. (王于玫等), 2014; Smith et al., 2009; Liu et al. (刘井元等), 2008; Pan et al. (潘清华等), 2007; Wilson and Reeder, 2005; Wang (王应祥), 2003; Motokaw and Lin, 2002; Zhang (张荣祖), 1997

80 / 台湾短尾鼩

Anourosorex yamashinai Kuroda, 1935

· Taiwanese Mole Shrew

▲ 分类地位 / Taxonomy

劳亚食虫目 Eulipotyphla / 鼩鼱科 Soricidae / 短尾鼩属 *Anourosorex*

科建立者及其文献 / Family Authority
G. Fischer, 1814

属建立者及其文献 / Genus Authority
Milne-Edwards, 1872

亚种 / Subspecies
无 None

模式标本产地 / Type Locality
中国
China, N Taiwan, Taiheizan, Taihoku-siu, 1676 m

▲ 其他名称 / Other Name(s)

其他中文名 / Other Chinese Name(s)
无 None

其他英文名 / Other English Name(s)
无 None

同物异名 / Synonym(s)
无 None

▲ 形态及生境 / Morphology and Habitat

形态特征 / Morphological Characteristics
齿式：3.0.1.3/2.0.1.3=26。头体长 51~ 98 mm。后足长 13~16 mm。尾长 7~13 mm。颅全长 23.5~26 mm。与微尾鼩相似，但体型略小。背腹黑色，尾和足背白色。上颌骨、下颌骨发育良好，骨质致密。下颌骨乳突和髁突非常发达。
Dental formula: 3.0.1.3/2.0.1.3=26. Head-body 51-98 mm, tail 7-13 mm, hindfoot 13-16 mm, condyloincisive length 23.5-26 mm. Morphologically similar to *Anourosorex squamipes* but smaller. Dorsal and ventral pelage is black, and feet and tail are white. Skull and mandible are robust. Mastoid and condylar processes are well developed.

生境 / Habitat
亚热带湿润山地森林、农田
Subtropical moist montane forest, farmland

▲ 地理分布 / Geographic Distribution

国内分布 / Domestic Distribution
台湾 Taiwan

全球分布 / World Distribution
中国 China

生物地理界 / Biogeographic Realm
印度马来界 Indomalaya

WWF 生物群系 / WWF Biome
热带和亚热带湿润阔叶林
Tropical & Subtropical Moist Broadleaf Forests

动物地理分布型 / Zoogeographic Distribution Type
J

分布标注 / Distribution Note
特有种 Endemic

▲ 濒危状况 / Threatened Status

中国生物多样性红色名录等级 / CB RL Category (2021)
数据缺乏 DD

IUCN 红色名录 / IUCN Red List (2021)
无危 LC

威胁因子 / Threats
未知 Unknown

▲ 法律保护地位 / Legal Protection Status

国家重点保护野生动物等级 / Category of National Key Protected Wild Animals (2021)
未列入 Not listed

"三有" 名录 / TWIESSV (2023)
未列入 Not listed

CITES 附录等级 / CITES Appendix (2023)
未列入 Not listed

迁徙物种公约附录 / CMS Appendix (2020)
未列入 Not listed

保护行动 / Conservation Action
尚无保护行动 No conservation action so far

▲ 参考文献 / References

Jiang et al. (蒋志刚等), 2021; Burgin et al., 2020; IUCN, 2020; Wilson and Mittermeier, 2018; Hutterer, 2005; Motokawa and Lin, 2002; Jameson and Jones, 1977; Pan et al. (潘清华等), 2007; Wilson and Reeder, 2005

81 / 喜马拉雅水鼩

Chimarrogale himalayica (Gray, 1842)

• Himalayan Water Shrew

▲ 分类地位 / Taxonomy

劳亚食虫目 Eulipotyphla / 鼩鼱科 Soricidae / 水麝鼩属 *Chimarrogale*

科建立者及其文献 / Family Authority
G. Fischer, 1814

属建立者及其文献 / Genus Authority
Anderson, 1877

亚种 / Subspecies
无共识 No consensus

模式标本产地 / Type Locality
印度
India, Punjab, Chamba

何锴 / 供图

▲ 其他名称 / Other Name(s)

其他中文名 / Other Chinese Name(s)
无 None

其他英文名 / Other English Name(s)
无 None

同物异名 / Synonym(s)
无 None

▲ 形态及生境 / Morphology and Habitat

形态特征 / Morphological Characteristics

齿式：3.1.1.3/2.0.1.3=28。头体长 111~132 mm。后足长 17~30 mm。尾长 79~88 mm。颅全长 25~28 mm。全身毛发短而细腻，芒毛长而稀疏，在臀部的位置较为集中。身体背腹异色，背部毛色棕黑色，腹部灰白色，背腹毛色没有明显的分界线。前后足侧边具有白色的流苏。尾双色，近四棱形，从根部到尾尖都覆盖短而密的白色绒毛。眼小，外耳退化，被毛发所遮盖。头骨骨质致密，脑颅两侧向外隆起。上单尖齿 3 枚，牙齿无色素沉积。

Dental formula: 3.1.1.3/2.0.1.3=28. Head-body 111-132 mm, tail 79-98 mm, hindfoot 17-30 mm. Body fur dense and lax. Dorsal pelage black, ventral surface ash gray, and the color gradually change. Dorsal pelage scattered white-tipped guard hairs especially on the lateral sides and on the rump. Eyes minute. Ears are reduced and covered by fur. Fringe of whitish hairs along margin of fore and hind feet and toes. Tail bicolored, the upperpart is blackish brown, and the underpart is white and covered by white hairs. A ventral keel is present on the tail. Braincase bony, flattened and enlarged in lateral. It has three upper unicuspids. Cusps of teeth are unpigmented.

生境 / Habitat
亚热带湿润低地森林
Subtropical moist lowland forest

▲ 地理分布 / Geographic Distribution

国内分布 / Domestic Distribution
云南、西藏
Yunnan, Tibet

全球分布 / World Distribution
中国、印度、老挝、缅甸、尼泊尔、越南
China, India, Laos, Myanmar, Nepal, Vietnam

生物地理界 / Biogeographic Realm
古北界、印度马来界 Palearctic, Indomalaya

WWF 生物群系 / WWF Biome
热带和亚热带湿润阔叶林、温带阔叶和混交林
Tropical & Subtropical Moist Broadleaf Forests, Temperate Broadleaf & Mixed Forests

动物地理分布型 / Zoogeographic Distribution Type
Sv

分布标注 / Distribution Note
非特有种 Non-Endemic

▲ 濒危状况 / Threatened Status

中国生物多样性红色名录等级 / CB RL Category (2021)
近危 NT

IUCN 红色名录 / IUCN Red List (2021)
无危 LC

威胁因子 / Threats
未知 Unknown

▲ 法律保护地位 / Legal Protection Status

国家重点保护野生动物等级 / Category of National Key Protected Wild Animals (2021)
未列入 Not listed

"三有" 名录 / TWIESSV (2023)
未列入 Not listed

CITES 附录等级 / CITES Appendix (2023)
未列入 Not listed

迁徙物种公约附录 / CMS Appendix (2020)
未列入 Not listed

保护行动 / Conservation Action
尚无保护行动 No conservation action so far

▲ 参考文献 / References

Jiang et al. (蒋志刚等), 2021; Burgin et al., 2020; IUCN, 2020; Wang et al. (汪巧云等), 2020; Wilson and Mittermeier, 2018; Deng et al. (邓可等), 2013; Smith et al., 2009; Qin et al. (秦岭等), 2007; Pan et al. (潘清华等), 2007; Wilson and Reeder, 2005; Wang (王应祥), 2003; Zhang (张荣祖), 1997

82 / 利安德水鼩

Chimarrogale leander Thomas, 1902

· Leander Water Shrew

▲ 分类地位 / Taxonomy

劳亚食虫目 Eulipotyphla / 鼩鼱科 Soricidae / 水麝鼩属 *Chimarrogale*

科建立者及其文献 / Family Authority
G. Fischer, 1814

属建立者及其文献 / Genus Authority
Anderson, 1877

亚种 / Subspecies
无 None

模式标本产地 / Type Locality
中国
Kuatun, N. W. Fokien. Altitude 1200 m

周佳俊 / 供图

▲ 其他名称 / Other Name(s)

其他中文名 / Other Chinese Name(s)
无 None

其他英文名 / Other English Name(s)
无 None

同物异名 / Synonym(s)
无 None

▲ 形态及生境 / Morphology and Habitat

形态特征 / Morphological Characteristics

齿式：3.1.1.3/2.0.1.3=28。体型较大的鼩鼱。头体长 80~130 mm。尾长 81~101 mm。后足 21~26 mm。与喜马拉雅水鼩外形十分相似，但毛色略浅。身体背部密布短而浓密的黑色毛发，腹部为深灰色。长有白色芒毛，但明显集中于臀部。尾近四棱形，背面深棕色，腹面暗灰色。头骨扁平，且较宽。牙齿无色素沉积。

Dental formula: 3.1.1.3/2.0.1.3=28. Head-body 80-130 mm, tail 81-101 mm, hindfoot 21-26 mm. Its external morphology is similar to *Chimarrogale himalayica* except that pelage color is less blackish grey. Dorsal pelage is black or blackish gray, and ventrum is dark gray. White-tipped guard hairs enriched around the rump. Tail is bicolored, the dorsal surface black and ventral white. Braincase is flattened and broad. Cusps of teeth are unpigmented.

生境 / Habitat

森林 Forest

▲ 地理分布 / Geographic Distribution

国内分布 / Domestic Distribution

浙江、安徽、福建、广东、广西、贵州、江苏、江西、湖南、湖北、
山西、重庆、四川、陕西、北京、河南、河北、台湾
Zhejiang, Anhui, Fujian, Guangdong, Guangxi, Guizhou, Jiangsu, Jiangxi, Hunan,
Hubei, Shanxi, Chongqing, Sichuan, Shaanxi, Beijing, Henan, Hebei, Taiwan

全球分布 / World Distribution

中国 China

生物地理界 / Biogeographic Realm

古北界、印度马来界 Palearctic, Indomalaya

WWF 生物群系 / WWF Biome

热带和亚热带湿润阔叶林、温带阔叶和混交林
Tropical & Subtropical Moist Broadleaf Forests, Temperate Broadleaf & Mixed
Forests

动物地理分布型 / Zoogeographic Distribution Type

Sa

分布标注 / Distribution Note

特有种 Endemic

▲ 濒危状况 / Threatened Status

中国生物多样性红色名录等级 / CB RL Category (2021)

数据缺乏 DD

IUCN 红色名录 / IUCN Red List (2021)

未评定 NE

威胁因子 / Threats

未知 Unknown

▲ 法律保护地位 / Legal Protection Status

国家重点保护野生动物等级 / Category of National Key Protected Wild Animals (2021)

未列入 Not listed

"三有" 名录 / TWIESSV (2023)

未列入 Not listed

CITES 附录等级 / CITES Appendix (2023)

未列入 Not listed

迁徙物种公约附录 / CMS Appendix (2020)

未列入 Not listed

保护行动 / Conservation Action

尚无保护行动 No conservation action so far

▲ 参考文献 / References

Jiang et al. (蒋志刚等), 2021; Wang et al. (汪巧云等), 2020; Wilson and Mittermeier, 2018; Yuan et al., 2013; Smith et al., 2009

83 / 斯氏水鼩

Chimarrogale styani de Winton, 1899

· Chinese Water Shrew

▲ 分类地位 / Taxonomy

劳亚食虫目 Eulipotyphla / 鼩鼱科 Soricidae / 水麝鼩属 *Chimarrogale*

科建立者及其文献 / Family Authority
G. Fischer, 1814

属建立者及其文献 / Genus Authority
Anderson, 1877

亚种 / Subspecies
无 None

模式标本产地 / Type Locality
中国
China, "Yang-liu-pa, N. W. Sechuen(Sichuan)"

刘洋 / 供图

▲ 其他名称 / Other Name(s)

其他中文名 / Other Chinese Name(s)
灰腹水鼩

其他英文名 / Other English Name(s)
Styan's Water Shrew

同物异名 / Synonym(s)
无 None

▲ 形态及生境 / Morphology and Habitat

形态特征 / Morphological Characteristics

齿式：3.1.1.3/2.0.1.3=28。头体长 95~110 mm。尾长 61~85 mm。后足 20~23 mm。外形特征与喜马拉雅水鼩十分相似，但斯氏水鼩腹部为白色。背部近黑色，体毛短而细腻，中间夹杂有较长的芒毛，集中分布于臀部。前后足白色。尾长明显小于头体长。牙齿无色素沉积。

Dental formula: 1.3.1.3 / 1.1.1.3=28. Head-body 95-110 mm, tail 61-85 mm, hind-foot 20-23 mm. Similar to *Chimarrogale himalayica* but smaller with a white belly. Dorsal and ventral pelage obviously bicolor. Ventral pelage slate black and covered with guard hairs which increases in length and numbers on the rump. Fore and hind feet are white. The tail is obviously shorter than the head-body. Cusps of teeth are unpigmented.

生境 / Habitat

湿润山地森林、溪流边
Subtropical moist montane forest, near stream

▲ 地理分布 / Geographic Distribution

国内分布 / Domestic Distribution
四川、云南、西藏、青海、甘肃
Sichuan, Yunnan, Tibet, Qinghai, Gansu

全球分布 / World Distribution
中国、缅甸
China, Myanmar

生物地理界 / Biogeographic Realm
古北界、印度马来界 Palearctic, Indomalaya

WWF 生物群系 / WWF Biome
热带和亚热带湿润阔叶林、温带阔叶和混交林
Tropical & Subtropical Moist Broadleaf Forests, Temperate Broadleaf & Mixed Forests

动物地理分布型 / Zoogeographic Distribution Type
Hc+Le

分布标注 / Distribution Note
非特有种 Non-Endemic

▲ 濒危状况 / Threatened Status

中国生物多样性红色名录等级 / CB RL Category (2021)
易危 VU

IUCN 红色名录 / IUCN Red List (2021)
无危 LC

威胁因子 / Threats
未知 Unknown

▲ 法律保护地位 / Legal Protection Status

国家重点保护野生动物等级 / Category of National Key Protected Wild Animals (2021)
未列入 Not listed

"三有"名录 / TWIESSV (2023)
未列入 Not listed

CITES 附录等级 / CITES Appendix (2023)
未列入 Not listed

迁徙物种公约附录 / CMS Appendix (2020)
未列入 Not listed

保护行动 / Conservation Action
尚无保护行动 No conservation action so far

▲ 参考文献 / References

Jiang et al. (蒋志刚等), 2021; Burgin et al., 2020; IUCN, 2020; Liu et al. (刘少英等), 2020; Wilson and Mittermeier, 2018; Zha et al. (扎史其等), 2014; Jiang et al. (姜雪松等), 2013; Smith et al., 2009; Pan et al. (潘清华等), 2007; Wilson and Reeder, 2005; Wang (王应祥), 2003; Zhang (张荣祖), 1997; Hoffmann, 1987

84 / 蹼足鼩

Nectogale elegans Milne-Edwards, 1870

· Elegant Water Shrew

▲ 分类地位 / Taxonomy

劳亚食虫目 Eulipotyphla / 鼩鼱科 Soricidae / 鼩属 *Nectogale*

科建立者及其文献 / Family Authority
G. Fischer, 1814

属建立者及其文献 / Genus Authority
Milne-Edwards, 1870

亚种 / Subspecies
无共识 No consensus

模式标本产地 / Type Locality
中国
China, Sichuan, Moupin (Baoxing)

▲ 其他名称 / Other Name(s)

其他中文名 / Other Chinese Name(s)
无 None

其他英文名 / Other English Name(s)
Web-footed Water Shrew

同物异名 / Synonym(s)
无 None

▲ 形态及生境 / Morphology and Habitat

形态特征 / Morphological Characteristics

齿式：3.1.1.3/2.0.1.3=28。体型较大的鼩鼱。头体长 90~115 mm。尾长 100~110 mm。后足 23~32 mm。毛发短而密集，似天鹅绒般光滑，中间夹杂着白色芒毛。背部黑色，腹部则为雪白色。眼很小，被毛发遮盖。外耳完全退化，耳洞被毛发遮盖。前后足均有蹼，且足外侧有白色梳状栉毛。尾横截面近四棱形，背面及两侧有龙骨状隆起。头骨扁平，牙齿没有色素沉积。

Dental formula: 3.1.1.3/2.0.1.3=28. Head-body 90-115 mm, tail 100-110 mm, hind-foot 23-32 mm. Body fur dense and lax. Dorsal and ventral sharply bicolored. Dorsal pelage nearly black with scattered white-tipped guard hairs especially on the rump, ventral pelage white. Eyes minute, about 1 mm in diameter, and covered by hairs. The external ears are completely reduced so that only a slim opening can be traced. Fore and hindlimbs are webbed to the base of the terminal phalanges and their edges are equipped with a fringe of stiff flattened hairs. The tail is equipped with one dorsal, two ventral, and two lateral keels of short stiff hairs. Braincase is flattened. Cusps of teeth are unpigmented.

生境 / Habitat
森林中的溪流边
Near streams in forest

▲ 地理分布 / Geographic Distribution

国内分布 / Domestic Distribution
四川、云南、西藏、陕西、甘肃、湖北
Sichuan, Yunnan, Tibet, Shaanxi, Gansu, Hubei

全球分布 / World Distribution
中国、印度、缅甸、尼泊尔、不丹
China, India, Myanmar, Nepal, Bhutan

生物地理界 / Biogeographic Realm
古北界、印度马来界 Palearctic, Indomalaya

WWF 生物群系 / WWF Biome
热带和亚热带湿润阔叶林、温带阔叶和混交林
Tropical & Subtropical Moist Broadleaf Forests, Temperate Broadleaf & Mixed Forests

动物地理分布型 / Zoogeographic Distribution Type
Hc

分布标注 / Distribution Note
非特有种 Non-Endemic

▲ 濒危状况 / Threatened Status

中国生物多样性红色名录等级 / CB RL Category (2021)
无危 LC

IUCN 红色名录 / IUCN Red List (2021)
无危 LC

威胁因子 / Threats
未知 Unknown

▲ 法律保护地位 / Legal Protection Status

国家重点保护野生动物等级 / Category of National Key Protected Wild Animals (2021)
未列入 Not listed

"三有" 名录 / TWIESSV (2023)
未列入 Not listed

CITES 附录等级 / CITES Appendix (2023)
未列入 Not listed

迁徙物种公约附录 / CMS Appendix (2020)
未列入 Not listed

保护行动 / Conservation Action
尚无保护行动 No conservation action so far

▲ 参考文献 / References

Fan et al., 2022; Jiang et al. (蒋志刚等), 2021; Burgin et al., 2020; IUCN, 2020; Wilson and Mittermeier, 2018; Deng et al. (邓可等), 2013; Jiang et al. (姜雪松等), 2013; Smith et al., 2009; Pan et al. (潘清华等), 2007; Wilson and Reeder, 2005; Wang (王应祥), 2003; Gong et al. (龚正达等), 2001; Zhang (张荣祖), 1997; Chu (褚新洛), 1989; Hoffmann, 1987; Feng et al. (冯祚建等), 1986

85 / 锡金蹼足鼩

Nectogale sikhimensis
(De Winton & Styan, 1899)

· Sikkim Water Shrew

▲ 分类地位 / Taxonomy

劳亚食虫目 Eulipotyphla / 鼩鼱科 Soricidae / 鼩属 *Nectogale*

科建立者及其文献 / Family Authority
G. Fischer, 1814

属建立者及其文献 / Genus Authority
Milne-Edwards, 1870

亚种 / Subspecies
无 None

模式标本产地 / Type Locality
印度
Sikkim India

▲ 其他名称 / Other Name(s)

其他中文名 / Other Chinese Name(s)
无 None

其他英文名 / Other English Name(s)
无 None

同物异名 / Synonym(s)
无 None

▲ 形态及生境 / Morphology and Habitat

形态特征 / Morphological Characteristics

齿式：3.1.1.3/2.0.1.3=28。体型较大，毛发短而密集，中间夹杂着白色芒毛。背部黑色，腹部则为雪白色。脑颅略微向上隆起，脑颅枕骨后缘底部近平直，乳突不向两侧凸起。锡金蹼足鼩的腭中缝呈弓形。锡金蹼足鼩的 M2 上次尖齿脊向前平缓延伸，不形成小尖结构。下颌单尖齿齿尖较低，牙齿无色素沉积。

Dental formula: 3.1.1.3/2.0.1.3=28. Body fur dense and lax. Dorsal and ventral sharply bicolored. Dorsal pelage nearly black with scattered white-tipped guard hairs, especially on the rump, ventral pelage white. The brain skull is slightly raised, and the posterior edge of the brain skull is less curved. The mastoid does not extend laterally. The palatal suture is arcuate-shaped. A cusp rests in the posterior of the hypocone and no cusp in the anterior of M2 in lingual view. The tip of the unicuspid is lower than in N. elegans. Cusps of teeth are unpigmented.

生境 / Habitat

热带和亚热带湿润阔叶林、温带阔叶和混交林
Tropical and subtropical moist broadleaf Forest, temperate broadleaf and mixed forest

▲ 地理分布 / Geographic Distribution

国内分布 / Domestic Distribution
西藏 Tibet

全球分布 / World Distribution
中国、印度、不丹、缅甸、尼泊尔
China, India, Bhutan, Myanmar, Nepal

生物地理界 / Biogeographic Realm
印度马来界 Indomalaya

WWF 生物群系 / WWF Biome
热带和亚热带湿润阔叶林、温带阔叶和混交林
Tropical & Subtropical Moist Broadleaf Forests, Temperate Broadleaf & Mixed Forests

动物地理分布型 / Zoogeographic Distribution Type
Hc

分布标注 / Distribution Note
非特有种 Non-Endemic

▲ 濒危状况 / Threatened Status

中国生物多样性红色名录等级 / CB RL Category (2021)
未评定 NE

IUCN 红色名录 / IUCN Red List (2021)
未评定 NE

威胁因子 / Threats
未知 Unknown

▲ 法律保护地位 / Legal Protection Status

国家重点保护野生动物等级 / Category of National Key Protected Wild Animals (2021)
未列入 Not listed

"三有"名录 / TWIESSV (2023)
列入 Listed

CITES 附录等级 / CITES Appendix (2023)
未列入 Not listed

迁徙物种公约附录 / CMS Appendix (2020)
未列入 Not listed

保护行动 / Conservation Action
尚无保护行动 No conservation action so far

▲ 参考文献 / References

Fan et al., 2022; De Winton and Styan, 1899

86 / 小臭鼩

Suncus etruscus (Savi, 1822)

· Etruscan Shrew

劳亚食虫目 Eulipotyphla / 鼩鼱科 Soricidae / 臭鼩属 *Suncus*

科建立者及其文献 / Family Authority
G. Fischer, 1814

属建立者及其文献 / Genus Authority
Ehrenberg, 1832

亚种 / Subspecies
无共识 No consensus

模式标本产地 / Type Locality
意大利
Italy, Pisa

▲ 其他名称 / Other Name(s)

其他中文名 / Other Chinese Name(s)
无 None

其他英文名 / Other English Name(s)
无 None

同物异名 / Synonym(s)
无 None

▲ 形态及生境 / Morphology and Habitat

形态特征 / Morphological Characteristics
齿式：3.1.2.3/2.0.1.3=30。上单尖齿 4 枚，比麝鼩属物种多 1 枚。头体长 33~50 mm。尾长 21~30 mm。耳长 4~6.2 mm。后足 7~7.5 mm。体重 1.2~2.7 g。体表毛发短而细腻，背部烟灰色，腹部银灰色。尾长通常为头体长的 60%。麝鼩亚科物种牙齿均无色素沉积。

Dental formula: 3.1.2.3/2.0.1.3=30. Four upper unicuspids, one more than that of a *Crocidura* species. Head-body 33-50 mm, tail 21-30 mm, hind-foot 7-7.5 mm, ear 4-6.2 mm, weight 1.2-2.7 g. Fur is short and soft, with no clear color border between belly and back. Back is smoky gray, and belly is slivery gray. Tail is usually 60% of the head-body length. Teeth are not pigmented as in all Crocidurinae species.

生境 / Habitat
草地、灌丛、森林、人工环境
Grassland, shrubland, forest, artificial environment

▲ 地理分布 / Geographic Distribution

国内分布 / Domestic Distribution

云南 Yunnan

全球分布 / World Distribution

阿富汗、阿尔巴尼亚、阿尔及利亚、阿塞拜疆、巴林、不丹、波斯尼亚和黑塞哥维那、保加利亚、中国、克罗地亚、塞浦路斯、埃及、埃塞俄比亚、法国、格鲁吉亚、希腊、几内亚、印度、伊朗、伊拉克、以色列、意大利、约旦、老挝、黎巴嫩、利比亚、马其顿、马来西亚、马耳他、黑山、摩洛哥、缅甸、尼泊尔、尼日利亚、阿曼、巴基斯坦、葡萄牙、斯洛文尼亚、西班牙、斯里兰卡、叙利亚、塔吉克斯坦、泰国、突尼斯、土耳其、土库曼斯坦、越南、也门

Afghanistan, Albania, Algeria, Azerbaijan, Bahrain, Bhutan, Bosnia and Herzegovina, Bulgaria, China, Croatia, Cyprus, Egypt, Ethiopia, France, Georgia, Greece, Guinea, India, Iran, Iraq, Israel, Italy, Jordan, Laos, Lebanon, Libya, Macedonia, Malaysia, Malta, Montenegro, Morocco, Myanmar, Nepal, Nigeria, Oman, Pakistan, Portugal, Slovenia, Spain, Sri Lanka, Syria, Tajikistan, Thailand, Tunisia, Turkey, Turkmenistan, Vietnam, Yemen

生物地理界 / Biogeographic Realm

古北界、印度马来界 Palearctic, Indomalaya

WWF 生物群系 / WWF Biome

热带和亚热带湿润阔叶林
Tropical & Subtropical Moist Broadleaf Forests

动物地理分布型 / Zoogeographic Distribution Type

Wa

分布标注 / Distribution Note

非特有种 Non-Endemic

▲ 濒危状况 / Threatened Status

中国生物多样性红色名录等级 / CB RL Category (2021)

易危 VU

IUCN 红色名录 / IUCN Red List (2021)

无危 LC

威胁因子 / Threats

未知 Unknown

▲ 法律保护地位 / Legal Protection Status

国家重点保护野生动物等级 / Category of National Key Protected Wild Animals (2021)
未列入 Not listed

"三有"名录 / TWIESSV (2023)
未列入 Not listed

CITES 附录等级 / CITES Appendix (2023)
未列入 Not listed

迁徙物种公约附录 / CMS Appendix (2020)
未列入 Not listed

保护行动 / Conservation Action
尚无保护行动 No conservation action so far

▲ 参考文献 / References

Jiang et al. (蒋志刚等), 2021; Burgin et al., 2020; IUCN, 2020; Liu et al. (刘少英等), 2020; Wilson and Mittermeier, 2018; Smith et al., 2009; Pan et al. (潘清华等), 2007; Pan et al. (潘清华等), 2007; Wang (王应祥), 2003; Wilson and Reeder, 2005; Wang (王应祥), 2003; Zhang (张荣祖), 1997; Wilson and Reeder, 1993

87 / 臭鼩

Suncus murinus Linnaeus, 1766

· House Shrew

▲ 分类地位 / Taxonomy

劳亚食虫目 Eulipotyphla / 鼩鼱科 Soricidae / 臭鼩属 *Suncus*

科建立者及其文献 / Family Authority
G. Fischer, 1814

属建立者及其文献 / Genus Authority
Ehrenberg, 1832

亚种 / Subspecies
指名亚种 *S. m. murinus* Linnaeus,1766
浙江、福建、台湾、广东、香港、澳门、广西、海南、贵州和云南
Zhejiang, Fujian, Taiwan, Guangdong, Hong Kong, Macao, Guangxi, Hainan, Guizhou and Yunnan

模式标本产地 / Type Locality
印度尼西亚
Indonesia, Java

曲利明 / 供图

▲ 其他名称 / Other Name(s)

其他中文名 / Other Chinese Name(s)
大臭鼩

其他英文名 / Other English Name(s)
无 None

同物异名 / Synonym(s)
无 None

▲ 形态及生境 / Morphology and Habitat

形态特征 / Morphological Characteristics

齿式：3.1.2.3/2.0.1.3=30。头体长 119~147 mm。尾长 60~85 mm。后足 19~22 mm。体色浅灰至深灰色，背腹颜色相近，腹部颜色略浅。外耳郭明显。尾巴肥大，直径远超麝鼩类，尾皮肤裸露，从根部到尖端都长有稀疏的针毛。

Dental formula: 3.1.2.3/2.0.1.3=30. Head-body length 119-147 mm, tail length 60-85 mm, hindfoot length 19-22 mm. The body color is whitish gray to dark gray, and the belly is slightly paler than the back. Ears are obvious. The tail is very thick, and naked, covered with sparse long hairs from base to the tip.

生境 / Habitat
耕地、池塘、灌丛、森林、沼泽
Arable land, pond, shrubland, forest, swamp

▲ 地理分布 / Geographic Distribution

国内分布 / Domestic Distribution
广西、福建、浙江、江西、湖南、广东、海南、四川、贵州、云南、
西藏、甘肃、台湾、香港、澳门
Guangxi, Fujian, Zhejiang, Jiangxi, Hunan, Guangdong, Hainan, Sichuan, Guizhou, Yunnan, Tibet, Gansu, Taiwan, Hong Kong, Macao

全球分布 / World Distribution
阿富汗、孟加拉国、不丹、文莱、柬埔寨、中国、印度、印度尼西亚、
老挝、马来西亚、缅甸、尼泊尔、巴基斯坦、新加坡、斯里兰卡、泰
国、越南
Afghanistan, Bengladesh, Bhutan, Brunei, Cambodia, China, India, Indonesia, Laos, Malaysia, Myanmar, Nepal, Pakistan, Singapore, Sri Lanka, Thailand, Vietnam

生物地理界 / Biogeographic Realm
印度马来界 Indomalaya

WWF 生物群系 / WWF Biome
热带和亚热带湿润阔叶林
Tropical & Subtropical Moist Broadleaf Forests

动物地理分布型 / Zoogeographic Distribution Type
Wd

分布标注 / Distribution Note
非特有种 Non-Endemic

▲ 濒危状况 / Threatened Status

中国生物多样性红色名录等级 / CB RL Category (2021)
无危 LC

IUCN 红色名录 / IUCN Red List (2021)
无危 LC

威胁因子 / Threats
未知 Unknown

▲ 法律保护地位 / Legal Protection Status

国家重点保护野生动物等级 / Category of National Key Protected Wild Animals (2021)
未列入 Not listed

"三有"名录 / TWIESSV (2023)
未列入 Not listed

CITES 附录等级 / CITES Appendix (2023)
未列入 Not listed

迁徙物种公约附录 / CMS Appendix (2020)
未列入 Not listed

保护行动 / Conservation Action
尚无保护行动 No conservation action so far

▲ 参考文献 / References

Jiang et al. (蒋志刚等), 2021; Burgin et al., 2020; IUCN, 2020; Wilson and Mittermeier, 2018; Smith et al., 2009; Pan et al. (潘清华等), 2007; Wilson and Reeder, 2005; Wang (王应祥), 2003; Zhang (张荣祖), 1997; Chen and Zhuge (陈水华和诸葛阳), 1993; Xin and Qiu (辛景禧和邱梦辞), 1990; Yang and Zhuge (杨士剑和诸葛阳), 1989; Xia (夏武平), 1988

88 / 安徽麝鼩

Crocidura anhuiensis
Zhang, Zhang & Li, 2020

• Anhui Shrew

▲ 分类地位 / Taxonomy

劳亚食虫目 Eulipotyphla / 鼩鼱科 Soricidae / 麝鼩属 *Crocidura*

科建立者及其文献 / Family Authority
G. Fischer, 1814

属建立者及其文献 / Genus Authority
Wagler, 1832

亚种 / Subspecies
无 None

模式标本产地 / Type Locality
中国
Anhui, China

▲ 其他名称 / Other Name(s)

其他中文名 / Other Chinese Name(s)
无 None

其他英文名 / Other English Name(s)
无 None

同物异名 / Synonym(s)
无 None

▲ 形态及生境 / Morphology and Habitat

形态特征 / Morphological Characteristics

齿式: 3.1.1.3/2.0.1.3=28。头体长 61~78 mm。尾长 48~59 mm。耳高 8.5~9.5 mm。体重 10.5~13.5 g。背部毛色棕灰色，腹部毛色略浅。尾靠近基部 50% 区域覆盖有稀疏的针毛。上单尖齿 3 枚，比臭鼩属物种少 1 枚。

Dental formula: 3.1.1.3/2.0.1.3=28. Head-body length 61-78 mm, tail length 48-59 mm, ears 8.5-9.5 mm. Weight 10.5-13.5 g. Dorsal pelage grayish brown and the ventral is paler. 50% of the tail near the base is covered with sparse long hairs. Three upper unicuspids, one less than that of a *Suncus* species.

生境 / Habitat
亚热带湿润山地森林
Subtropical moist montane forest

▲ 地理分布 / Geographic Distribution

国内分布 / Domestic Distribution
安徽、江西、浙江、福建、湖南
Anhui, Jiangxi, Zhejiang, Fujian, hunan

全球分布 / World Distribution
中国 China

生物地理界 / Biogeographic Realm
古北界 Palearctic

WWF 生物群系 / WWF Biome
热带和亚热带湿润阔叶林
Tropical & Subtropical Moist Broadleaf Forests

动物地理分布型 / Zoogeographic Distribution Type
Sc

分布标注 / Distribution Note
特有种 Endemic

▲ 濒危状况 / Threatened Status

中国生物多样性红色名录等级 / CB RL Category (2021)
数据缺乏 DD

IUCN 红色名录 / IUCN Red List (2021)
未评定 NE

威胁因子 / Threats
未知 Unknown

▲ 法律保护地位 / Legal Protection Status

国家重点保护野生动物等级 / Category of National Key Protected Wild Animals (2021)
未列入 Not listed

"三有" 名录 / TWIESSV (2023)
未列入 Not listed

CITES 附录等级 / CITES Appendix (2023)
未列入 Not listed

迁徙物种公约附录 / CMS Appendix (2020)
未列入 Not listed

保护行动 / Conservation Action
尚无保护行动 No conservation action so far

▲ 参考文献 / References

Zhang et al., 2020

89 / 灰麝鼩

Crocidura attenuata Milne-Edwards, 1872

· Grey Shrew

▲ 分类地位 / Taxonomy

劳亚食虫目 Eulipotyphla / 鼩鼱科 Soricidae / 麝鼩属 *Crocidura*

科建立者及其文献 / Family Authority
G. Fischer, 1814

属建立者及其文献 / Genus Authority
Wagler, 1832

亚种 / Subspecies
指名亚种 *C. a. attenuata* Milne-Edwards,1872
安徽、浙江、江西、福建、广东、广西、海南、贵州、湖南、湖北、
四川和云南
Anhui, Zhejiang, Jiangxi, Fujian, Guangdong, Guangxi, Hainan, Guizhou, Hunan, Hubei, Sichuan and Yunnan

喜马拉雅亚种 *C. a. rubricosa* Anderson, 1877
云南
Yunnan

模式标本产地 / Type Locality
中国
China, Szechuan, Moupin (Sichuan, Baoxing)

刘洋 / 供图

▲ 其他名称 / Other Name(s)

其他中文名 / Other Chinese Name(s)
无 None

其他英文名 / Other English Name(s)
无 None

同物异名 / Synonym(s)
无 None

▲ 形态及生境 / Morphology and Habitat

形态特征 / Morphological Characteristics
齿式：3.1.1.3/2.0.1.3=28。头体长 66~89 mm。尾长 41~60 mm。后足 13~16 mm。背部棕灰色，向腹部逐渐过渡为暗灰色。夏季毛发较短，但毛色更深。尾巴背面毛色较深，腹面略浅。

Dental formula: 3.1.1.3/2.0.1.3=28. Head-body length 66-89 mm, tail length 41-60 mm, hind foot length 13-16 mm. Dorsal pelage brownish-gray, gradually transit to dark gray on the belly. In summer, hairs are shorter but darker. Tail is darker on the back and lighter on the underside.

生境 / Habitat
热带亚热带湿润低地山地森林、草地、灌丛
Tropical subtropical moist lowland montane forest, grassland, shrubland

▲ 地理分布 / Geographic Distribution

国内分布 / Domestic Distribution

广西、云南、湖南、陕西、江苏、浙江、安徽、福建、江西、湖北、
广东、海南、四川、贵州、西藏、甘肃、香港、重庆
Guangxi, Yunnan, Hunan, Shaanxi, Jiangsu, Zhejiang, Anhui, Fujian, Jiangxi,
Hubei, Guangdong, Hainan, Sichuan, Guizhou, Tibet, Gansu, Hong Kong,
Chongqing

全球分布 / World Distribution

柬埔寨、中国、印度、老挝、马来西亚、缅甸、菲律宾、泰国、
越南
Cambodia, China, India, Laos, Malaysia, Myanmar, Philippines, Thailand,
Vietnam

生物地理界 / Biogeographic Realm

古北界、印度马来界 Palearctic, Indomalaya

WWF 生物群系 / WWF Biome

热带和亚热带湿润阔叶林
Tropical & Subtropical Moist Broadleaf Forests

动物地理分布型 / Zoogeographic Distribution Type

Sd

分布标注 / Distribution Note

非特有种 Non-Endemic

▲ 濒危状况 / Threatened Status

中国生物多样性红色名录等级 / CB RL Category (2021)

无危 LC

IUCN 红色名录 / IUCN Red List (2021)

无危 LC

威胁因子 / Threats

未知 Unknown

▲ 法律保护地位 / Legal Protection Status

国家重点保护野生动物等级 / Category of National Key Protected Wild Animals (2021)

未列入 Not listed

"三有"名录 / TWIESSV (2023)

未列入 Not listed

CITES 附录等级 / CITES Appendix (2023)

未列入 Not listed

迁徙物种公约附录 / CMS Appendix (2020)

未列入 Not listed

保护行动 / Conservation Action

尚无保护行动 No conservation action so far

▲ 参考文献 / References

Jiang et al. (蒋志刚等), 2021; Burgin et al., 2020; Chen et al. 2020; IUCN, 2020; Liu et al. (刘少英等), 2020; Li et al., 2019; Wilson and Mittermeier, 2018; Xie et al. (谢文华等), 2014; Smith et al., 2009; Liu et al. (刘井元等), 2008; Wang et al. (王淯等), 2006; Pan et al. (潘清华等), 2007; Wilson and Reeder, 2005; Wang (王应祥), 2003; Zhang (张荣祖), 1997

90 / 东阳江麝鼩

Crocidura donyangjiangensis
Liu Y, Chen SD, Liu SY, 2020

· Dongyangjiang Shrew

▲ 分类地位 / Taxonomy

劳亚食虫目 Eulipotyphla / 鼩鼱科 Soricidae / 麝鼩属 *Crocidura*

科建立者及其文献 / Family Authority
G. Fischer, 1814

属建立者及其文献 / Genus Authority
Wagler, 1832

亚种 / Subspecies
无 None

模式标本产地 / Type Locality
中国
Zhejiang(Dongyang), China

▲ 其他名称 / Other Name(s)

其他中文名 / Other Chinese Name(s)
黄山麝鼩

其他英文名 / Other English Name(s)
Huangshan Shrew

同物异名 / Synonym(s)
无 None

▲ 形态及生境 / Morphology and Habitat

形态特征 / Morphological Characteristics
齿式：3.1.1.3/2.0.1.3=28。头体长 48~61 mm。耳长 7~8 mm。后足长 11~13 mm。尾长 35~49 mm。体重 2.5~4.5 g。由刘洋等（2020）和 Yang et al.（2020）同时命名发表。背部毛发灰褐色，腹部浅灰色。尾巴背面毛色较深，尾巴腹面略浅。

Dental formula: 3.1.1.3/2.0.1.3=28. Head-body length 48-61 mm, ear length 7-8 mm, hindfoot 11-13 mm, tail length 35-49 mm, weight 2.5-4.5 g. Discovered and named by Liu et al. (2020) and Yang et al. (2020) simultaneously. Dorsal pelage brownish gray and ventrum is slightly paler. Tail is darker on the upperpart and lighter on the underside.

生境 / Habitat
亚热带湿润山地森林
Subtropical moist montane forest

▲ 地理分布 / Geographic Distribution

国内分布 / Domestic Distribution
浙江、安徽、湖南、江西、广东
Zhejiang, Anhui, Hunan, Jiangxi, Guangdong

全球分布 / World Distribution
中国 China

生物地理界 / Biogeographic Realm
古北界 Palearctic

WWF 生物群系 / WWF Biome
热带和亚热带湿润阔叶林、温带阔叶和混交林
Tropical & Subtropical Moist, Broadleaf Forests, Temperate Broadleaf & Mixed
Forests

动物地理分布型 / Zoogeographic Distribution Type
Sc

分布标注 / Distribution Note
特有种 Endemic

▲ 濒危状况 / Threatened Status

中国生物多样性红色名录等级 / CB RL Category (2021)
数据缺乏 DD

IUCN 红色名录 / IUCN Red List (2021)
未评定 NE

威胁因子 / Threats
未知 Unknown

▲ 法律保护地位 / Legal Protection Status

国家重点保护野生动物等级 / Category of National Key Protected Wild Animals (2021)
未列入 Not listed

"三有"名录 / TWIESSV (2023)
未列入 Not listed

CITES 附录等级 / CITES Appendix (2023)
未列入 Not listed

迁徙物种公约附录 / CMS Appendix (2020)
未列入 Not listed

保护行动 / Conservation Action
尚无保护行动 No conservation action so far

▲ 参考文献 / References

Chen et al., 2021; Liu et al. (刘洋 等), 2020; Yang et al., 2020

91 / 白尾梢麝鼩

Crocidura dracula Blyth, 1855

· Dracula's Shrew

▲ 分类地位 / Taxonomy

劳亚食虫目 Eulipotyphla / 鼩鼱科 Soricidae / 麝鼩属 *Crocidura*

科建立者及其文献 / Family Authority
G. Fischer, 1814

属建立者及其文献 / Genus Authority
Wagler, 1832

亚种 / Subspecies
无共识 No consensus

模式标本产地 / Type Locality
中国
"Probably Mengzi" of Yunnan, China

阳德才 / 供图

▲ 其他名称 / Other Name(s)

其他中文名 / Other Chinese Name(s)
无 None

其他英文名 / Other English Name(s)
White-toothed Shrew

同物异名 / Synonym(s)
无 None

▲ 形态及生境 / Morphology and Habitat

形态特征 / Morphological Characteristics

齿式：3.1.1.3/2.0.1.3=28。头体长 79~104 mm。尾长 62~89 mm。后足长 15~19 mm。背部棕褐色，逐渐向腹部过渡为暗灰色。前后足背部灰白色。尾有稀疏针毛，背面棕色，腹面颜色较浅，部分个体的尾尖为明显白色。脑颅部分较扁，头骨吻部较宽。

Dental formula: 3.1.1.3/2.0.1.3=28. Head-body length 79-104 mm, tail length 62-89 mm, hindfoot length 15-19 mm, ear length 10 mm, weight 13 g. Dorsal pelage smoky brown to dark blackish gray, gradually merging into dark gray on the ventral. Feet are dull whitish. Tail dark brown above and lighter below, usually longer than 80% of the head-body length. Skull and braincase are low, and rostrum is broad.

生境 / Habitat
湿润低地森林
Moist lowland forest

▲ 地理分布 / Geographic Distribution

国内分布 / Domestic Distribution
四川、湖北、云南、广西、贵州、西藏、陕西、重庆
Sichuan, Hubei, Yunnan, Guangxi, Guizhou, Tibet, Shaanxi, Chongqing

全球分布 / World Distribution
中国、缅甸、越南
China, Myanmar, Vietnam

生物地理界 / Biogeographic Realm
古北界，印度马来界 Palearctic, Indomalaya

WWF 生物群系 / WWF Biome
热带和亚热带湿润阔叶林、温带阔叶和混交林
Tropical & Subtropical Moist Broadleaf Forests, Temperate Broadleaf & Mixed Forests

动物地理分布型 / Zoogeographic Distribution Type
Sd

分布标注 / Distribution Note
非特有种 Non-Endemic

▲ 濒危状况 / Threatened Status

中国生物多样性红色名录等级 / CB RL Category (2021)
无危 LC

IUCN 红色名录 / IUCN Red List (2021)
无危 LC

威胁因子 / Threats
无 None

▲ 法律保护地位 / Legal Protection Status

国家重点保护野生动物等级 / Category of National Key Protected Wild Animals (2021)
未列入 Not listed

"三有" 名录 / TWIESSV (2023)
未列入 Not listed

CITES 附录等级 / CITES Appendix (2023)
未列入 Not listed

迁徙物种公约附录 / CMS Appendix (2020)
未列入 Not listed

保护行动 / Conservation Action
尚无保护行动 No conservation action so far

▲ 参考文献 / References

Jiang et al. (蒋志刚等), 2021; Burgin et al., 2020; IUCN, 2020; Liu et al. (刘少英等), 2020; Wilson and Mittermeier, 2018; Cui et al. (崔茂欢等), 2014; Zhang et al. (张斌等), 2014; Smith et al., 2009; Hutterer, 2005; Pan et al. (潘清华等), 2007; Wilson and Reeder, 2005; Wang (王应祥), 2003

92 / 印支小麝鼩

Crocidura indochinensis
Robinson & Kloss, 1922

· Indochinese Shrew

▲ 分类地位 / Taxonomy

劳亚食虫目 Eulipotyphla / 鼩鼱科 Soricidae / 麝鼩属 *Crocidura*

科建立者及其文献 / Family Authority
G. Fischer, 1814

属建立者及其文献 / Genus Authority
Wagler, 1832

亚种 / Subspecies
无共识 No consensus

模式标本产地 / Type Locality
越南
Vietnam, Langbian Plateau, Dalat

何锴 / 供图

▲ 其他名称 / Other Name(s)

其他中文名 / Other Chinese Name(s)
无 None

其他英文名 / Other English Name(s)
无 None

同物异名 / Synonym(s)
无 None

▲ 形态及生境 / Morphology and Habitat

形态特征 / Morphological Characteristics
齿式：3.1.1.3/2.0.1.3=28。体型较小，头体长 53~61 mm。尾长 40~50 mm。后足 10~13 mm。背部银灰略带污黄，腹部毛色略浅，偏灰白色。尾细，近基部三分之一的部分皮肤完全裸露，后三分之二的部分覆盖有极短的白色绒毛，只在近臀部的位置有稀疏的针毛。
Dental formula: 3.1.1.3/2.0.1.3=28. Head-body 53-61 mm, tail 40-50 mm, hind-foot 10-13 mm. Pelage dark brownish gray, belly is paler and grayish-white. Tail is slender, and has few bristles on the basal part of the tail. The basal 1/3 part of the tail is naked and the other 2/3 of the tail is covered with short white hairs.

生境 / Habitat
亚热带湿润山地森林
Subtropical moist montane forest

▲ 地理分布 / Geographic Distribution

国内分布 / Domestic Distribution
云南、贵州
Yunnan, Guizhou

全球分布 / World Distribution
中国、老挝、缅甸、泰国、越南
China, Laos, Myanmar, Thailand, Vietnam

生物地理界 / Biogeographic Realm
古北界、印度马来界 Palearctic, Indomalaya

WWF 生物群系 / WWF Biome
热带和亚热带湿润阔叶林
Tropical & Subtropical Moist Broadleaf Forests

动物地理分布型 / Zoogeographic Distribution Type
Wa

分布标注 / Distribution Note
非特有种 Non-Endemic

▲ 濒危状况 / Threatened Status

中国生物多样性红色名录等级 / CB RL Category (2021)
近危 NT

IUCN 红色名录 / IUCN Red List (2021)
无危 LC

威胁因子 / Threats
未知 Unknown

▲ 法律保护地位 / Legal Protection Status

国家重点保护野生动物等级 / Category of National Key Protected Wild Animals (2021)
未列入 Not listed

"三有" 名录 / TWIESSV (2023)
未列入 Not listed

CITES 附录等级 / CITES Appendix (2023)
未列入 Not listed

迁徙物种公约附录 / CMS Appendix (2020)
未列入 Not listed

保护行动 / Conservation Action
尚无保护行动 No conservation action so far

▲ 参考文献 / References

Jiang et al. (蒋志刚等), 2021; Burgin et al., 2020; IUCN, 2020; Wilson and Mittermeier, 2018; Deng et al. (邓可等), 2013; Wang (王应祥), 2003; Gong et al. (龚正达等), 2001; Pan et al. (潘清华等), 2007; Wilson and Reeder, 2005; Wang (王应祥), 2003; Zhang (张荣祖), 1997

93 / 大麝鼩

Crocidura lasiura Dobson, 1890

· Ussuri Shrew

▲ 分类地位 / Taxonomy

劳亚食虫目 Eulipotyphla / 鼩鼱科 Soricidae / 麝鼩属 *Crocidura*

科建立者及其文献 / Family Authority
G. Fischer, 1814

属建立者及其文献 / Genus Authority
Wagler, 1832

亚种 / Subspecies
指名亚种 *C. l. lasiura* Dobson, 1890
黑龙江、吉林和内蒙古东北部
Heilongjiang, Jilin and Inner Mongolia (northeastern part)

华东亚种 *C. l. campuslincolnensis* Soweiby, 1945
江苏、上海和四川东北部（剑阁和南充）
Jiangsu, Shanghai and Sichuan (northeastem parts-Jiange and Nanchong)

模式标本产地 / Type Locality
中国
NE China, (Manchuria), Ussuri River

▲ 其他名称 / Other Name(s)

其他中文名 / Other Chinese Name(s)
无 None

其他英文名 / Other English Name(s)
无 None

同物异名 / Synonym(s)
无 None

▲ 形态及生境 / Morphology and Habitat

形态特征 / Morphological Characteristics
齿式：3.1.1.3/2.0.1.3=28。头体长 66~100 mm。尾长 29~41 mm。体重 8.8~26.7 g。体型较大的鼩鼱。背部毛色深棕近黑色，腹部毛色略浅，背腹近乎同色。

Dental formula: 3.1.1.3/2.0.1.3=28. Head-body length 66-100 mm, tail length 29-41 mm, weight 8.8-26.7 g. A large-sized shrew with a short tail. Dorsal pelage dark brown and belly somewhat lighter.

生境 / Habitat
森林、沼泽、草地、灌丛、耕地
Forest, swamp, grassland, shrub land, arable land

▲ 地理分布 / Geographic Distribution

国内分布 / Domestic Distribution
黑龙江、吉林、辽宁、内蒙古、上海、江苏、浙江、山东
Heilongjiang, Jilin, Liaoning, Inner Mongolia, Shanghai, Jiangsu, Zhejiang, Shandong

全球分布 / World Distribution
中国、朝鲜、韩国、俄罗斯
China, Democratic People's Republic of Korea, Republic of Korea, Russia

生物地理界 / Biogeographic Realm
古北界 Palearctic

WWF 生物群系 / WWF Biome
温带阔叶和混交林
Temperate Broadleaf & Mixed Forests

动物地理分布型 / Zoogeographic Distribution Type
O

分布标注 / Distribution Note
非特有种 Non-Endemic

▲ 濒危状况 / Threatened Status

中国生物多样性红色名录等级 / CB RL Category (2021)
近危 NT

IUCN 红色名录 / IUCN Red List (2021)
无危 LC

威胁因子 / Threats
无 None

▲ 法律保护地位 / Legal Protection Status

国家重点保护野生动物等级 / Category of National Key Protected Wild Animals (2021)
未列入 Not listed

"三有"名录 / TWIESSV (2023)
未列入 Not listed

CITES 附录等级 / CITES Appendix (2023)
未列入 Not listed

迁徙物种公约附录 / CMS Appendix (2020)
未列入 Not listed

保护行动 / Conservation Action
尚无保护行动 No conservation action so far

▲ 参考文献 / References

Jiang et al. (蒋志刚等), 2021; Burgin et al., 2020; IUCN, 2020; Liu et al. (刘少英等), 2020; Wilson and Mittermeier, 2018; Smith et al., 2009; Pan et al. (潘清华等), 2007; Wilson and Reeder, 2005; Ohdachi et al., 2004; Axel, 2002; Wang (王应祥), 2003; Motokawa et al., 2000; Zhang (张荣祖), 1997

94 / 华南中麝鼩

Crocidura rapax G. M. Allen, 1923

· Chinese White-toothed Shrew

▲ 分类地位 / Taxonomy

劳亚食虫目 Eulipotyphla / 鼩鼱科 Soricidae / 麝鼩属 *Crocidura*

科建立者及其文献 / Family Authority
G. Fischer, 1814

属建立者及其文献 / Genus Authority
Wagler, 1832

亚种 / Subspecies
指名亚种 *C. r. rapax* G. Allen, 1929
云南、广西、海南、贵州和广东
Yunnan, Guangxi, Hainan, Guizhou and Guangdong

台湾亚种 *C. r. kurodai* Tekuda et Kano, 1936
台湾 Taiwan

台湾绿岛亚种 *C. r. lutaoensis* Fang et Lee, 2002
台湾 Taiwan

台湾兰屿亚种 *C. r. tadae* Tokuda et Kano, 1936
台湾兰屿
Taiwan (Lanyu Island)

模式标本产地 / Type Locality
中国
China, Yunnan, Mekong River, Yinpankai

▲ 其他名称 / Other Name(s)

其他中文名 / Other Chinese Name(s)
无 None

其他英文名 / Other English Name(s)
无 None

同物异名 / Synonym(s)
无 None

▲ 形态及生境 / Morphology and Habitat

形态特征 / Morphological Characteristics
齿式：3.1.1.3/2.0.1.3=28。上门齿钩状，向前端伸出。第一上单尖齿约为第二上单尖齿的2倍大小。头体长56~70 mm。尾长38~47 mm。后足长11~13 mm。颅全长17.4~18.3 mm。背部棕色，腹部浅棕灰色。尾不明显双色，背面棕色，腹面毛色略浅，尾长是头体长的63%~69%，尾部靠近基部的一半长度有长针毛。

Dental formula: 3.1.1.3/2.0.1.3=28. Upper incisors hook-like projecting anteriorly. The first upper unicuspid is large and high, twice as long as the second one. Head-body 56-70 mm, tail 38-47 mm, hind-foot 11-13 mm, condyloincisive length 17.5-18.3 mm. Dorsal pelage brown, ventral pelage pale grayish brown. Tail length ranges between 63%-69% of the head-body length. Tail is faintly bicolored, brown above, pale below. Long bristle hair extends along 50% of the proximal portion of the tail.

生境 / Habitat
温带森林
Temperate forest

▲ 地理分布 / Geographic Distribution

国内分布 / Domestic Distribution
云南、湖南、广西、海南、四川、贵州、台湾
Yunnan, Hunan, Guangxi, Hainan, Sichuan, Guizhou, Taiwan

全球分布 / World Distribution
中国、越南、缅甸
China, Vietnam, Myanmar

生物地理界 / Biogeographic Realm
古北界、印度马来界 Palearctic, Indomalaya

WWF 生物群系 / WWF Biome
热带和亚热带湿润阔叶林、温带阔叶和混交林
Tropical & Subtropical Moist Broadleaf Forests, Temperate Broadleaf & Mixed Forests

动物地理分布型 / Zoogeographic Distribution Type
Sc+J

分布标注 / Distribution Note
非特有种 Non-Endemic

▲ 濒危状况 / Threatened Status

中国生物多样性红色名录等级 / CB RL Category (2021)
近危 NT

IUCN 红色名录 / IUCN Red List (2021)
数据缺乏 DD

威胁因子 / Threats
无 None

▲ 法律保护地位 / Legal Protection Status

国家重点保护野生动物等级 / Category of National Key Protected Wild Animals (2021)
未列入 Not listed

"三有"名录 / TWIESSV (2023)
未列入 Not listed

CITES 附录等级 / CITES Appendix (2023)
未列入 Not listed

迁徙物种公约附录 / CMS Appendix (2020)
未列入 Not listed

保护行动 / Conservation Action
尚无保护行动 No conservation action so far

▲ 参考文献 / References

Jiang et al. (蒋志刚等), 2021; Burgin et al., 2020; Chen et al., 2020; IUCN, 2020; Liu et al. (刘少英等), 2020; Wilson and Mittermeier, 2018; Liu et al. (刘应雄等), 2014; Yu et al. (余国睿等), 2014; Smith et al., 2009; Pan et al. (潘清华等), 2007; Wilson and Reeder, 2005; Wang (王应祥), 2003

95 / 山东小麝鼩

Crocidura shantungensis Miller, 1901

· Shantung White-toothed Shrew

▲ 分类地位 / Taxonomy

劳亚食虫目 Eulipotyphla / 鼩鼱科 Soricidae / 麝鼩属 *Crocidura*

科建立者及其文献 / Family Authority
G. Fischer, 1814

属建立者及其文献 / Genus Authority
Wagler, 1832

亚种 / Subspecies
华北亚种 *C. s. shantungensis* Miller, 1901
河北、北京、山东、山西、安徽、浙江和江苏
Hebei, Beijing, Shandong, Shanxi, Anhui, Zhejiang and Jiangsu

东北亚种 *C. s. orientis* Ognev, 1921
黑龙江（乌苏里江地区）、辽宁和吉林
Heilongjiang (Ussuri River area), Liaoning and Jilin

西南亚种 *C. s. phaeopus* G. Allen, 1923
四川、云南、贵州、湖北、陕西和甘肃
Sichuan, Yunnan, Guizhou, Hebei, Shaanxi and Gansu

台湾亚种 *C. s. hosletti* Jameson et Jones, 1977
台湾
Taiwan

模式标本产地 / Type Locality
中国
China, Shantung (Shandong), Chimeh

乔轶伦 / 供图

▲ 其他名称 / Other Name(s)

其他中文名 / Other Chinese Name(s)
无 None

其他英文名 / Other English Name(s)
Shantung Shrew, Asian Lesser White-toothed Shrew, Manchurian White-toothed Shrew

同物异名 / Synonym(s)
无 None

▲ 形态及生境 / Morphology and Habitat

形态特征 / Morphological Characteristics
齿式：3.1.1.3/2.0.1.3=28。头体长 51~74 mm。尾长 26~40 mm。后足 10~13 mm。体重 6.2~8.2 g。背部毛色银灰色，腹部毛色略浅于背部毛色。
Dental formula: 3.1.1.3/2.0.1.3=28. Head-body 51-74 mm, tail 26-40 mm, hind-foot 10-13 mm, weight 6.2-8.2 g. Dorsal pelage dull grayish brown, ventral pelage light gray.

生境 / Habitat
草地、森林 Grassland, forest

▲ 地理分布 / Geographic Distribution

国内分布 / Domestic Distribution
黑龙江、辽宁、吉林、内蒙古、四川、贵州、湖北、陕西、甘肃、青海、河北、北京、山东、山西、安徽、浙江、江苏、宁夏、台湾
Heilongjiang, Liaoning, Jilin, Inner Mongolia, Sichuan, Guizhou, Hubei, Shaanxi, Gansu, Qinghai, Hebei, Beijing, Shandong, Shanxi, Anhui, Zhejiang, Jiangsu, Ningxia, Taiwan

全球分布 / World Distribution
中国、日本、朝鲜、俄罗斯、蒙古国
China, Japan, Democratic People's Republic of Korea, Russia, Mongolia

生物地理界 / Biogeographic Realm
古北界 Palearctic

WWF 生物群系 / WWF Biome
温带阔叶和混交林、热带和亚热带湿润阔叶林
Temperate Broadleaf & Mixed Forests, Tropical & Subtropical Moist Broadleaf Forests

动物地理分布型 / Zoogeographic Distribution Type
Ba

分布标注 / Distribution Note
非特有种 Non-Endemic

▲ 濒危状况 / Threatened Status

中国生物多样性红色名录等级 / CB RL Category (2021)
无危 LC

IUCN 红色名录 / IUCN Red List (2021)
无危 LC

威胁因子 / Threats
未知 Unknown

▲ 法律保护地位 / Legal Protection Status

国家重点保护野生动物等级 / Category of National Key Protected Wild Animals (2021)
未列入 Not listed

"三有" 名录 / TWIESSV (2023)
未列入 Not listed

CITES 附录等级 / CITES Appendix (2023)
未列入 Not listed

迁徙物种公约附录 / CMS Appendix (2020)
未列入 Not listed

保护行动 / Conservation Action
尚无保护行动 No conservation action so far

▲ 参考文献 / References

Jiang et al. (蒋志刚等), 2021; Burgin et al., 2020; IUCN, 2020; Liu et al. (刘少英等), 2020; Wilson and Mittermeier, 2018; Sun et al. (孙治宇等), 2013; Bannikova et al., 2009; Simth et al., 2009; Qin et al. (秦岭等), 2007; Pan et al. (潘清华等), 2007; Wilson and Reeder, 2005; Wang (王应祥), 2003

96 / 西伯利亚麝鼩

Crocidura sibirica Dukelsky, 1930

· Siberian Shrew

▲ 分类地位 / Taxonomy

劳亚食虫目 Eulipotyphla / 鼩鼱科 Soricidae / 麝鼩属 *Crocidura*

科建立者及其文献 / Family Authority
G. Fischer, 1814

属建立者及其文献 / Genus Authority
Wagler, 1832

亚种 / Subspecies
无 None

模式标本产地 / Type Locality
俄罗斯
Russia, Siberia, S Krasnoyarsky Krai, upper Yenisei River, 96 km S of Minusinsk, Oznatchenoie

▲ 其他名称 / Other Name(s)

其他中文名 / Other Chinese Name(s)
无 None

其他英文名 / Other English Name(s)
Siberian White-toothed Shrew

同物异名 / Synonym(s)
无 None

▲ 形态及生境 / Morphology and Habitat

形态特征 / Morphological Characteristics
齿式：3.1.1.3/2.0.1.3=28。头体长 77~86 mm。尾长 31~38 mm。后足 10~13 mm。体重 7.5~9.4 g。毛色整体背部灰褐色，腹部灰白色，背部腹部颜色分明。尾棕灰色。

Dental formula: 3.1.1.3/2.0.1.3=28. Head-body length 77-86 mm, tail 31-38 mm, hind foot 10-13 mm, weight 7.5-9.4 g. Dorsal pelage is dark brown or grayish brown, and venter is light gray. Tail is unicolored and grayish brown.

生境 / Habitat
森林、沼泽、溪流边
Forest, swamp, near streams

▲ 地理分布 / Geographic Distribution

国内分布 / Domestic Distribution
新疆 Xinjiang

全球分布 / World Distribution
中国、哈萨克斯坦、吉尔吉斯斯坦、蒙古国、俄罗斯
China, Kazakhstan, Kyrgyzstan, Mongolia, Russia

生物地理界 / Biogeographic Realm
古北界 Palearctic

WWF 生物群系 / WWF Biome
温带针叶树森林，温带草原、热带稀树草原和灌木地
Temperate Conifer Forests, Temperate Grasslands, Savannas & Shrublands

动物地理分布型 / Zoogeographic Distribution Type
Dc

分布标注 / Distribution Note
非特有种 Non-Endemic

▲ 濒危状况 / Threatened Status

中国生物多样性红色名录等级 / CB RL Category (2021)
近危 NT

IUCN 红色名录 / IUCN Red List (2021)
无危 LC

威胁因子 / Threats
过度放牧、污染 Over-grazing, pollution

▲ 法律保护地位 / Legal Protection Status

国家重点保护野生动物等级 / Category of National Key Protected Wild Animals (2021)
未列入 Not listed

"三有"名录 / TWIESSV (2023)
未列入 Not listed

CITES 附录等级 / CITES Appendix (2023)
未列入 Not listed

迁徙物种公约附录 / CMS Appendix (2020)
未列入 Not listed

保护行动 / Conservation Action
尚无保护行动 No conservation action so far

▲ 参考文献 / References

Jiang et al. (蒋志刚等), 2021; Burgin et al., 2020; IUCN, 2020; Wilson and Mittermeier, 2018; Duan et al. (段海生等), 2011; Smith et al., 2009; Ohdachi et al., 2004; Pan et al. (潘清华等), 2007; Wilson and Reeder, 2005; Wang (王应祥), 2003

97 / 小麝鼩

Crocidura suaveolens (Pallas, 1811)

· Lesser Shrew

▲ 分类地位 / Taxonomy

劳亚食虫目 Eulipotyphla / 鼩鼱科 Soricidae / 麝鼩属 *Crocidura*

科建立者及其文献 / Family Authority
G. Fischer, 1814

属建立者及其文献 / Genus Authority
Wagler, 1832

亚种 / Subspecies
伊犁亚种 *C. s. ilensis* (Miller, 1901)

模式标本产地 / Type Locality
俄罗斯
Russia, Crimea, Khersones, near Sevastopol

刘晔 / 供图

▲ 其他名称 / Other Name(s)

其他中文名 / Other Chinese Name(s)
无 None

其他英文名 / Other English Name(s)
无 None

同物异名 / Synonym(s)
无 None

▲ 形态及生境 / Morphology and Habitat

形态特征 / Morphological Characteristics

齿式：3.1.1.3/2.0.1.3=28。头体长 47~80 mm。尾长 25~40 mm。体重 6.5~9.4 g。背部毛色浅褐色至棕褐色，腹部毛色浅灰色至灰白色。背腹明显异色，冬季色差更为明显。前后足足背覆盖白色短毛。尾通常背腹异色，背面为深棕色，腹面毛色较浅；尾基部至尾长 3/4 处有较长的针毛，尾尖有一簇短毛。

Dental formula: 3.1.1.3/2.0.1.3=28. Head-body length 47-80 mm, tail length 25-40 mm, weight 6.5-9.4 g. Dorsal pelage brownish gray to light gray, and ventral is white or light gray. In winter, dorsum is darker and the ventrum is lighter. Tail is usually bicolored. Long bristle hair extends along 75% of the proximal portion of the tail, a tuft of hair present on the tip.

生境 / Habitat
草地、林地、牧场
Grassland, forest, pastureland

▲ 地理分布 / Geographic Distribution

国内分布 / Domestic Distribution
陕西、安徽、内蒙古、甘肃、黑龙江、山西、山东、江苏、辽宁、宁夏、
浙江、湖北、新疆、四川
Shaanxi, Anhui, Inner Mongolia, Gansu, Heilongjiang, Shanxi, Shandong, Jiangsu,
Liaoning, Ningxia, Zhejiang, Hubei, Xinjiang, Sichuan

全球分布 / World Distribution
西欧延伸至中国北方，广泛分布
Widely distributed from the coast of Western Europe to Eastern Siberia, including
northern China

生物地理界 / Biogeographic Realm
古北界 Palearctic

WWF 生物群系 / WWF Biome
温带草原，热带稀树草原和灌木地、北方森林 / 针叶林
Temperate Grasslands, Savannas & Shrublands, Boreal Forests/Coniferous Forests

动物地理分布型 / Zoogeographic Distribution Type
Dc

分布标注 / Distribution Note
非特有种 Non-Endemic

▲ 濒危状况 / Threatened Status

中国生物多样性红色名录等级 / CB RL Category (2021)
无危 LC

IUCN 红色名录 / IUCN Red List (2021)
无危 LC

威胁因子 / Threats
未知 Unknown

▲ 法律保护地位 / Legal Protection Status

国家重点保护野生动物等级 / Category of National Key Protected Wild Animals (2021)
未列入 Not listed

"三有"名录 / TWIESSV (2023)
未列入 Not listed

CITES 附录等级 / CITES Appendix (2023)
未列入 Not listed

迁徙物种公约附录 / CMS Appendix (2020)
未列入 Not listed

保护行动 / Conservation Action
尚无保护行动 No conservation action so far

▲ 参考文献 / References

Burgin et al., 2020; IUCN, 2020; Wilson and Mittermeier, 2019; Simth et al., 2009

98 | 台湾灰麝鼩

Crocidura tanakae Kuroda, 1938

· Taiwanese Gray Shrew

▲ 分类地位 / Taxonomy

劳亚食虫目 Eulipotyphla / 鼩鼱科 Soricidae / 麝鼩属 *Crocidura*

科建立者及其文献 / Family Authority
G. Fischer, 1814

属建立者及其文献 / Genus Authority
Wagler, 1832

亚种 / Subspecies
无共识 No consensus

模式标本产地 / Type Locality
中国
"Shohosha, Horigai, Taichusiu" = Xiaopushe, Pulijie, Taichung, Taiwan

何鍇 / 供图

▲ 其他名称 / Other Name(s)

其他中文名 / Other Chinese Name(s)
台湾长尾麝鼩

其他英文名 / Other English Name(s)
Taiwanese Gray White-toothed Shrew

同物异名 / Synonym(s)
无 None

▲ 形态及生境 / Morphology and Habitat

形态特征 / Morphological Characteristics
齿式：3.1.1.3/2.0.1.3=28。头体长 69~86 mm。颅全长 20~22 mm。耳长
8~10 mm。后足长 12~15 mm。尾长 47~63 mm。与灰麝鼩外形相似，
但体型略小。背部毛色浅灰色，腹部毛色略浅。足背灰色，后足脚掌较
宽。尾长约为头体长的 68%，靠近基部的 2/3 部分长有针毛。脑颅较低，
下颌骨上升支发育良好，喙突较高。
Dental formula: 3.1.1.3/2.0.1.3=28. Head-body 69-86 mm, tail 47-63 mm, hind-foot
12-14.5 mm. It is morphologically similar to *S. attenuata*, but smaller. The dorsal
pelage is pale gray and the ventral is slightly lighter. Upperside of feet gray. Hindfeet
larger and broader with a rounder thenar (plantar) pad and hypothenar pad which are
located close to each other. Tail is about 68% of the head-body in length, and bristle
hair is distributed on the first two-third of the tail. The braincase is low. Ascending
ramus is robust, and the coronoid process is well developed and high.

生境 / Habitat
草地、次生林、牧场
Grassland, secondary forest, pastureland

▲ 地理分布 / Geographic Distribution

国内分布 / Domestic Distribution
台湾、浙江、福建、广东、海南、安徽、江西、湖北、湖南、四川、
重庆、贵州、广西、云南、河南
Taiwan, Zhejiang, Fujian, Guangdong, Hainan, Anhui, Jiangxi, Hubei, Hunan,
Sichuan, Chongqing, Guizhou, Guangxi, Yunnan, Henan

全球分布 / World Distribution
中国、菲律宾、老挝、越南
China, Philippines, Vietnam, Laos

生物地理界 / Biogeographic Realm
印度马来界 Indomalaya

WWF 生物群系 / WWF Biome
热带和亚热带湿润阔叶林、温带阔叶和混交林
Tropical & Subtropical Moist Broadleaf Forests, Temperate Broadleaf & Mixed
Forests

动物地理分布型 / Zoogeographic Distribution Type
S+J

分布标注 / Distribution Note
非特有种 Non-Endemic

▲ 濒危状况 / Threatened Status

中国生物多样性红色名录等级 / CB RL Category (2021)
无危 LC

IUCN 红色名录 / IUCN Red List (2021)
无危 LC

威胁因子 / Threats
无 None

▲ 法律保护地位 / Legal Protection Status

国家重点保护野生动物等级 / Category of National Key Protected Wild Animals (2021)
未列入 Not listed

"三有"名录 / TWIESSV (2023)
未列入 Not listed

CITES 附录等级 / CITES Appendix (2023)
未列入 Not listed

迁徙物种公约附录 / CMS Appendix (2020)
未列入 Not listed

保护行动 / Conservation Action
尚无保护行动 No conservation action so far

▲ 参考文献 / References

Jiang et al. (蒋志刚等), 2021; Burgin et al., 2020; Chen et al., 2020; IUCN, 2020; Liu et al. (刘少英等), 2020; Li et al., 2019; Wilson and
Mittermeier, 2018; Esselstynand Oliveros, 2010; Smith et al., 2009; Pan et al. (潘清华等), 2007; Wilson and Reeder, 2005; Wang (王应祥), 2003;
Fang and Lee, 2002; Motokawa et al., 2001

99 / 西南中麝鼩

Crocidura vorax G. M. Allen, 1923

· Voracious Shrew

▲ 分类地位 / Taxonomy

劳亚食虫目 Eulipotyphla / 鼩鼱科 Soricidae / 麝鼩属 *Crocidura*

科建立者及其文献 / Family Authority
G. Fischer, 1814

属建立者及其文献 / Genus Authority
Wagler, 1832

亚种 / Subspecies
无 None

模式标本产地 / Type Locality
中国
China, Yunnan, Li-kiang (Lijiang) Valley

▲ 其他名称 / Other Name(s)

其他中文名 / Other Chinese Name(s)
无 None

其他英文名 / Other English Name(s)
Voracious White-toothed Shrew

同物异名 / Synonym(s)
无 None

▲ 形态及生境 / Morphology and Habitat

形态特征 / Morphological Characteristics
齿式：3.1.1.3/2.0.1.3=28。头骨粗壮，上门齿较长，与脑颅方向垂直。头体长 54~90 mm。后足长 11~14 mm。尾长 41~51 mm。与灰麝鼩相似，但体型更小，毛色更偏棕色。背毛淡灰棕色，腹毛浅棕灰色。尾双色，背面深色，腹面浅灰色，尾长靠近臀部 80% 长有稀疏的针毛。

Dental formula: 3.1.1.3/2.0.1.3=28. Skull is robust. The upper incisor is obviously long, extending straightly downward. Head-body 54-90 mm, tail 41-51 mm, hind-foot 11-14 mm. Similar to *C. attenuata* but smaller and with a more brownish pelage. Dorsal pelage is pale grayish brown, ventrum is light brownish gray. Tail is sharply bicolored, upperpart dark and underpart light gray. Bristle hair extends over 80% of the proximal portion of the tail.

生境 / Habitat
温带森林、亚热带湿润低地森林
Temperate forest, subtropical moist lowland forest

▲ 地理分布 / Geographic Distribution

国内分布 / Domestic Distribution
四川、贵州、云南、湖南
Sichuan, Guizhou, Yunnan, Hunan

全球分布 / World Distribution
中国、老挝、泰国
China, Laos, Thailand

生物地理界 / Biogeographic Realm
古北界、印度马来界 Palearctic, Indomalaya

WWF 生物群系 / WWF Biome
热带和亚热带湿润阔叶林、温带阔叶和混交林
Tropical & Subtropical Moist Broadleaf Forests, Temperate Broadleaf & Mixed Forests

动物地理分布型 / Zoogeographic Distribution Type
Hm

分布标注 / Distribution Note
非特有种 Non-Endemic

▲ 濒危状况 / Threatened Status

中国生物多样性红色名录等级 / CB RL Category (2021)
近危 NT

IUCN 红色名录 / IUCN Red List (2021)
无危 LC

威胁因子 / Threats
未知 Unknown

▲ 法律保护地位 / Legal Protection Status

国家重点保护野生动物等级 / Category of National Key Protected Wild Animals (2021)
未列入 Not listed

"三有" 名录 / TWIESSV (2023)
未列入 Not listed

CITES 附录等级 / CITES Appendix (2023)
未列入 Not listed

迁徙物种公约附录 / CMS Appendix (2020)
未列入 Not listed

保护行动 / Conservation Action
尚无保护行动 No conservation action so far

▲ 参考文献 / References

Jiang et al. (蒋志刚等), 2021; Burgin et al., 2020; IUCN, 2020; Wilson and Mittermeier, 2018; Tu et al. (涂飞云等), 2012; Qin et al. (秦岭等), 2007; Pan et al. (潘清华等), 2007; Wilson and Reeder, 2005; Wang (王应祥), 2003

100 / 五指山小麝鼩

Crocidura wuchihensis Wang, 1966

· Hainan Island Shrew

▲ 分类地位 / Taxonomy

劳亚食虫目 Eulipotyphla / 鼩鼱科 Soricidae / 麝鼩属 *Crocidura*

科建立者及其文献 / Family Authority
G. Fischer, 1814

属建立者及其文献 / Genus Authority
Wagler, 1832

亚种 / Subspecies
无 None

模式标本产地 / Type Locality
中国
China, Hainan Isl, Mt. Wuchih

▲ 其他名称 / Other Name(s)

其他中文名 / Other Chinese Name(s)
无 None

其他英文名 / Other English Name(s)
Wuchi Shrew

同物异名 / Synonym(s)
无 None

▲ 形态及生境 / Morphology and Habitat

形态特征 / Morphological Characteristics

齿式：3.1.1.3/2.0.1.3=28。头体长 50~60.5 mm。尾长 34~39 mm。后足长 10~10.5 mm。耳长 8 mm。体重 4.5~5.5 g。来自广西和越南的标本体型更大，但其分类地位有待于验证。背部暗灰色，夹杂白色的毛发，腹部毛色略淡。足背覆盖有白色的毛发。尾长约为头体长的 67%，颜色与体毛相似，尾近臀部 20% 的区域有稀疏的针毛。头骨较窄，脑颅较低。牙齿与台湾分布的华南灰麝鼩相似。

Dental formula: 3.1.1.3/2.0.1.3=28. head-body 50-60.5 mm, tail 34-39 mm, hindfoot 10-10.5 mm. The specimens from Guangxi and Vietnam are slightly larger, but their taxonomic states need to be revised. Pelage is not obviously bicolor, the dorsum is dark gray to grayish brown, and the ventrum is lighter. Upperside of hindfeet covered with light hairs. Tail is about 67% of the head-body length, and is similar to the pelage in color. Bristle hair extends along 20% of the proximal portion of the tail. Skull is narrow and the braincase is low. Teeth were described as similar to *Crocidura rapax kurodai* from Taiwan.

生境 / Habitat
热带、亚热带森林
Tropical, subtropical forest

▲ 地理分布 / Geographic Distribution

国内分布 / Domestic Distribution
海南、云南、广西
Hainan, Yunnan, Guangxi

全球分布 / World Distribution
中国、越南
China, Vietnam

生物地理界 / Biogeographic Realm
印度马来界 Indomalaya

WWF 生物群系 / WWF Biome
热带和亚热带湿润阔叶林
Tropical & Subtropical Moist Broadleaf Forests

动物地理分布型 / Zoogeographic Distribution Type
S+J

分布标注 / Distribution Note
非特有种 Non-Endemic

▲ 濒危状况 / Threatened Status

中国生物多样性红色名录等级 / CB RL Category (2021)
近危 NT

IUCN 红色名录 / IUCN Red List (2021)
数据缺乏 DD

威胁因子 / Threats
未知 Unknown

▲ 法律保护地位 / Legal Protection Status

国家重点保护野生动物等级 / Category of National Key Protected Wild Animals (2021)
未列入 Not listed

"三有"名录 / TWIESSV (2023)
未列入 Not listed

CITES 附录等级 / CITES Appendix (2023)
未列入 Not listed

迁徙物种公约附录 / CMS Appendix (2020)
未列入 Not listed

保护行动 / Conservation Action
尚无保护行动 No conservation action so far

▲ 参考文献 / References

Jiang et al. (蒋志刚等), 2021; Burgin et al., 2020; Chen et al., 2020; IUCN, 2020; Liu et al. (刘少英等), 2020; Wilson and Mittermeier, 2018; Duan et al. (段海生等), 2011; Smith et al., 2009; Pan et al. (潘清华等), 2007; Wilson and Reeder, 2005; Tan (谭邦杰), 1992

101 / 北树鼩

Tupaia belangeri (Wagner, 1841)

• Northern Tree Shrew

▲ 分类地位 / Taxonomy

攀鼩目 Scandentia / 树鼩科 Tupaiidae / 树鼩属 *Tupaia*

科建立者及其文献 / Family Authority
Gray, 1825

属建立者及其文献 / Genus Authority
Raffles, 1821

亚种 / Subspecies

海南亚种 *T. b. modesta* J. Mien, 1906 海南 Hainan

瑶山亚种 *T. b. yaoshanensis* Wang, 1987 广西 Guangxi

越北亚种 *T. b. tonquinia* Thomas, 1925 广西 Guangxi

滇东南亚种 *T. b. yunalis* Thomas, 1914 云南和贵州 Yunnan and Guizhou

滇西亚种 *T. b. chinensis* Anderson, 1879 云南和四川 Yunnan and Sichuan

高黎贡山亚种 *T. b. gaohgongemis* Wang, 1987 云南 Yunnan

藏东南亚种 *T. b. versurae* Thomas, 1922 西藏 Tibet

藏南亚种 *T. b. lepcha* Thomas, 1922 西藏 Tibet

模式标本产地 / Type Locality
缅甸
Burma (Myanmar), Pegu, Siriam (near Yangon)

许明岗 / 供图

邢睿 / 供图

▲ 其他名称 / Other Name(s)

其他中文名 / Other Chinese Name(s)
树鼩、中缅树鼩

其他英文名 / Other English Name(s)
Chinese Treeshrew

同物异名 / Synonym(s)
无 None

▲ 形态及生境 / Morphology and Habitat

形态特征 / Morphological Characteristics

齿式：2.1.3.3/1.3.3.4=20。吻部钝圆，拉长，灰色，橄榄色，比松鼠的吻部长。头体长 160~230 mm。尾长 150~200 mm。耳长 15~20 mm。成年雄性眼睛周围有一圈白毛。雌性有 3 对乳头。背部毛发暗棕色或脏黄色。下腹部毛发通常淡黄色。云南种群毛色多为橄榄褐色，海南种群毛色多为深红色。尾巴扁平，被覆稀疏长毛，不像松鼠那样蓬松多毛。

Dental formula: 2.1.3.3/1.3.3.4. Snout blunt and round, elongated, greyish, olive color, longer than that of a squirrel. Head and body length 160-230 mm. Tail length 150-200 mm. Ear length 15-20 mm. A ring of white hair around the eye in adult males. Females have 3 pairs of nipples. Dorsal hairs dirty brown or dirty yellow. The underbelly is usually pale yellow. The hair color of the Yunnan subspecies is more olive brown, while that of Hainan subspecies is more scarlet. Tail flat, sparsely covered with long hairs, unlike the puffy tai of squirrels.

生境 / Habitat

热带亚热带森林、喀斯特地区
Tropical subtropical forest, karst landscape

▲ 地理分布 / Geographic Distribution

国内分布 / Domestic Distribution
云南、四川、西藏、海南、广东、广西、贵州
Yunnan, Sichuan, Tibet, Hainan, Guangdong, Guangxi, Guizhou

全球分布 / World Distribution
中国、泰国、缅甸、印度、柬埔寨、老挝、越南
China, Thailand, Myanmar, India, Cambodia, Laos, Vietnam

生物地理界 / Biogeographic Realm
印度马来界 Indomalaya

WWF 生物群系 / WWF Biome
热带和亚热带湿润阔叶林
Tropical & Subtropical Moist Broadleaf Forests

动物地理分布型 / Zoogeographic Distribution Type
Wb

分布标注 / Distribution Note
非特有种 Non-Endemic

▲ 濒危状况 / Threatened Status

中国生物多样性红色名录等级 / CB RL Category (2021)
无危 LC

IUCN 红色名录 / IUCN Red List (2021)
无危 LC

威胁因子 / Threats
未知 Unknown

▲ 法律保护地位 / Legal Protection Status

国家重点保护野生动物等级 / Category of National Key Protected Wild Animals (2021)
未列入 Not listed

"三有" 名录 / TWIESSV (2023)
列入 Listed

CITES 附录等级 / CITES Appendix (2023)
II

迁徙物种公约附录 / CMS Appendix (2020)
未列入 Not listed

保护行动 / Conservation Action
尚无保护行动 No conservation action so far

▲ 参考文献 / References

Jiang et al. (蒋志刚等), 2021; Burgin et al., 2020; IUCN, 2020; Wilson and Mittermeier, 2018; Liu and Yao, 2013; Xu et al. (许凌等), 2013; Pan et al. (潘清华等), 2007; Helgen, 2005; Wang (王应祥), 2003; Zhang (张荣祖), 1997; Xia (夏武平), 1988

102 / 抱尾果蝠

Rousettus amplexicaudatus
(É. Geoffroy Saint-Hilaire, 1810)

• Geoffroy's Rousette

▲ 分类地位 / Taxonomy

翼手目 Chiroptera / 狐蝠科 Pteropodidae / 果蝠属 *Rousettus*

科建立者及其文献 / Family Authority
Gray, 1821

属建立者及其文献 / Genus Authority
Gray, 1821

亚种 / Subspecies
无 None

模式标本产地 / Type Locality
印度尼西亚
Indonesia, Lesser Sunda Isls, Timor Isl

▲ 其他名称 / Other Name(s)

其他中文名 / Other Chinese Name(s)
无 None

其他英文名 / Other English Name(s)
Common Rousette

同物异名 / Synonym(s)
无 None

▲ 形态及生境 / Morphology and Habitat

形态特征 / Morphological Characteristics
齿式：2.1.3.2 /2.1.3.3 = 34。头体长 105~115 mm。耳长 18~20 mm。前臂长 79~87 mm。后足长 20~23 mm。尾长 15~17 mm。颅全长 35~40 mm。形态上与棕果蝠 *Rousettus leschenaultii* 相似。体色暗灰褐色。头颈部浅灰色，腹部灰褐色。翅膀深棕色。

Dental formula: 2.1.3.2 /2.1.3.3 = 34. Head and body length 105-115 mm. Ear length 18-20 mm. Forearm 79-87 mm. Hind foot length 20-23 mm. Tail length 15-17 mm. Greatest skull length 35-40 mm. Morphologically similar to Leschenault's Rousette *Rousettus leschenaultii*. Body color is dull grayish brown. The underparts are gray-brown, the neck is pale gray, and the wings are uniformly dark brown.

生境 / Habitat
森林、洞穴、内陆岩石区域
Forest, cave, inland rocky area

▲ 地理分布 / Geographic Distribution

国内分布 / Domestic Distribution
云南 Yunnan

全球分布 / World Distribution
文莱、柬埔寨、中国、印度尼西亚、老挝、马来西亚、缅甸、巴布亚新几内亚（俾斯麦群岛）、菲律宾、新加坡、所罗门群岛、泰国、东帝汶、越南
Brunei, Cambodia, China, Indonesia, Laos, Malaysia, Myanmar, Papua New Guinea(Bismarck Archipelaago), Philippines, Singapore, Solomon Islands, Thailand, Timor-Leste, Vietnam

生物地理界 / Biogeographic Realm
印度马来界、大洋洲界 Indomalaya, Oceanian

WWF 生物群系 / WWF Biome
热带和亚热带湿润阔叶林
Tropical & Subtropical Moist Broadleaf Forests

动物地理分布型 / Zoogeographic Distribution Type
Wa

分布标注 / Distribution Note
非特有种 Non-Endemic

▲ 濒危状况 / Threatened Status

中国生物多样性红色名录等级 / CB RL Category (2021)
易危 VU

IUCN 红色名录 / IUCN Red List (2021)
无危 LC

威胁因子 / Threats
未知 Unknown

▲ 法律保护地位 / Legal Protection Status

国家重点保护野生动物等级 / Category of National Key Protected Wild Animals (2021)
未列入 Not listed

"三有"名录 / TWIESSV (2023)
列入 Listed

CITES 附录等级 / CITES Appendix (2023)
未列入 Not listed

迁徙物种公约附录 / CMS Appendix (2020)
未列入 Not listed

保护行动 / Conservation Action
尚无保护行动 No conservation action so far

▲ 参考文献 / References

Jiang et al. (蒋志刚等), 2021; IUCN, 2021; Burgin et al., 2020; Liu et al. (刘少英等), 2020; Wilson and Mittermeier, 2019; Smith and Xie, 2009; Pan et al. (潘清华等), 2007; Wang (王应祥), 2003

103 / 棕果蝠

Rousettus leschenaultii (Desmarest, 1820)

· Leschenault's Rousette

▲ 分类地位 / Taxonomy

翼手目 Chiroptera / 狐蝠科 Pteropodidae / 果蝠属 *Rousettus*

科建立者及其文献 / Family Authority
Gray, 1821

属建立者及其文献 / Genus Authority
Gray, 1821

亚种 / Subspecies
指名亚种 *R. l. leschenaidtii* (Desmarest, 1820)
西藏、云南、四川、贵州、广西、海南、广东、香港、福建和江西
Tibet, Yunnan, Sichuan, Guizhou, Guangxi, Hainan, Guangdong, Hong Kong, Fujian and Jiangxi

模式标本产地 / Type Locality
印度
India, Pondicherry

赵江波 / 供图

▲ 其他名称 / Other Name(s)

其他中文名 / Other Chinese Name(s)
果蝠、列氏果蝠

其他英文名 / Other English Name(s)
Fulvous Fruit Bat, Shortridge's Rousette

同物异名 / Synonym(s)
无 None

▲ 形态及生境 / Morphology and Habitat

形态特征 / Morphological Characteristics
齿式 2.1.3.2 /2.1.3.3 = 34。头体长 95~120 mm。耳长 18~24 mm。前臂长 80~99 mm。第一指长 23~31 mm。第三指的第二指骨长 41~51 mm。后足长 19~24 mm。尾长 10~18 mm。颅全长 34~40 mm。吻部似犬吻部。眼睛大。耳呈椭圆形，无耳屏。体背毛深褐至黑色，颈背及体腹呈灰褐色。
Dental formula: 2.1.3.2 /2.1.3.3 = 34. Head and body length 95-120 mm. Ear length 18 -24 mm. Forearm length 80-99 mm. Tail length 10-18 mm. The first finger length 23-31 mm, and the length of the second phalanx of the third finger is 41-51 mm. Hind foot length 19-24 mm. Greatest skull length 34-40 mm. Snout is similar in resemblance to a dog. Eyes large. The ears are elliptic without tragus. Dorsal hairs are dark brown to black, the neck back and the body abdomen are grayish brown.

生境 / Habitat
森林、人造建筑、洞穴
Forest, man-made buildings, cave

▲ 地理分布 / Geographic Distribution

国内分布 / Domestic Distribution
江西、贵州、广西、广东、海南、福建、云南、香港、澳门、四川、西藏
Jiangxi, Guizhou, Guangxi, Guangdong, Hainan, Fujian, Yunnan, Hong Kong, Macao, Sichuan, Tibet

全球分布 / World Distribution
孟加拉国、不丹、柬埔寨、中国、印度、印度尼西亚、老挝、马来西亚、缅甸、尼泊尔、巴基斯坦、斯里兰卡、泰国、越南
Bangladesh, Bhutan, Cambodia, China, India, Indonesia, Laos, Malaysia, Myanmar, Nepal, Pakistan, Sri Lanka, Thailand, Vietnam

生物地理界 / Biogeographic Realm
印度马来界、 古北界 Indomalaya, Palearctic

WWF 生物群系 / WWF Biome
热带和亚热带湿润阔叶林
Tropical & Subtropical Moist Broadleaf Forests

动物地理分布型 / Zoogeographic Distribution Type
Wa

分布标注 / Distribution Note
非特有种 Non-Endemic

▲ 濒危状况 / Threatened Status

中国生物多样性红色名录等级 / CB RL Category (2021)
近危 NT

IUCN 红色名录 / IUCN Red List (2021)
无危 LC

威胁因子 / Threats
旅游 、采石、狩猎及采集陆生动物
Tourism, quarrying, hunting or collection

▲ 法律保护地位 / Legal Protection Status

国家重点保护野生动物等级 / Category of National Key Protected Wild Animals (2021)
未列入 Not listed

"三有"名录 / TWIESSV (2023)
列入 Listed

CITES 附录等级 / CITES Appendix (2023)
II

迁徙物种公约附录 / CMS Appendix (2020)
未列入 Not listed

保护行动 / Conservation Action
尚无保护行动 No conservation action so far

▲ 参考文献 / References

Jiang et al. (蒋志刚等), 2021; Burgin et al., 2020; IUCN, 2020; Liu et al. (刘少英等), 2020; Wilson and Mittermeier, 2019; Huang et al. (黄继展等), 2013; Pan et al. (潘清华等), 2007; Shao et al. (邵伟伟等), 2007; Zhu (朱光剑), 2007; Chen et al. (陈忠等), 2005; Tang et al. (唐占辉等), 2005; Wilson and Reeder, 2005; Wang (王应祥), 2003; Zhang (张荣祖), 1997; Xia (夏武平), 1988

104 / 琉球狐蝠

Pteropus dasymallus (Temminck, 1825)

- Ryukyu Flying Fox

▲ 分类地位 / Taxonomy

翼手目 Chiroptera / 狐蝠科 Pteropodidae / 狐蝠属 *Pteropus*

科建立者及其文献 / Family Authority
Gray, 1821

属建立者及其文献 / Genus Authority
Brisson, 1762

亚种 / Subspecies
台湾亚种 *P. d. formosus* Sclater, 1873
台湾
Taiwan

模式标本产地 / Type Locality
琉球群岛
Ryukyu Isls, Kuchinoerabu Isl (restricted by Kuroda, 1933)

顏振晖 / 供图

▲ 其他名称 / Other Name(s)

其他中文名 / Other Chinese Name(s)
台湾狐蝠

其他英文名 / Other English Name(s)
Daito Flying Fox, Erabu Flying Fox,
Taiwanese Flying Fox, Orii's Flying Fox,
Yaeyama Flying Fox

同物异名 / Synonym(s)
无 None

▲ 形态及生境 / Morphology and Habitat

形态特征 / Morphological Characteristics
齿式：2.1.3.2 /2.1.3.3=34。头体长 95~120 mm。耳长 18~24 mm。前臂
长 80~99 mm。第一指长 23~31 mm，第三指的第二指骨长 41~51 mm。
后足长 19~24 mm。尾长 10~18 mm。颅全长 34~40 mm。吻部似犬吻部。
眼睛大。耳呈椭圆形，无耳屏。体背毛深褐至黑色，颈背及体腹呈灰褐色。
Dental formula: 2.1.3.2 /2.1.3.3=34. Head and body length 95-120 mm. Hind foot
length 19-24 mm. Ear length 18-24 mm. Forearm length 80-99 mm. Greatest skull
length 34-40 mm. Dark brown hairs on the head and abdomen, yellow or white hairs
on the nape. Snout is prominent, like that of a dog, the auricle is round and short, and
there is no tragus or nose leaf. Wings are brown and the belly and back are yellow.
Dorsal side is brown.

生境 / Habitat
未知 Unknown

▲ 地理分布 / Geographic Distribution

国内分布 / Domestic Distribution
台湾 Taiwan

全球分布 / World Distribution
日本、菲律宾、中国
Japan, Philippines, China

生物地理界 / Biogeographic Realm
印度马来界 Indomalaya

WWF 生物群系 / WWF Biome
热带和亚热带湿润阔叶林
Tropical & Subtropical Moist Broadleaf Forests

动物地理分布型 / Zoogeographic Distribution Type
Wb

分布标注 / Distribution Note
非特有种 Non-Endemic

▲ 濒危状况 / Threatened Status

中国生物多样性红色名录等级 / CB RL Category (2021)
近危 NT

IUCN 红色名录 / IUCN Red List (2021)
无危 LC

威胁因子 / Threats
狩猎及采集
Hunting or collection

▲ 法律保护地位 / Legal Protection Status

国家重点保护野生动物等级 / Category of National Key Protected Wild Animals (2021)
未列入 Not listed

"三有" 名录 / TWIESSV (2023)
未列入 Not listed

CITES 附录等级 / CITES Appendix (2023)
II

迁徙物种公约附录 / CMS Appendix (2020)
未列入 Not listed

保护行动 / Conservation Action
政策性措施和立法保护、生境及栖息地保护
Policy-based actions and legistation, habitat protection

▲ 参考文献 / References

Jiang et al. (蒋志刚等), 2021; IUCN, 2021; Burgin et al., 2020; Liu et al. (刘少英等), 2020; Wilson and Mittermeier, 2019; Pan et al. (潘清华等), 2007; Wilson and Reeder, 2005; Wang (王应祥), 2003; Lin et al. (林良恭等), 2002; Zhang (张荣祖), 1997; Chen (陈兼善), 1969

105 / 印度大狐蝠

Pteropus giganteus (Brünnich, 1782)

• Indian Flying Fox

0294

▲ 分类地位 / Taxonomy

翼手目 Chiroptera / 狐蝠科 Pteropodidae / 狐蝠属 *Pteropus*

科建立者及其文献 / Family Authority
Gray, 1821

属建立者及其文献 / Genus Authority
Brisson, 1762

亚种 / Subspecies
无共识 No consensus

模式标本产地 / Type Locality
印度
India, Bengal

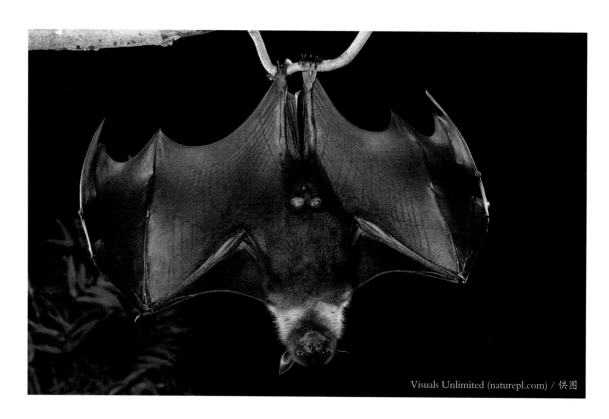

Visuals Unlimited (naturepl.com) / 供图

▲ 其他名称 / Other Name(s)

其他中文名 / Other Chinese Name(s)
无 None

其他英文名 / Other English Name(s)
无 None

同物异名 / Synonym(s)
无 None

▲ 形态及生境 / Morphology and Habitat

形态特征 / Morphological Characteristics
齿式：2.1.3.2 /2.1.3.3=34。头体长 20~23 cm，后足长 4.3~5.8 cm。耳长 3.3~4.5 cm。前臂长 15.2~18.6 cm。颅全长 6.3~7.8 cm。耳长而尖。尾翼膜宽度 2.8 cm。翼展 1.2~1.5 m。体重 600~1600 g。身体被毛颜色深棕色、灰色或黑色，颈背部被毛淡黄色。下颌、喉部和上胸部被毛呈黑棕色。腹部毛色较背部浅。

Dental formula: 2.1.3.2 /2.1.3.3=34. Head-body length 20-23 cm, Hind foot length 4.3-5.8 cm. Ears length 3.3-4.5 cm. Forearm length 15.2-18.6 cm, Greatest skull length 6.3-7.8 cm. Ears long and sharp; breadth of uropatagium 2.8 cm. Wingspan may range from 1.2-1.5 m. Body mass 600-1600 g. Dark brown, gray, or black body hair color with a contrasting yellowish mantle head. Chin, throat, and upper chest blackish brown. Ventral hairs are paler than dorsum.

生境 / Habitat
未知
Unknown

▲ 地理分布 / Geographic Distribution

国内分布 / Domestic Distribution
青海（仅 1 例标本）
Qinghai

全球分布 / World Distribution
孟加拉国、不丹、中国、印度、马尔代夫、缅甸、尼泊尔、巴基斯坦、斯里兰卡
Bangladesh, Bhutan, China, India, Maldives, Myanmar, Nepal, Pakistan, Sri Lanka

生物地理界 / Biogeographic Realm
印度马来界 Indomalaya

WWF 生物群系 / WWF Biome
热带和亚热带湿润阔叶林
Tropical & Subtropical Moist Broadleaf Forests

动物地理分布型 / Zoogeographic Distribution Type
We

分布标注 / Distribution Note
非特有种 Non-Endemic

▲ 濒危状况 / Threatened Status

中国生物多样性红色名录等级 / CB RL Category (2021)
数据缺乏 DD

IUCN 红色名录 / IUCN Red List (2021)
无危 LC

威胁因子 / Threats
未知 Unknown

▲ 法律保护地位 / Legal Protection Status

国家重点保护野生动物等级 / Category of National Key Protected Wild Animals (2021)
未列入 Not listed

"三有" 名录 / TWIESSV (2023)
未列入 Not listed

CITES 附录等级 / CITES Appendix (2023)
II

迁徙物种公约附录 / CMS Appendix (2020)
未列入 Not listed

保护行动 / Conservation Action
尚无保护行动 No conservation action so far

▲ 参考文献 / References

Jiang et al. (蒋志刚等), 2021; Burgin et al., 2020; IUCN, 2021; Liu et al. (刘少英等), 2000; Smith et al., 2009; Pan et al. (潘清华等), 2007; Wilson and Reeder, 2005; Wang (王应祥), 2003; Zhang (张荣祖), 1997; Wang and Wang, 1962

106 / 短耳犬蝠

Cynopterus brachyotis (Müller, 1838)

• Lesser Dog-faced Fruit Bat

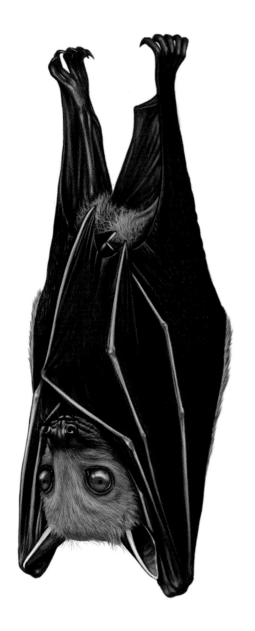

▲ 分类地位 / Taxonomy

翼手目 Chiroptera / 狐蝠科 Pteropodidae / 犬蝠属 *Cynopterus*

科建立者及其文献 / Family Authority
Gray, 1821

属建立者及其文献 / Genus Authority
F. Cuvier, 1824

亚种 / Subspecies
越北亚种 *C. b. hoffeti* Bourret, 1944
云南 Yunnan

模式标本产地 / Type Locality
婆罗洲
Borneo, Dewei (Dewai) River

▲ 其他名称 / Other Name(s)

其他中文名 / Other Chinese Name(s)
无 None

其他英文名 / Other English Name(s)
Common Short-nosed Fruit Bat, Lesser Dog-faced Fruit Bat, Sunda Short-nosed Fruit Bat

同物异名 / Synonym(s)
无 None

▲ 形态及生境 / Morphology and Habitat

形态特征 / Morphological Characteristics
齿式：2.1.3.1/2.1.3.2=30。头体长 54~72 mm。后足长 13~15 mm。耳长 13~18 mm。颅全长 27~31 mm。体重 30~100 g。头部像狐狸，眼睛大、黑色。毛色从浅灰色到深棕色或亮棕色。成年个体喉部和肩部有橙色或黄色毛发。掌指颜色略浅，与深棕色翅膜形成对比。

Dental formula: 2.1.3.1/2.1.3.2=30. Head and body length 54-72 mm. Hind foot length 13-15 mm. Ear length 13-18 mm. Greatest skull length 27-31 mm. Fur color ranges from light grayish through dark or bright brown. Breeding adults may have orange or yellow hairs on throat and shoulders. Fingers reddish in contrast to dark brown wing membranes. It should be noted that *C. brachyotis* has been split into two species (forest and Sunda) based on phylogenetics, but these two species have not been officially resolved.

生境 / Habitat
森林、乡村种植园、种植园、城市
Forest, rural garden, plantation, urban area

▲ 地理分布 / Geographic Distribution

国内分布 / Domestic Distribution
云南、西藏、广东
Yunnan, Tibet, Guangdong

全球分布 / World Distribution
柬埔寨、中国、印度、印度尼西亚、老挝、马来西亚、缅甸、新加坡、
斯里兰卡、泰国、东帝汶、越南
Cambodia, China, India, Indonesia, Laos, Malaysia, Myanmar, Singapore, Sri Lanka,
Thailand, Timor-Leste, Vietnam

生物地理界 / Biogeographic Realm
印度马来界
Indomalaya

WWF 生物群系 / WWF Biome
热带和亚热带湿润阔叶林
Tropical & Subtropical Moist Broadleaf Forests

动物地理分布型 / Zoogeographic Distribution Type
Wb

分布标注 / Distribution Note
非特有种 Non-Endemic

▲ 濒危状况 / Threatened Status

中国生物多样性红色名录等级 / CB RL Category (2021)
易危 LC

IUCN 红色名录 / IUCN Red List (2021)
无危 LC

威胁因子 / Threats
未知 Unknown

▲ 法律保护地位 / Legal Protection Status

国家重点保护野生动物等级 / Category of National Key Protected Wild Animals (2021)
未列入 Not listed

"三有" 名录 / TWIESSV (2023)
未列入 Not listed

CITES 附录等级 / CITES Appendix (2023)
未列入 Not listed

迁徙物种公约附录 / CMS Appendix (2020)
未列入 Not listed

保护行动 / Conservation Action
尚无保护行动 No conservation action so far

▲ 参考文献 / References

Jiang et al. (蒋志刚等), 2021; IUCN, 2021; Burgin et al., 2020; Wilson and Mittermeier, 2019; Pan et al. (潘清华等), 2007; Wilson and Reeder, 2005; Wang (王应祥), 2003; Zhang (张荣祖), 1997; Corbet and Hill, 1992; Heideman and Heaney, 1989

107 / 犬蝠

Cynopterus sphinx (Vahl, 1797)

• Greater Shortnosed Fruit Bat

▲ 分类地位 / Taxonomy

翼手目 Chiroptera / 狐蝠科 Pteropodidae / 犬蝠属 *Cynopterus*

科建立者及其文献 / Family Authority
Gray, 1821

属建立者及其文献 / Genus Authority
F. Cuvier, 1824

亚种 / Subspecies
指名亚种 *C. s. sphinx* (Vahl, 1799)
西藏、海南、福建
Tibet, Hainan, Fujian
泰国亚种 *C. s. angulatus* Miller 1898
广西、云南、海南、广东、香港、澳门和福建
Guangxi, Yunnan, Hainan, Guangdong, Hong Kong, Macao and Fujian

模式标本产地 / Type Locality
印度
India, Madras, Tranquebar

吴毅 / 供图

▲ 其他名称 / Other Name(s)

其他中文名 / Other Chinese Name(s)
短鼻果蝠、短吻果蝠、犬面果蝠

其他英文名 / Other English Name(s)
Short-nosed Indian Fruit Bat, Sphinx Fruit Bat

同物异名 / Synonym(s)
无 None

▲ 形态及生境 / Morphology and Habitat

形态特征 / Morphological Characteristics

齿式：2.1.3.1/2.1.3.2=30。头体长 80~90 mm。耳长 18~21 mm。前臂长 66~83 mm。后足长 13~15 mm。尾长 7~12 mm。翼展 380 mm。颅全长 30~38 mm。头部侧面观似犬，头骨前额部外侧凸起，中间有明显凹陷。耳壳边缘为白色。背毛橄榄棕色，体侧毛色红褐色，腹部毛色锈黄色到绿棕色。

Dental formula: 2.1.3.1/2.1.3.2=30. Head and body length 80-90 mm. Ear length 18-21 mm. Forearm length 66-83 mm. Hind foot length 13-15 mm. Tail length 7-12 mm. Wingspan 380 mm. Greatest skull length 30-38 mm. Dorsal hairs olive-brown. Nostrils protrude like tubes. Ears with pale edges. Lateral part of the anterior frontal part of the skull is raised, with a marked depression in the middle. Body sides reddish-brown ventral surface rusty yellow to greenish-brown.

生境 / Habitat

森林、城市林地蒲葵（或棕榈）树叶
Forest, leaves of bulrush (or palm) forest in urban area

▲ 地理分布 / Geographic Distribution

国内分布 / Domestic Distribution
福建、广西、云南、海南、西藏、广东、香港、澳门
Fujian, Guangxi, Yunnan, Hainan, Tibet, Guangdong, Hong Kong, Macao

全球分布 / World Distribution
孟加拉国、不丹、柬埔寨、中国、印度、印度尼西亚、老挝、马来西亚、
缅甸、尼泊尔、巴基斯坦、斯里兰卡、泰国、越南
Bangladesh, Bhutan, Cambodia, China, India, Indonesia, Laos, Malaysia,
Myanmar, Nepal, Pakistan, Sri Lanka, Thailand, Vietnam

生物地理界 / Biogeographic Realm
印度马来界、古北界、大洋洲界
Indomalaya, Palearctic, Oceanian

WWF 生物群系 / WWF Biome
热带和亚热带湿润阔叶林
Tropical & Subtropical Moist Broadleaf Forests

动物地理分布型 / Zoogeographic Distribution Type
Wb

分布标注 / Distribution Note
非特有种 Non-Endemic

▲ 濒危状况 / Threatened Status

中国生物多样性红色名录等级 / CB RL Category (2021)
近危 NT

IUCN 红色名录 / IUCN Red List (2021)
无危 LC

威胁因子 / Threats
未知 Unknown

▲ 法律保护地位 / Legal Protection Status

国家重点保护野生动物等级 / Category of National Key Protected Wild Animals (2021)
未列入 Not listed

"三有" 名录 / TWIESSV (2023)
未列入 Not listed

CITES 附录等级 / CITES Appendix (2023)
未列入 Not listed

迁徙物种公约附录 / CMS Appendix (2020)
未列入 Not listed

保护行动 / Conservation Action
尚无保护行动 No conservation action so far

▲ 参考文献 / References

Jiang et al. (蒋志刚等), 2021; IUCN, 2021; Burgin et al., 2020; Liu et al. (刘少英等), 2020; Wilson and Mittermeier, 2019; Huang et al. (黄继展等), 2013; Smith and Xie, 2009; Zhang (张伟), 2008; Zhu (朱光剑), 2007; Feng et al. (冯祚建等), 1984; Institute of Entomology/ Guangdong Province and Department of Biology/ Sun Yat-sen University, 1983; Pan et al. (潘清华等), 2007; Wilson and Reeder, 2005; Wang (王应祥), 2003; Zhang (张荣祖), 1997

108 / 球果蝠

Sphaerias blanfordi (Thomas, 1891)

• Blandford's Fruit Bat

▲ 分类地位 / Taxonomy

翼手目 Chiroptera / 狐蝠科 Pteropodidae / 球果蝠属 *Sphaerias*

科建立者及其文献 / Family Authority
Gray, 1821

属建立者及其文献 / Genus Authority
Miller, 1906

亚种 / Subspecies
指名亚种 *S. b. blanfordi* (Thomas. 1891)
云南 Yunnan

墨脱亚种 *S. b. medogensis* Cai et Zhang
西藏 Tibet

模式标本产地 / Type Locality
缅甸
Burma (Myanmar), Karin Hills, Cheba, Leito

吴毅 / 供图

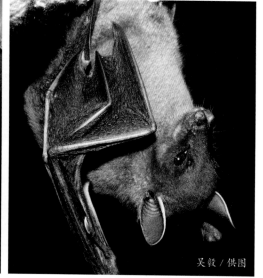
吴毅 / 供图

▲ 其他名称 / Other Name(s)

其他中文名 / Other Chinese Name(s)
球鼻果蝠、山果蝠

其他英文名 / Other English Name(s)
无 None

同物异名 / Synonym(s)
无 None

▲ 形态及生境 / Morphology and Habitat

形态特征 / Morphological Characteristics
齿式：2.1.3.1/2.1.3.2=30。头体长 80~90 mm。耳长 16~20 mm。前臂长 50~58 mm。后足长 11~12 mm。颅全长 27~28 mm。身体暗灰棕色。毛皮长而浓密，皮毛延伸到胫骨和前臂下侧。耳缘浅白色。无足距。股间膜狭窄。无尾。
Dental formula: 2.1.3.1/2.1.3.2=30. Head and body length 80-90 mm. Ear length 16-20 mm. Forearm length 50-58 mm. Hind foot length 11-12 mm. Greatest skull length 27-28 mm. Body dull grayish brown; pelage long and dense; fur extends onto tibia and underside of forearm. Anterior margin of ears edged in white. Antitragus is small and triangular. Uropatagium reduced. Tail absent.

生境 / Habitat
热带和亚热带湿润山地森林
Tropical and subtropical moist montane forest

▲ 地理分布 / Geographic Distribution

国内分布 / Domestic Distribution
西藏、云南
Tibet, Yunnan

全球分布 / World Distribution
不丹、中国、印度、缅甸、尼泊尔、泰国、越南
Bhutan, China, India, Myanmar, Nepal, Thailand, Vietnam

生物地理界 / Biogeographic Realm
印度马来界、古北界 Indomalaya, Palearctic

WWF 生物群系 / WWF Biome
热带和亚热带湿润阔叶林
Tropical & Subtropical Moist Broadleaf Forests

动物地理分布型 / Zoogeographic Distribution Type
Wb

分布标注 / Distribution Note
非特有种 Non-Endemic

▲ 濒危状况 / Threatened Status

中国生物多样性红色名录等级 / CB RL Category (2021)
易危 VU

IUCN 红色名录 / IUCN Red List (2021)
无危 LC

威胁因子 / Threats
未知 Unknown

▲ 法律保护地位 / Legal Protection Status

国家重点保护野生动物等级 / Category of National Key Protected Wild Animals (2021)
未列入 Not listed

"三有" 名录 / TWIESSV (2023)
列入 Listed

CITES 附录等级 / CITES Appendix (2023)
未列入 Not listed

迁徙物种公约附录 / CMS Appendix (2020)
未列入 Not listed

保护行动 / Conservation Action
尚无保护行动 No conservation action so far

▲ 参考文献 / References

Jiang et al. (蒋志刚等), 2021; IUCN, 2021; Burgin et al., 2020; Liu et al. (刘少英等), 2020; Wilson and Mittermeier, 2019; Cheng et al. (程志营等), 2011; Feng et al. (冯庆等), 2008; Pan et al. (潘清华等), 2007; Wilson and Reeder, 2005; Wang (王应祥), 2003; Zhang (张荣祖), 1997; Li (李思华), 1989; Feng et al. (冯祚建等), 1986

109 / 长舌果蝠

Eonycteris spelaea (Dobson, 1871)

• Dawn Bat

▲ 分类地位 / Taxonomy

翼手目 Chiroptera / 狐蝠科 Pteropodidae / 长舌果蝠属 *Eonycteris*

科建立者及其文献 / Family Authority
Gray, 1821

属建立者及其文献 / Genus Authority
Dobson, 1873

亚种 / Subspecies
无 None

模式标本产地 / Type Locality
缅甸
Burma (Myanmar), Tenasserim, Moulmein, Farm Caves

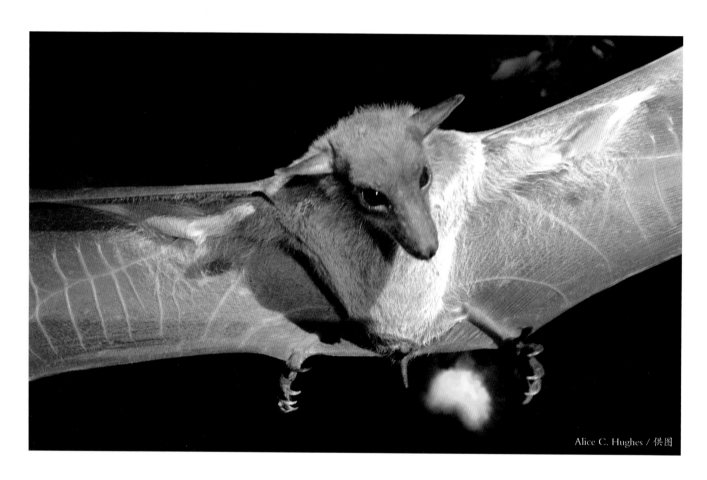

Alice C. Hughes / 供图

▲ 其他名称 / Other Name(s)

其他中文名 / Other Chinese Name(s)
大长舌果蝠、花蕊果蝠

其他英文名 / Other English Name(s)
Cave Nectar Bat, Common Dawn Bat,
Common Nectar Bat, Dobson's Long-
tongued Fruit Bat, Dawn Cave Bat

同物异名 / Synonym(s)
无 None

▲ 形态及生境 / Morphology and Habitat

形态特征 / Morphological Characteristics
齿式：2.1.3.2/2.1.3.3=34。前臂长 65~72 mm。吻部较长，上唇前有一沟槽，
鼻孔不突出。舌长而尖，可伸展。颊齿较粗壮，颊齿排列紧密。第二指
无爪。翼膜止于踝部。尾与后足等长。

Dental formula: 2.1.3.2/2.1.3.3=34. Forearm length 65-72 mm. Snout is long, with a
groove in front of the upper lip and no prominent nostrils. Tongue is long, pointed,
and extensible. Buccal teeth are stout and the teeth are closely arranged. The second
finger has no claws. The pterygoid membrane rests at the ankle. Tail is the same length
as the hind foot.

生境 / Habitat
洞穴、森林、种植园
Caves, forest, plantation

▲ 地理分布 / Geographic Distribution

国内分布 / Domestic Distribution
广西、云南
Guangxi, Yunnan

全球分布 / World Distribution
文莱、柬埔寨、中国、印度、印度尼西亚、老挝、马来西亚、缅甸、
菲律宾、新加坡、泰国、东帝汶、越南
Brunei, Cambodia, China, India, Indonesia, Laos, Malaysia, Myanmar, Philippines,
Singapore, Thailand, Timor-Leste, Vietnam

生物地理界 / Biogeographic Realm
印度马来界、古北界、大洋洲界
Indomalaya, Palearctic, Oceanian

WWF 生物群系 / WWF Biome
热带和亚热带湿润阔叶林
Tropical & Subtropical Moist Broadleaf Forests

动物地理分布型 / Zoogeographic Distribution Type
Wb

分布标注 / Distribution Note
非特有种 Non-Endemic

▲ 濒危状况 / Threatened Status

中国生物多样性红色名录等级 / CB RL Category (2021)
易危 VU

IUCN 红色名录 / IUCN Red List (2021)
无危 LC

威胁因子 / Threats
森林砍伐、耕种、狩猎及采集
Logging, farming, hunting or collection

▲ 法律保护地位 / Legal Protection Status

国家重点保护野生动物等级 / Category of National Key Protected Wild Animals (2021)
未列入 Not listed

"三有" 名录 / TWIESSV (2023)
列入 Listed

CITES 附录等级 / CITES Appendix (2023)
未列入 Not listed

迁徙物种公约附录 / CMS Appendix (2020)
未列入 Not listed

保护行动 / Conservation Action
尚无保护行动 No conservation action so far

▲ 参考文献 / References

Jiang et al. (蒋志刚等), 2021; IUCN, 2021; Burgin et al., 2020; Liu et al. (刘少英等), 2020; Wilson and Mittermeier, 2019; Smith and Xie, 2009; Pan et al. (潘清华等), 2007; Wilson and Reeder, 2005; Wang (王应祥), 2003; Zhang (张荣祖), 1997; Yang et al. (杨德华等), 1993; Xia(夏武平), 1988

110 / 安氏长舌果蝠

Macroglossus sobrinus K. Andersen, 1911

· Hill Long-tongued Fruit Bat

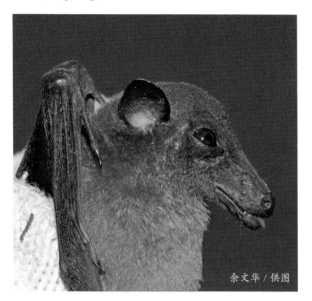

余文华 / 供图

▲ 分类地位 / Taxonomy

翼手目 Chiroptera / 狐蝠科 Pteropodidae /
小长舌果蝠属 *Macroglossus*

科建立者及其文献 / Family Authority
Gray, 1821

属建立者及其文献 / Genus Authority
F. Cuvier, 1824

亚种 / Subspecies
无 None

模式标本产地 / Type Locality
马来西亚
Malaysia, Perak, Gunong Igari (Mt Igari), 610 m

▲ 其他名称 / Other Name(s)

其他中文名 / Other Chinese Name(s)
无 None

其他英文名 / Other English Name(s)
Greater Long-nosed Blossom Bat, Greater Long-nosed Fruit
Bat, Greater Long-tongued Fruit Bat, Greater Nectar Bat, Hill
Long-tongued Blossom Bat, Hill Long-tongued Fruit Bat

同物异名 / Synonym(s)
无 None

▲ 形态及生境 / Morphology and Habitat

形态特征 / Morphological Characteristics
齿式：2.1.3.2/2.1.3.3=34。头体长 70~89 mm。耳长
14~19 mm。前臂长 50~58 mm。后足长 70~89 mm。
尾长 0~6 mm。颅全长 28~29 mm。牙齿细窄，齿间
隙宽。吻长而细窄，上唇前缘无沟槽。舌非常长，有
丝状乳头构成的羽状尖端，用于摄入花粉。鼻口长而
窄，上唇前缘无槽。毛皮柔软细，背面均匀地呈泥褐
色，腹部呈棕黄色。尾短，隐于毛被之内。
Dental formula: 2.1.3.2/2.1.3.3=34. Head and body length 70-
89 mm. Ear length 14-19 mm. Forearm length 50-58 mm.
Length of hind foot 70-89 mm. Tail length 0-6 mm. Greatest
skull length 28-29 mm. Teeth are narrow and the gaps between
teeth are wide. Snout is long and narrow, and the leading edge of
the upper lip is ungrooved. The tip of the toungue tip is feathery
and is made of filiform papillae and is used in feeding on pollen.
Pelage soft and fine, uniformly clay brown dorsally, and buffy
brown ventrally. Tail short, hidden under the coat.

生境 / Habitat
森林、次生林、种植园、人造建筑
Forest, secondary forest, plantation, man-made building

▲ 地理分布 / Geographic Distribution

国内分布 / Domestic Distribution
云南 Yunnan

全球分布 / World Distribution
柬埔寨、中国、印度、印度尼西亚、老挝、马来西亚、缅甸、泰国、越南
Cambodia, China, India, Indonesia, Laos, Malaysia, Myanmar, Thailand, Vietnam

生物地理界 / Biogeographic Realm
印度马来界 Indomalaya

WWF 生物群系 / WWF Biome
热带和亚热带湿润阔叶林
Tropical & Subtropical Moist Broadleaf Forests

动物地理分布型 / Zoogeographic Distribution Type
WB

分布标注 / Distribution Note
非特有种 Non-Endemic

▲ 濒危状况 / Threatened Status

中国生物多样性红色名录等级 / CB RL Category (2021)
濒危 EN

IUCN 红色名录 / IUCN Red List (2021)
无危 LC

威胁因子 / Threats
森林砍伐、耕种、狩猎及采集
Logging, farming, hunting or collection

▲ 法律保护地位 / Legal Protection Status

国家重点保护野生动物等级 / Category of National Key Protected Wild Animals (2021)
未列入 Not listed

"三有" 名录 / TWIESSV (2023)
列入 Listed

CITES 附录等级 / CITES Appendix (2023)
未列入 Not listed

迁徙物种公约附录 / CMS Appendix (2020)
未列入 Not listed

保护行动 / Conservation Action
尚无保护行动 No conservation action so far

▲ 参考文献 / References

Jiang et al. (蒋志刚等), 2021; IUCN, 2021; Burgin et al., 2020; Liu et al. (刘少英等), 2020; Wilson and Mittermeier, 2019; Zhang et al., 2010; Feng et al. (冯庆等), 2007; Pan et al. (潘清华等), 2007; Wang (王应祥), 2003; Lekagul and McNeely, 1977

111 / 无尾果蝠

Megaerops ecaudatus (Temminck, 1837)

· Temminck's Tailless Fruit Bat

▲ 分类地位 / Taxonomy

翼手目 Chiroptera / 狐蝠科 Pteropodidae / 无尾果蝠属 *Megaerops*

科建立者及其文献 / Family Authority
Gray, 1821

属建立者及其文献 / Genus Authority
Peters, 1865

亚种 / Subspecies
无 None

模式标本产地 / Type Locality
印度尼西亚
Indonesia, W Sumatra, Padang

▲ 其他名称 / Other Name(s)

其他中文名 / Other Chinese Name(s)
无 None

其他英文名 / Other English Name(s)
Sunda Tailless Fruit Bat, Temminck's Fruit Bat

同物异名 / Synonym(s)
无 None

▲ 形态及生境 / Morphology and Habitat

形态特征 / Morphological Characteristics
齿式：2.1.3.1/1.1.3.2=28。体毛较短而蓬松，棕色，眼睛深褐色，大而突出。耳壳小而黑，鼻吻部相当短、粗壮，鼻尖明显分叉，鼻孔周围有一圈皮肤。其生态学尚待研究。

Dental formula: 2.1.3.1/1.1.3.2=28. Fur is short, fluffy, brown color, with dark brown eyes, large and bulging. The ears are small and black, and the nose is rather short and robust, with a clearly forked tip and a ring of skin around the nostrils. Until now, the ecology of tailless fruit bats has been barely studied.

生境 / Habitat
热带亚热带湿润低地森林、次生林
Tropical subtropical moist lowland forest and secondary forest

▲ 地理分布 / Geographic Distribution

国内分布 / Domestic Distribution
云南 Yunnan

全球分布 / World Distribution
中国、文莱、印度尼西亚、马来西亚、泰国
China, Brunei, Indonesia, Malaysia, Thailand

生物地理界 / Biogeographic Realm
印度马来界 Indomalaya

WWF 生物群系 / WWF Biome
热带和亚热带湿润阔叶林
Tropical & Subtropical Moist Broadleaf Forests

动物地理分布型 / Zoogeographic Distribution Type
Hc

分布标注 / Distribution Note
非特有种 Non-Endemic

▲ 濒危状况 / Threatened Status

中国生物多样性红色名录等级 / CB RL Category (2021)
数据缺乏 DD

IUCN 红色名录 / IUCN Red List (2021)
无危 LC

威胁因子 / Threats
未知 Unknown

▲ 法律保护地位 / Legal Protection Status

国家重点保护野生动物等级 / Category of National Key Protected Wild Animals (2021)
未列入 Not listed

"三有"名录 / TWIESSV (2023)
列入 Listed

CITES 附录等级 / CITES Appendix (2023)
未列入 Not listed

迁徙物种公约附录 / CMS Appendix (2020)
未列入 Not listed

保护行动 / Conservation Action
尚无保护行动 No conservation action so far

▲ 参考文献 / References

Jiang et al. (蒋志刚等), 2021; IUCN, 2021; Burgin et al., 2020; Wilson and Mittermeier, 2019; Pan et al. (潘清华等) 2007; Feng et al. (冯庆等), 2006

112 / 泰国无尾果蝠

Megaerops niphanae Yenbutra & Felten, 1983

• Ratanaworabhan's Fruit Bat

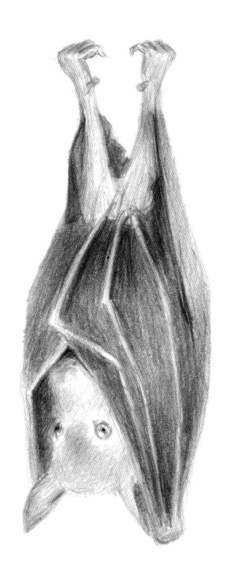

▲ 分类地位 / Taxonomy

翼手目 Chiroptera / 狐蝠科 Pteropodidae / 无尾果蝠属 *Megaerops*

科建立者及其文献 / Family Authority
Gray, 1821

属建立者及其文献 / Genus Authority
Peters, 1865

亚种 / Subspecies
无 None

模式标本产地 / Type Locality
泰国
Thailand, Nakhon Ratchasima Province, Amphoe Pak Thong Chai, Sakaerat Environmental Research Station

▲ 其他名称 / Other Name(s)

其他中文名 / Other Chinese Name(s)
无 None

其他英文名 / Other English Name(s)
无 None

同物异名 / Synonym(s)
无 None

▲ 形态及生境 / Morphology and Habitat

形态特征 / Morphological Characteristics
齿式：2.1.3.1/1.1.3.2=28。前臂长 50~58 mm。后足长 77 mm。耳长 16 mm。背毛呈均匀一致的棕灰色，腹毛中灰色，侧面毛色稍深。鼻孔略管状。耳边缘深色，耳尖宽圆形。第二指远端有爪。无尾。

Dental formula: 2.1.3.1/1.1.3.2=28. Forearm length 50-58 mm. Length of hind foot 77 mm. Ear length 16 mm. Nostrils slightly tubular. Dorsal hairs are uniformly brownish-grey. Ventral hairs are mid-grey with slightly darker flanks. Ears are simple with broadly rounded tips and dark margins. Claw present on distal extremity of the second digit. tail absent.

生境 / Habitat
落叶林、针叶林、竹林和亚热带混交林
Deciduous forests, coniferous forest, bamboo forest and subtropical mixed forest

▲ 地理分布 / Geographic Distribution

国内分布 / Domestic Distribution
云南 Yunnan

全球分布 / World Distribution
中国、柬埔寨、印度、老挝、泰国、越南
China, Cambodia, India, Laos, Thailand, Vietnam

生物地理界 / Biogeographic Realm
印度马来界 Indomalaya

WWF 生物群系 / WWF Biome
热带和亚热带湿润阔叶林
Tropical & Subtropical Moist Broadleaf Forests

动物地理分布型 / Zoogeographic Distribution Type
Hc

分布标注 / Distribution Note
非特有种 Non-Endemic

▲ 濒危状况 / Threatened Status

中国生物多样性红色名录等级 / CB RL Category (2021)
数据缺乏 DD

IUCN 红色名录 / IUCN Red List (2021)
无危 LC

威胁因子 / Threats
未知 Unknown

▲ 法律保护地位 / Legal Protection Status

国家重点保护野生动物等级 / Category of National Key Protected Wild Animals (2021)
未列入 Not listed

"三有"名录 / TWIESSV (2023)
列入 Listed

CITES 附录等级 / CITES Appendix (2023)
未列入 Not listed

迁徙物种公约附录 / CMS Appendix (2020)
未列入 Not listed

保护行动 / Conservation Action
尚无保护行动 No conservation action so far

▲ 参考文献 / References

Jiang et al. (蒋志刚等), 2021; IUCN, 2021; Burgin et al., 2020; Pan et al. (潘清华等), 2007; Feng et al. (冯庆等), 2006; Simmons, 2005

113 / 黑髯墓蝠

Taphozous melanopogon Temminck, 1841

· Black-bearded Tomb Bat

▲ 分类地位 / Taxonomy

翼手目 Chiroptera / 鞘尾蝠科 Emballonuridae / 墓蝠属 *Taphozous*

科建立者及其文献 / Family Authority
Gervais, 1855

属建立者及其文献 / Genus Authority
Geoffroy, 1818

亚种 / Subspecies
马来亚种 *T. m. fretensis* (Thomas, 1916)
云南、贵州、广西、海南、广东、香港和澳门
Yunnan, Guizhou, Guangxi, Hainan, Guangdong, Hong Kong and Macao
北京亚种 *T. m. solifer* (Hollister, 1913)
北京 Beijing

模式标本产地 / Type Locality
印度尼西亚
Indonesia, W Java, Bantam

吴毅 / 供图

▲ 其他名称 / Other Name(s)

其他中文名 / Other Chinese Name(s)
黑髯鞘尾蝠、墓蝠

其他英文名 / Other English Name(s)
无 None

同物异名 / Synonym(s)
无 None

▲ 形态及生境 / Morphology and Habitat

形态特征 / Morphological Characteristics
齿式：1.1.2.3/2.1.2.3=30。前臂长 50~58 mm。头体长 67~86 mm。耳长 16~23 mm。后足长 8~15 mm。尾长 20~32 mm。颅全长 19.4~21.9 mm。眼大。下唇有一肉质裂片。下颌上有一小簇黑胡须。翼膜附在胫骨脚踝以上。背部和腹部毛为黑棕色，毛基部白色。尾端部自股间膜背面穿出，故名"鞘尾"。
Dental formula: 1.1.2.3/2.1.2.3=30. Forearm length 50-58 mm. Head and body length 67-86 mm. Ear length 16-23 mm. Hind foot length 8-15 mm. Tail length 20-32 mm. Greatest skull length 19.4-21.9 mm. Big eyes. A fleshy lobe on the lower lip and a small tuft of black beard on the chin. Membranous wing attached to the tibia above the ankle. The back and belly hairs are black and brown, and the base of the hair is white. Tail comes out from the dorsal side of the interfemoral membrane.

生境 / Habitat
热带亚热带湿润低地山地森林、洞穴
Tropical subtropical moist lowland montane forest, cave

▲ 地理分布 / Geographic Distribution

国内分布 / Domestic Distribution
广西、海南、贵州、云南、香港、澳门、北京、广东
Guangxi, Hainan, Guizhou, Yunnan, Hong Kong, Macao, Beijing, Guangdong

全球分布 / World Distribution
文莱、柬埔寨、中国、印度、印度尼西亚、老挝、马来西亚、缅甸、菲律宾、新加坡、斯里兰卡、泰国、东帝汶、越南
Brunei, Cambodia, China, India, Indonesia, Laos, Malaysia, Myanmar, Philippines, Singapore, Sri Lanka, Thailand, Timor-Leste, Vietnam

生物地理界 / Biogeographic Realm
印度马来界、大洋洲界
Indomalaya, Oceanian

WWF 生物群系 / WWF Biome
热带和亚热带湿润阔叶林
Tropical & Subtropical Moist Broadleaf Forests

动物地理分布型 / Zoogeographic Distribution Type
Sb

分布标注 / Distribution Note
非特有种 Non-Endemic

▲ 濒危状况 / Threatened Status

中国生物多样性红色名录等级 / CB RL Category (2021)
无危 LC

IUCN 红色名录 / IUCN Red List (2021)
无危 LC

威胁因子 / Threats
无 None

▲ 法律保护地位 / Legal Protection Status

国家重点保护野生动物等级 / Category of National Key Protected Wild Animals (2021)
未列入 Not listed

"三有"名录 / TWIESSV (2023)
未列入 Not listed

CITES 附录等级 / CITES Appendix (2023)
未列入 Not listed

迁徙物种公约附录 / CMS Appendix (2020)
未列入 Not listed

保护行动 / Conservation Action
尚无保护行动 No conservation action so far

▲ 参考文献 / References

Jiang et al. (蒋志刚等), 2021; IUCN, 2021; Burgin et al., 2020; Liu et al. (刘少英等), 2020; Wilson and Mittermeier, 2019; Huang et al. (黄继展等), 2013; Pan et al. (潘清华等), 2007; Wei(韦力), 2007; Wilson and Reeder, 2005; Wang (王应祥), 2003; Zhang (张荣祖), 1997; Luo et al. (罗蓉等), 1993; Chu(褚新洛), 1989; Institute of Entomology/ Guangdong Province and Department of Biology/ Sun Yat-sen University (广东省昆虫研究所动物室, 中山大学生物系), 1983; Gao et al. (高耀亭等), 1962

114 / 大墓蝠

Taphozous theobaldi Dobson, 1872

· Theobold's Bat

▲ 分类地位 / Taxonomy

翼手目 Chiroptera / 鞘尾蝠科 Emballonuridae /
墓蝠属 *Taphozous*

科建立者及其文献 / Family Authority
Gervais, 1855

属建立者及其文献 / Genus Authority
Geoffroy, 1818

亚种 / Subspecies
指名亚种 *T. th. theobaldi* Dobson, 1872
云南南部
Yunnan (southern part)

模式标本产地 / Type Locality
缅甸
Burma (Myanmar), Tenasserim

▲ 其他名称 / Other Name(s)

其他中文名 / Other Chinese Name(s)
无 None

其他英文名 / Other English Name(s)
Theobald's Bat

同物异名 / Synonym(s)
无 None

▲ 形态及生境 / Morphology and Habitat

形态特征 / Morphological Characteristics
齿式：1.1.2.3/2.1.2.3=30。前臂长 70~76 mm。头体长
88~95 mm。耳长 21~26 mm。后足长 11~18 mm。尾
长 25~35 mm。颅全长 22~24 mm。吻部近乎裸露。成
年雄性有红棕色胡须。鼻孔内缘抬高。耳壳较大，耳
屏小。背侧毛深棕色，腹侧浅灰色。尾膜无毛。翼膜
具发育良好的膜袋。尾端部可见几根长毛。
Dental formula: 1.1.2.3/2.1.2.3=30. Head and body length 88-95
mm. Ear length 21-26 mm. Forearm length 70-76 mm. Hind foot
length 11-18 mm. Tail Length 25-35 mm. Greatest skull length
22-24 mm. Muzzle nearly naked. Adult males have reddish-
brown whiskers. Inner margin of the nostril is elevated. Ears are
large and the tragus small. Dorsal hairs are dark brown, ventral
hairs light gray. Uropatagium hairless. Wings with well-developed
wing pouch. There are a few long hairs on the tip of the tail.

生境 / Habitat
森林、洞穴
Forest, caves

▲ 地理分布 / Geographic Distribution

国内分布 / Domestic Distribution
广东、云南
Guangdong, Yunnan

全球分布 / World Distribution
中国、柬埔寨、印度、缅甸、泰国、越南
China, Cambodia, India, Myanmar, Thailand, Vietnam

生物地理界 / Biogeographic Realm
印度马来界 Indomalaya

WWF 生物群系 / WWF Biome
热带和亚热带湿润阔叶林
Tropical & Subtropical Moist Broadleaf Forests

动物地理分布型 / Zoogeographic Distribution Type
Sb

分布标注 / Distribution Note
非特有种 Non-Endemic

▲ 濒危状况 / Threatened Status

中国生物多样性红色名录等级 / CB RL Category (2021)
近危 NT

IUCN 红色名录 / IUCN Red List (2021)
无危 LC

威胁因子 / Threats
狩猎及采集
Hunting and collection

▲ 法律保护地位 / Legal Protection Status

国家重点保护野生动物等级 / Category of National Key Protected Wild Animals (2021)
未列入 Not listed

"三有"名录 / TWIESSV (2023)
未列入 Not listed

CITES 附录等级 / CITES Appendix (2023)
未列入 Not listed

迁徙物种公约附录 / CMS Appendix (2020)
未列入 Not listed

保护行动 / Conservation Action
尚无保护行动 No conservation action so far

▲ 参考文献 / References

Jiang et al. (蒋志刚等), 2021; Burgin et al., 2020; IUCN, 2020; Wilson and Mittermeier, 2019; Zhou et al. (周全等), 2012; Pan et al. (潘清华等), 2007; Wang (王应祥), 2003

115 / 印度假吸血蝠

Megaderma lyra É. Geoffroy, 1810

• Greater False Vampire

翼手目 Chiroptera / 假吸血蝠科 Megadermatidae / 假吸血蝠属 *Megaderma*

科建立者及其文献 / Family Authority
Allen, 1864

属建立者及其文献 / Genus Authority
É. Geoffroy, 1810

亚种 / Subspecies
华南亚种 *M. l. sinensis* (Anderson et Wroughton, 1907)
西藏、云南、四川、重庆、贵州、湖南、广西、广东、海南和福建
Tibet, Yunnan, Sichuan, Chongqing, Guizhou, Hunan, Guangxi, Guangdong, Hainan
and Fujian

模式标本产地 / Type Locality
印度
India, Madras

赵江波 / 供图

▲ 其他名称 / Other Name(s)

其他中文名 / Other Chinese Name(s)
大耳蝠、大巨耳蝠、亚洲假吸血蝠

其他英文名 / Other English Name(s)
无 None

同物异名 / Synonym(s)
无 None

▲ 形态及生境 / Morphology and Habitat

形态特征 / Morphological Characteristics
齿式：0.1.2.3/2.1.2.3=28。头体长 70~95 mm。耳长 31~45 mm。前臂长
56~72 mm。后足长 14~20 mm。颅全长 30 mm。耳大，呈卵圆形，两耳
内缘在额头连接。鼻叶大，突起，卵形，结构简单，背部毛色鼠棕色。
腹部被毛基部深灰色，毛尖白色，整体毛色呈苍白色。胫骨长于前臂长
度的一半。尾仅余痕迹或缺失。

Dental formula: 0.1.2.3/2.1.2.3=28. Head and body length 70-95 mm. Ear length 31-
45 mm. Forearm length 56-72 mm. Hind foot length 14-20 mm. Greatest skull length
30 mm. Ears are large, oval, and the inner edges of the ears connects to the forehead.
Nose leaves are large, protruded, ovate, simple in structure, and back coat color is
mouse brown. Belly coat is dark gray at the base of hairs and white at the hair tips,
and the belly coat looks pale. Tibia is longer than half the length of the forearm. Tail
remains only vestigial or absence.

生境 / Habitat
森林、洞穴、人造建筑
Forest, cave, man-made building

▲ 地理分布 / Geographic Distribution

国内分布 / Domestic Distribution
湖南、四川、广西、云南、广东、福建、贵州、重庆、海南、西藏
Hunan, Sichuan, Guangxi, Yunnan, Guangdong, Fujian, Guizhou, Chongqing, Hainan, Tibet

全球分布 / World Distribution
阿富汗、孟加拉国、柬埔寨、中国、印度、老挝、马来西亚、缅甸、尼泊尔、巴基斯坦、斯里兰卡、泰国、越南
Afghanistan, Bangladesh, Cambodia, China, India, Laos, Malaysia, Myanmar, Nepal, Pakistan, Sri Lanka, Thailand, Vietnam

生物地理界 / Biogeographic Realm
印度马来界、古北界
Indomalaya, Palearctic

WWF 生物群系 / WWF Biome
热带和亚热带湿润阔叶林
Tropical & Subtropical Moist Broadleaf Forests

动物地理分布型 / Zoogeographic Distribution Type
Wb

分布标注 / Distribution Note
非特有种 Non-Endemic

▲ 濒危状况 / Threatened Status

中国生物多样性红色名录等级 / CB RL Category (2021)
易危 VU

IUCN 红色名录 / IUCN Red List (2021)
无危 LC

威胁因子 / Threats
采石、森林砍伐、火灾、耕种
Quarrying, logging, fire, farming

▲ 法律保护地位 / Legal Protection Status

国家重点保护野生动物等级 / Category of National Key Protected Wild Animals (2021)
未列入 Not listed

"三有"名录 / TWIESSV (2023)
未列入 Not listed

CITES 附录等级 / CITES Appendix (2023)
未列入 Not listed

迁徙物种公约附录 / CMS Appendix (2020)
未列入 Not listed

保护行动 / Conservation Action
尚无保护行动 No conservation action so far

▲ 参考文献 / References

Jiang et al. (蒋志刚等), 2021; IUCN, 2021; Burgin et al., 2020; Liu et al. (刘少英等), 2020; Zhang et al. (张礼标等), 2007; Luo et al. (罗蓉等), 1993; Yang et al. (杨德华等), 1993; Chu (褚新洛), 1989; Shi and Zhao (施白南和赵尔宓), 1980

116 / 马来假吸血蝠

Megaderma spasma (Linnaeus, 1758)

· Lesser False Vampire

▲ 分类地位 / Taxonomy

翼手目 Chiroptera / 假吸血蝠科 Megadermatidae / 假吸血蝠属 *Megaderma*

科建立者及其文献 / Family Authority
Allen, 1864

属建立者及其文献 / Genus Authority
É. Geoffroy, 1810

亚种 / Subspecies
无 None

模式标本产地 / Type Locality
印度尼西亚
Indonesia, Molucca Isls, Ternate

Alice C. Hughes / 供图

Alice C. Hughes / 供图

▲ 其他名称 / Other Name(s)

其他中文名 / Other Chinese Name(s)
无 None

其他英文名 / Other English Name(s)
无 None

同物异名 / Synonym(s)
无 None

▲ 形态及生境 / Morphology and Habitat

形态特征 / Morphological Characteristics

齿式：0.1.2.3/2.1.2.3=28。前臂长约 60 mm。体重约 21 g。耳长 37 mm。卵圆形鼻叶突起，鼻叶椭圆形，顶部钝圆、两侧缘外隆、中央具纵形隆嵴。耳呈椭圆形，两耳在前额上方相连，耳屏双叉状。上体毛淡灰褐色，下体毛色较浅淡。无尾。

Dental formula: 0.1.2.3/2.1.2.3=28. Forearm length 60 mm. Body mass 21 g. Ear length 37 mm. Ovoid nasal lobe protruding, the posterior nasal lobe is elliptic, the top is blunt, two sides are convex, and the center has a longitudinal crest. Ears are oval, the bases of the two ears are joined above the forehead, and the tragus is bifurcated. Upper body hairs are grayish brown, and the lower body hairs are lighter. Tailless.

生境 / Habitat

森林 Forest

▲ 地理分布 / Geographic Distribution

国内分布 / Domestic Distribution
云南 Yunnan

全球分布 / World Distribution
中国、斯里兰卡、印度、东帝汶、越南、老挝、泰国、马来西亚、
印度尼西亚、文莱、菲律宾、新加坡、柬埔寨、缅甸
China, Sri Lanka, India, Timor-Leste, Vietnam, Laos, Thailand, Malaysia,
Indonesia, Brunei, Philippines, Singapore, Cambodia, Myanmar

生物地理界 / Biogeographic Realm
印度马来界 Indomalaya

WWF 生物群系 / WWF Biome
热带和亚热带湿润阔叶林
Tropical & Subtropical Moist Broadleaf Forests

动物地理分布型 / Zoogeographic Distribution Type
Wa

分布标注 / Distribution Note
非特有种 Non-Endemic

▲ 濒危状况 / Threatened Status

中国生物多样性红色名录等级 / CB RL Category (2021)
数据缺乏 DD

IUCN 红色名录 / IUCN Red List (2021)
无危 LC

威胁因子 / Threats
未知 Unknown

▲ 法律保护地位 / Legal Protection Status

国家重点保护野生动物等级 / Category of National Key Protected Wild Animals (2021)
未列入 Not listed

"三有" 名录 / TWIESSV (2023)
未列入 Not listed

CITES 附录等级 / CITES Appendix (2023)
未列入 Not listed

迁徙物种公约附录 / CMS Appendix (2020)
未列入 Not listed

保护行动 / Conservation Action
尚无保护行动 No conservation action so far

▲ 参考文献 / References

Jiang et al. (蒋志刚等), 2021; IUCN, 2021; Burgin et al., 2020; Liu et al. (刘少英等), 2020; Zhang et al. (张礼标等), 2010

117 / 中菊头蝠

Rhinolophus affinis Horsfield, 1823

• Intermediate Horseshoe Bat

▲ 分类地位 / Taxonomy

翼手目 Chiroptera / 菊头蝠科 Rhinolophidae / 菊头蝠属 *Rhinolophus*

科建立者及其文献 / Family Authority
Gray, 1825

属建立者及其文献 / Genus Authority
Lacépède, 1799

亚种 / Subspecies
喜马拉雅亚种 *R. a. himalayanus* (Anderson, 1905)
云南、贵州、四川、陕西和湖北
Yunnan, Guizhou, Sichuan, Shaanxi and Hubei

华南亚种 *R. a. macrurus* Anderson, 1905
江苏、浙江、安徽、福建、江西、湖南、广东、香港和广西
Jiangsu, Zhejiang, Anhui, Fujian, Jiangxi, Hunan and Guangdong, Hong Kong and Guangxi

海南亚种 *R. a. hainanus* J. Allen, 1905
海南 Hainan

模式标本产地 / Type Locality
印度尼西亚
Indonesia, Java

余文华 / 供图

杜晌 / 供图

▲ 其他名称 / Other Name(s)

其他中文名 / Other Chinese Name(s)
无 None

其他英文名 / Other English Name(s)
无 None

同物异名 / Synonym(s)
无 None

▲ 形态及生境 / Morphology and Habitat

形态特征 / Morphological Characteristics
齿式：1.1.2.3/2.1.3.3=32。头体长 58~63 mm。尾长 20~35 mm。后足长 11~13 mm。耳长 15~21 mm。前臂长 46~56 mm。颅全长 22~24 mm。鞍状叶两侧缘向中间凹入。联接叶顶端低圆，与鞍状叶之间有凹陷。第三、四、五指掌骨近等长。体背毛深暗褐色，腹毛淡。尾短，与股间膜接近平行。

Dental formula: 1.1.2.3/2.1.3.3=32. Head and body length 58-63 mm. Ear length 15-21 mm. Forearm length 46-56 mm. Hind foot length 11-13 mm. Tail length 20-35 mm. Greatest skull length 22-24 mm. Both sides of the saddle leaflet margin concave to the middle, fiddle-shaped. The tip of the coupling leaflet is low round and concave with the saddle leaf. The 3rd, 4th, 5th metacarpal bones are of subequal length. The body back hairs are dark brown, and the abdominal hairs are light. The tail is short, nearly parallel to the interfemoral membrane.

生境 / Habitat
洞穴、种植园、耕地、森林
Caves, plantations, arable land, forest

▲ 地理分布 / Geographic Distribution

国内分布 / Domestic Distribution
陕西、江苏、浙江、安徽、湖南、江西、四川、贵州、云南、广西、
海南、福建、香港、重庆、广东
Shaanxi, Jiangsu, Zhejiang, Anhui, Hunan, Jiangxi, Sichuan, Guizhou, Yunnan,
Guangxi, Hainan, Fujian, Hong Kong, Chongqing, Guangdong

全球分布 / World Distribution
孟加拉国、不丹、柬埔寨、中国、印度、印度尼西亚、老挝、马来西亚、
缅甸、尼泊尔、新加坡、泰国、越南
Bangladesh, Bhutan, Cambodia, China, India, Indonesia, Laos, Malaysia,
Myanmar, Nepal, Singapore, Thailand, Vietnam

生物地理界 / Biogeographic Realm
印度马来界、古北界、大洋洲界
Indomalaya, Palearctic, Oceanian

WWF 生物群系 / WWF Biome
热带和亚热带湿润阔叶林
Tropical & Subtropical Moist Broadleaf Forests

动物地理分布型 / Zoogeographic Distribution Type
Wc

分布标注 / Distribution Note
非特有种 Non-Endemic

▲ 濒危状况 / Threatened Status

中国生物多样性红色名录等级 / CB RL Category (2021)
无危 LC

IUCN 红色名录 / IUCN Red List (2021)
无危 LC

威胁因子 / Threats
无 None

▲ 法律保护地位 / Legal Protection Status

国家重点保护野生动物等级 / Category of National Key Protected Wild Animals (2021)
未列入 Not listed

"三有" 名录 / TWIESSV (2023)
未列入 Not listed

CITES 附录等级 / CITES Appendix (2023)
未列入 Not listed

迁徙物种公约附录 / CMS Appendix (2020)
未列入 Not listed

保护行动 / Conservation Action
尚无保护行动 No conservation action so far

▲ 参考文献 / References

Jiang et al. (蒋志刚等), 2021; IUCN, 2021; Burgin et al., 2020; Wilson and Mittermeier, 2019; Mao et al., 2010; Zhang et al., 2009; Wu et al., 2009; Niu et al., 2007; Pan et al. (潘清华等), 2007; Liu et al. (刘延德等), 2006; Wilson and Reeder, 2005; Zhou et al. (周昭敏等), 2005; Wang (王应祥), 2003; Gu et al. (谷晓明等), 2003; Zhang (张荣祖), 1997

118 / 马铁菊头蝠

Rhinolophus ferrumequinum
(Schreber, 1774)

• Greater Horseshoe Bat

吴毅 / 供图

周佳俊 / 供图

▲ 分类地位 / Taxonomy

翼手目 Chiroptera / 菊头蝠科 Rhinolophidae / 菊头蝠属 *Rhinolophus*

科建立者及其文献 / Family Authority
Gray, 1825

属建立者及其文献 / Genus Authority
Lacépède, 1799

亚种 / Subspecies
尼泊尔亚种 *R. f. tragatus* (Hodgeson, 1835)
云南和贵州
Yunnan and Guizhou

日本亚种 *R. f. nippon* Temminick, 1835
吉林、辽宁、河北、北京、河南、山东、山西、江苏、上海、浙江、安徽、福建、湖南、广西、贵州、云南、四川、湖北、陕西、甘肃和宁夏
Jilin, Liaoning, Hebei, Beijing, Henan, Shandong, Shanxi, Jiangsu, Shanghai, Zhejiang, Anhui, Fujian, Guangxi, Hunan, Guizhou, Yunnan, Sichuan, Hubei, Shaanxi, Gansu and Ningxia

模式标本产地 / Type Locality
法国
France

▲ 其他名称 / Other Name(s)

其他中文名 / Other Chinese Name(s)
暗褐菊头蝠、大菊头蝠

其他英文名 / Other English Name(s)
Larger Horseshoe Bat

同物异名 / Synonym(s)
无 None

▲ 形态及生境 / Morphology and Habitat

形态特征 / Morphological Characteristics
齿式：1.1.2.3/2.1.3.3=32。头体长 56~79 mm。耳长 18~29 mm。前臂长 53~64 mm。后足长 10~14 mm。尾长 25~44 mm。颅全长 21~25 mm。马蹄叶较宽，附叶小，鞍状叶两侧内凹，联接叶较低而圆，顶叶近三角形。耳大，无耳屏，有对耳屏。背毛浅棕褐色，毛基淡灰棕色。腹毛淡灰棕色。股间膜发达。翼和股间膜棕褐色。尾长，呈锥状。
Dental formula: 1.1.2.3/2.1.3.3=32. Head and body length 56–79 mm. Tail length 25–44 mm. Hind foot length 10-14 mm. Ear length 18-29 mm. Forearm length 53-64 mm. Greatest skull length 21-25 mm. Leaves of horseshoe are wide, appendages are small, saddle leaves are concave on both sides, connecting leaves are low and round, and parietal leaves are nearly triangular. Ears are large, without tragus, but with antitragus. Undercoat is light tan with a grayish-brown base. Abdominal hairs are grayish brown. Interfemoral membrane is well developed. Wings and interfemoral membranes are brown. Tail is very long and conical.

生境 / Habitat
牧场、地中海灌木植被、温带森林、洞穴、人造建筑
Pastureland, mediterranean-type shrubby vegetation, temperate forest, caves, man-made building

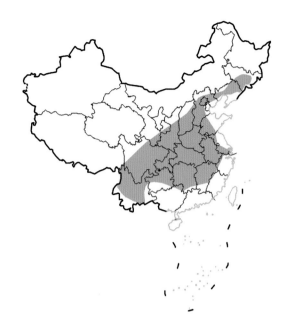

▲ 地理分布 / Geographic Distribution

国内分布 / Domestic Distribution

吉林、辽宁、河北、北京、山东、河南、陕西、山西、上海、浙江、安徽、江西、四川、甘肃、贵州、云南、广西、福建、重庆

Jilin, Liaoning, Hebei, Beijing, Shandong, Henan, Shaanxi, Shanxi, Shanghai, Zhejiang, Anhui, Jiangxi, Sichuan, Gansu, Guizhou, Yunnan, Guangxi, Fujian, Chongqing

全球分布 / World Distribution

阿富汗、阿尔巴尼亚、阿尔及利亚、安道尔、亚美尼亚、奥地利、阿塞拜疆、孟加拉国、不丹、波斯尼亚和黑塞哥维那、保加利亚、中国、克罗地亚、塞浦路斯、捷克、法国、格鲁吉亚、德国、直布罗陀、希腊、匈牙利、印度、伊朗、伊拉克、以色列、意大利、日本、约旦、哈萨克斯坦、朝鲜、韩国、黎巴嫩、列支敦士登、卢森堡、马其顿、摩尔多瓦、摩纳哥、黑山、摩洛哥、尼泊尔、巴基斯坦、巴勒斯坦、波兰、葡萄牙、罗马尼亚、俄罗斯、圣马力诺、沙特阿拉伯、塞尔维亚、斯洛伐克、斯洛文尼亚、西班牙、瑞士、叙利亚、塔吉克斯坦、突尼斯、土耳其、土库曼斯坦、乌克兰、英国、乌兹别克斯坦

Afghanistan, Albania, Algeria, Andorra, Armenia, Austria, Azerbaijan, Bangladesh, Bhutan, Bosnia and Herzegovina, Bulgaria, China, Croatia, Cyprus, Czech, France, Georgia, Germany, Gibraltar, Greece, Hungary, India, Iran, Iraq, Israel, Italy, Japan, Jordan, Kazakhstan, Democratic People's Republic of Korea, Republic of Korea, Lebanon, Liechtenstein, Luxembourg, Macedonia, Moldova, Monaco, Montenegro, Morocco, Nepal, Pakistan, Palestine, Poland, Portugal, Romania, Russian, San Marino, Saudi Arabia, Serbia, Slovakia, Slovenia, Spain, Switzerland, Syria, Tajikistan, Tunisia, Turkey, Turkmenistan, Ukraine, United Kingdom, Uzbekistan

生物地理界 / Biogeographic Realm

古北界 Palearctic

WWF 生物群系 / WWF Biome

热带和亚热带湿润阔叶林
Tropical & Subtropical Moist Broadleaf Forests

动物地理分布型 / Zoogeographic Distribution Type

Ug

分布标注 / Distribution Note

非特有种 Non-Endemic

▲ 濒危状况 / Threatened Status

中国生物多样性红色名录等级 / CB RL Category (2021)

无危 LC

IUCN 红色名录 / IUCN Red List (2021)

无危 LC

威胁因子 / Threats

无 None

▲ 法律保护地位 / Legal Protection Status

国家重点保护野生动物等级 / Category of National Key Protected Wild Animals (2021)

未列入 Not listed

"三有" 名录 / TWIESSV (2023)

未列入 Not listed

CITES 附录等级 / CITES Appendix (2023)

未列入 Not listed

迁徙物种公约附录 / CMS Appendix (2020)

未列入 Not listed

保护行动 / Conservation Action

尚无保护行动 No conservation action so far

▲ 参考文献 / References

Jiang et al. (蒋志刚等), 2021; IUCN, 2021; Burgin et al., 2020; Wilson and Mittermeier, 2019; He (何新焕), 2011; Luo (罗丽), 2011; Li (李国红), 2010; Wang et al. (王静等), 2010; Wang(王新华), 2008; Pan et al. (潘清华等), 2007; Gao (高晶), 2006; Liu et al. (刘延德等), 2006; Peng and Zhong (彭基泰和钟祥清), 2005; Wilson and Reeder, 2005; Wang(王应祥), 2003; Zhang (张荣祖), 1997; Xia (夏武平), 1988

119 / 台湾菊头蝠

Rhinolophus formosae Sanborn, 1939

• Taiwan Woolly Horseshoe Bat

▲ 分类地位 / Taxonomy

翼手目 Chiroptera / 菊头蝠科 Rhinolophidae / 菊头蝠属 *Rhinolophus*

科建立者及其文献 / Family Authority
Gray, 1825

属建立者及其文献 / Genus Authority
Lacépède, 1799

亚种 / Subspecies
无 None

模式标本产地 / Type Locality
中国台湾
Taiwan, China

周政翰 / 供图

▲ 其他名称 / Other Name(s)

其他中文名 / Other Chinese Name(s)
台湾大蹄鼻蝠

其他英文名 / Other English Name(s)
无 None

同物异名 / Synonym(s)
无 None

▲ 形态及生境 / Morphology and Habitat

形态特征 / Morphological Characteristics

齿式：1.1.2.3/2.1.3.3=32。头体长 90 mm。耳长 28~33 mm。前臂长 56~61 mm。后足长 16~17 mm。尾长 28~39 mm。颅全长 24~28 mm。耳郭宽大，对耳屏宽而短。背毛深褐色到黑色。翼膜宽圆，向后延伸到趾基部。

Dental formula: 1.1.2.3/2.1.3.3=32. Head and body length 90 mm. Ear length 28-33 mm. Forearm length 56-61 mm. Hind foot length 16-17 mm. Tail length 28-39 mm. Greatest skull length 24-28 mm. Auricle broad antitragus short. Dorsal hairs dark brown to black. Membrane is broadly circular and extends backward to the base of the toe.

生境 / Habitat
洞穴、人造建筑、森林
Caves, man-made building, forest

▲ 地理分布 / Geographic Distribution

国内分布 / Domestic Distribution
台湾 Taiwan

全球分布 / World Distribution
中国 China

生物地理界 / Biogeographic Realm
印度马来界 Indomalaya

WWF 生物群系 / WWF Biome
热带和亚热带湿润阔叶林
Tropical & Subtropical Moist Broadleaf Forests

动物地理分布型 / Zoogeographic Distribution Type
J

分布标注 / Distribution Note
特有种 Endemic

▲ 濒危状况 / Threatened Status

中国生物多样性红色名录等级 / CB RL Category (2021)
近危 NT

IUCN 红色名录 / IUCN Red List (2021)
无危 LC

威胁因子 / Threats
住宅区及商业发展、公路 铁路、耕种
Residential and commercial development, roads and railroads, farming

▲ 法律保护地位 / Legal Protection Status

国家重点保护野生动物等级 / Category of National Key Protected Wild Animals (2021)
未列入 Not listed

"三有" 名录 / TWIESSV (2023)
未列入 Not listed

CITES 附录等级 / CITES Appendix (2023)
未列入 Not listed

迁徙物种公约附录 / CMS Appendix (2020)
未列入 Not listed

保护行动 / Conservation Action
尚无保护行动 No conservation action so far

▲ 参考文献 / References

Jiang et al. (蒋志刚等), 2021; Burgin et al., 2020; IUCN, 2021; Wilson and Mittermeier, 2019; Pan et al. (潘清华等), 2007; Wilson and Reeder, 2005; Wang (王应祥), 2003; Lin et al. (林良恭等), 1997; Chen (陈兼善), 1969; Sanborn, 1939

120 / 短翼菊头蝠

Rhinolophus lepidus Blyth, 1844

• Blyth's Horseshoe Bat

翼手目 Chiroptera / 菊头蝠科 Rhinolophidae /
菊头蝠属 *Rhinolophus*

科建立者及其文献 / Family Authority
Gray, 1825

属建立者及其文献 / Genus Authority
Lacépède, 1799

亚种 / Subspecies
缅甸亚种 *R. l. shortridgei* Anderson, 1918
云南、四川、贵州、湖南、湖北、广西、海南、广
东和福建
Yunnan, Sichuan, Guizhou, Hunan, Hubei, Guangxi, Hainan,
Guangdong and Fujian

模式标本产地 / Type Locality
印度
India, Bengal, Calcutta (uncertain); see Das (1986)

▲ 其他名称 / Other Name(s)

其他中文名 / Other Chinese Name(s)
无 None

其他英文名 / Other English Name(s)
无 None

同物异名 / Synonym(s)
无 None

▲ 形态及生境 / Morphology and Habitat

形态特征 / Morphological Characteristics
齿式：1.1.2.3/2.1.3.3=32。头体长 51~59 mm。尾长
18~25 mm。后足长 9 mm。耳长 16~20 mm。前臂
长 39~43 mm。颅全长 16.8~18.7 mm。鼻孔未完全
被鼻叶覆盖，马蹄叶中间有凹痕。两侧各有一个小
凹陷。鞍状叶基部宽于顶部，中部稍凹入；下唇有
三个垂直凹槽；头部毛发颜色棕色。尾略长于胫骨。
Dental formula: 1.1.2.3/2.1.3.3=32. Head and body length
51-59 mm. Tail length 18-25 mm. Hind foot length 9 mm.
Ear length 16-20 mm. Forearm length 39-43 mm. Greatest
skull length 16.8-18.7 mm. Nostril is not completely covered
by a nose leaf, which possesses an indentation in the middle.
with a small depression on either side; basal sellar leaflet wider
than the top, middle part slightly recessed; lower lip with three
vertical grooves; hair color brown, Tail slightly longer than the
tibia.

生境 / Habitat
洞穴、人造建筑、热带湿润低地森林
Caves, man-made building, tropical moist lowland forest

▲ 地理分布 / Geographic Distribution

国内分布 / Domestic Distribution
浙江、安徽、江西、四川、云南、广西、贵州、湖南、福建、重庆、广东
Zhejiang, Anhui, Jiangxi, Sichuan, Yunnan, Guangxi, Guizhou, Hunan, Fujian, Chongqing, Guangdong

全球分布 / World Distribution
阿富汗、孟加拉国、柬埔寨、中国、印度、印度尼西亚、马来西亚、缅甸、尼泊尔、巴基斯坦、泰国、越南
Afghanistan, Bangladesh, Cambodia, China, India, Indonesia, Malaysia, Myanmar, Nepal, Pakistan, Thailand, Vietnam

生物地理界 / Biogeographic Realm
印度马来界、古北界
Indomalaya, Palearctic

WWF 生物群系 / WWF Biome
热带和亚热带湿润阔叶林
Tropical & Subtropical Moist Broadleaf Forests

动物地理分布型 / Zoogeographic Distribution Type
Wc

分布标注 / Distribution Note
非特有种 Non-Endemic

▲ 濒危状况 / Threatened Status

中国生物多样性红色名录等级 / CB RL Category (2021)
近危 NT

IUCN 红色名录 / IUCN Red List (2021)
无危 LC

威胁因子 / Threats
旅游 Tourism

▲ 法律保护地位 / Legal Protection Status

国家重点保护野生动物等级 / Category of National Key Protected Wild Animals (2021)
未列入 Not listed

"三有" 名录 / TWIESSV (2023)
未列入 Not listed

CITES 附录等级 / CITES Appendix (2023)
未列入 Not listed

迁徙物种公约附录 / CMS Appendix (2020)
未列入 Not listed

保护行动 / Conservation Action
尚无保护行动 No conservation action so far

▲ 参考文献 / References

Jiang et al. (蒋志刚等), 2021; IUCN, 2021; Burgin et al., 2020; Wilson and Mittermeier, 2019; Pan et al. (潘清华等), 2007; Wilson and Reeder, 2005; Wang (王应祥), 2003; Luo and Gao (罗键和高红英), 2002; Zhang (张荣祖), 1997

121 / 大菊头蝠

Rhinolophus luctus Temminck, 1834

· Great Woolly Horsehoe Bat

▲ 分类地位 / Taxonomy

翼手目 Chiroptera / 菊头蝠科 Rhinolophidae / 菊头蝠属 *Rhinolophus*

科建立者及其文献 / Family Authority
Gray, 1825

属建立者及其文献 / Genus Authority
Lacépède, 1799

亚种 / Subspecies
喜马拉雅亚种 *R. l. pernmiger* Hodgson, 1843
云南 Yunnan

华南亚种 *R. l. lanous* Anderson, 1905
浙江、安徽、福建、江西、广东、广西、贵州和四川
Zhejiang, Anhui, Fujian, Jiangxi, Guangdong, Guangxi, Guizhou and Sichuan

海南亚种 *R. l. spurcus* G. Allen, 1928
海南 Hainan

模式标本产地 / Type Locality
印度尼西亚
Indonesia, Java, Tapos

吴毅 / 供图   周佳俊 / 供图

▲ 其他名称 / Other Name(s)

其他中文名 / Other Chinese Name(s)
绒毛菊头蝠、毛面菊头蝠、羊毛蝠

其他英文名 / Other English Name(s)
Typical Woolly Horseshoe Bat

同物异名 / Synonym(s)
无 None

▲ 形态及生境 / Morphology and Habitat

形态特征 / Morphological Characteristics

齿式：1.1.2.3/2.1.3.3=32。头体长 75~95 mm。尾长 36~61 mm。后足长 16~18 mm。耳长 28~44 mm。前臂长 58~81 mm。颅全长 28~33 mm。鼻叶中马蹄叶发达，覆盖鼻吻部，两侧无附小叶。体毛长而密，略显卷曲。背毛棕褐或灰褐色，毛尖隐约有灰白色，腹毛颜色稍淡。

Dental formula: 1.1.2.3/2.1.3.3=32. Head and body length 75-95 mm. Tail length 36-61 mm. Hind foot length 16-18 mm. Ear length 28-44 mm. Forearm length 58-81 mm. Greatest skull length 28-33 mm. Horseshoe noseleafs developed, covering the snout of the nose, without attached lobules on both sides. Dorsal hairs are brown or grayish brown, hair tip is faintly grayish-white; the abdominal hair color is slightly lighter.

生境 / Habitat
森林、人造建筑
Forest, man-made building

▲ 地理分布 / Geographic Distribution

国内分布 / Domestic Distribution

安徽、浙江、江西、四川、贵州、广东、海南、广西、福建、台湾、重庆、香港、澳门

Anhui, Zhejiang, Jiangxi, Sichuan, Guizhou, Guangdong, Hainan, Guangxi, Fujian, Taiwan, Chongqing, Hongkong, Macao

全球分布 / World Distribution

孟加拉国、柬埔寨、中国、印度、印度尼西亚、老挝、马来西亚、缅甸、尼泊尔、新加坡、泰国、越南

Bangladesh, Cambodia, China, India, Indonesia, Laos, Malaysia, Myanmar, Nepal, Singapore, Thailand, Vietnam

生物地理界 / Biogeographic Realm

印度马来界、古北界、大洋洲界

Indomalaya, Palearctic, Oceanian

WWF 生物群系 / WWF Biome

热带和亚热带湿润阔叶林

Tropical & Subtropical Moist Broadleaf Forests

动物地理分布型 / Zoogeographic Distribution Type

Wc

分布标注 / Distribution Note

非特有种 Non-Endemic

▲ 濒危状况 / Threatened Status

中国生物多样性红色名录等级 / CB RL Category (2021)

近危 NT

IUCN 红色名录 / IUCN Red List (2021)

无危 LC

威胁因子 / Threats

森林砍伐、火灾、耕种、狩猎及采集

Logging, fire, farming, hunting and collection

▲ 法律保护地位 / Legal Protection Status

国家重点保护野生动物等级 / Category of National Key Protected Wild Animals (2021)

未列入 Not listed

"三有" 名录 / TWIESSV (2023)

未列入 Not listed

CITES 附录等级 / CITES Appendix (2023)

未列入 Not listed

迁徙物种公约附录 / CMS Appendix (2020)

未列入 Not listed

保护行动 / Conservation Action

尚无保护行动 No conservation action so far

▲ 参考文献 / References

Jiang et al. (蒋志刚等), 2021; IUCN, 2021; Burgin et al., 2020; Wilson and Mittermeier, 2019; Zhang et al. (张婵等), 2013; Xu (徐海龙), 2012; Pei (裴俊峰), 2011; Yang et al. (杨锐等), 2010; Zhang et al., 2009; Zhang et al. (张佑祥等), 2008; Pan et al. (潘清华等), 2007; Wilson and Reeder, 2005; Wang (王应祥), 2003; Zhang (张荣祖), 1997; Liang and Dong, 1984

122 / 大耳菊头蝠

Rhinolophus macrotis Blyth, 1844

· Big-eared Horseshoe Bat

▲ 分类地位 / Taxonomy

翼手目 Chiroptera / 菊头蝠科 Rhinolophidae / 菊头蝠属 *Rhinolophus*

科建立者及其文献 / Family Authority
Gray, 1825

属建立者及其文献 / Genus Authority
Lacépède, 1799

亚种 / Subspecies
四川亚种 *R. m. episcopus* G. Allen, 1923
四川和陕西
Sichuan and Shaanxi

福建亚种 *R. m. caldwelli* G. Allen, 1923
浙江、福建、江西、广东、广西和贵州
Zhejiang, Fujian, Jiangxi, Guangdong, Guangxi and Guizhou

模式标本产地 / Type Locality
尼泊尔
Nepal

周佳俊 / 供图   周佳俊 / 供图

▲ 其他名称 / Other Name(s)

其他中文名 / Other Chinese Name(s)
无 None

其他英文名 / Other English Name(s)
Great-eared Horseshoe Bat

同物异名 / Synonym(s)
无 None

▲ 形态及生境 / Morphology and Habitat

形态特征 / Morphological Characteristics
齿式：1.1.2.3/2.1.3.3=32。头体长 47~51 mm。尾长 12~32 mm。后足长 9~10 mm。耳长 18~27 mm。前臂长 39~48 mm。颅全长 20 mm。马蹄叶宽，中央具明显缺刻，两侧各具一小型附叶。头骨矢状嵴不明显。耳大，耳屏不发达。背毛毛基灰白色，毛尖暗褐色，腹毛色浅淡。

Dental formula: 1.1.2.3/2.1.3.3=32. Head and body length 47-51 mm. Tail length 12-32 mm. Hind foot length 9-10 mm. Ear length 18-27 mm. Forearm length 39-48 mm. Greatest skull length 20 mm. Leaves of the horseshoe are wide, with a marked notch in the center and a small appendage on each side. Ears are large and the antitragus is underdeveloped, though the anti-tragus is large. Base of dorsal hairs gray, hair tip dark brown, middle part light color.

生境 / Habitat
洞穴、人造建筑
Caves, man-made building

▲ 地理分布 / Geographic Distribution

国内分布 / Domestic Distribution
陕西、浙江、江西、四川、贵州、广西、云南、福建、重庆、广东
Shaanxi, Zhejiang, Jiangxi, Sichuan, Guizhou, Guangxi, Yunnan, Fujian,
Chongqing, Guangdong

全球分布 / World Distribution
孟加拉国、中国、印度、印度尼西亚、老挝、马来西亚、缅甸、尼泊尔、
巴基斯坦、菲律宾、泰国、越南
Bangladesh, China, India, Indonesia, Laos, Malaysia, Myanmar, Nepal, Pakistan,
Philippines, Thailand, Vietnam

生物地理界 / Biogeographic Realm
印度马来界、古北界
Indomalaya, Palearctic

WWF 生物群系 / WWF Biome
热带和亚热带湿润阔叶林
Tropical & Subtropical Moist Broadleaf Forests

动物地理分布型 / Zoogeographic Distribution Type
Wc

分布标注 / Distribution Note
非特有种 Non-Endemic

▲ 濒危状况 / Threatened Status

中国生物多样性红色名录等级 / CB RL Category (2021)
无危 LC

IUCN 红色名录 / IUCN Red List (2021)
无危 LC

威胁因子 / Threats
无 None

▲ 法律保护地位 / Legal Protection Status

国家重点保护野生动物等级 / Category of National Key Protected Wild Animals (2021)
未列入 Not listed

"三有" 名录 / TWIESSV (2023)
未列入 Not listed

CITES 附录等级 / CITES Appendix (2023)
未列入 Not listed

迁徙物种公约附录 / CMS Appendix (2020)
未列入 Not listed

保护行动 / Conservation Action
尚无保护行动 No conservation action so far

▲ 参考文献 / References

Jiang et al. (蒋志刚等), 2021; Burgin et al., 2020; IUCN, 2020; Wilson and Mittermeier, 2019; Wei (魏学文), 2013; Li et al. (李艳丽等), 2012; Wang et al. (王延校等), 2012; Liu et al., 2011; Zhang et al., 2009; Feng et al., 2008; Pan et al. (潘清华等), 2007; Wilson and Reeder, 2005; Wang (王应祥), 2003; Zhang (张荣祖), 1997

123 / 马来菊头蝠

Rhinolophus malayanus Bonhote, 1903

· Malayan Horseshoe Bat

▲ 分类地位 / Taxonomy

翼手目 Chiroptera / 菊头蝠科 Rhinolophidae /
菊头蝠属 *Rhinolophus*

科建立者及其文献 / Family Authority
Gray, 1825

属建立者及其文献 / Genus Authority
Lacépède, 1799

亚种 / Subspecies
无 None

模式标本产地 / Type Locality
泰国
Thailand, Jalor, Biserat

▲ 其他名称 / Other Name(s)

其他中文名 / Other Chinese Name(s)
无 None

其他英文名 / Other English Name(s)
无 None

同物异名 / Synonym(s)
无 None

▲ 形态及生境 / Morphology and Habitat

形态特征 / Morphological Characteristics
齿 式：1.1.2.3/2.1.3.3=32。头 体 长 42~45
mm。耳长 14~16 mm。前臂长 42~43 mm。
后足长 8~8.5 mm。尾长 20~23 mm。体重
6.5~6.7 g。体型与小褐菊头蝠 *Rhinolophus
stheno* 很相似。很难分辨形态差异。建议进
行分子生物学分析鉴别。
Dental formula: 1.1.2.3/2.1.3.3=32. Head and body
length 42-45 mm. Ear length 14-16 mm. Forearm
length 42-43 mm. Hind length 8-8.5 mm. Tail length
20-23 mm. Weight 6.5-6.7 g. Body size similar to the
small brown horseshoe bat *Rhinolophus stheno*.
Morphological differences are hard to discern.
Molecular biological analysis is recommended for
identification.

生境 / Habitat
洞穴、人造建筑、森林
Caves, man-made building, forest

▲ 地理分布 / Geographic Distribution

国内分布 / Domestic Distribution
云 南 Yunnan

全球分布 / World Distribution
中 国 China

生物地理界 / Biogeographic Realm
印度马来界 Indomalaya

WWF 生物群系 / WWF Biome
热带和亚热带湿润阔叶林
Tropical & Subtropical Moist Broadleaf Forests

动物地理分布型 / Zoogeographic Distribution Type
Wa

分布标注 / Distribution Note
非特有种 Non-Endemic

▲ 濒危状况 / Threatened Status

中国生物多样性红色名录等级 / CB RL Category (2021)
未评定 NE

IUCN 红色名录 / IUCN Red List (2021)
无危 LC

威胁因子 / Threats
未知 Unknown

▲ 法律保护地位 / Legal Protection Status

国家重点保护野生动物等级 / Category of National Key Protected Wild Animals (2021)
未列入 Not listed

"三有" 名录 / TWIESSV (2023)
未列入 Not listed

CITES 附录等级 / CITES Appendix (2023)
未列入 Not listed

迁徙物种公约附录 / CMS Appendix (2020)
未列入 Not listed

保护行动 / Conservation Action
尚无保护行动 No conservation action so far

▲ 参考文献 / References

Burgin et al., 2020; IUCN, 2020; Wilson and Mittermeier, 2019

124 / 马氏菊头蝠

Rhinolophus marshalli Thonglongya, 1973

· Marshall's Horseshoe Bat

翼手目 Chiroptera / 菊头蝠科 Rhinolophidae / 菊头蝠属 *Rhinolophus*

科建立者及其文献 / Family Authority
Gray, 1825

属建立者及其文献 / Genus Authority
Lacépède, 1799

亚种 / Subspecies
无 None

模式标本产地 / Type Locality
泰国
Thailand, Chantaburi, Amphoe Pong Nam Ron, foothills of Khao Soi Dao Thai

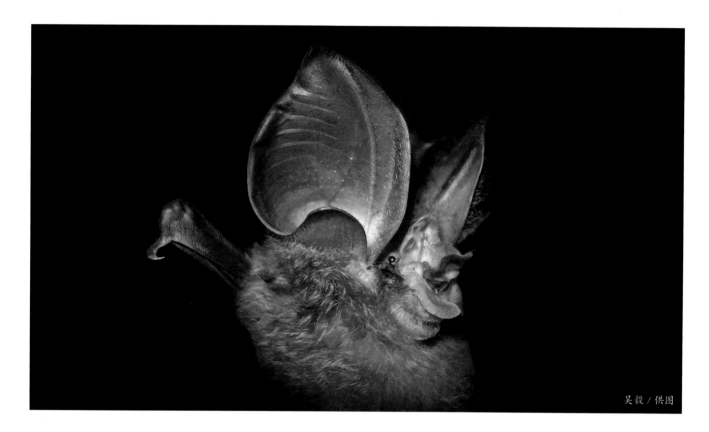

吴毅 / 供图

▲ 其他名称 / Other Name(s)

其他中文名 / Other Chinese Name(s)
无 None

其他英文名 / Other English Name(s)
无 None

同物异名 / Synonym(s)
无 None

▲ 形态及生境 / Morphology and Habitat

形态特征 / Morphological Characteristics
齿式：1.1.2.3/2.1.3.3=32。前臂长 41~47 mm。马蹄叶大，前缘光滑，后缘锯齿状；鞍状叶舌状，质薄而宽短，顶端圆弧形。耳大，对耳屏发达。背毛暗褐色，腹毛毛色较浅淡。
Dental formula: 1.1.2.3/2.1.3.3=32. Forearm length 41-47 mm. Horseshoe noseleafs are large, the leading edge is smooth, and the trailing edge is serrated. Saddle-shaped leaves ligulate, thin and short, round arc at the top, the base extends to both sides, and cup leaves to form a butterfly-shaped pterygoid process. Ears are large and the antitragus well developed. Dorsal hairs are dark brown and the abdominal hairs are light.

生境 / Habitat
石灰岩洞穴岩石裂缝，能耐受高度扰动的生境。
Limestone caves, and rock crevices, and highly disturbed tolerant habitat

▲ 地理分布 / Geographic Distribution

国内分布 / Domestic Distribution
广西、云南
Guangxi, Yunnan

全球分布 / World Distribution
中国、老挝、马来西亚、缅甸、泰国、越南
China, Laos, Malaysia, Myanmar, Thailand, Vietnam

生物地理界 / Biogeographic Realm
印度马来界 Indomalaya

WWF 生物群系 / WWF Biome
热带和亚热带湿润阔叶林
Tropical & Subtropical Moist Broadleaf Forests

动物地理分布型 / Zoogeographic Distribution Type
Wb

分布标注 / Distribution Note
非特有种 Non-Endemic

▲ 濒危状况 / Threatened Status

中国生物多样性红色名录等级 / CB RL Category (2021)
近危 NT

IUCN 红色名录 / IUCN Red List (2021)
无危 LC

威胁因子 / Threats
森林砍伐、耕种、火灾、采矿、采石、旅游
Logging, farming, fire, mining, quarrying, tourism

▲ 法律保护地位 / Legal Protection Status

国家重点保护野生动物等级 / Category of National Key Protected Wild Animals (2021)
未列入 Not listed

"三有"名录 / TWIESSV (2023)
未列入 Not listed

CITES 附录等级 / CITES Appendix (2023)
未列入 Not listed

迁徙物种公约附录 / CMS Appendix (2020)
未列入 Not listed

保护行动 / Conservation Action
尚无保护行动 No conservation action so far

▲ 参考文献 / References

Jiang et al. (蒋志刚等), 2021; Burgin et al., 2020; IUCN, 2020; Liu et al. (刘少英等), 2020; Pan et al. (潘清华等), 2007; Simmons, 2005; Zhang et al. (张礼标等), 2005; Wu et al. (吴毅等), 2004

125 / 单角菊头蝠

Rhinolophus monoceros K. Andersen, 1905

• Taiwan Least Horseshoe Bat

▲ 分类地位 / Taxonomy

翼手目 Chiroptera / 菊头蝠科 Rhinolophidae / 菊头蝠属 *Rhinolophus*

科建立者及其文献 / Family Authority
Gray, 1825

属建立者及其文献 / Genus Authority
Lacépède, 1799

亚种 / Subspecies
无 None

模式标本产地 / Type Locality
中国
China, Taiwan, Baksa

周政翰 / 供图

▲ 其他名称 / Other Name(s)

其他中文名 / Other Chinese Name(s)
台湾小蹄鼻蝠

其他英文名 / Other English Name(s)
Taiwan Horseshoe Bat

同物异名 / Synonym(s)
无 None

▲ 形态及生境 / Morphology and Habitat

形态特征 / Morphological Characteristics
齿式：1.1.2.3/2.1.3.3=32。头体长 40~50 mm。尾长 15~27 mm。后足长 7~9 mm。耳长 16~17 mm。前臂长 34~40 mm。颅全长 14~16 mm。鼻叶中的连接叶顶端尖细，微微向下弯曲。头部毛发呈淡褐色。耳郭宽，对耳屏短。翼膜宽圆，连接到趾基部。

Dental formula: 1.1.2.3/2.1.3.3=32. Head and body length 40-50 mm. Tail length 15-27 mm. Hind foot length 7-9 mm. Ear length 16-17 mm. Forearm length 34-40 mm. Greatest skull length 14-16 mm. The conjunction in the nasal lobes is tapered at the apex and slightly curved downwards. Hairs on the head are light brown. Auricle broad, and antitragus short. Pterygoid membrane is broadly circular and connected to the base of the toe.

生境 / Habitat
洞穴、亚热带湿润低地森林
Caves, subtropical moist lowland forest

▲ 地理分布 / Geographic Distribution

国内分布 / Domestic Distribution
台湾 Taiwan

全球分布 / World Distribution
中国 China

生物地理界 / Biogeographic Realm
印度马来界 Indomalaya

WWF 生物群系 / WWF Biome
热带和亚热带湿润阔叶林
Tropical & Subtropical Moist Broadleaf Forests

动物地理分布型 / Zoogeographic Distribution Type
Sb

分布标注 / Distribution Note
特有种 Endemic

▲ 濒危状况 / Threatened Status

中国生物多样性红色名录等级 / CB RL Category (2021)
易危 VU

IUCN 红色名录 / IUCN Red List (2021)
无危 LC

威胁因子 / Threats
森林砍伐、耕种、采矿、采石、旅游
Logging, farming, mining, quarrying, tourism

▲ 法律保护地位 / Legal Protection Status

国家重点保护野生动物等级 / Category of National Key Protected Wild Animals (2021)
未列入 Not listed

"三有"名录 / TWIESSV (2023)
未列入 Not listed

CITES 附录等级 / CITES Appendix (2023)
未列入 Not listed

迁徙物种公约附录 / CMS Appendix (2020)
未列入 Not listed

保护行动 / Conservation Action
尚无保护行动 No conservation action so far

▲ 参考文献 / References

Jiang et al. (蒋志刚等), 2021; Burgin et al., 2020; IUCN, 2020; Wilson and Mittermeier, 2019; Smith and Xie, 2009; Xu et al. (许立杰等), 2008; Wang (王应祥), 2003; Lin (林良恭), 2000

126 / 丽江菊头蝠

Rhinolophus osgoodi Sanborn, 1939

· Osgood's Horseshoe Bat

▲ 分类地位 / Taxonomy

翼手目 Chiroptera / 菊头蝠科 Rhinolophidae /
菊头蝠属 *Rhinolophus*

科建立者及其文献 / Family Authority
Gray, 1825

属建立者及其文献 / Genus Authority
Lacépède, 1799

亚种 / Subspecies
无 None

模式标本产地 / Type Locality
中国云南
China, Yunnan, N of Likiang, Nguluko, (27°5'N, 100°5'E)

▲ 其他名称 / Other Name(s)

其他中文名 / Other Chinese Name(s)
奥氏菊头幅

其他英文名 / Other English Name(s)
无 None

同物异名 / Synonym(s)
无 None

▲ 形态及生境 / Morphology and Habitat

形态特征 / Morphological Characteristics
齿式：1.1.2.3/2.1.3.3=32。头体长 52~54 mm。
尾长 17~21 mm。后足长 8~9 mm。耳长 12~20
mm。前臂长 41~46 mm。颅全长 15~16 mm。背
侧毛浅棕色，有基于灰色的毛；腹面的毛呈浅
灰色。

Dental formula: 1.1.2.3/2.1.3.3=32. Head and body length
52-54 mm. Tail length 17-21 mm. Hind foot length 8-9
mm. Ear length 12-20 mm. Forearm length 41-46 mm.
Greatest skull length 15-16 mm. Dorsal pelage pale brown
with hairs gray-based; pelage paler gray below.

生境 / Habitat
未知 Unknown

▲ 地理分布 / Geographic Distribution

国内分布 / Domestic Distribution
云南 Yunnan

全球分布 / World Distribution
中国、越南
China, Vietnam

生物地理界 / Biogeographic Realm
古北界 Palearctic

WWF 生物群系 / WWF Biome
热带和亚热带湿润阔叶林
Tropical & Subtropical Moist Broadleaf Forests

动物地理分布型 / Zoogeographic Distribution Type
Sa

分布标注 / Distribution Note
非特有种 Non-Endemic

▲ 濒危状况 / Threatened Status

中国生物多样性红色名录等级 / CB RL Category (2021)
数据缺乏 DD

IUCN 红色名录 / IUCN Red List (2021)
数据缺乏 DD

威胁因子 / Threats
未知 Unknown

▲ 法律保护地位 / Legal Protection Status

国家重点保护野生动物等级 / Category of National Key Protected Wild Animals (2021)
未列入 Not listed

"三有" 名录 / TWIESSV (2023)
未列入 Not listed

CITES 附录等级 / CITES Appendix (2023)
未列入 Not listed

迁徙物种公约附录 / CMS Appendix (2020)
未列入 Not listed

保护行动 / Conservation Action
尚无保护行动 No conservation action so far

▲ 参考文献 / References

Jiang et al. (蒋志刚等), 2021; Burgin et al., 2020; IUCN, 2020; Liu et al. (刘少英等), 2020; Wilson and Mittermeier, 2019; Zhang et al., 2009; Sanborn, 1939; Pan et al. (潘清华等), 2007; Wilson and Reeder, 2005; Wang (王应祥), 2003

127 / 皮氏菊头蝠

Rhinolophus pearsonii Horsfield, 1851

· Pearson's Horseshoe Bat

▲ 分类地位 / Taxonomy

翼手目 Chiroptera / 菊头蝠科 Rhinolophidae / 菊头蝠属 *Rhinolophus*

科建立者及其文献 / Family Authority
Gray, 1825

属建立者及其文献 / Genus Authority
Lacépède, 1799

亚种 / Subspecies
指名亚种 *R. p. pearsoni* Horsfield, 1851
西藏、四川、贵州、云南、陕西和湖北
Tibet, Sichuan, Guizhou, Yunnan, Shaanxi and Hubei

华南亚种 *R. p. chinensis* Anderson, 1905
浙江、安徽、福建、江西、广东、广西和湖南
Zhejiang, Anhui, Fujian, Jiangxi, Guangdong, Guangxi and Hunan

模式标本产地 / Type Locality
印度
India, W Bengal, Darjeeling

黄耀华 / 供图

吴毅 / 供图

▲ 其他名称 / Other Name(s)

其他中文名 / Other Chinese Name(s)
毕氏菊头蝠、绒毛菊头蝠

其他英文名 / Other English Name(s)
无 None

同物异名 / Synonym(s)
无 None

▲ 形态及生境 / Morphology and Habitat

形态特征 / Morphological Characteristics
齿式：1.1.2.3/2.1.3.3=32。头体长 61~68 mm。耳长 23~29 mm。前臂长 47~56 mm。后足长 12~13 mm。尾长 16~29 mm。颅全长 24 mm。鼻叶复杂，马蹄叶宽大，覆盖上唇。翼膜止于胫基，股间膜后缘较平，不呈锥状。体毛长而柔密，背毛暗褐色或棕褐色，腹毛稍淡。
Dental formula: 1.1.2.3/2.1.3.3=32. Head and body length 61-68 mm. Ear length 23-29 mm. Forearm length 47-56 mm. Hind foot length 12-13 mm. Tail length 16-29 mm. Greatest skull length 24 mm. Nasal leaves are complex, and the horseshoe noseleafs are broad, covering the upper lip. Pterygoid membrane stops at the tibial base, and the posterior margin of the interfemoral membrane is flat and not conical. Body hairs are long and dense, and the back hairs are dark brown or tan, and the abdominal hairs are slightly lighter color.

生境 / Habitat
洞穴、森林、种植园
Caves, forest, plantations

▲ 地理分布 / Geographic Distribution

国内分布 / Domestic Distribution

浙江、陕西、西藏、安徽、湖南、江西、四川、贵州、云南、广西、广东、福建、重庆

Zhejiang, Shaanxi, Tibet, Anhui, Hunan, Jiangxi, Sichuan, Guizhou, Yunnan, Guangxi, Guangdong, Fujian, Chongqing

全球分布 / World Distribution

孟加拉国、不丹、中国、印度、老挝、马来西亚、缅甸、尼泊尔、泰国、越南

Bangladesh, Bhutan, China, India, Laos, Malaysia, Myanmar, Nepal, Thailand, Vietnam

生物地理界 / Biogeographic Realm

印度马来界、古北界

Indomalaya, Palearctic

WWF 生物群系 / WWF Biome

热带和亚热带湿润阔叶林

Tropical & Subtropical Moist Broadleaf Forests

动物地理分布型 / Zoogeographic Distribution Type

Wc

分布标注 / Distribution Note

非特有种 Non-Endemic

▲ 濒危状况 / Threatened Status

中国生物多样性红色名录等级 / CB RL Category (2021)

无危 LC

IUCN 红色名录 / IUCN Red List (2021)

无危 LC

威胁因子 / Threats

无 None

▲ 法律保护地位 / Legal Protection Status

国家重点保护野生动物等级 / Category of National Key Protected Wild Animals (2021)

未列入 Not listed

"三有" 名录 / TWIESSV (2023)

未列入 Not listed

CITES 附录等级 / CITES Appendix (2023)

未列入 Not listed

迁徙物种公约附录 / CMS Appendix (2020)

未列入 Not listed

保护行动 / Conservation Action

尚无保护行动 No conservation action so far

▲ 参考文献 / References

Jiang et al. (蒋志刚等), 2021; Burgin et al., 2020; IUCN, 2020; Liu et al. (刘少英等), 2020; Wilson and Mittermeier, 2019; Mao (毛秀光), 2010; Song (宋华), 2009; Lin et al. (林爱青等), 2009; Niu et al. (牛红星等), 2008; Peng and Zhong (彭基泰和钟祥清), 2005; Zhou et al. (周江等), 2002

128 / 小菊头蝠

Rhinolophus pusillus Temminck, 1834

• Least Horseshoe Bat

▲ 分类地位 / Taxonomy

翼手目 Chiroptera / 菊头蝠科 Rhinolophidae / 菊头蝠属 *Rhinolophus*

科建立者及其文献 / Family Authority
Gray, 1825

属建立者及其文献 / Genus Authority
Lacépède, 1799

亚种 / Subspecies
四川亚种 *R. p. szechwanus* Anderson, 1918
西藏、四川和贵州
Tibet, Sichuan and Guizhou

福建亚种 *R. p. calidus* G. Allen, 1923
福建、广东、贵州和广西
Fujian, Guangdong, Guizhou and Guangxi

海南亚种 *R. p. parcus* G. Allen, 1923
海南 Hainan

清迈亚种 *R. p. lakkhanae* Yoshiyuki, 1990
云南 Yunnan

模式标本产地 / Type Locality
印度尼西亚
Indonesia, Java

吴毅 / 供图

周佳俊 / 供图

▲ 其他名称 / Other Name(s)

其他中文名 / Other Chinese Name(s)
无 None

其他英文名 / Other English Name(s)
无 None

同物异名 / Synonym(s)
无 None

▲ 形态及生境 / Morphology and Habitat

形态特征 / Morphological Characteristics
齿式：1.1.2.3/2.1.3.3=32。头体长 38~42 mm。耳长 13~20 mm。前臂长 33~40 mm。后足长 6~8 mm。尾长 13~26 mm。颅全长 15~17 mm。鞍状叶上窄下宽，与联接叶之间具明显凹陷。耳郭基部被毛。体背毛呈茶褐色，毛基部灰白色，腹毛肉桂色。喉部毛色较浅。

Dental formula: 1.1.2.3/2.1.3.3=32. Head and body length 38-42 mm. Ear length 13-20 mm. Forearm length 33-40 mm. Hind foot length 6-8 mm. Tail length 13-26 mm. Greatest skull length 15-17 mm. Saddle-shaped leaves are narrow at the top and wide at the bottom, with obvious depressions between them and the connecting leaves. Hairy at the base of the auricle. Dorsal hairs are brown, bases of the hairs are grayish-white, and the abdominal hairs are cinnamon color. Throat hairs are lighter colored.

生境 / Habitat
洞穴、森林、人造建筑、喀斯特地貌
Caves, forest, man-made building, karst

▲ 地理分布 / Geographic Distribution

国内分布 / Domestic Distribution

湖北、湖南、浙江、安徽、江西、贵州、四川、云南、广西、海南、
福建、重庆、广东、澳门
Hubei, Hunan, Zhejiang, Anhui, Jiangxi, Guizhou, Sichuan, Yunnan, Guangxi,
Hainan, Fujian, Chongqing, Guangdong, Macao

全球分布 / World Distribution

柬埔寨、中国、印度、印度尼西亚、日本、老挝、马来西亚、缅甸、
尼泊尔、泰国、越南
Cambodia, China, India, Indonesia, Japan, Laos, Malaysia, Myanmar, Nepal,
Thailand, Vietnam

生物地理界 / Biogeographic Realm

印度马来界、古北界
Indomalaya, Palearctic

WWF 生物群系 / WWF Biome

热带和亚热带湿润阔叶林
Tropical & Subtropical Moist Broadleaf Forests

动物地理分布型 / Zoogeographic Distribution Type

Wc

分布标注 / Distribution Note

非特有种 Non-Endemic

▲ 濒危状况 / Threatened Status

中国生物多样性红色名录等级 / CB RL Category (2021)

无危 LC

IUCN 红色名录 / IUCN Red List (2021)

无危 LC

威胁因子 / Threats

无 None

▲ 法律保护地位 / Legal Protection Status

国家重点保护野生动物等级 / Category of National Key Protected Wild Animals (2021)

未列入 Not listed

"三有"名录 / TWIESSV (2023)

未列入 Not listed

CITES 附录等级 / CITES Appendix (2023)

未列入 Not listed

迁徙物种公约附录 / CMS Appendix (2020)

未列入 Not listed

保护行动 / Conservation Action

尚无保护行动 No conservation action so far

▲ 参考文献 / References

Jiang et al. (蒋志刚等), 2021; Burgin et al., 2020; IUCN, 2020; Liu et al. (刘少英等), 2020; Wilson and Mittermeier, 2019; Xu et al. (许立杰等),
2008; Pan et al. (潘清华等), 2007; Wilson and Reeder, 2005; Wang (王应祥), 2003

129 / 贵州菊头蝠

Rhinolophus rex G. M. Allen, 1923

· King Horseshoe Bat

翼手目 Chiroptera / 菊头蝠科 Rhinolophidae / 菊头蝠属 *Rhinolophus*

科建立者及其文献 / Family Authority
Gray, 1825

属建立者及其文献 / Genus Authority
Lacépède, 1799

亚种 / Subspecies
无 None

模式标本产地 / Type Locality
中国
China, Szechwan, Wanhsien

余文华 / 供图

▲ 其他名称 / Other Name(s)

其他中文名 / Other Chinese Name(s)
无 None

其他英文名 / Other English Name(s)
Bourret's Horseshoe Bat

同物异名 / Synonym(s)
无 None

▲ 形态及生境 / Morphology and Habitat

形态特征 / Morphological Characteristics
齿式：1.1.2.3/2.1.3.3=32。头体长 45~50 mm，耳长 30~34 mm，尾长 32~38 mm。前臂长 55~63 mm。颅全长 23~24 mm。耳郭特别大，超过体长之半。背毛可达 15 mm 长，棕色，腹毛浅色。
Dental formula: 1.1.2.3/2.1.3.3=32. Head and body length 50 mm. Tail length 32-38 mm. Forearm length 55-63 mm. Greatest skull length 23-24 mm. The auricle is very large. Ear length is about half the length of the head and body. Dorsal hairs up to 15 mm long, brown, abdominal hairs light color.

生境 / Habitat
洞穴、人造建筑、森林
Caves, man-made building, forest

▲ 地理分布 / Geographic Distribution

国内分布 / Domestic Distribution
广东、重庆、四川、贵州、广西
Guangdong, Chongqing, Sichuan, Guizhou, Guangxi

全球分布 / World Distribution
中国 China

生物地理界 / Biogeographic Realm
古北界 Palearctic

WWF 生物群系 / WWF Biome
热带和亚热带湿润阔叶林
Tropical & Subtropical Moist Broadleaf Forests

动物地理分布型 / Zoogeographic Distribution Type
Wb

分布标注 / Distribution Note
特有种 Endemic

▲ 濒危状况 / Threatened Status

中国生物多样性红色名录等级 / CB RL Category (2021)
近危 NT

IUCN 红色名录 / IUCN Red List (2021)
无危 LC

威胁因子 / Threats
森林砍伐、耕种、火灾、采矿、采石、旅游
Logging, farming, fire, mining, quarrying, tourism

▲ 法律保护地位 / Legal Protection Status

国家重点保护野生动物等级 / Category of National Key Protected Wild Animals (2021)
未列入 Not listed

"三有" 名录 / TWIESSV (2023)
未列入 Not listed

CITES 附录等级 / CITES Appendix (2023)
未列入 Not listed

迁徙物种公约附录 / CMS Appendix (2020)
未列入 Not listed

保护行动 / Conservation Action
尚无保护行动 No conservation action so far

▲ 参考文献 / References

Jiang et al. (蒋志刚等), 2021; Burgin et al., 2020; IUCN, 2020; Liu et al. (刘少英等), 2020; Wilson and Mittermeier, 2019; Wu et al., 2012; Jiang et al., 2010; Chen and Peng, 2010; Pan et al. (潘清华等), 2007; Wilson and Reeder, 2005; Wang (王应祥), 2003; Wu and Harada, 2005; Gu et al. (谷晓明等), 2003; Zhou (周江), 2001; Zhang (张荣祖), 1997

130 / 施氏菊头蝠

Rhinolophus schnitzleri Wu & Thong, 2011

· Schnitzler's Horseshoe Bat

▲ 分类地位 / Taxonomy

翼手目 Chiroptera / 菊头蝠科 Rhinolophidae / 菊头蝠属 *Rhinolophus*

科建立者及其文献 / Family Authority
Gray, 1825

属建立者及其文献 / Genus Authority
Lacépède, 1799

亚种 / Subspecies
无 None

模式标本产地 / Type Locality
中国
Xiao-dong Cave, Gengjiaying Commune, Yi-liang County, Kunming City, Yunnan, China

李锋 / 供图

▲ 其他名称 / Other Name(s)

其他中文名 / Other Chinese Name(s)
无 None

其他英文名 / Other English Name(s)
无 None

同物异名 / Synonym(s)
无 None

▲ 形态及生境 / Morphology and Habitat

形态特征 / Morphological Characteristics

齿式：1.1.2.3/2.1.3.3=32。颅全长 22 mm。前臂长 58 mm。马蹄叶宽大，无附鼻叶。耳郭、对耳屏均发达，形态特征与贵州菊头蝠相似。背毛呈栗色，腹毛浅棕色。股间膜和翼膜均为棕色。脚踝被膜，翼膜无毛。尾尖突出股间膜。

Dental formula: 1.1.2.3/2.1.3.3=32. Cranial length 22 mm. Forearm length 58 mm. Horseshoe noseleafs wide, without nasal leaves. Auricle and antitragus are well developed, the morphological characteristics are similar to those of Guizhou Horseshoe Bat. Dorsal hairs are chestnut colored and the belly hairs are light brown. Interfemoral and pterygoid membranes are brown. Ankle covered with a membrane, which is glabrous. Caudal tip protrudes the interfemoral membrane.

生境 / Habitat
洞穴 Caves

▲ 地理分布 / Geographic Distribution

国内分布 / Domestic Distribution
云南 Yunnan

全球分布 / World Distribution
中国 China

生物地理界 / Biogeographic Realm
印度马来界 Indomalaya

WWF 生物群系 / WWF Biome
热带和亚热带湿润阔叶林
Tropical & Subtropical Moist Broadleaf Forests

动物地理分布型 / Zoogeographic Distribution Type
Sa

分布标注 / Distribution Note
特有种 Endemic

▲ 濒危状况 / Threatened Status

中国生物多样性红色名录等级 / CB RL Category (2021)
数据缺乏 DD

IUCN 红色名录 / IUCN Red List (2021)
未评定 NE

威胁因子 / Threats
未知 Unknown

▲ 法律保护地位 / Legal Protection Status

国家重点保护野生动物等级 / Category of National Key Protected Wild Animals (2021)
未列入 Not listed

"三有" 名录 / TWIESSV (2023)
未列入 Not listed

CITES 附录等级 / CITES Appendix (2023)
未列入 Not listed

迁徙物种公约附录 / CMS Appendix (2020)
未列入 Not listed

保护行动 / Conservation Action
尚无保护行动 No conservation action so far

▲ 参考文献 / References

Jiang et al. (蒋志刚等), 2021; Burgin et al., 2020; IUCN, 2020; Liu et al. (刘少英等), 2020; Wilson and Mittermeier, 2019; Wu and Thong, 2011

131 / 清迈菊头蝠

Rhinolophus siamensis Glydenstolpe, 1917

· Thai Horseshoe Bat

▲ 分类地位 / Taxonomy

翼手目 Chiroptera / 菊头蝠科 Rhinolophidae / 菊头蝠属 *Rhinolophus*

科建立者及其文献 / Family Authority
Gray, 1825

属建立者及其文献 / Genus Authority
Lacépède, 1799

亚种 / Subspecies
无 None

模式标本产地 / Type Locality
泰国
Thailand, NW Thailand, Doi Par Sakang

▲ 其他名称 / Other Name(s)

其他中文名 / Other Chinese Name(s)
无 None

其他英文名 / Other English Name(s)
无 None

同物异名 / Synonym(s)
无 None

▲ 形态及生境 / Morphology and Habitat

形态特征 / Morphological Characteristics
齿式：1.1.2.3/2.1.3.3=32。头体长 38 mm。耳长 19~22 mm。前臂长 36~41 mm。后足长 8~9 mm。尾长 14 mm。颅全长 17~18 mm。鞍状叶宽度与高度几乎相等。耳大，对耳屏小。基部侧翼与鼻孔内缘侧叶连成较浅的杯状叶，联接叶起始于鞍状叶的亚顶端。背毛毛基灰白色，毛尖暗褐色，腹毛色浅白色。
Dental formula: 1.1.2.3/2.1.3.3=32. Head and body length 38 mm. Ear length 19-22 mm. Forearm length 36-41 mm. Hind foot length 8-9 mm. Tail length 14 mm. Greatest skull length 17-18 mm. Saddle leaflet is almost equal in width and height. Ears are large, the antitragus is small. Basal flank is connected with the lateral lobe of the inner margin of the nostril to form a shallow cup leaf. Connecting leaflet starts from the subtip of the saddle leaf. Bases of dorsal hairs gray, and hair tips dark brown, and abdominal hair color light white.

生境 / Habitat
人造建筑、洞穴
Man-made building, cave

▲ 地理分布 / Geographic Distribution

国内分布 / Domestic Distribution
广东、广西、江西、云南
Guangdong, Guangxi, Jiangxi, Yunnan

全球分布 / World Distribution
中国、老挝、越南、泰国
China, Laos, Vietnam, Thailand

生物地理界 / Biogeographic Realm
印度马来界 Indomalaya

WWF 生物群系 / WWF Biome
热带和亚热带湿润阔叶林
Tropical & Subtropical Moist Broadleaf Forests

动物地理分布型 / Zoogeographic Distribution Type
Wb

分布标注 / Distribution Note
非特有种 Non-Endemic

▲ 濒危状况 / Threatened Status

中国生物多样性红色名录等级 / CB RL Category (2021)
近危 NT

IUCN 红色名录 / IUCN Red List (2021)
未评定 NE

威胁因子 / Threats
森林砍伐、耕种、火灾、采矿、采石、旅游
Logging, farming, fire, mining, quarrying, tourism

▲ 法律保护地位 / Legal Protection Status

国家重点保护野生动物等级 / Category of National Key Protected Wild Animals (2021)
未列入 Not listed

"三有" 名录 / TWIESSV (2023)
未列入 Not listed

CITES 附录等级 / CITES Appendix (2023)
未列入 Not listed

迁徙物种公约附录 / CMS Appendix (2020)
未列入 Not listed

保护行动 / Conservation Action
尚无保护行动 No conservation action so far

▲ 参考文献 / References

Jiang et al. (蒋志刚等), 2021; Liu et al. (刘少英等), 2020; Wilson and Mittermeier, 2019; Zhang et al., 2009; Wu et al., 2008

132 / 中华菊头蝠

Rhinolophus sinicus K. Andersen, 1905

· Chinese Horseshoe Bat

▲ 分类地位 / Taxonomy

翼手目 Chiroptera / 菊头蝠科 Rhinolophidae / 菊头蝠属 *Rhinolophus*

科建立者及其文献 / Family Authority
Gray, 1825

属建立者及其文献 / Genus Authority
Lacépède, 1799

亚种 / Subspecies
无 None

模式标本产地 / Type Locality
中国
China, Anhwei (Anhui), Chinteh

吴毅 / 供图　　周佳俊 / 供图

▲ 其他名称 / Other Name(s)

其他中文名 / Other Chinese Name(s)
无 None

其他英文名 / Other English Name(s)
Chinese Horseshoe Bat, Little Nepalese
Horseshoe Bat

同物异名 / Synonym(s)
无 None

▲ 形态及生境 / Morphology and Habitat

形态特征 / Morphological Characteristics

齿式：1.1.2.3/2.1.3.3=32。头体长 43~53 mm。耳长 15~20 mm。前臂长 45~52 mm。后足长 7~10 mm。尾长 21~30 mm。颅全长 19~23 mm。鼻叶中马蹄叶两侧下缘各具一片小型附叶，鞍状叶左右呈平行状。背毛毛尖栗色，毛基灰白色，腹毛赭褐色。

Dental formula: 1.1.2.3/2.1.3.3=32. Head and body length 43-53 mm. Ear length 15-20 mm. Forearm length 45-52 mm. Hind foot length 7-10 mm. Tail length 21-30 mm. Greatest skull length 19-23 mm. There is small leaflets on both sides of the lower margin of the horseshoe leaflet in the noseleaf. Saddle-shaped leaves are parallel to each other. Dorsal hairs are chestnut colored, bases of hairs are grayish-white and the abdominal hairs are ochre brown.

生境 / Habitat
热带和亚热带湿润山地森林、洞穴、人造建筑
Tropical and subtropical moist montane forest, caves, man-made building

▲ 地理分布 / Geographic Distribution

国内分布 / Domestic Distribution

福建、江苏、浙江、安徽、江西、湖北、湖南、广东、广西、四川、贵州、云南、西藏、陕西、甘肃、香港、海南、重庆
Fujian, Jiangsu, Zhejiang, Anhui, Jiangxi, Hubei, Hunan, Guangdong, Guangxi, Sichuan, Guizhou, Yunnan, Tibet, Shaanxi, Gansu, Hong Kong, Hainan, Chongqing

全球分布 / World Distribution

中国、印度、缅甸、尼泊尔、越南
China, India, Myanmar, Nepal, Vietnam

生物地理界 / Biogeographic Realm

印度马来界、古北界
Indomalaya, Palearctic

WWF 生物群系 / WWF Biome

热带和亚热带湿润阔叶林
Tropical & Subtropical Moist Broadleaf Forests

动物地理分布型 / Zoogeographic Distribution Type

Wd

分布标注 / Distribution Note

非特有种 Non-Endemic

▲ 濒危状况 / Threatened Status

中国生物多样性红色名录等级 / CB RL Category (2021)

无危 LC

IUCN 红色名录 / IUCN Red List (2021)

无危 LC

威胁因子 / Threats

无 None

▲ 法律保护地位 / Legal Protection Status

国家重点保护野生动物等级 / Category of National Key Protected Wild Animals (2021)

未列入 Not listed

"三有"名录 / TWIESSV (2023)

未列入 Not listed

CITES 附录等级 / CITES Appendix (2023)

未列入 Not listed

迁徙物种公约附录 / CMS Appendix (2020)

未列入 Not listed

保护行动 / Conservation Action

尚无保护行动 No conservation action so far

▲ 参考文献 / References

Jiang et al. (蒋志刚等), 2021; Wilson and Mittermeier, 2019; Zhou et al. (周昭敏等), 2009; Pan et al. (潘清华等), 2007; Wilson and Reeder, 2005

133 / 小褐菊头蝠

Rhinolophus stheno K. Andersen, 1905

· Lesser Brown Horseshoe Bat

Alice C. Hughes / 供图

▲ 分类地位 / Taxonomy

翼手目 Chiroptera / 菊头蝠科 Rhinolophidae / 菊头蝠属 *Rhinolophus*

科建立者及其文献 / Family Authority
Gray, 1825

属建立者及其文献 / Genus Authority
Lacépède, 1799

亚种 / Subspecies
无 None

模式标本产地 / Type Locality
马来西亚
Malaysia, Selangor

▲ 其他名称 / Other Name(s)

其他中文名 / Other Chinese Name(s)
无 None

其他英文名 / Other English Name(s)
无 None

同物异名 / Synonym(s)
无 None

▲ 形态及生境 / Morphology and Habitat

形态特征 / Morphological Characteristics
齿式：1.1.2.3/2.1.3.3=32。前臂长 43 mm，颅全长约 18.3 mm。鞍状叶两侧几近平行，上端圆弧形。顶叶凹入或侧缘平行，顶端延长，呈楔形。连接叶突出，前端钝圆，上着生少许尖毛。背毛棕褐色，腹毛色浅。翼膜黑褐色，其上无毛附着。
Dental formula: 1.1.2.3/2.1.3.3=32. Forearm length is about 43 mm, and the cranial length is about 18.3 mm. Both sides of the saddle leaflets are nearly parallel, with a round upper end. Parietal leaf concave or lateral margin parallel, apex elongated, cuneate. Connecting leaves protruding, blunt round front end, with a little sharp hairs. Dorsal hairs are brown and the belly hairs are light-colored. Pterygoid membrane is dark brown and glabrous.

生境 / Habitat
森林、热带亚热带严重退化的森林、喀斯特地貌
Forest, tropical and subtropical seriously degraded forest, karst landforms

▲ 地理分布 / Geographic Distribution

国内分布 / Domestic Distribution
云南 Yunnan

全球分布 / World Distribution
中国、印度尼西亚、老挝、马来西亚、缅甸、泰国、越南
China, Indonesia, Laos, Malaysia, Myanmar, Thailand, Vietnam

生物地理界 / Biogeographic Realm
印度马来界 Indomalaya

WWF 生物群系 / WWF Biome
热带和亚热带湿润阔叶林
Tropical & Subtropical Moist Broadleaf Forests

动物地理分布型 / Zoogeographic Distribution Type
Wa

分布标注 / Distribution Note
非特有种 Non-Endemic

▲ 濒危状况 / Threatened Status

中国生物多样性红色名录等级 / CB RL Category (2021)
近危 NT

IUCN 红色名录 / IUCN Red List (2021)
无危 LC

威胁因子 / Threats
未知 Unknown

▲ 法律保护地位 / Legal Protection Status

国家重点保护野生动物等级 / Category of National Key Protected Wild Animals (2021)
未列入 Not listed

"三有" 名录 / TWIESSV (2023)
未列入 Not listed

CITES 附录等级 / CITES Appendix (2023)
未列入 Not listed

迁徙物种公约附录 / CMS Appendix (2020)
未列入 Not listed

保护行动 / Conservation Action
尚无保护行动 No conservation action so far

▲ 参考文献 / References

Jiang et al. (蒋志刚等), 2021; Burgin et al., 2020; IUCN, 2020; Liu et al. (刘少英等), 2020; Mao et al., 2013; Song et al., 2009; Zhang et al. (张劲硕等), 2005; Pan et al. (潘清华等), 2007

134 / 葛氏菊头蝠

Rhinolophus subbadius Blyth, 1844

· Geoffroy's Horseshoe Bat

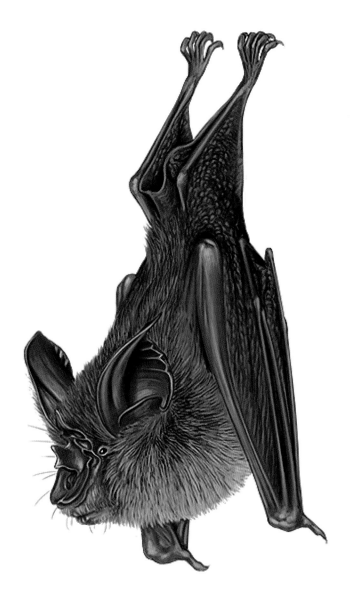

▲ 分类地位 / Taxonomy

翼手目 Chiroptera / 菊头蝠科 Rhinolophidae / 菊头蝠属 *Rhinolophus*

科建立者及其文献 / Family Authority
Gray, 1825

属建立者及其文献 / Genus Authority
Lacépède, 1799

亚种 / Subspecies
无 None

模式标本产地 / Type Locality
尼泊尔
Nepal

▲ 其他名称 / Other Name(s)

其他中文名 / Other Chinese Name(s)
无 None

其他英文名 / Other English Name(s)
Little Nepalese Horseshoe Bat

同物异名 / Synonym(s)
无 None

▲ 形态及生境 / Morphology and Habitat

形态特征 / Morphological Characteristics
齿式：1.1.2.3/2.1.3.3=32。头体长 35~37 mm。颅全长 14~15 mm。耳长 14~18 mm。前臂长 33~38 mm。后足长 7~8 mm。尾长 16~19 mm。下唇有 3 个凹槽。第三掌骨略短于第四和第五掌骨。背毛肉桂棕色，毛具灰色的基部和带棕色的尖端；腹部毛色略苍白。
Dental formula: 1.1.2.3/2.1.3.3=32. Head and body length 35-37 mm. Greatest skull length 14-15 mm. Ear length 14-18 mm. Forearm length 33-38 mm. Hind foot length 7-8 mm. Tail length 16-19 mm. Lower lip with three grooves. Third metacarpal is slightly shorter than fourth and fifth. Dorsal Pelage cinnamon-brown. Hairs with grayish bases and brownish tips; underparts slightly paler.

生境 / Habitat
有竹丛的密林地带
Dense forest with bamboo grove

▲ 地理分布 / Geographic Distribution

国内分布 / Domestic Distribution
西藏、云南 Tibet, Yunnan

全球分布 / World Distribution
中国、孟加拉国、印度、缅甸、尼泊尔
China, Bangladesh, India, Myanmar, Nepal

生物地理界 / Biogeographic Realm
印度马来界 Indomalaya

WWF 生物群系 / WWF Biome
热带和亚热带湿润阔叶林
Tropical & Subtropical Moist Broadleaf Forests

动物地理分布型 / Zoogeographic Distribution Type
Wa

分布标注 / Distribution Note
非特有种 Non-Endemic

▲ 濒危状况 / Threatened Status

中国生物多样性红色名录等级 / CB RL Category (2021)
数据缺乏 DD

IUCN 红色名录 / IUCN Red List (2021)
无危 LC

威胁因子 / Threats
未知 Unknown

▲ 法律保护地位 / Legal Protection Status

国家重点保护野生动物等级 / Category of National Key Protected Wild Animals (2021)
未列入 Not listed

"三有"名录 / TWIESSV (2023)
未列入 Not listed

CITES 附录等级 / CITES Appendix (2023)
未列入 Not listed

迁徙物种公约附录 / CMS Appendix (2020)
未列入 Not listed

保护行动 / Conservation Action
尚无保护行动 No conservation action so far

▲ 参考文献 / References

Jiang et al. (蒋志刚等), 2021; Burgin et al., 2020; IUCN, 2020; Wilson and Mittermeier, 2019; Zhang et al., 2009

135 / 托氏菊头蝠

Rhinolophus thomasi K. Andersen, 1905

• Thomas's Horseshoe Bat

▲ 分类地位 / Taxonomy

翼手目 Chiroptera / 菊头蝠科 Rhinolophidae / 菊头蝠属 *Rhinolophus*

科建立者及其文献 / Family Authority
Gray, 1825

属建立者及其文献 / Genus Authority
Lacépède, 1799

亚种 / Subspecies
滇北亚种 *R. t. septentrionalis* Sanbom, 1939
云南 Yunnan

越北亚种 *R. t. latifolius* Sanbom, 1939
云南、贵州和广西
Yunnan, Guizhou and Guangxi

模式标本产地 / Type Locality
缅甸
Burma (Myanmar), Karin Hills

▲ 其他名称 / Other Name(s)

其他中文名 / Other Chinese Name(s)
无 None

其他英文名 / Other English Name(s)
无 None

同物异名 / Synonym(s)
无 None

▲ 形态及生境 / Morphology and Habitat

形态特征 / Morphological Characteristics
齿式：1.1.2.3/2.1.3.3=32。头体长 48~50 mm。耳长 16~20 mm。前臂长 40~48 mm。后足长 8~10 mm。尾长 18~28 mm。颅全长 18~20 mm。鼻叶中马蹄形裂片，宽 7.2~8.9 mm。顶端叶短。鞍状叶侧缘平行。掌骨短，第三指第二指骨的长度超过第一指骨长度的一半。
Dental formula: 1.1.2.3/2.1.3.3=32. Head and body length 48-50 mm. Ear length 16-20 mm. Forearm length 40-48 mm. Hind foot length 8-10 mm. Tail length 18-28 mm. Greatest skull length 18-20 mm. Horseshoe lobe of nose leaflet medium, breadth 7.2-8.9 mm. Apical lobe short. Lateral margins of sellar leaflets parallel. Metacarpal bones short. The second phalanx of the third digit is longer than half the length of the first phalanx.

生境 / Habitat
喀斯特地貌、洞穴
Karst, caves

▲ 地理分布 / Geographic Distribution

国内分布 / Domestic Distribution
云南、贵州、广西、四川
Yunnan, Guizhou, Guangxi, Sichuan

全球分布 / World Distribution
中国、老挝、越南、泰国、缅甸
China, Laos, Vietnam, Thailand, Myanmar

生物地理界 / Biogeographic Realm
印度马来界、古北界
Indomalaya, Palearctic

WWF 生物群系 / WWF Biome
热带和亚热带湿润阔叶林
Tropical & Subtropical Moist Broadleaf Forests

动物地理分布型 / Zoogeographic Distribution Type
Wa

分布标注 / Distribution Note
非特有种 Non-Endemic

▲ 濒危状况 / Threatened Status

中国生物多样性红色名录等级 / CB RL Category (2021)
近危 NT

IUCN 红色名录 / IUCN Red List (2021)
无危 LC

威胁因子 / Threats
未知 Unknown

▲ 法律保护地位 / Legal Protection Status

国家重点保护野生动物等级 / Category of National Key Protected Wild Animals (2021)
未列入 Not listed

"三有" 名录 / TWIESSV (2023)
未列入 Not listed

CITES 附录等级 / CITES Appendix (2023)
未列入 Not listed

迁徙物种公约附录 / CMS Appendix (2020)
未列入 Not listed

保护行动 / Conservation Action
尚无保护行动 No conservation action so far

▲ 参考文献 / References

Jiang et al. (蒋志刚等), 2021; Burgin et al., 2020; IUCN, 2020; Liu et al. (刘少英等), 2020; Wilson and Mittermeier, 2019; Pan et al. (潘清华等), 2007; Wang (王应祥), 2003; Zhang (张荣祖), 1997

136 / 锲鞍菊头蝠

Rhinolophus xinanzhongguoensis
Zou, Guillén-Servent, Lim, Eger, Wang & Jiang, 2009

· Xinan Horseshoe Bat

▲ 分类地位 / Taxonomy

翼手目 Chiroptera / 菊头蝠科 Rhinolophidae /
菊头蝠属 *Rhinolophus*

科建立者及其文献 / Family Authority
Gray, 1825

属建立者及其文献 / Genus Authority
Lacépède, 1799

亚种 / Subspecies
无 None

模式标本产地 / Type Locality
中国
Wumulong, Yongde County, Yunnan Province, China.
24°22'N, 99°39'E 1980 m above sea level

▲ 其他名称 / Other Name(s)

其他中文名 / Other Chinese Name(s)
无 None

其他英文名 / Other English Name(s)
South-western China Horseshoe Bat, Wedge-sella
Horseshoe Bat

同物异名 / Synonym(s)
无 None

▲ 形态及生境 / Morphology and Habitat

形态特征 / Morphological Characteristics
齿式：1.1.2.3/2.1.3.3=32。头体长 59~70 mm。
前臂长 58.7~60.4 mm。尾长 30~39 mm。耳长
21~22 mm。体重20~26 g。耳小，棕色，部分透明。
背毛暗棕色，腹毛色浅。翼膜深褐色。
Dental formula: 1.1.2.3/2.1.3.3=32. Head and body
length 59-70 mm. Forearm length 58.7-60.4 mm. Tail
length 30-39 mm. Ear length 21-22 mm. Body weight
20-26 g. Ears brown, small, partly transparent. Dorsal
fur dull medium brown, ventral fur paler in color. Flight
membranes are dark brown.

生境 / Habitat
洞穴 Caves

▲ 地理分布 / Geographic Distribution

国内分布 / Domestic Distribution
云南、贵州
Yunnan, Guizhou

全球分布 / World Distribution
中国 China

生物地理界 / Biogeographic Realm
古北界 Palearctic

WWF 生物群系 / WWF Biome
热带和亚热带湿润阔叶林
Tropical & Subtropical Moist Broadleaf Forests

动物地理分布型 / Zoogeographic Distribution Type
Sd

分布标注 / Distribution Note
特有种 Endemic

▲ 濒危状况 / Threatened Status

中国生物多样性红色名录等级 / CB RL Category (2021)
数据缺乏 DD

IUCN 红色名录 / IUCN Red List (2021)
未评定 NE

威胁因子 / Threats
未知 Unknown

▲ 法律保护地位 / Legal Protection Status

国家重点保护野生动物等级 / Category of National Key Protected Wild Animals (2021)
未列入 Not listed

"三有" 名录 / TWIESSV (2023)
未列入 Not listed

CITES 附录等级 / CITES Appendix (2023)
未列入 Not listed

迁徙物种公约附录 / CMS Appendix (2020)
未列入 Not listed

保护行动 / Conservation Action
尚无保护行动 No conservation action so far

▲ 参考文献 / References

Jiang et al. (蒋志刚等), 2021; Burgin et al., 2020; IUCN, 2020; Wilson and Mittermeier, 2019

137 / 云南菊头蝠

Rhinolophus yunanensis Dobson, 1872

· Dobson's Horseshoe Bat

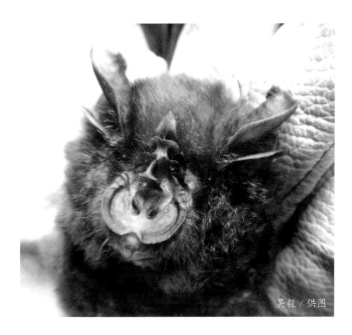

吴毅 / 供图

▲ 分类地位 / Taxonomy

翼手目 Chiroptera / 菊头蝠科 Rhinolophidae / 菊头蝠属 *Rhinolophus*

科建立者及其文献 / Family Authority
Gray, 1825

属建立者及其文献 / Genus Authority
Lacépède, 1799

亚种 / Subspecies
无 None

模式标本产地 / Type Locality
中国
China, Yunnan, Hotha

▲ 其他名称 / Other Name(s)

其他中文名 / Other Chinese Name(s)
无 None

其他英文名 / Other English Name(s)
无 None

同物异名 / Synonym(s)
无 None

▲ 形态及生境 / Morphology and Habitat

形态特征 / Morphological Characteristics
齿式：1.1.2.3/2.1.3.3=32。头体长 60~68 mm。耳长 23~32 mm。前臂长 54~60 mm。后足长 12~14 mm。尾长 18~26 mm。颅全长 24~28 mm。马蹄叶较宽，尾短于胫长。体毛长而密，棕褐色。颧宽略大于后头宽。
Dental formula: 1.1.2.3/2.1.3.3=32. Head and body length 60-68 mm. Hind foot length 12-14 mm. Ear length 23-32 mm. Forearm length 54-60 mm. Tail length 18-26 mm. Greatest skull length 24-28 mm. Horseshoe noseleafs are wider, tail shorter than shin length. Dorsal hairs are long and dense and brown. Zygomatic width is slightly larger than the posterior width.

生境 / Habitat
竹林、人造建筑
Bamboo forest, man-made building

▲ 地理分布 / Geographic Distribution

国内分布 / Domestic Distribution
云南、四川
Yunnan, Sichuan

全球分布 / World Distribution
中国、印度、缅甸、泰国
China, India, Myanmar, Thailand

生物地理界 / Biogeographic Realm
印度马来界 Indomalaya

WWF 生物群系 / WWF Biome
热带和亚热带湿润阔叶林
Tropical & Subtropical Moist Broadleaf Forests

动物地理分布型 / Zoogeographic Distribution Type
Wb

分布标注 / Distribution Note
非特有种 Non-Endemic

▲ 濒危状况 / Threatened Status

中国生物多样性红色名录等级 / CB RL Category (2021)
易危 VU

IUCN 红色名录 / IUCN Red List (2021)
无危 LC

威胁因子 / Threats
未知 Unknown

▲ 法律保护地位 / Legal Protection Status

国家重点保护野生动物等级 / Category of National Key Protected Wild Animals (2021)
未列入 Not listed

"三有" 名录 / TWIESSV (2023)
未列入 Not listed

CITES 附录等级 / CITES Appendix (2023)
未列入 Not listed

迁徙物种公约附录 / CMS Appendix (2020)
未列入 Not listed

保护行动 / Conservation Action
尚无保护行动 No conservation action so far

▲ 参考文献 / References

Jiang et al. (蒋志刚等), 2021; Wilson and Mittermeier, 2019; Pan et al. (潘清华等), 2007; Peng and Zhong (彭基泰和钟祥清), 2005; Wilson and Reeder, 2005; Wang (王应祥), 2003

138 / 大蹄蝠

Hipposideros armiger (Hodgson, 1835)

· Great Leaf-nosed Bat

周佳俊 / 供图

余文华 / 供图

▲ 分类地位 / Taxonomy

翼手目 Chiroptera / 蹄蝠科 Hipposideridae / 蹄蝠属 *Hipposideros*

科建立者及其文献 / Family Authority
Lydekker, 1891

属建立者及其文献 / Genus Authority
Gray, 1831

亚种 / Subspecies
指名亚种 *H. a. armiger* (Hodgson, 1835)
四川、云南、贵州和陕西
Sichuan, Yunnan, Guizhou and Shaanxi

华东亚种 *H. a. swinhoei* (Prtere, 1871)
浙江、安徽、江苏、福建、江西、湖南、广东、香港、澳门、广西和海南
Zhejiang, Anhui, Jiangsu, Fujian, Jiangxi, Hunan, Guangdong, Hong Kong, Macao, Guangxi and Hainan

闽南亚种 *H. a. fujianensis* Zhen, 1987
福建 Fujian

模式标本产地 / Type Locality
尼泊尔
Nepal

▲ 其他名称 / Other Name(s)

其他中文名 / Other Chinese Name(s)
无 None

其他英文名 / Other English Name(s)
无 None

同物异名 / Synonym(s)
无 None

▲ 形态及生境 / Morphology and Habitat

形态特征 / Morphological Characteristics

齿式：1.1.2.3/2.1.2.3=30。头体长 80~110 mm。尾长 48~70 mm。后足长 13~17 mm。耳长 26~35 mm。前臂长 82~99 mm。颅全长 31~33 mm。上颌第一小前白齿位于齿列外。耳大，呈三角形。额部中间有一腺囊开口。头骨吻部由前向后逐渐升高。矢状嵴发达。上体棕褐色，下体毛黄褐色，毛基暗褐或栗褐。

Dental formula: 1.1.2.3/2.1.2.3=30. Head and body length 80-110 mm. Tail length 48-70 mm. Hind foot length 13-17 mm. Ear length 26-35 mm. Forearm length 82-99 mm. Greatest skull length 31-33 mm. The maxillary 1st minor premolar is located outside the dentition. Ears are large and triangular in shape. There is a glandular sac opening in the middle of the forehead. Snout of the skull rises gradually from front to back. Sagittal crest is well developed. Dorsal part is brown, the lower body hairs are yellow-brown. The bases of hairs are dark brown or chestnut brown.

生境 / Habitat
洞穴、人造建筑
Caves, man-made building

▲ 地理分布 / Geographic Distribution

国内分布 / Domestic Distribution
陕西、江苏、浙江、安徽、湖南、江西、四川、贵州、云南、广西、
广东、海南、台湾、香港、重庆、澳门
Shaanxi, Jiangsu, Zhejiang, Anhui, Hunan, Jiangxi, Sichuan, Guizhou, Yunnan,
Guangxi, Guangdong, Hainan, Taiwan, Hong Kong, Chongqing, Macao

全球分布 / World Distribution
柬埔寨、中国、印度、老挝、马来西亚、缅甸、尼泊尔、泰国、越南
Cambodia, China, India, Laos, Malaysia, Myanmar, Nepal, Thailand, Vietnam

生物地理界 / Biogeographic Realm
印度马来界、古北界
Indomalaya, Palearctic

WWF 生物群系 / WWF Biome
热带和亚热带湿润阔叶林
Tropical & Subtropical Moist Broadleaf Forests

动物地理分布型 / Zoogeographic Distribution Type
Wb

分布标注 / Distribution Note
非特有种 Non-Endemic

▲ 濒危状况 / Threatened Status

中国生物多样性红色名录等级 / CB RL Category (2021)
无危 LC

IUCN 红色名录 / IUCN Red List (2021)
无危 LC

威胁因子 / Threats
无 None

▲ 法律保护地位 / Legal Protection Status

国家重点保护野生动物等级 / Category of National Key Protected Wild Animals (2021)
未列入 Not listed

"三有" 名录 / TWIESSV (2023)
列入 Listed

CITES 附录等级 / CITES Appendix (2023)
未列入 Not listed

迁徙物种公约附录 / CMS Appendix (2020)
未列入 Not listed

保护行动 / Conservation Action
尚无保护行动 No conservation action so far

▲ 参考文献 / References

Jiang et al. (蒋志刚等), 2021; Burgin et al., 2020; IUCN, 2020; Mao (毛秀光), 2010; Pan et al. (潘清华等), 2007; Wilson and Reeder, 2005; Wang (王应祥), 2003; Chen (陈敏), 2003; Zhang (张荣祖), 1997

139 | 灰小蹄蝠

Hipposideros cineraceus Blyth, 1853

• Least Leaf-nosed Bat

▲ 分类地位 / Taxonomy

翼手目 Chiroptera / 蹄蝠科 Hipposideridae / 蹄蝠属 *Hipposideros*

科建立者及其文献 / Family Authority
Lydekker, 1891

属建立者及其文献 / Genus Authority
Gray, 1831

亚种 / Subspecies
无 None

模式标本产地 / Type Locality
巴基斯坦
Pakistan, Punjab, Salt Range, near Pind Dadan Khan

余文华 / 供图

▲ 其他名称 / Other Name(s)

其他中文名 / Other Chinese Name(s)
无 None

其他英文名 / Other English Name(s)
Ashy Leaf-nosed Bat, Ashy Roundleaf Bat,
Least Roundleaf Bat

同物异名 / Synonym(s)
无 None

▲ 形态及生境 / Morphology and Habitat

形态特征 / Morphological Characteristics
齿式：1.1.2.3/2.1.2.3=30。前臂长 36 mm 以下。马蹄叶前端无缺刻和附小叶。鼻间隔膜膨胀隆起，类似肾形肉垂。耳壳黄褐色，前端钝尖，相对较大，往前折略微超出吻端，对耳屏不到耳长 1/3。翼膜及股间膜褐色，止于踝部。

Dental formula: 1.1.2.3/2.1.2.3=30. Forearm length less than 36 mm. Front-end of the horseshoe leaflet is unbroken with the attached leaflet. Nasal septum bulges, creating a kidney-like lobes. Auricle is yellowish-brown, blunt at the front, relatively large, folded forward slightly beyond the snout, and the antitragus is less than 1/3 the length of the ear. Pterygoid membrane and interfemoral membrane brown, attached to the ankle.

生境 / Habitat
洞穴、森林 Caves, forest

▲ 地理分布 / Geographic Distribution

国内分布 / Domestic Distribution
云南、广西
Yunnan, Guangxi

全球分布 / World Distribution
中国、印度、印度尼西亚、老挝、马来西亚、缅甸、巴基斯坦、泰国、
越南
China, India, Indonesia, Laos, Malaysia, Myanmar, Pakistan, Thailand, Vietnam

生物地理界 / Biogeographic Realm
印度马来界 Indomalaya

WWF 生物群系 / WWF Biome
热带和亚热带湿润阔叶林
Tropical & Subtropical Moist Broadleaf Forests

动物地理分布型 / Zoogeographic Distribution Type
Hc

分布标注 / Distribution Note
非特有种 Non-Endemic

▲ 濒危状况 / Threatened Status

中国生物多样性红色名录等级 / CB RL Category (2021)
近危 NT

IUCN 红色名录 / IUCN Red List (2021)
无危 LC

威胁因子 / Threats
未知 Unknown

▲ 法律保护地位 / Legal Protection Status

国家重点保护野生动物等级 / Category of National Key Protected Wild Animals (2021)
未列入 Not listed

"三有"名录 / TWIESSV (2023)
未列入 Not listed

CITES 附录等级 / CITES Appendix (2023)
未列入 Not listed

迁徙物种公约附录 / CMS Appendix (2020)
未列入 Not listed

保护行动 / Conservation Action
尚无保护行动 No conservation action so far

▲ 参考文献 / References

Jiang et al. (蒋志刚等), 2021; IUCN, 2021; Burgin et al., 2020; Liu et al. (刘少英等), 2020; Li and Wu, 2011; Zhen, 1987; Liang and Dong, 1984; Phillips and Wilson, 1968; Shaw et al., 1966

140 / 大耳小蹄蝠

Hipposideros fulvus Gray, 1838

• Fulvus Leaf-nosed Bat

▲ 分类地位 / Taxonomy

翼手目 Chiroptera / 蹄蝠科 Hipposideridae / 蹄蝠属 *Hipposideros*

科建立者及其文献 / Family Authority
Lydekker, 1891

属建立者及其文献 / Genus Authority
Gray, 1831

亚种 / Subspecies
无 None

模式标本产地 / Type Locality
印度
India, Karnatika, Dharwar

▲ 其他名称 / Other Name(s)

其他中文名 / Other Chinese Name(s)
无 None

其他英文名 / Other English Name(s)
Fulvus Roundleaf Bat

同物异名 / Synonym(s)
无 None

▲ 形态及生境 / Morphology and Habitat

形态特征 / Morphological Characteristics
齿 式：1.1.2.3/2.1.2.3=30。头体长 43 mm。耳长 20~23 mm。尾长 25~29 mm。体重 8~10 g。平均翼展 130 mm。背毛红棕色、暗黄色、暗棕色、浅灰色或金橙色，腹毛从乳白色到淡灰色。翼膜苍白色。
Dental formula: 1.1.2.3/2.1.2.3=30. Head and body length 43 mm. Ear length 20-23 mm. Tail length 25-29 mm. Body mass 8-10 g. Average wingspan 130 mm. Dorsal fur reddish-brown, dull yellow, dull brown, light gray or golden orange. Ventral fur ranges from creamy white to pale gray. Members are pale.

生境 / Habitat
洞穴、人造建筑、森林、草地
Caves, man-made building, forest, grassland

▲ 地理分布 / Geographic Distribution

国内分布 / Domestic Distribution
云南 Yunnan

全球分布 / World Distribution
中国、阿富汗、孟加拉国、印度、巴基斯坦、斯里兰卡
China, Afghanistan, Bangladesh, India, Pakistan, Sri Lanka

生物地理界 / Biogeographic Realm
印度马来界 Indomalaya

WWF 生物群系 / WWF Biome
热带和亚热带湿润阔叶林
Tropical & Subtropical Moist Broadleaf Forests

动物地理分布型 / Zoogeographic Distribution Type
Sa

分布标注 / Distribution Note
非特有种 Non-Endemic

▲ 濒危状况 / Threatened Status

中国生物多样性红色名录等级 / CB RL Category (2021)
数据缺乏 DD

IUCN 红色名录 / IUCN Red List (2021)
无危 LC

威胁因子 / Threats
未知 Unknown

▲ 法律保护地位 / Legal Protection Status

国家重点保护野生动物等级 / Category of National Key Protected Wild Animals (2021)
未列入 Not listed

"三有"名录 / TWIESSV (2023)
未列入 Not listed

CITES 附录等级 / CITES Appendix (2023)
未列入 Not listed

迁徙物种公约附录 / CMS Appendix (2020)
未列入 Not listed

保护行动 / Conservation Action
尚无保护行动 No conservation action so far

▲ 参考文献 / References

Jiang et al. (蒋志刚等), 2021; Burgin et al., 2 021; IUCN, 2020; Wilson and Mittermeier, 2019; Tan et al. (谭敏等), 2009; Wang (王应祥), 2003; Koopman, 1993

141 / 中蹄蝠

Hipposideros larvatus (Horsfield, 1823)

• Horsfield's Leaf-nosed Bat

吴毅 / 供图

▲ 分类地位 / Taxonomy

翼手目 Chiroptera / 蹄蝠科 Hipposideridae /
蹄蝠属 *Hipposideros*

科建立者及其文献 / Family Authority
Lydekker, 1891

属建立者及其文献 / Genus Authority
Gray, 1831

亚种 / Subspecies
海南亚种 *H. l. poutensis* J. Allen, 1906
海南、广东、广西和贵州
Hainan, Guangdong, Guangxi and Guizhou

越北亚种 *H. l. alongensis* Bourret, 1942
云南 Yunnan

缅甸亚种 *H. l. grandis* Allen, 1936
云南 Yunnan

模式标本产地 / Type Locality
印度尼西亚
Indonesia, Java

▲ 其他名称 / Other Name(s)

其他中文名 / Other Chinese Name(s)
无 None

其他英文名 / Other English Name(s)
Large Asian Roundleaf Bat

同物异名 / Synonym(s)
无 None

▲ 形态及生境 / Morphology and Habitat

形态特征 / Morphological Characteristics
齿式：1.1.2.3/2.1.2.3=30。前臂长 54~61 mm。
颅全长约 23 mm。具额腺囊，开口于两耳之间，
有一束长毛从囊口伸出。耳大且呈三角形。体
毛有暗色型（海南亚种）和淡色型（缅甸亚种）
区分，前者具有褐白色三角形肩斑，后者为淡
褐色肩斑。
Dental formula: 1.1.2.3/2.1.2.3=30. Forearm length 54-
61 mm. A frontal glandular sac opening between the
ears with a strand of long hairs protruding from the sac.
Ears are large and triangular in shape. Body hairs are
divided into dark type (Hainan subspecies) and light
type (Burmese subspecies), the former with brown-white
triangular spots on the shoulders, the latter with light
brown spots on the shoulders.

生境 / Habitat
洞穴、人造建筑
Caves, man-made building

▲ 地理分布 / Geographic Distribution

国内分布 / Domestic Distribution
贵州、海南、广西、广东
Guizhou, Hainan, Guangxi, Guangdong

全球分布 / World Distribution
孟加拉国、柬埔寨、中国、印度、印度尼西亚、老挝、马来西亚、缅甸、
泰国、越南
Bangladesh, Cambodia, China, India, Indonesia, Laos, Malaysia, Myanmar,
Thailand, Vietnam

生物地理界 / Biogeographic Realm
印度马来界 Indomalaya

WWF 生物群系 / WWF Biome
热带和亚热带湿润阔叶林
Tropical & Subtropical Moist Broadleaf Forests

动物地理分布型 / Zoogeographic Distribution Type
Wa

分布标注 / Distribution Note
非特有种 Non-Endemic

▲ 濒危状况 / Threatened Status

中国生物多样性红色名录等级 / CB RL Category (2021)
无危 LC

IUCN 红色名录 / IUCN Red List (2021)
无危 LC

威胁因子 / Threats
无 None

▲ 法律保护地位 / Legal Protection Status

国家重点保护野生动物等级 / Category of National Key Protected Wild Animals (2021)
未列入 Not listed

"三有"名录 / TWIESSV (2023)
未列入 Not listed

CITES 附录等级 / CITES Appendix (2023)
未列入 Not listed

迁徙物种公约附录 / CMS Appendix (2020)
未列入 Not listed

保护行动 / Conservation Action
尚无保护行动 No conservation action so far

▲ 参考文献 / References

Jiang et al. (蒋志刚等), 2021; Burgin et al., 2020; IUCN, 2020; Wilson and Mittermeier, 2019; Wang et al., 2013; Smith and Xie, 2009;
Pan et al. (潘清华等), 2007; Wilson and Reeder, 2005; Wang (王应祥), 2003; Zhang (张荣祖), 1997; Koopman, 1993

142 / 莱氏蹄蝠

Hipposideros lylei Thomas, 1913

• Shield-faced Leaf-nosed Bat

▲ 分类地位 / Taxonomy

翼手目 Chiroptera / 蹄蝠科 Hipposideridae / 蹄蝠属 *Hipposideros*

科建立者及其文献 / Family Authority
Lydekker, 1891

属建立者及其文献 / Genus Authority
Gray, 1831

亚种 / Subspecies
无 None

模式标本产地 / Type Locality
泰国
Thailand, 50 mi. (80 km) N Chiengmai (Chiang Mai), Chiengdao Cave, 350 m

▲ 其他名称 / Other Name(s)

其他中文名 / Other Chinese Name(s)
鞘面蹄蝠

其他英文名 / Other English Name(s)
Shield-faced Roundleaf Bat

同物异名 / Synonym(s)
无 None

▲ 形态及生境 / Morphology and Habitat

形态特征 / Morphological Characteristics
齿式：1.1.2.3/2.1.2.3=30。头体长 72~95 mm。耳长 30 mm。前臂长 78~84 mm。后足长 16~21 mm。尾长 48~55 mm。颅全长 28~29 mm。耳郭宽且呈三角形，无对耳屏。雄性额囊具翅状突起，且与鼻叶后侧缘连续。鼻叶呈带褐色的粉红色。耳郭和翅膀淡棕色，略带褐色或灰色，腹部淡黄色。
Dental formula: 1.1.2.3/2.1.2.3=30. Head and body length 72-95 mm. Tail length 48-55 mm. Hind foot length 16-21 mm. Ear length 30 mm. Forearm length 78-84 mm. Greatest skull length 28-29 mm. Ears wide and triangular, without antitragus. Frontal sac with winglike projections, prominent in males. Lateral margins of anterior and posterior nose leave continuous. Nose leaflet brownish pink. Ears and wings pale brown. Color brownish or grayish, with buffy venter.

生境 / Habitat
洞穴、耕地、热带和亚热带森林
Cave, arable land, subtropical tropical forest

▲ 地理分布 / Geographic Distribution

国内分布 / Domestic Distribution
云南 Yunnan

全球分布 / World Distribution
中国、马来西亚、缅甸、泰国、越南
China, Malaysia, Myanmar, Thailand, Vietnam

生物地理界 / Biogeographic Realm
印度马来界 Indomalaya

WWF 生物群系 / WWF Biome
热带和亚热带湿润阔叶林
Tropical & Subtropical Moist Broadleaf Forests

动物地理分布型 / Zoogeographic Distribution Type
Sa

分布标注 / Distribution Note
非特有种 Non-Endemic

▲ 濒危状况 / Threatened Status

中国生物多样性红色名录等级 / CB RL Category (2021)
易危 VU

IUCN 红色名录 / IUCN Red List (2021)
无危 LC

威胁因子 / Threats
旅游、采矿、采石
Tourism, mining, quarrying

▲ 法律保护地位 / Legal Protection Status

国家重点保护野生动物等级 / Category of National Key Protected Wild Animals (2021)
未列入 Not listed

"三有"名录 / TWIESSV (2023)
未列入 Not listed

CITES 附录等级 / CITES Appendix (2023)
未列入 Not listed

迁徙物种公约附录 / CMS Appendix (2020)
未列入 Not listed

保护行动 / Conservation Action
尚无保护行动 No conservation action so far

▲ 参考文献 / References

Jiang et al. (蒋志刚等), 2021; Burgin et al., 2020; IUCN, 2020; Wilson and Mittermeier, 2019; He et al. (贺新平等), 2014; Yuan et al., 2012; Wei et al. (韦力等), 2011; Jiang et al., 2010; Pan et al. (潘清华等), 2007; Wang (王应祥), 2003

143 / 小蹄蝠

Hipposideros pomona K. Andersen, 1918

• Andersen's Leaf-nosed Bat

黄泰 / 供图

▲ 其他名称 / Other Name(s)

其他中文名 / Other Chinese Name(s)
大耳双色蹄蝠、果树蹄蝠

其他英文名 / Other English Name(s)
Pomona Roundleaf Bat

同物异名 / Synonym(s)
无 None

▲ 形态及生境 / Morphology and Habitat

形态特征 / Morphological Characteristics

齿式：1.1.2.3/2.1.2.3=30。头体长 36~52 mm。耳长 18~25 mm。前臂长 38~43 mm。后足长 6~9 mm。尾长 28~35 mm。颅全长 17~18 mm。鼻叶复杂，马蹄叶中央缺刻，两侧无附小叶，中叶不发达，顶叶有两纵隔。雌雄均具额囊腺。耳宽而圆，对耳屏低且与耳壳全部相连。体毛柔软，背毛棕褐色，毛基灰白色，腹毛呈灰白色。

Dental formula: 1.1.2.3/2.1.2.3=30. Head and body length 36-52 mm. Ear length 18-25 mm. Forearm length 38-43 mm. Hind foot length 6-9 mm. Tail length 28-35 mm. Greatest skull length 17-18 mm. Nasal leaves complex. Middle of the Horseshoe noseleafs are absent, and there are no attached lobules on both sides. Middle leaves are not developed. Parietal lobe has two mediastinums. Both males and females have frontal cystic glands. Ear is broad and round, low to the antitragus, and fully connected with the auricle. Body hairs are soft. Dorsal hairs brown, with gray and white hair bases. Abdominal hairs are gray and white.

生境 / Habitat

地下洞穴和裂缝、栖息在高度改变的栖息地，甚至栖息在城区
It roosts in small colonies of a few individuals in caves and crevices in subterranean habitats and is considered to be tolerant modified habitats and can even occur in urban areas

▲ 地理分布 / Geographic Distribution

国内分布 / Domestic Distribution
福建、湖南、广西、广东、海南、四川、贵州、云南、香港
Fujian, Hunan, Guangxi, Guangdong, Hainan, Sichuan, Guizhou, Yunnan,
Hong Kong

全球分布 / World Distribution
孟加拉国、柬埔寨、中国、印度、老挝、马来西亚、缅甸、尼
泊尔、泰国、越南
Bangladesh, Cambodia, China, India, Laos, Malaysia, Myanmar, Nepal,
Thailand, Vietnam

生物地理界 / Biogeographic Realm
印度马来界、古北界
Indomalaya, Palearctic

WWF 生物群系 / WWF Biome
热带和亚热带湿润阔叶林
Tropical & Subtropical Moist Broadleaf Forests

动物地理分布型 / Zoogeographic Distribution Type
Wa

分布标注 / Distribution Note
非特有种 Non-Endemic

▲ 濒危状况 / Threatened Status

中国生物多样性红色名录等级 / CB RL Category (2021)
无危 LC

IUCN 红色名录 / IUCN Red List (2021)
无危 LC

威胁因子 / Threats
无 None

▲ 法律保护地位 / Legal Protection Status

国家重点保护野生动物等级 / Category of National Key Protected Wild Animals (2021)
未列入 Not listed

"三有"名录 / TWIESSV (2023)
未列入 Not listed

CITES 附录等级 / CITES Appendix (2023)
未列入 Not listed

迁徙物种公约附录 / CMS Appendix (2020)
未列入 Not listed

保护行动 / Conservation Action
尚无保护行动 No conservation action so far

▲ 参考文献 / References

Jiang et al. (蒋志刚等), 2021; Burgin et al., 2020; IUCN, 2020; Liu et al. (刘少英等), 2020; Wilson and Mittermeier, 2019; Yang et al., 2012; Yuan et al., 2012; Hong et al., 2011; Tan et al. (谭敏等), 2009; Zhang et al., 2009; Pan et al. (潘清华等), 2007; Wilson and Reeder, 2005; Wang (王应祥), 2003; Lu et al. (陆长坤等), 1965

144 / 普氏蹄蝠

Hipposideros pratti Thomas, 1891

- Pratt's Leaf-nosed Bat

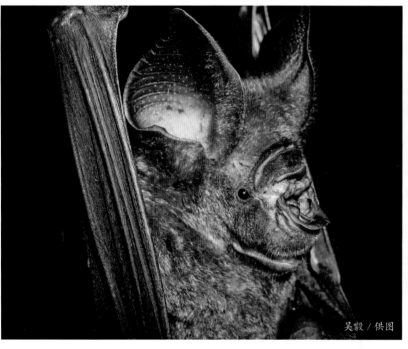

吴毅 / 供图

吴毅 / 供图

▲ 其他名称 / Other Name(s)

其他中文名 / Other Chinese Name(s)
柏氏蹄蝠、黄大蹄蝠

其他英文名 / Other English Name(s)
Pratt's Roundleaf Bat

同物异名 / Synonym(s)
无 None

▲ 形态及生境 / Morphology and Habitat

形态特征 / Morphological Characteristics

齿式：1.1.2.3/2.1.2.3=30。头体长 91~110 mm。耳长 33~38 mm。前臂长 75~90 mm。后足长 15~22 mm。尾长 50~62 mm。颅全长 28~35 mm。耳大而宽。马蹄叶左右侧各有 2 片附小叶，中间有凹。顶部裂片中部高于侧面。顶叶后面有 2 个大的皮叶，中间具有 1 束直立的长毛。背毛颜色棕色或深烟棕色。腹侧毛发色淡棕色。

Dental formula: 1.1.2.3/2.1.2.3=30. Head and body length 91-110 mm. Ear length 33-38 mm. Forearm length 75-90 mm. Hind foot length 15-22 mm. Tail length 50-62 mm. Greatest skull length 28-35 mm. Ears large and wide. Antitragi is low and small. Horseshoe anterior lobe of nose leaflet with only two leaflets on either side and with a concavity in middle. Middle of apical lobe higher than sides. Behind apical lobe are two large leaflets, with a bundle of long and straight hair, especially developed in adult males. Pelage color brown or dark smoky brown. Ventral surface light.

生境 / Habitat
洞穴 Caves

▲ 地理分布 / Geographic Distribution

国内分布 / Domestic Distribution

陕西、江苏、浙江、安徽、湖南、江西、四川、贵州、云南、广西、福建、重庆、广东、海南

Shaanxi, Jiangsu, Zhejiang, Anhui, Hunan, Jiangxi, Sichuan, Guizhou, Yunnan, Guangxi, Fujian, Chongqing, Guangdong, Hainan

全球分布 / World Distribution

中国、越南

China, Vietnam

生物地理界 / Biogeographic Realm

印度马来界、古北界

Indomalaya, Palearctic

WWF 生物群系 / WWF Biome

热带和亚热带湿润阔叶林

Tropical & Subtropical Moist Broadleaf Forests

动物地理分布型 / Zoogeographic Distribution Type

Wc

分布标注 / Distribution Note

非特有种 Non-Endemic

▲ 濒危状况 / Threatened Status

中国生物多样性红色名录等级 / CB RL Category (2021)

近危 NT

IUCN 红色名录 / IUCN Red List (2021)

无危 LC

威胁因子 / Threats

采矿、采石

Mining, quarrying

▲ 法律保护地位 / Legal Protection Status

国家重点保护野生动物等级 / Category of National Key Protected Wild Animals (2021)

未列入 Not listed

"三有" 名录 / TWIESSV (2023)

未列入 Not listed

CITES 附录等级 / CITES Appendix (2023)

未列入 Not listed

迁徙物种公约附录 / CMS Appendix (2020)

未列入 Not listed

保护行动 / Conservation Action

尚无保护行动 No conservation action so far

▲ 参考文献 / References

Jiang et al. (蒋志刚等), 2021; Burgin et al., 2020; IUCN, 2020; Liu et al. (刘少英等), 2020; Wilson and Mittermeier, 2019; Zeng (曾峰), 2012; Liu (刘文超), 2009; Wang (王婉莹), 2007; Chen et al. (陈敏等), 2002; Pan et al. (潘清华等), 2007; Shek and Lau, 2006; Peng and Zhong (彭基泰和钟祥清), 2005; Wilson and Reeder, 2005; Wang (王应祥), 2003; Zhang (张荣祖), 1997

145 / 三叶蹄蝠

Aselliscus stoliczkanus (Dobson, 1871)

• Stoliczka's Asian Trident Bat

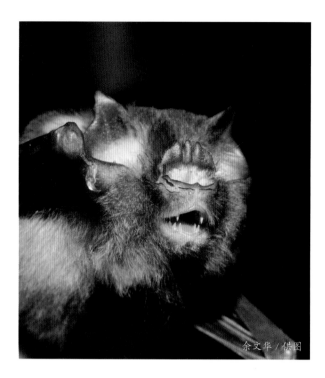

余文华 / 供图

▲ 分类地位 / Taxonomy

翼手目 Chiroptera / 蹄蝠科 Hipposideridae /
三叶蹄蝠属 *Aselliscus*

科建立者及其文献 / Family Authority
Lydekker, 1891

属建立者及其文献 / Genus Authority
Tate, 1941

亚种 / Subspecies
无 None

模式标本产地 / Type Locality
马来西亚
Malaysia, West, Penang Isl

▲ 其他名称 / Other Name(s)

其他中文名 / Other Chinese Name(s)
无 None

其他英文名 / Other English Name(s)
Stoliczka's Trident Bat

同物异名 / Synonym(s)
无 None

▲ 形态及生境 / Morphology and Habitat

形态特征 / Morphological Characteristics

齿式：1.1.2.3/2.1.2.3=30。头体长 40~50 mm。耳长
10~14 mm。前臂长 39~44 mm。后足长 9~10 mm。
尾长 30~40 mm。颅全长 16 mm。前鼻叶两侧各有
2 片小型附叶，后鼻叶被 2 条纵沟分成 3 叶状，中
间一叶细长。雄性胸前有 2 个腺体，胸毛白色长。
耳小，无对耳屏。体色棕褐，体毛长而绒软，背毛
毛尖棕褐，毛基淡白，腹毛浅棕褐色，从前胸到腹
部有淡色区，两侧有棕褐斑。尾明显突出于尾间膜。
Dental formula: 1.1.2.3/2.1.2.3=30. Head and body length
40-50 mm. Ear length 10-14 mm. Forearm length 39-44 mm.
Hind foot length 9-10 mm. Tail length 30-40 mm. Greatest
skull length 16 mm. Two small appendages on each side of the
anterior nasal lobe. Posterior nasal lobe is divided into 3 lobes
by 2 longitudinal furrows, the middle lobe is long and thin.
Males have two glands on their chests along with long white
"fluffy" chest hairs. Ears are small, without antitragus. Body
color is brown, the body hairs are long and soft, the tips of
dorsal hairs are brown, with pale white hair bases. Abdominal
hairs are light brown. Light-colored hairs extend from the
chest to the abdomen. There are brown spots on both sides.
Caudate clearly protrudes from the intercaudal membrane.

生境 / Habitat

洞穴、耕地、亚热带热带退化森林
Caves, arable land, subtropical tropical degraded forest

▲ 地理分布 / Geographic Distribution

国内分布 / Domestic Distribution
云南、广东、广西、贵州
Yunnan, Guangdong, Guangxi, Guizhou

全球分布 / World Distribution
中国、老挝、马来西亚、缅甸、泰国、越南
China, Laos, Malaysia, Myanmar, Thailand, Vietnam

生物地理界 / Biogeographic Realm
印度马来界、古北界
Indomalaya, Palearctic

WWF 生物群系 / WWF Biome
热带和亚热带湿润阔叶林
Tropical & Subtropical Moist Broadleaf Forests

动物地理分布型 / Zoogeographic Distribution Type
Wb

分布标注 / Distribution Note
非特有种 Non-Endemic

▲ 濒危状况 / Threatened Status

中国生物多样性红色名录等级 / CB RL Category (2021)
近危 NT

IUCN 红色名录 / IUCN Red List (2021)
无危 LC

威胁因子 / Threats
采矿、采石
Mining, quarrying

▲ 法律保护地位 / Legal Protection Status

国家重点保护野生动物等级 / Category of National Key Protected Wild Animals (2021)
未列入 Not listed

"三有"名录 / TWIESSV (2023)
未列入 Not listed

CITES 附录等级 / CITES Appendix (2023)
未列入 Not listed

迁徙物种公约附录 / CMS Appendix (2020)
未列入 Not listed

保护行动 / Conservation Action
尚无保护行动 No conservation action so far

▲ 参考文献 / References

Jiang et al. (蒋志刚等), 2021; Burgin et al., 2020; IUCN, 2020; Wilson and Mittermeier, 2019; Wang (王婉莹), 2007; Pan et al. (潘清华等), 2007; Wilson and Reeder, 2005; Wang (王应祥), 2003; Zhang (张荣祖), 1997

146 / 无尾蹄蝠

Coelops frithii Blyth, 1848

· Tail-less Leaf-nosed Bat

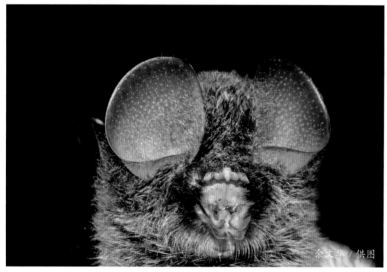

余文华 / 供图

余文华 / 供图

▲ 其他名称 / Other Name(s)

其他中文名 / Other Chinese Name(s)
无 None

其他英文名 / Other English Name(s)
East Asian Tailless Roundleaf Bat, Tailless
Leaf-nosed Bat

同物异名 / Synonym(s)
无 None

▲ 形态及生境 / Morphology and Habitat

形态特征 / Morphological Characteristics
齿式：1.1.2.3/2.1.2.3=30。头体长 38~50 mm。耳长 11~15 mm。前臂长
35~42 mm。后足长 5~9 mm。颅全长 16 mm。鼻叶分前、中、后鼻叶，
后鼻叶两侧各有 1 片小叶。耳大而圆。体毛背腹各异，背毛基部黑褐色，
毛尖赤褐色，腹毛基部灰褐色，毛尖灰白色。无尾。
Dental formula: 1.1.2.3/2.1.2.3=30. Head and body length 38-50 mm. Ear length 11-
15 mm. Forearm length 35-42 mm. Hind foot length 5-9 mm. Greatest skull length
16 mm. Nasal lobe is divided into the anterior, middle, and posterior nasal lobes and
each side of the posterior nasal lobes has a lobule. Big ears. Dorsal and abdominal hairs
are different, dorsal hair bases black-brown, hair tips russet, abdominal hair bases gray-
brown, with gray-white hair tips. Tailless.

生境 / Habitat
洞穴、耕地、亚热带热带严重退化森林
Caves, arable land, subtropical tropical degraded forest

▲ 地理分布 / Geographic Distribution

国内分布 / Domestic Distribution
广东、江西、四川、海南、福建、台湾、广西、云南、重庆
Guangdong, Jiangxi, Sichuan, Hainan, Fujian, Taiwan, Guangxi, Yunnan,
Chongqing

全球分布 / World Distribution
孟加拉国、中国、印度、印度尼西亚、老挝、马来西亚、缅甸、泰国、
越南
Bangladesh, China, India, Indonesia, Laos, Malaysia, Myanmar, Thailand, Vietnam

生物地理界 / Biogeographic Realm
印度马来界 Indomalaya

WWF 生物群系 / WWF Biome
热带和亚热带湿润阔叶林
Tropical & Subtropical Moist Broadleaf Forests

动物地理分布型 / Zoogeographic Distribution Type
Wc

分布标注 / Distribution Note
非特有种 Non-Endemic

▲ 濒危状况 / Threatened Status

中国生物多样性红色名录等级 / CB RL Category (2021)
易危 VU

IUCN 红色名录 / IUCN Red List (2021)
无危 LC

威胁因子 / Threats
森林砍伐、耕种
Logging, farming

▲ 法律保护地位 / Legal Protection Status

国家重点保护野生动物等级 / Category of National Key Protected Wild Animals (2021)
未列入 Not listed

"三有" 名录 / TWIESSV (2023)
未列入 Not listed

CITES 附录等级 / CITES Appendix (2023)
未列入 Not listed

迁徙物种公约附录 / CMS Appendix (2020)
未列入 Not listed

保护行动 / Conservation Action
尚无保护行动 No conservation action so far

▲ 参考文献 / References

Jiang et al. (蒋志刚等), 2021; Burgin et al., 2020; IUCN, 2020; Liu et al. (刘少英等), 2020; Wilson and Mittermeier, 2019; Xu et al. (徐忠鲜), 2013; Wang et al., 2013; Wang (王延校), 2012; Mao et al., 2010; Liu et al. (刘森等), 2008; Zhu (朱斌良), 2008; Pan et al. (潘清华等), 2007; Wilson and Reeder, 2005; Wu and Peng, 2005; Wang (王应祥), 2003; Feng et al. (冯江等), 2002; Zhou (周江), 2001; He(何晓瑞), 1999; Zhang (张荣祖), 1997

147 / 宽耳犬吻蝠

Tadarida insignis Blyth, 1862

• East Asian Free-tailed Bat

张礼标 / 供图

▲ 分类地位 / Taxonomy

翼手目 Chiroptera / 犬吻蝠科 Molossidae /
犬吻蝠属 *Tadarida*

科建立者及其文献 / Family Authority
Gervais, 1855

属建立者及其文献 / Genus Authority
Rafinesque, 1814

亚种 / Subspecies
无共识
Non consensus

模式标本产地 / Type Locality
中国
China, Fukien (Fujian), Amoy

▲ 其他名称 / Other Name(s)

其他中文名 / Other Chinese Name(s)
欧亚皱唇蝠、皱唇蝠

其他英文名 / Other English Name(s)
Oriental Free-tailed Bat

同物异名 / Synonym(s)
无 None

▲ 形态及生境 / Morphology and Habitat

形态特征 / Morphological Characteristics
齿式：1.1.2.3/3.1.2.3=32。头体长 84~94 mm。耳长 31~34 mm。前臂长 57~65 mm。后足长 10~15 mm。尾长 48~60 mm。颅全长 22~24 mm。上唇肥厚，鼻部突出。吻部覆盖密毛，具 8~10 条皱褶。耳内具 9~10 条横纹，耳后缘中部具凹刻。背毛呈土褐色，毛基色淡呈苍白色，腹毛色淡。后足超过胫骨长之半，脚趾附曲毛。翼狭长，似革质。尾后端从尾膜后缘伸出，呈游离状。

Dental formula: 1.1.2.3/3.1.2.3=32. Head and body length 84-94 mm. Greatest skull length 22-24 mm. Ear length 31-34 mm. Forearm length 57-65 mm. Hind foot length 10-15 mm. Tail length 48-60 mm. Upper lip is hypertrophic and the nose is prominent. Snout densely hairy, with 8-10 folds. There are 9-10 horizontal stripes in the ear and a concave inscription in the middle of the posterior margin of the ear. Dorsal hairs are drab brown. Primary hair color is pale. Ventral hair color light. Hind foot is more than half the length of the tibia, and the toes are furnished with curved hairs. The wings are long and narrow, like leather. Tail protrudes from the posterior margin of the caudate membrane.

生境 / Habitat

洞穴 Caves

▲ 地理分布 / Geographic Distribution

国内分布 / Domestic Distribution
安徽、湖北、云南、广西、福建、四川、贵州、台湾
Anhui, Hubei, Yunnan, Guangxi, Fujian, Sichuan, Guizhou, Taiwan

全球分布 / World Distribution
中国、朝鲜、韩国
China, Democratic People's Republic of Korea, Republic of Korea

生物地理界 / Biogeographic Realm
印度马来界 Indomalaya

WWF 生物群系 / WWF Biome
温带阔叶和混交林
Temperate Broadleaf & Mixed Forests

动物地理分布型 / Zoogeographic Distribution Type
Ed

分布标注 / Distribution Note
非特有种 Non-Endemic

▲ 濒危状况 / Threatened Status

中国生物多样性红色名录等级 / CB RL Category (2021)
近危 NT

IUCN 红色名录 / IUCN Red List (2021)
数据缺乏 DD

威胁因子 / Threats
采矿、采石
Mining, quarrying

▲ 法律保护地位 / Legal Protection Status

国家重点保护野生动物等级 / Category of National Key Protected Wild Animals (2021)
未列入 Not listed

"三有" 名录 / TWIESSV (2023)
未列入 Not listed

CITES 附录等级 / CITES Appendix (2023)
未列入 Not listed

迁徙物种公约附录 / CMS Appendix (2020)
II

保护行动 / Conservation Action
尚无保护行动 No conservation action so far

▲ 参考文献 / References

Jiang et al. (蒋志刚等), 2021; Burgin et al., 2020; IUCN, 2020; Liu et al. (刘少英等), 2020; Wilson and Mittermeier, 2019; You et al. (由玉岩等), 2009; Pan et al. (潘清华等), 2007; Wilson and Reeder, 2005; Wang (王应祥), 2003; Zhang (张维道), 1985; Xu et al. (徐亚君等), 1982

148 / 华北犬吻蝠

Tadarida latouchei Thomas, 1920

· La Touche's Free-tailed Bat

▲ 分类地位 / Taxonomy

翼手目 Chiroptera / 犬吻蝠科 Molossidae / 犬吻蝠属 *Tadarida*

科建立者及其文献 / Family Authority
Gervais, 1855

属建立者及其文献 / Genus Authority
Rafinesque, 1814

亚种 / Subspecies
无 None

模式标本产地 / Type Locality
中国
China, NE coast of Hopei (Hebei), Ching-wang Tao (Qinhuangdao)

▲ 其他名称 / Other Name(s)

其他中文名 / Other Chinese Name(s)
皱唇犬吻蝠

其他英文名 / Other English Name(s)
Touche's Free-tailed Bat

同物异名 / Synonym(s)
无 None

▲ 形态及生境 / Morphology and Habitat

形态特征 / Morphological Characteristics
齿式：1.1.2.3/2.1.2.3=30。头体长 67~72 mm。耳长 22~25 mm。后足长
12~13 mm。尾长 41~46 mm。颅全长 20~22 mm。耳郭前端基部相连。
背毛柔软浓密，颜色淡黑棕色，腹毛基部近白色。体侧毛尖端颜色浅。
翼膜向后延伸附着于胫骨基三分之一处。

Dental formula: 1.1.2.3/2.1.2.3=30. Head and body length 67-72 mm. Ear length 22-
25 mm. Hind foot length 12-13 mm. Tail length 41-46 mm. Greatest skull length 20-
22 mm. Ears joined at front. Fur soft and dense. Pelage color blackish brown, basal
portion of hairs nearly white, tips of ventral hairs relatively light. Plagiopatagium
attached to a basal third of tibia.

生境 / Habitat
森林、洞穴
Forest, caves

▲ 地理分布 / Geographic Distribution

国内分布 / Domestic Distribution
河北、北京、内蒙古、黑龙江、辽宁
Hebei, Beijing, Inner Mongolia, Heilongjiang, Liaoning

全球分布 / World Distribution
中国、老挝、泰国
China, Laos, Thailand

生物地理界 / Biogeographic Realm
古北界 Palearctic

WWF 生物群系 / WWF Biome
温带阔叶和混交林
Temperate Broadleaf & Mixed Forests

动物地理分布型 / Zoogeographic Distribution Type
Wd

分布标注 / Distribution Note
非特有种 Non-Endemic

▲ 濒危状况 / Threatened Status

中国生物多样性红色名录等级 / CB RL Category (2021)
近危 NT

IUCN 红色名录 / IUCN Red List (2021)
数据缺乏 DD

威胁因子 / Threats
旅游、采矿、采石
Tourism, mining, quarrying

▲ 法律保护地位 / Legal Protection Status

国家重点保护野生动物等级 / Category of National Key Protected Wild Animals (2021)
未列入 Not listed

"三有" 名录 / TWIESSV (2023)
未列入 Not listed

CITES 附录等级 / CITES Appendix (2023)
未列入 Not listed

迁徙物种公约附录 / CMS Appendix (2020)
II

保护行动 / Conservation Action
尚无保护行动 No conservation action so far

▲ 参考文献 / References

Jiang et al. (蒋志刚等), 2021; Burgin et al., 2020; IUCN, 2020; Wilson and Mittermeier, 2019; Zhou(周现召), 2012; Pan et al. (潘清华等), 2007; Wilson and Reeder, 2005; Gao et al. (高武等), 1996

149 / 小犬吻蝠

Chaerephon plicatus (Buchanan, 1800)

• Wrinkle-lipped Bat

▲ 分类地位 / Taxonomy

翼手目 Chiroptera / 犬吻蝠科 Molossidae / 小犬吻蝠属 *Chaerephon*

科建立者及其文献 / Family Authority
Gervais, 1856

属建立者及其文献 / Genus Authority
Dobson, 1874

亚种 / Subspecies
指名亚种 *C. p plicata*
西藏、甘肃、云南、贵州、广西、海南、广东和香港
Tibet, Gansu, Yunnan, Guizhou, Guangxi, Hainan, Guangdong and Hong Kong

模式标本产地 / Type Locality
印度
India, Bengal, Puttahaut (restricted to Puttahaut by G. M. Allen, 1939)

吴毅 / 供图   吴毅 / 供图

▲ 其他名称 / Other Name(s)

其他中文名 / Other Chinese Name(s)
无 None

其他英文名 / Other English Name(s)
无 None

同物异名 / Synonym(s)
无 None

▲ 形态及生境 / Morphology and Habitat

形态特征 / Morphological Characteristics
齿式：1.1.2.3/2.1.2.3=30。前臂长54~65 mm。吻部突出。上唇较下唇宽大，具纵行褶皱。耳宽阔，几呈方形，其内缘加厚。股间膜窄。尾长，有一半以上从股间膜后缘穿出，呈游离状。足趾外缘具硬毛。体毛短，上体毛暗褐色或灰黑色；下体毛色较浅，翼膜浅褐色。
Dental formula: 1.1.2.3/2.1.2.3=30. Forearm length 54-65 mm. Snout protrudes. Upper lip wider than the lower lip, with longitudinal folds. The ears are broad, almost square, with a thickened inner margin. The interfemoral membrane is narrow. Tail is long, more than half from the posterior margin of the interfemoral membrane through, is free. Toe outer margin hirsute. Short body hairs, and upper body hairs dark brown or grayish black. Lower body hair color is lighter, and pterygium light brown.

生境 / Habitat
洞穴、人造建筑、内陆岩石区域
Caves, man-made building, inland rocky area

▲ 地理分布 / Geographic Distribution

国内分布 / Domestic Distribution
云南、广东、香港、海南、贵州、甘肃、宁夏、广西
Yunnan, Guangdong, Hong Kong, Hainan, Guizhou, Gansu, Ningxia, Guangxi

全球分布 / World Distribution
柬埔寨、中国、印度、老挝、马来西亚、菲律宾、斯里兰卡、越南
Cambodia, China, India, Laos, Malaysia, Philippines, Sri Lanka, Vietnam

生物地理界 / Biogeographic Realm
印度马来界 Indomalaya

WWF 生物群系 / WWF Biome
热带和亚热带湿润阔叶林
Tropical & Subtropical Moist Broadleaf Forests

动物地理分布型 / Zoogeographic Distribution Type
Wd

分布标注 / Distribution Note
非特有种 Non-Endemic

▲ 濒危状况 / Threatened Status

中国生物多样性红色名录等级 / CB RL Category (2021)
无危 LC

IUCN 红色名录 / IUCN Red List (2021)
无危 LC

威胁因子 / Threats
无 None

▲ 法律保护地位 / Legal Protection Status

国家重点保护野生动物等级 / Category of National Key Protected Wild Animals (2021)
未列入 Not listed

"三有" 名录 / TWIESSV (2023)
未列入 Not listed

CITES 附录等级 / CITES Appendix (2023)
未列入 Not listed

迁徙物种公约附录 / CMS Appendix (2020)
未列入 Not listed

保护行动 / Conservation Action
尚无保护行动 No conservation action so far

▲ 参考文献 / References

Jiang et al. (蒋志刚等), 2021; Burgin et al., 2020; IUCN, 2020; Liu et al. (刘少英等), 2020; Wilson and Mittermeier, 2019; Smith and Xie, 2009; Simmons, 2005; Thomas, 1920

150 / 盘足蝠

Eudiscopus denticulus (Osgood, 1932)

· Disk-footed Bat

▲ 分类地位 / Taxonomy

翼手目 Chiroptera / 蝙蝠科 Vespertilionidae / 盘足蝠属 *Eudiscopus*

科建立者及其文献 / Family Authority
Gervais, 1856

属建立者及其文献 / Genus Authority
Conisbee, 1953

亚种 / Subspecies
无 None

模式标本产地 / Type Locality
老挝
Laos, Phong Saly, 4,000 ft. (1,219 m)

▲ 其他名称 / Other Name(s)

其他中文名 / Other Chinese Name(s)
无 None

其他英文名 / Other English Name(s)
无 None

同物异名 / Synonym(s)
无 None

▲ 形态及生境 / Morphology and Habitat

形态特征 / Morphological Characteristics

齿式：1.1.2.3/2.1.2.3=30。前臂长 40~50 mm。上犬齿与鼠耳蝠相似。第一上门齿比第二上门齿高。头骨宽且扁平。吻端向前向上相对延长。耳向前弯曲时能达到口鼻端，耳屏直而狭窄，末端钝。皮毛浓密柔软，背毛为红褐色。与 *Tylonycteris* 属和 *Glischropus* 属的蝙蝠相似，足腹面有明显盘状垫。

Dental formula: 1.1.2.3/2.1.2.3=30. Forearm length 40-50 mm. Upper canine is Myotis-like. First upper incisor is higher than the second. The skull is relatively broad and strikingly flattened; rostrum is elongated and relatively long and upturned anteriorly. The ears reach the tip of the muzzle when laid forward, and the tragus is straight and distinctly narrowing, ending in a blunt tip. The fur is dense and soft, reddish-brown at the dorsum. Noticeable disklike adhesive pads are present on the feet, similar to those in *Tylonycteris* and *Glischropus*.

生境 / Habitat
森林、洞穴
Forest, caves

▲ 地理分布 / Geographic Distribution

国内分布 / Domestic Distribution
云南 Yunnan

全球分布 / World Distribution
中国，老挝，缅甸，泰国，越南
China, Laos, Myanmar, Thailand, Vietnam

生物地理界 / Biogeographic Realm
印度马来界 Indomalaya

WWF 生物群系 / WWF Biome
热带和亚热带湿润阔叶林
Tropical & Subtropical Moist Broadleaf Forests

动物地理分布型 / Zoogeographic Distribution Type
Wa

分布标注 / Distribution Note
非特有种 Non-Endemic

▲ 濒危状况 / Threatened Status

中国生物多样性红色名录等级 / CB RL Category (2021)
未评定 NE

IUCN 红色名录 / IUCN Red List (2021)
无危 LC

威胁因子 / Threats
未知 Unknown

▲ 法律保护地位 / Legal Protection Status

国家重点保护野生动物等级 / Category of National Key Protected Wild Animals (2021)
未列入 Not listed

"三有" 名录 / TWIESSV (2023)
未列入 Not listed

CITES 附录等级 / CITES Appendix (2023)
未列入 Not listed

迁徙物种公约附录 / CMS Appendix (2020)
未列入 Not listed

保护行动 / Conservation Action
尚无保护行动 No conservation action so far

▲ 参考文献 / References

Yu et al., 2021; Burgin et al., 2020; IUCN, 2020

151 / 西南鼠耳蝠

Myotis altarium Thomas, 1911

· Szechwan Mouse-eared Bat

▲ 分类地位 / Taxonomy

翼手目 Chiroptera / 蝙蝠科 Vespertilionidae / 鼠耳蝠属 *Myotis*

科建立者及其文献 / Family Authority
Gray, 1821

属建立者及其文献 / Genus Authority
Kaup, 1829

亚种 / Subspecies
无 None

模式标本产地 / Type Locality
中国
China, Szechwan, Omi San (Omei Shan)

吴毅 / 供图

▲ 其他名称 / Other Name(s)

其他中文名 / Other Chinese Name(s)
峨眉鼠耳蝠、四川鼠耳蝠

其他英文名 / Other English Name(s)
South-western Mouse-eared Bat,
Szechwan Myotis

同物异名 / Synonym(s)
无 None

▲ 形态及生境 / Morphology and Habitat

形态特征 / Morphological Characteristics

齿式：2.1.3.3/3.1.3.3=38。头体长 55~60 mm。耳长 22~24 mm。前臂长 42~46 mm。后足长 11~12 mm。尾长 48~50 mm。颅全长 15~16 mm。头骨吻短，矢状嵴和人字嵴不发达。鼻额部略凹，耳壳狭长，前折超过吻端 5~7 mm，耳屏尖长。第三到第五掌骨近等长。体毛较长而柔和，背毛棕褐色，腹毛颜色近似背毛色。

Dental formula: 2.1.3.3/3.1.3.3=38. Head and body length 55-60 mm. Ear length 22-24 mm. Forearm length 42-46 mm. Hind foot length 11-12 mm. Tail length 48-50 mm. Greatest skull length 15-16 mm. Snout of the skull is short, and the sagittal crest and herrings are not well developed. Nose forehead is slightly concave. Auricle is long and narrow, the front fold exceeds the snout, and the tip of the tragus is long. 3-5 metacarpal nearly equal in length. Body hairs are long and soft, the dorsal hairs are brown, and the abdominal hairs are similar in color.

生境 / Habitat

喀斯特洞穴 Karst caves

▲ 地理分布 / Geographic Distribution

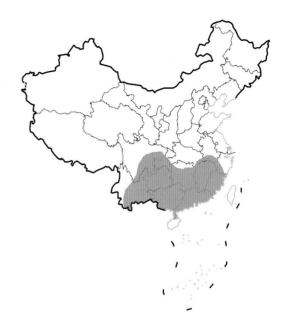

国内分布 / Domestic Distribution
安徽、江西、四川、贵州、重庆、湖北、湖南、云南、广西、广东、
福建
Anhui, Jiangxi, Sichuan, Guizhou, Chongqing, Hubei, Hunan, Yunnan, Guangxi, Guangdong, Fujian

全球分布 / World Distribution
中国、泰国
China, Thailand

生物地理界 / Biogeographic Realm
印度马来界、古北界
Indomalaya, Palearctic

WWF 生物群系 / WWF Biome
热带和亚热带湿润阔叶林
Tropical & Subtropical Moist Broadleaf Forests

动物地理分布型 / Zoogeographic Distribution Type
Wb

分布标注 / Distribution Note
非特有种 Non-Endemic

▲ 濒危状况 / Threatened Status

中国生物多样性红色名录等级 / CB RL Category (2021)
近危 NT

IUCN 红色名录 / IUCN Red List (2021)
无危 LC

威胁因子 / Threats
采矿、采石、旅游
Mining, quarrying, tourism

▲ 法律保护地位 / Legal Protection Status

国家重点保护野生动物等级 / Category of National Key Protected Wild Animals (2021)
未列入 Not listed

"三有"名录 / TWIESSV (2023)
未列入 Not listed

CITES 附录等级 / CITES Appendix (2023)
未列入 Not listed

迁徙物种公约附录 / CMS Appendix (2020)
未列入 Not listed

保护行动 / Conservation Action
尚无保护行动 No conservation action so far

▲ 参考文献 / References

Jiang et al. (蒋志刚等), 2021; Burgin et al., 2020; IUCN, 2020; Wilson and Mittermeier, 2019; Pei (裴俊峰), 2012; Fu et al. (符丹凤等), 2010; Zhang et al. (张燕均等), 2010; Pan et al. (潘清华等), 2007; Sun et al. (孙振国等), 2006; Wang (王应祥), 2003; Wang and Hu (王酉之和胡锦矗), 1999; Zhang (张荣祖), 1997; Chen et al. (陈延熹等), 1989; Chen et al. (陈延熹等), 1987; Liang and Dong (梁仁济和董永文), 1984; Xu et al. (徐亚君等), 1984

152 / 缺齿鼠耳蝠

Myotis annectans (Dobson, 1871)

• Hairy-faced Mouse-eared Bat

▲ 分类地位 / Taxonomy

翼手目 Chiroptera / 蝙蝠科 Vespertilionidae /
鼠耳蝠属 *Myotis*

科建立者及其文献 / Family Authority
Gray, 1821

属建立者及其文献 / Genus Authority
Kaup, 1829

亚种 / Subspecies
无 None

模式标本产地 / Type Locality
印度
India, NE India, Assam, Naga Hills

▲ 其他名称 / Other Name(s)

其他中文名 / Other Chinese Name(s)
无 None

其他英文名 / Other English Name(s)
Hairy-faced Bat, Intermediate Bat

同物异名 / Synonym(s)
无 None

▲ 形态及生境 / Morphology and Habitat

形态特征 / Morphological Characteristics
齿式：2.1.3.3/3.1.3.3=38。头体长 45~55 mm。耳长
14~16 mm。前臂长 45~48 mm。后足长 10 mm。
尾长 39~48 mm。颅全长 15~17 mm。耳郭较狭长，
耳屏较宽。毛发长、浓密、柔软。脸部毛发深棕色。
背部毛发深棕色。腹部被毛基部黑褐色，毛尖浅
灰黄色，下腹部有一毛尖为鲜棕色的肚斑。
Dental formula: 2.1.3.3/3.1.3.3=38. Head and body
length 45-55 mm. Ear length 14-16 mm. Forearm length
45-48 mm. Hind foot length 10 mm. Tail length 39-48
mm. Greatest skull length 15-17 mm. Ears relative long
and narrow, with relatively broad tragus. Pelage long,
dense, and soft. Face covered with dark brown hairs.
Dorsal hairs are dark brown. The bases of the abdominal
hairs are dark brown, and the tips of the hairs are light
gray yellow, and there is a bright brown spot on the lower
belly.

生境 / Habitat
喀斯特洞穴 Karst caves

▲ 地理分布 / Geographic Distribution

国内分布 / Domestic Distribution
云南 Yunnan

全球分布 / World Distribution
柬埔寨、中国、印度、老挝、泰国、越南
Cambodia, China, India, Laos, Thailand, Vietnam

生物地理界 / Biogeographic Realm
印度马来界 Indomalaya

WWF 生物群系 / WWF Biome
热带和亚热带湿润阔叶林
Tropical & Subtropical Moist Broadleaf Forests

动物地理分布型 / Zoogeographic Distribution Type
Wa

分布标注 / Distribution Note
特有种 Endemic

▲ 濒危状况 / Threatened Status

中国生物多样性红色名录等级 / CB RL Category (2021)
近危 NT

IUCN 红色名录 / IUCN Red List (2021)
无危 LC

威胁因子 / Threats
旅游 、采矿、采石
Tourism, mining, quarrying

▲ 法律保护地位 / Legal Protection Status

国家重点保护野生动物等级 / Category of National Key Protected Wild Animals (2021)
未列入 Not listed

"三有" 名录 / TWIESSV (2023)
未列入 Not listed

CITES 附录等级 / CITES Appendix (2023)
未列入 Not listed

迁徙物种公约附录 / CMS Appendix (2020)
未列入 Not listed

保护行动 / Conservation Action
尚无保护行动 No conservation action so far

▲ 参考文献 / References

Jiang et al. (蒋志刚等), 2021; Burgin et al., 2020; IUCN, 2020; Wilson and Mittermeier, 2019; Luo (罗一宁), 1987

153 / 栗鼠耳蝠

Myotis badius
Tiunov, Kruskop & Feng, 2011

• Bay Mouse-eared Bat

▲ 分类地位 / Taxonomy

翼手目 Chiroptera / 蝙蝠科 Vespertilionidae / 鼠耳蝠属 *Myotis*

科建立者及其文献 / Family Authority
Gray, 1821

属建立者及其文献 / Genus Authority
Kaup, 1829

亚种 / Subspecies
无 None

模式标本产地 / Type Locality
中国
Dashi Cave, near Shicaohe village, Shuanghe Town, Yunnan Province, China

Gabor Csorba/ 供图

▲ 其他名称 / Other Name(s)

其他中文名 / Other Chinese Name(s)
无 None

其他英文名 / Other English Name(s)
Chestnut Myotis

同物异名 / Synonym(s)
无 None

▲ 形态及生境 / Morphology and Habitat

形态特征 / Morphological Characteristics

齿式：2.1.3.3/3.1.3.3=38。头体长 39~45.3 mm。耳长 13~17 mm。前臂长 35~37 mm。后足长 6.1~12.5 mm。体重 4~6 g。颅基长 12~12.6 mm。外耳短，顶部钝圆。耳屏短，前端尖而直。背部毛发褐色或栗色，腹毛呈浅褐色。翼膜较宽，第 5 掌骨为第 3 掌骨的 4/5。尾膜始于第一趾跖骨。尾尖游离于尾膜外 0.45 mm。

Dental formula: 2.1.3.3/3.1.3.3=38. Head and body length 39-45.3 mm. Ear length 13-17 mm. Forearm length 35-37 mm. Hind Foot length 6.1-12.5 mm, Body mass 4-6 g. Condylobasal length of skull 12-12.6 mm. Outer ears short, tips blunt round. Tragus short, sharp, and straight front. Dorsal hairs brown or chestnut, abdominal hairs light brown. Pterygoid membrane is wide, and the 5th metacarpal bone is 4/5 of the length of the 3rd metacarpal bone. Caudal membrane begins at the metatarsal of the first toe. Tail tip 0.45 mm outside the caudal membrane.

生境 / Habitat

洞穴 Caves

▲ 地理分布 / Geographic Distribution

国内分布 / Domestic Distribution
云南 Yunnan

全球分布 / World Distribution
中国 China

生物地理界 / Biogeographic Realm
印度马来界 Indomalaya

WWF 生物群系 / WWF Biome
热带和亚热带湿润阔叶林
Tropical & Subtropical Moist Broadleaf Forests

动物地理分布型 / Zoogeographic Distribution Type
Sa

分布标注 / Distribution Note
特有种 Endemic

▲ 濒危状况 / Threatened Status

中国生物多样性红色名录等级 / CB RL Category (2021)
数据缺乏 DD

IUCN 红色名录 / IUCN Red List (2021)
未评定 NE

威胁因子 / Threats
未知 Unknown

▲ 法律保护地位 / Legal Protection Status

国家重点保护野生动物等级 / Category of National Key Protected Wild Animals (2021)
未列入 Not listed

"三有" 名录 / TWIESSV (2023)
未列入 Not listed

CITES 附录等级 / CITES Appendix (2023)
未列入 Not listed

迁徙物种公约附录 / CMS Appendix (2020)
未列入 Not listed

保护行动 / Conservation Action
尚无保护行动 No conservation action so far

▲ 参考文献 / References

Jiang et al. (蒋志刚等), 2021; Burgin et al., 2020; IUCN, 2020; Liu et al. (刘少英等), 2020; Wilson and Mittermeier, 2019; Tiunov et al., 2011
Smith and Xie, 2009

154 / 狭耳鼠耳蝠

Myotis blythii (Tomes, 1857)

• Lesser Mouse-eared Bat

▲ 分类地位 / Taxonomy

翼手目 Chiroptera / 蝙蝠科 Vespertilionidae / 鼠耳蝠属 *Myotis*

科建立者及其文献 / Family Authority
Gray, 1821

属建立者及其文献 / Genus Authority
Kaup, 1829

亚种 / Subspecies
新疆亚种 *M. b. omari* Thomas, 1906
新疆和内蒙古
Xinjiang and Inner Mongolia

陕西亚种 *M. b. ancilla* Thomas, 1910
陕西、北京、山西、重庆、贵州、广东和广西
Shaanxi, Beijing, Shanxi, Chongqing, Guizhou, Guangdong and Guangxi

模式标本产地 / Type Locality
印度
India, Rajasthan, Nasirabad

▲ 其他名称 / Other Name(s)

其他中文名 / Other Chinese Name(s)
尖耳鼠耳蝠

其他英文名 / Other English Name(s)
Lesser Mouse-eared Myotis

同物异名 / Synonym(s)
无 None

▲ 形态及生境 / Morphology and Habitat

形态特征 / Morphological Characteristics
齿式：2.1.3.3/3.1.3.3=38。头体长 65~89 mm。耳长 19~22 mm。前臂长 53~70 mm。后足长 11~17 mm。尾长 53~81 mm。体重 20~45 g。翅展 365~450 mm。颅全长 21~23 mm。耳长而窄，耳屏细长。背毛黑棕色或灰褐色，腹面棕灰色或灰白色。毛尖端灰色毛。翼膜附着在脚踝处。
Dental formula: 2.1.3.3/3.1.3.3=38. Head and body length 65-89 mm. Ear length 19-22 mm. Forearm length 53-70 mm. Hind foot length 11-17 mm. Tail length 53-81 mm. Body mass 20-45 g. Wingspan 365-450 mm. Greatest skull length 21-23 mm. Ears long and narrow, tragus slender. Dorsal hairs black-brown or taupe, ventral surface brown-gray or off white, with gray hair tips. Wing membrane attached at ankles.

生境 / Habitat
洞穴、人造建筑、耕地、乡村种植园、灌丛、草地
Caves, man-made building, arable land, rural garden, shrubland, grassland

▲ 地理分布 / Geographic Distribution

国内分布 / Domestic Distribution

内蒙古、新疆、陕西、北京、广东、广西、山西、重庆、贵州

Inner Mongolia, Xinjiang, Shaanxi, Beijing, Guangdong, Guangxi, Shanxi, Chongqing, Guizhou

全球分布 / World Distribution

阿富汗、阿尔巴尼亚、阿尔及利亚、安道尔、安哥拉、亚美尼亚、奥地利、阿塞拜疆、孟加拉国、不丹、波斯尼亚和黑塞哥维那、保加利亚、中国、克罗地亚、塞浦路斯、捷克、法国、格鲁吉亚、德国、直布罗陀、希腊、梵蒂冈、匈牙利、印度、伊朗、伊拉克、以色列、意大利、约旦、哈萨克斯坦、吉尔吉斯斯坦、黎巴嫩、利比亚、马其顿、摩尔多瓦、摩纳哥、黑山、摩洛哥、尼泊尔、巴基斯坦、波兰、葡萄牙、罗马尼亚、俄罗斯、圣马力诺、塞尔维亚、斯洛伐克、斯洛文尼亚、西班牙、瑞士、叙利亚、塔吉克斯坦、土耳其、土库曼斯坦、乌克兰

Afghanistan, Albania, Algeria, Andorra, Angola, Armenia, Austria, Azerbaijan, Bangladesh, Bhutan, Bosnia and Herzegovina, Bulgaria, China, Croatia, Cyprus, Czech, France, Georgia, Germany, Gibraltar, Greece, Vatican, Hungary, India, Iran, Iraq, Israel, Italy, Jordan, Kazakhstan, Kyrgyzstan, Lebanon, Libya, Macedonia, Moldova, Monaco, Montenegro, Morocco, Nepal, Pakistan, Poland, Portugal, Romania, Russia, San Marino, Serbia, Slovakia, Slovenia, Spain, Switzerland, Syria, Tajikistan, Turkey, Turkmenistan, Ukraine

生物地理界 / Biogeographic Realm

古北界 Palearctic

WWF 生物群系 / WWF Biome

热带和亚热带湿润阔叶林

Tropical & Subtropical Moist Broadleaf Forests

动物地理分布型 / Zoogeographic Distribution Type

Ud

分布标注 / Distribution Note

非特有种 Non-Endemic

▲ 濒危状况 / Threatened Status

中国生物多样性红色名录等级 / CB RL Category (2021)

近危 DD

IUCN 红色名录 / IUCN Red List (2021)

无危 LC

威胁因子 / Threats

旅游、住房及城市建设、采矿、采石、耕种

Tourism, housing and urban areas, mining, quarrying, farming

▲ 法律保护地位 / Legal Protection Status

国家重点保护野生动物等级 / Category of National Key Protected Wild Animals (2021)

未列入 Not listed

"三有"名录 / TWIESSV (2023)

未列入 Not listed

CITES 附录等级 / CITES Appendix (2023)

未列入 Not listed

迁徙物种公约附录 / CMS Appendix (2020)

未列入 Not listed

保护行动 / Conservation Action

尚无保护行动 No conservation action so far

▲ 参考文献 / References

Jiang et al. (蒋志刚等), 2021; Liu et al. (刘少英等), 2020; Wilson and Mittermeier, 2019; Zhou and Yang(周江和杨天友), 2012; Pan et al. (潘清华等), 2007; Wilson and Reeder, 2005; Wang (王应祥), 2003; Liu et al. (刘少英等), 2001; Zhang (张荣祖), 1997

155 / 远东鼠耳蝠

Myotis bombinus Thomas, 1906

• Far Eastern Mouse-eared Bat

翼手目 Chiroptera / 蝙蝠科 Vespertilionidae /
鼠耳蝠属 *Myotis*

科建立者及其文献 / Family Authority
Gray, 1821

属建立者及其文献 / Genus Authority
Kaup, 1829

亚种 / Subspecies
指名亚种 *M. b. bombinus* Thomas, 1906
黑龙江、吉林
Heilongjiang, Jilin

模式标本产地 / Type Locality
日本
Japan, Kiushiu, Miyasaki Ken, Tano

▲ 其他名称 / Other Name(s)

其他中文名 / Other Chinese Name(s)
无 None

其他英文名 / Other English Name(s)
无 None

同物异名 / Synonym(s)
无 None

▲ 形态及生境 / Morphology and Habitat

形态特征 / Morphological Characteristics
齿式：2.1.3.3/3.1.3.3=38。头体长 41~52 mm。耳
长 14~19 mm。前臂长 37~42 mm。后足长 8~12
mm。尾长 38~45 mm。颅全长 14 mm。耳长，尖窄。
耳屏狭窄和明显下弯，耳屏游离边缘上有直或微
弯的硬刚毛。毛皮柔软，背毛深色，背毛基部褐
黄色，毛尖苍白，腹部颜色苍白。翅膜附在脚趾
基部。

Dental formula: 2.1.3.3/3.1.3.3=38. Head and body
length 41-52 mm. Ear length 14-19 mm. Forearm length
37-42 mm. Hind foot length 8-12 mm. Tail length 38-45
mm. Greatest skull length 14 mm. Ears long and narrow,
with narrow tips. Tragus narrow and markedly recurved,
border of stiff bristles on the free edge of uropatagium
straight or very slightly curved. Fur soft, woolly, and dark,
with the basal part of dorsal hairs fuscous, tips paler,
ventral coloration paler. Wings attached to the base of
toes.

生境 / Habitat
森林、洞穴、人造建筑
Forest, caves, man-made building

▲ 地理分布 / Geographic Distribution

国内分布 / Domestic Distribution
吉林、黑龙江
Jilin, Heilongjiang

全球分布 / World Distribution
中国、日本、朝鲜、俄罗斯、蒙古国
China, Japan, Democratic People's Republic of Korea, Russia, Mongolia

生物地理界 / Biogeographic Realm
古北界 Palearctic

WWF 生物群系 / WWF Biome
北方森林 / 针叶林
Boreal Forests/Coniferous Forests

动物地理分布型 / Zoogeographic Distribution Type
Xa

分布标注 / Distribution Note
非特有种 Non-Endemic

▲ 濒危状况 / Threatened Status

中国生物多样性红色名录等级 / CB RL Category (2021)
近危 NT

IUCN 红色名录 / IUCN Red List (2021)
近危 NT

威胁因子 / Threats
人类活动干扰
Human disturbance

▲ 法律保护地位 / Legal Protection Status

国家重点保护野生动物等级 / Category of National Key Protected Wild Animals (2021)
未列入 Not listed

"三有"名录 / TWIESSV (2023)
未列入 Not listed

CITES 附录等级 / CITES Appendix (2023)
未列入 Not listed

迁徙物种公约附录 / CMS Appendix (2020)
未列入 Not listed

保护行动 / Conservation Action
尚无保护行动 No conservation action so far

▲ 参考文献 / References

Jiang et al. (蒋志刚等), 2021; Burgin et al., 2020; IUCN, 2020; Wilson and Mittermeier, 2019; Pan et al. (潘清华等), 2007; Wilson and Reeder, 2005; Wang (王应祥), 2003

156 / 布氏鼠耳蝠

Myotis brandtii (Eversmann, 1845)

· Brandt's Mouse-eared Bat

▲ 分类地位 / Taxonomy

翼手目 Chiroptera / 蝙蝠科 Vespertilionidae /
鼠耳蝠属 *Myotis*

科建立者及其文献 / Family Authority
Gray, 1821

属建立者及其文献 / Genus Authority
Kaup, 1829

亚种 / Subspecies
无 None

模式标本产地 / Type Locality
俄罗斯
Russia, Orenburgsk. Obl., S. Ural, Bolshoi-Ik River,
Spasskoie. Foothills of the Ural Mountains

▲ 其他名称 / Other Name(s)

其他中文名 / Other Chinese Name(s)
无 None

其他英文名 / Other English Name(s)
Brandt's Myotis

同物异名 / Synonym(s)
无 None

▲ 形态及生境 / Morphology and Habitat

形态特征 / Morphological Characteristics
齿式：2.1.3.3/3.1.3.3=38。头体长 39~51 mm。耳
长 12~17 mm。前臂长 33~39 mm。后足长 7~9
mm。尾长 32~44 mm。颅全长 14 mm。耳屏窄而尖，
约为耳长的一半。体毛长、淡棕色，有金色光泽，
浅灰色或淡黄色。翼膜附着在外趾的基部。
Dental formula: 2.1.3.3/3.1.3.3=38. Head and body
length 39-51 mm. Ear length 12-17 mm. Forearm length
33-39 mm. Hind foot length 7-9 mm. Tail length 32-44
mm. Greatest skull length 14 mm. Tragus narrow and
pointed, about half as long as the ear. Hair long, dorsum
pale brown with a golden sheen; venter paler gray, or
yellowish tinge. Wing membrane attaches at the base of
the outer toe.

生境 / Habitat
森林、人造建筑、洞穴
Forest, man-made building, caves

▲ 地理分布 / Geographic Distribution

国内分布 / Domestic Distribution
内蒙古、辽宁、吉林、黑龙江、西藏
Inner Mongolia, Liaoning, Heilongjiang, Jilin, Tibet

全球分布 / World Distribution
阿尔巴尼亚、奥地利、白俄罗斯、比利时、保加利亚、中国、克罗地亚、捷克、丹麦、爱沙尼亚、芬兰、法国、德国、希腊、匈牙利、意大利、日本、哈萨克斯坦、朝鲜、拉脱维亚、列支敦士登、立陶宛、卢森堡、摩尔多瓦、蒙古国、黑山、荷兰、挪威、波兰、罗马尼亚、俄罗斯、塞尔维亚、斯洛伐克、斯洛文尼亚、西班牙、瑞典、瑞士、土耳其、乌克兰、英国
Albania, Austria, Belarus, Belgium, Bulgaria, China, Croatia, Czech, Denmark, Estonia, Finland, France, Germany, Greece, Hungary, Italy, Japan, Kazakhstan, Democratic People's Republic of Korea, Latvia, Liechtenstein, Lithuania, Luxembourg, Moldova, Mongolia, Montenegro, Netherlands, Norway, Poland, Romania, Russia, Serbia, Slovakia, Slovenia, Spain, Sweden, Switzerland, Turkey, Ukraine, United Kingdom

生物地理界 / Biogeographic Realm
古北界 Palearctic

WWF 生物群系 / WWF Biome
温带针叶树森林
Temperate Conifer Forests

动物地理分布型 / Zoogeographic Distribution Type
Ud

分布标注 / Distribution Note
非特有种 Non-Endemic

▲ 濒危状况 / Threatened Status

中国生物多样性红色名录等级 / CB RL Category (2021)
近危 NT

IUCN 红色名录 / IUCN Red List (2021)
无危 NT

威胁因子 / Threats
未知 Unknown

▲ 法律保护地位 / Legal Protection Status

国家重点保护野生动物等级 / Category of National Key Protected Wild Animals (2021)
未列入 Not listed

"三有"名录 / TWIESSV (2023)
未列入 Not listed

CITES 附录等级 / CITES Appendix (2023)
未列入 Not listed

迁徙物种公约附录 / CMS Appendix (2020)
未列入 Not listed

保护行动 / Conservation Action
尚无保护行动 No conservation action so far

▲ 参考文献 / References

Jiang et al. (蒋志刚等), 2021; Burgin et al., 2020; IUCN, 2020; Liu et al. (刘少英等), 2020; Wilson and Mittermeier, 2019; Seim et al., 2013; Smith and Xie, 2009; Kawai et al., 2003; Wang (王应祥), 2003; Horácek et al., 2000; Yoon, 1990; Horácek and Hanák, 1984

157 / 中华鼠耳蝠

Myotis chinensis (Tomes, 1857)

· Large Mouse-eared Bat

▲ 分类地位 / Taxonomy

翼手目 Chiroptera / 蝙蝠科 Vespertilionidae / 鼠耳蝠属 *Myotis*

科建立者及其文献 / Family Authority
Gray, 1821

属建立者及其文献 / Genus Authority
Kaup, 1829

亚种 / Subspecies
华南亚种 *M. c. chinensis* (Thomas, 1857)
湖南、安徽、福建、江苏、江西、浙江、广东、广西、海南、香港
Hunan, Anhui, Fujian, Jiangsu, Jiangxi, Zhejiang, Guangdong, Guangxi, Hainan and Hong Kong

四川亚种 *M. c. luctuosus* G. Allen, 1923
四川、重庆、贵州和云南
Sichuan, Chongqing, Guizhou and Yunnan

模式标本产地 / Type Locality
中国
"Southern China"

▲ 其他名称 / Other Name(s)

其他中文名 / Other Chinese Name(s)
大鼠耳蝠

其他英文名 / Other English Name(s)
无 None

同物异名 / Synonym(s)
无 None

▲ 形态及生境 / Morphology and Habitat

形态特征 / Morphological Characteristics
齿式：2.1.3.3/3.1.3.3=38。头体长 91~97 mm。耳长 20~23 mm。前臂长 64~69 mm。后足长 16~18 mm。尾长 53~58 mm。颅全长 23 mm。耳狭长，向前折可达吻端。耳屏长而直，约为耳长之半。背部毛黑褐色，毛尖灰褐色。腹部毛暗灰色。翼膜止于外趾基部。

Dental formula: 2.1.3.3/3.1.3.3=38. Head and body length 91-97 mm. Forearm length 64-69 mm. Hind foot length 16-18 mm. Tail length 53-58 mm. Greatest skull length 23 mm. Ear length 20-23 mm. Ears narrow, long, when folded forward to reach the snout. Tragus is long and straight, about half the length of the ear. Dorsal hairs black brown, and hair tips grayish-brown. Ventral hairs are dark gray. Pterygoid membrane attaches at the toe base.

生境 / Habitat
喀斯特洞穴 Karst caves

▲ 地理分布 / Geographic Distribution

国内分布 / Domestic Distribution
内蒙古、重庆、贵州、四川、云南，湖南、安徽、福建、江苏、江西、
浙江、广东、广西、海南、香港
Inner Mongolia, Chongqing, Guizhou, Sichuan, Yunnan, Hunan, Anhui, Fujian,
Jiangsu, Jiangxi, Zhejiang, Guangdong, Guangxi, Hainan, Hong Kong

全球分布 / World Distribution
中国、缅甸、泰国、越南
China, Myanmar, Thailand, Vietnam

生物地理界 / Biogeographic Realm
印度马来界、古北界
Indomalaya, Palearctic

WWF 生物群系 / WWF Biome
热带和亚热带湿润阔叶林
Tropical & Subtropical Moist Broadleaf Forests

动物地理分布型 / Zoogeographic Distribution Type
Wd

分布标注 / Distribution Note
非特有种 Non-Endemic

▲ 濒危状况 / Threatened Status

中国生物多样性红色名录等级 / CB RL Category (2021)
近危 NT

IUCN 红色名录 / IUCN Red List (2021)
无危 NT

威胁因子 / Threats
未知 Unknown

▲ 法律保护地位 / Legal Protection Status

国家重点保护野生动物等级 / Category of National Key Protected Wild Animals (2021)
未列入 Not listed

"三有" 名录 / TWIESSV (2023)
未列入 Not listed

CITES 附录等级 / CITES Appendix (2023)
未列入 Not listed

迁徙物种公约附录 / CMS Appendix (2020)
未列入 Not listed

保护行动 / Conservation Action
尚无保护行动 No conservation action so far

▲ 参考文献 / References

Jiang et al. (蒋志刚等), 2021; Burgin et al., 2020; IUCN, 2020; Liu et al. (刘少英等), 2020; Wilson and Mittermeier, 2019; Liu et al. (刘昊等), 2012; Tian and Jin, 2012; Ma et al., 2008; Wilson and Reeder, 2005; Ma et al., 2004; Wang (王应祥), 2003; Zhang (张荣祖), 1997; Zhang (张维道), 1984

158 / 沼泽鼠耳蝠

Myotis dasycneme (Boie, 1825)

· Pond Mouse-eared Bat

翼手目 Chiroptera / 蝙蝠科 Vespertilionidae / 鼠耳蝠属 *Myotis*

科建立者及其文献 / Family Authority
Gray, 1821

属建立者及其文献 / Genus Authority
Kaup, 1829

亚种 / Subspecies
无 None

模式标本产地 / Type Locality
丹麦
Denmark, Jutland, Dagbieg (near Wiborg)

▲ 其他名称 / Other Name(s)

其他中文名 / Other Chinese Name(s)
无 None

其他英文名 / Other English Name(s)
Pond Bat

同物异名 / Synonym(s)
无 None

▲ 形态及生境 / Morphology and Habitat

形态特征 / Morphological Characteristics

齿式：2.1.3.3/3.1.3.3=38。头体长 57~67 mm。耳长 17~18 mm。前臂长 43~49 mm。后足长 11~12 mm。尾长 46~51 mm。颅全长 17.2~18.1 mm。耳屏短，长度不及耳长的一半。背色被毛黄褐色。腹部被毛石板黑色。

Dental formula: 2.1.3.3/3.1.3.3=38. Head and body length 57-67 mm. Ear length 17-18 mm. Forearm length 43-49 mm. Hind foot length 11-12 mm. Tail length 46-51 mm. Greatest skull length 17.2-18.1 mm. Tragus is short, less than half length of the ear. Dorsal hairs yellowish brown. Ventral hairs slate black.

生境 / Habitat
溪流边、淡水湖、洞穴、人造建筑
Streams, fresh water lake, caves, man-made structure

▲ 地理分布 / Geographic Distribution

国内分布 / Domestic Distribution
山东 Shandong

全球分布 / World Distribution
白俄罗斯、比利时、保加利亚、中国、捷克、丹麦、爱沙尼亚、芬兰、法国、德国、匈牙利、哈萨克斯坦、拉脱维亚、立陶宛、卢森堡、摩尔多瓦、黑山、荷兰、波兰、罗马尼亚、俄罗斯、塞尔维亚、斯洛伐克、瑞典、乌克兰

Belarus, Belgium, Bulgaria, China, Czech, Denmark, Estonia, Finland, France, Germany, Hungary, Kazakhstan, Latvia, Lithuania, Luxembourg, Moldova, Montenegro, Netherlands, Poland, Romania, Russia, Serbia, Slovakia, Sweden, Ukraine

生物地理界 / Biogeographic Realm
古北界 Palearctic

WWF 生物群系 / WWF Biome
温带阔叶和混交林
Temperate Broadleaf & Mixed Forests

动物地理分布型 / Zoogeographic Distribution Type
Ud

分布标注 / Distribution Note
非特有种 Non-Endemic

▲ 濒危状况 / Threatened Status

中国生物多样性红色名录等级 / CB RL Category (2021)
无危 LC

IUCN 红色名录 / IUCN Red List (2021)
近危 NT

威胁因子 / Threats
未知 Unknown

▲ 法律保护地位 / Legal Protection Status

国家重点保护野生动物等级 / Category of National Key Protected Wild Animals (2021)
未列入 Not listed

"三有"名录 / TWIESSV (2023)
未列入 Not listed

CITES 附录等级 / CITES Appendix (2023)
未列入 Not listed

迁徙物种公约附录 / CMS Appendix (2020)
未列入 Not listed

保护行动 / Conservation Action
尚无保护行动 No conservation action so far

▲ 参考文献 / References

Jiang et al. (蒋志刚等), 2021; Burgin et al., 2020; IUCN, 2020; Wilson and Mittermeier, 2019

159 / 大卫鼠耳蝠

Myotis davidii Peters, 1869

· David's Mouse-eared Bat

▲ 分类地位 / Taxonomy

翼手目 Chiroptera / 蝙蝠科 Vespertilionidae / 鼠耳蝠属 *Myotis*

科建立者及其文献 / Family Authority
Gray, 1821

属建立者及其文献 / Genus Authority
Kaup, 1829

亚种 / Subspecies
无 None

模式标本产地 / Type Locality
中国
China, Hopei, Peiping

▲ 其他名称 / Other Name(s)

其他中文名 / Other Chinese Name(s)
无 None

其他英文名 / Other English Name(s)
无 None

同物异名 / Synonym(s)
无 None

▲ 形态及生境 / Morphology and Habitat

形态特征 / Morphological Characteristics
齿式：2.1.3.3/3.1.3.3=38。头体长 41~44 mm。耳长 12~15 mm。前臂长 31~36 mm。后足长 7~9 mm。尾长 30~43 mm。颅全长 11.8~14.3 mm。背毛深棕色，毛梢色稍浅。腹毛颜色与背毛相同，毛梢灰色。后足长超过胫骨长的一半。翅膜附着于足趾基部。距长，其长度略大于至尾尖距离的一半。
Dental formula: 2.1.3.3/3.1.3.3=38. Head and body length 41-44 mm. Ear length 12-15 mm. Forearm length 31-36 mm. Hind foot length 7-9 mm. Tail length 30-43 mm. Greatest skull length 11.8-14.3 mm. Dorsal hairs dark brown, hair tips a little paler. Ventral hairs color same as that of dorsal hairs, but tips grayish. Length of hind foot is more than half of the tibia length. Wing membrane attaches to the basal part of the toe. Calcar long, slightly half the length of distance to tail tip.

生境 / Habitat
未知 Unknown

▲ 地理分布 / Geographic Distribution

国内分布 / Domestic Distribution
北京、河北、山西、内蒙古、甘肃、江西、贵州、海南、广东、广西、重庆、陕西、云南、湖南、安徽、江苏、浙江、香港
Beijing, Hebei, Shanxi, Inner Mongolia, Gansu, Jiangxi, Guizhou, Hainan, Guangdong, Guangxi, Chongqing, Shaanxi, Yunnan, Hunan, Anhui, Jiangsu, Zhejiang, HongKong

全球分布 / World Distribution
中国 China

生物地理界 / Biogeographic Realm
古北界 Palearctic

WWF 生物群系 / WWF Biome
温带草原、热带稀树草原和灌木地
Temperate Grasslands, Savannas & Shrublands

动物地理分布型 / Zoogeographic Distribution Type
Ue

分布标注 / Distribution Note
特有种 Endemic

▲ 濒危状况 / Threatened Status

中国生物多样性红色名录等级 / CB RL Category (2021)
无危 LC

IUCN 红色名录 / IUCN Red List (2021)
无危 LC

威胁因子 / Threats
未知 Unknown

▲ 法律保护地位 / Legal Protection Status

国家重点保护野生动物等级 / Category of National Key Protected Wild Animals (2021)
未列入 Not listed

"三有"名录 / TWIESSV (2023)
未列入 Not listed

CITES 附录等级 / CITES Appendix (2023)
未列入 Not listed

迁徙物种公约附录 / CMS Appendix (2020)
未列入 Not listed

保护行动 / Conservation Action
尚无保护行动 No conservation action so far

▲ 参考文献 / References

Jiang et al. (蒋志刚等), 2021; Burgin et al., 2020; IUCN, 2020; Liu et al. (刘少英等), 2020; Wilson and Mittermeier, 2019; Ren et al. (任锐君等), 2017; You (由玉岩), 2013; Seim et al., 2013; Jiang and Feng (江廷磊和冯江), 2011; Yin et al. (尹皓等), 2011; You and Du (由玉岩和杜江峰), 2011; You et al., 2010; Zhang et al., 2009; Smith et al., 2009; Pan et al. (潘清华等), 2007; Wilson and Reeder, 2005; Wang (王应祥), 2003; Zhang (张荣祖), 1997; Phillips and Wilson, 1968

160 / 毛腿鼠耳蝠

Myotis fimbriatus (Peters, 1871)

• Fringed Long-footed Mouse-eared Bat

▲ 分类地位 / Taxonomy

翼手目 Chiroptera / 蝙蝠科 Vespertilionidae / 鼠耳蝠属 *Myotis*

科建立者及其文献 / Family Authority
Gray, 1821

属建立者及其文献 / Genus Authority
Kaup, 1829

亚种 / Subspecies
无 None

模式标本产地 / Type Locality
中国
China, Fujian, Amoy

张礼标 / 供图

▲ 其他名称 / Other Name(s)

其他中文名 / Other Chinese Name(s)
无 None

其他英文名 / Other English Name(s)
Hairy-legged Myotis

同物异名 / Synonym(s)
无 None

▲ 形态及生境 / Morphology and Habitat

形态特征 / Morphological Characteristics

齿式：2.1.3.3/3.1.3.3=38。头体长 42~52 mm。耳长 14~16 mm。前臂长
37~40 mm。后足长 8~10 mm。尾长 37~48 mm。颅全长 15 mm。上颌
第三枚前白齿位于齿列中或偏向内侧。身体覆毛短而密，背毛黑色、棕
色及灰色，腹毛灰褐色，显灰白色调。尾基白。后足长超过胫骨长之半。
胫外缘具毛。翼膜附着于踝部。

Dental formula: 2.1.3.3/3.1.3.3=38. Head and body length 42-52 mm. Ear length 14-
16 mm. Forearm length 37-40 mm. Hind foot length 8-10 mm. Tail length 37-48 mm.
Greatest skull length 15 mm. Body coat is short and dense. Dorsal hairs are black,
brown, and gray. Hairs on the abdominal part are grayish brown, showing a grayish-
white tone. Tail white. Hind foot is more than half the length of the tibia. Tibia outer
margin covered with hair. Pterygoid membrane is attached to the ankle.

生境 / Habitat

喀斯特洞穴 Karst caves

▲ 地理分布 / Geographic Distribution

国内分布 / Domestic Distribution
江苏、浙江、安徽、福建、江西、四川、贵州、云南、香港
Jiangsu, Zhejiang, Anhui, Fujian, Jiangxi, Sichuan, Guizhou, Yunnan, Hong Kong

全球分布 / World Distribution
中国 China

生物地理界 / Biogeographic Realm
古北界 Palearctic

WWF 生物群系 / WWF Biome
热带和亚热带湿润阔叶林
Tropical & Subtropical Moist Broadleaf Forests

动物地理分布型 / Zoogeographic Distribution Type
Ue

分布标注 / Distribution Note
特有种 Endemic

▲ 濒危状况 / Threatened Status

中国生物多样性红色名录等级 / CB RL Category (2021)
近危 NT

IUCN 红色名录 / IUCN Red List (2021)
无危 LC

威胁因子 / Threats
采石、旅游
Quarrying, tourism

▲ 法律保护地位 / Legal Protection Status

国家重点保护野生动物等级 / Category of National Key Protected Wild Animals (2021)
未列入 Not listed

"三有" 名录 / TWIESSV (2023)
未列入 Not listed

CITES 附录等级 / CITES Appendix (2023)
未列入 Not listed

迁徙物种公约附录 / CMS Appendix (2020)
未列入 Not listed

保护行动 / Conservation Action
尚无保护行动 No conservation action so far

▲ 参考文献 / References

Jiang et al. (蒋志刚等), 2021; Burgin et al., 2020; IUCN, 2020; Wilson and Mittermeier, 2019; Wang et al. (王会等), 2009; Pan et al. (潘清华等), 2007; Wu and Harada, 2006; Wilson and Reeder, 2005; Liu et al. (刘颖等), 2003; Wang (王应祥), 2003; Wang and Hu (王酉之和胡锦矗), 1999; Dong et al. (董聿茂等), 1989; Chen et al. (陈延熹等), 1987; Xu et al. (徐亚君等), 1984; Zhang et al. (张维道等), 1983

161 / 金黄鼠耳蝠

Myotis formosus (Hodgson, 1835)

· Hodgson's Bat

▲ 分类地位 / Taxonomy

翼手目 Chiroptera / 蝙蝠科 Vespertilionidae / 鼠耳蝠属 *Myotis*

科建立者及其文献 / Family Authority
Gray, 1821

属建立者及其文献 / Genus Authority
Kaup, 1829

亚种 / Subspecies
指名亚种 *M. f. formosus* G. Allen, 1923
江西、西藏、湖南
Jiangxi, Tibet, Hunan
台湾亚种 *M. f. formosus* Ognev, 1927
台湾 Taiwan

模式标本产地 / Type Locality
尼泊尔
Nepal

周政翰 / 供图

周政翰 / 供图

▲ 其他名称 / Other Name(s)

其他中文名 / Other Chinese Name(s)
无 None

其他英文名 / Other English Name(s)
Hodgson's Myotis

同物异名 / Synonym(s)
无 None

▲ 形态及生境 / Morphology and Habitat

形态特征 / Morphological Characteristics
齿式：2.1.3.3/3.1.3.3=38。头体长 50~57 mm。耳长 16~17 mm。前臂长 45~50 mm。后足长 10~12 mm。尾长 43~52 mm。颅全长 17.4~19.5 mm。吻端突出。耳长大于耳宽，耳端较尖，耳屏长而窄。全身毛发金黄色，腹面毛颜色浅。翼膜呈褐色或深紫色，翼膜向两侧延伸，连接到趾基部。

Dental formula: 2.1.3.3/3.1.3.3=38. Head and body length 50-57 mm. Ear length 16-17 mm. Forearm length 45-50 mm. Hind foot length 10-12 mm. Tail length 43-52 mm. Greatest skull length 17.4-19.5 mm. Snout protrudes. Ear length is larger than the ear width. Tips of the ears pointed, tragus long and narrow. Dorsal hairs are golden yellow, abdominal hairs are light in color. Pterygium brown or dark purple, extends laterally to the base of the toe.

生境 / Habitat
森林、竹林、耕地、牧场
Forest, bamboo forest, arable land, pastureland

▲ 地理分布 / Geographic Distribution

国内分布 / Domestic Distribution
江西、西藏、湖南、台湾
Jiangxi, Tibet, Hunan, Taiwan

全球分布 / World Distribution
阿富汗、巴基斯坦、中国、印度、尼泊尔、不丹、孟加拉国
Afghanistan, Pakistan, China, India, Nepal, Bhutan, Bangladesh

生物地理界 / Biogeographic Realm
印度马来界、古北界
Indomalaya, Palearctic

WWF 生物群系 / WWF Biome
热带和亚热带湿润阔叶林
Tropical & Subtropical Moist Broadleaf Forests

动物地理分布型 / Zoogeographic Distribution Type
Uh

分布标注 / Distribution Note
非特有种 Non-Endemic

▲ 濒危状况 / Threatened Status

中国生物多样性红色名录等级 / CB RL Category (2021)
易危 VU

IUCN 红色名录 / IUCN Red List (2021)
近危 NT

威胁因子 / Threats
基础建设及交通、堤坝建设，生境退化、丧失
Infrastructure and transportation, dams and water management use, habitat degradation loss

▲ 法律保护地位 / Legal Protection Status

国家重点保护野生动物等级 / Category of National Key Protected Wild Animals (2021)
未列入 Not listed

"三有"名录 / TWIESSV (2023)
未列入 Not listed

CITES 附录等级 / CITES Appendix (2023)
未列入 Not listed

迁徙物种公约附录 / CMS Appendix (2020)
未列入 Not listed

保护行动 / Conservation Action
尚无保护行动 No conservation action so far

▲ 参考文献 / References

Jiang et al. (蒋志刚等), 2021; Burgin et al., 2020; IUCN, 2020; Liu et al. (刘少英等), 2020; Wilson and Mittermeier, 2019; Feng et al. (冯磊等), 2019; Jiang et al., 2010; Pan et al. (潘清华等), 2007; Wilson and Reeder, 2005; Wang (王应祥), 2003; Zhang (张荣祖), 1997

162 / 长尾鼠耳蝠

Myotis frater G. M. Allen, 1923

· Fraternal Mouse-eared Bat

▲ 分类地位 / Taxonomy

翼手目 Chiroptera / 蝙蝠科 Vespertilionidae / 鼠耳蝠属 *Myotis*

科建立者及其文献 / Family Authority
Gray, 1821

属建立者及其文献 / Genus Authority
Kaup, 1829

亚种 / Subspecies
无 None

模式标本产地 / Type Locality
中国
China, SE China, Fukien (Fujian), Yenping

Gabor Csorba/ 供图

▲ 其他名称 / Other Name(s)

其他中文名 / Other Chinese Name(s)
长腿鼠耳蝠、长胫鼠耳蝠

其他英文名 / Other English Name(s)
Long-tailed Whiskered Bat

同物异名 / Synonym(s)
无 None

▲ 形态及生境 / Morphology and Habitat

形态特征 / Morphological Characteristics
齿式：2.1.3.3/3.1.3.3=38。头体长 43~57 mm。耳长 11~14 mm。前臂长 36~42 mm。后足长 7~12 mm。尾长 38~47 mm。颅全长 13.5 mm。吻鼻背中央具浅的纵沟。耳短，前折不达吻端。胫骨长为后足长的 2 倍。背毛长而蓬松，毛基暗褐，毛尖浅沙黄色带光泽。腹毛较背毛短，毛基3/4 黑褐色，毛尖浅褐色。翼膜起于外趾的基部。翼、股间膜颜色较背毛色略深。胫部及股间膜背腹面均无毛。

Dental formula: 2.1.3.3/3.1.3.3=38. Head and body length 43-57 mm. Ear length 11-14 mm. Forearm length 36-42 mm. Hind foot length 7-12 mm. Tail length 38-47 mm. Greatest skull length 13.5 mm. Rostral dorsum has a shallow longitudinal groove in the center. Ears short, folded forward not to reach the snout. Tibia is twice as long as the hind foot. Dorsal hairs long and fluffy, the hair bases are dark brown, the hair tips are light sand yellow with luster. Abdominal hairs are shorter than the dorsal hairs, hair bases 3/4 black brown, hair tips light brown. Pterygoid membrane arises from the base of the outer toe. Color of the wing and thigh membrane is slightly darker than the color of the back coat. Tibia and interfemoral membrane dorsal ventral surface hairless.

生境 / Habitat
森林、岩洞
Forest, caves

▲ 地理分布 / Geographic Distribution

国内分布 / Domestic Distribution
黑龙江、吉林、内蒙古、安徽、江西、福建、四川、台湾
Heilongjiang, Jilin, Inner Mongolia, Anhui, Jiangxi, Fujian, Sichuan, Taiwan

全球分布 / World Distribution
中国、日本、朝鲜、韩国、俄罗斯
China, Japan, Democratic People's Republic of Korea, Republic of Korea, Russia

生物地理界 / Biogeographic Realm
古北界 Palearctic

WWF 生物群系 / WWF Biome
热带和亚热带湿润阔叶林
Tropical & Subtropical Moist Broadleaf Forests

动物地理分布型 / Zoogeographic Distribution Type
Ud

分布标注 / Distribution Note
非特有种 Non-Endemic

▲ 濒危状况 / Threatened Status

中国生物多样性红色名录等级 / CB RL Category (2021)
数据缺乏 DD

IUCN 红色名录 / IUCN Red List (2021)
数据缺乏 DD

威胁因子 / Threats
未知 Unknown

▲ 法律保护地位 / Legal Protection Status

国家重点保护野生动物等级 / Category of National Key Protected Wild Animals (2021)
未列入 Not listed

"三有" 名录 / TWIESSV (2023)
未列入 Not listed

CITES 附录等级 / CITES Appendix (2023)
未列入 Not listed

迁徙物种公约附录 / CMS Appendix (2020)
未列入 Not listed

保护行动 / Conservation Action
尚无保护行动 No conservation action so far

▲ 参考文献 / References

Jiang et al. (蒋志刚等), 2021; Burgin et al., 2020; IUCN, 2020; Liu et al. (刘少英等), 2020; Wilson and Mittermeier, 2019; Ruedi et al., 2015; Liu (刘伟), 2012; Zhen et al. (张桢珍等), 2008; Pan et al. (潘清华等), 2007; Luo and Gao (罗键和高红英), 2006; Peng and Zhong (彭基泰和钟祥清), 2005; Wilson and Reeder, 2005; Wang (王应祥), 2003; Chen et al. (陈昌笃等), 2000; Zhang et al. (张树义等), 2000; Zhang (张荣祖), 1997; Dong et al. (董聿茂等), 1989; Liang and Dong (梁仁济和董永文), 1984

163 / 小巨足鼠耳蝠

Myotis hasseltii (Temminck, 1840)

- Lesser Large-footed Mouse-eared Bat

张礼标 / 供图

▲ 分类地位 / Taxonomy

翼手目 Chiroptera / 蝙蝠科 Vespertilionidae / 鼠耳蝠属 *Myotis*

科建立者及其文献 / Family Authority
Gray, 1821

属建立者及其文献 / Genus Authority
Kaup, 1829

亚种 / Subspecies
无 None

模式标本产地 / Type Locality
印度尼西亚
Indonesia, Java, Bantam

▲ 其他名称 / Other Name(s)

其他中文名 / Other Chinese Name(s)
无 None

其他英文名 / Other English Name(s)
Lesser Hasselt's Large-footed Myotis

同物异名 / Synonym(s)
无 None

▲ 形态及生境 / Morphology and Habitat

形态特征 / Morphological Characteristics
齿式：2.1.3.3/3.1.3.3=38。头体长 53~58 mm。前臂长 38~43 mm。头骨矢状嵴缺失，人字嵴发达。背毛灰褐泛白，毛尖灰色，毛基暗褐色至黑色。腹毛颜色稍浅，毛尖灰白，毛基暗黑。翼膜及股间膜黑褐色，无毛。股间膜宽大，后端锐尖。翼膜止于踝部。后足较大，其长超过胫长之半，趾背具稀疏针毛。
Dental formula: 2.1.3.3/3.1.3.3=38. Head and body length 53-58 mm. Forearm length 38-43 mm. Sagittal crest of the skull missing and the herringbone crest are developed. Dorsal hairs grey and brown, with grey hair tips and dark brown to black hair bases. Ventral hair color slightly light, with gray hair tips and dark hair bases. Pterygoid and interfemoral membranes are dark brown, hairless. Interfemoral membrane wide and the posterior end acute. Pterygoid membrane rests at the ankle. Hind foot larger, its length exceeds half of the shin length, back of the claw has sparse bristles.

生境 / Habitat
热带亚热带干旱森林、红树林、竹林、人造建筑、海岸、喀斯特地貌、洞穴
Tropical subtropical dry forest, mangrove, bamboo forest, man-made building, coast, karst, caves

▲ 地理分布 / Geographic Distribution

国内分布 / Domestic Distribution
云南 Yunnan

全球分布 / World Distribution
中国、柬埔寨、印度、印度尼西亚、马来西亚、缅甸、斯里兰卡、泰国、越南
China, Cambodia, India, Indonesia, Malaysia, Myanmar, Sri Lanka, Thailand, Vietnam

生物地理界 / Biogeographic Realm
印度马来界 Indomalaya

WWF 生物群系 / WWF Biome
热带和亚热带湿润阔叶林
Tropical & Subtropical Moist Broadleaf Forests

动物地理分布型 / Zoogeographic Distribution Type
Wb

分布标注 / Distribution Note
非特有种 Non-Endemic

▲ 濒危状况 / Threatened Status

中国生物多样性红色名录等级 / CB RL Category (2021)
易危 VU

IUCN 红色名录 / IUCN Red List (2021)
无危 LC

威胁因子 / Threats
森林砍伐、耕种
Logging, farming

▲ 法律保护地位 / Legal Protection Status

国家重点保护野生动物等级 / Category of National Key Protected Wild Animals (2021)
未列入 Not listed

"三有"名录 / TWIESSV (2023)
未列入 Not listed

CITES 附录等级 / CITES Appendix (2023)
未列入 Not listed

迁徙物种公约附录 / CMS Appendix (2020)
未列入 Not listed

保护行动 / Conservation Action
尚无保护行动 No conservation action so far

▲ 参考文献 / References

Jiang et al. (蒋志刚等), 2021; Burgin et al., 2020; IUCN, 2020; Zhang et al. (张礼标等), 2004; Pan et al. (潘清华等), 2007; Wilson and Reeder, 2005

164 / 霍氏鼠耳蝠

Myotis horsfieldii Temminck, 1840

· Horsfield's Mouse-eared Bat

▲ 分类地位 / Taxonomy

翼手目 Chiroptera / 蝙蝠科 Vespertilionidae / 鼠耳蝠属 *Myotis*

科建立者及其文献 / Family Authority
Gray, 1821

属建立者及其文献 / Genus Authority
Kaup, 1829

亚种 / Subspecies
泰国亚种 *M. h. deignani* Shamei, 1942
广东、海南、江西和香港
Guangdong, Hainan, Jiangxi and Hong Kong

模式标本产地 / Type Locality
印度尼西亚
Indonesia, Java, Mount Gede

张礼标 / 供图

▲ 其他名称 / Other Name(s)

其他中文名 / Other Chinese Name(s)
无 None

其他英文名 / Other English Name(s)
Common Asiatic Myotis, Horsfield's Bat,
Lesser Large-tooth Bat

同物异名 / Synonym(s)
无 None

▲ 形态及生境 / Morphology and Habitat

形态特征 / Morphological Characteristics

齿式：2.1.3.3/3.1.3.3=38。头体长 49~59 mm。耳长 13~15 mm。前臂长 36~42 mm。后足长 7~11 mm。尾长 34~42 mm。颅全长 14.8 mm。耳郭突出，耳屏短。背毛黑褐色，腹毛深褐色，近尾部有浅灰色毛尖。后足长超过胫骨之半，胫骨外缘无毛。翼膜附着于外跖部。

Dental formula: 2.1.3.3/3.1.3.3=38. Head and body length 49-59 mm. Ear length 13-15 mm. Forearm length 36-42 mm. Hind foot length 7-11 mm. Tail length 34-42 mm. Greatest skull length 14.8 mm. Auricle is prominent, and the tragus is short. Dorsal hairs dark brown, abdominal hairs deep brown, hairs near the tail base with light gray tips. Hind foot longer than half tibia, tibia outer margin glabrous. Pterygoid membrane is attached to the outer plantar.

生境 / Habitat
森林、人造建筑、种植园、溪流边
Forest, man-made building, plantations, streams

▲ 地理分布 / Geographic Distribution

国内分布 / Domestic Distribution
香港、广东、海南、江西
Hong Kong, Guangdong, Hainan, Jiangxi

全球分布 / World Distribution
文莱、柬埔寨、中国、印度、印度尼西亚、老挝、马来西亚、缅甸、
菲律宾、泰国、越南
Brunei, Cambodia, China, India, Indonesia, Laos, Malaysia, Myanmar, Philippines,
Thailand, Vietnam

生物地理界 / Biogeographic Realm
印度马来界、大洋洲界
Indomalaya, Oceanian

WWF 生物群系 / WWF Biome
热带和亚热带湿润阔叶林
Tropical & Subtropical Moist Broadleaf Forests

动物地理分布型 / Zoogeographic Distribution Type
Wa

分布标注 / Distribution Note
非特有种 Non-Endemic

▲ 濒危状况 / Threatened Status

中国生物多样性红色名录等级 / CB RL Category (2021)
无危 LC

IUCN 红色名录 / IUCN Red List (2021)
无危 LC

威胁因子 / Threats
无 None

▲ 法律保护地位 / Legal Protection Status

国家重点保护野生动物等级 / Category of National Key Protected Wild Animals (2021)
未列入 Not listed

"三有"名录 / TWIESSV (2023)
未列入 Not listed

CITES 附录等级 / CITES Appendix (2023)
未列入 Not listed

迁徙物种公约附录 / CMS Appendix (2020)
未列入 Not listed

保护行动 / Conservation Action
尚无保护行动 No conservation action so far

▲ 参考文献 / References

Jiang et al. (蒋志刚等), 2021; Burgin et al., 2020; IUCN, 2020; Liu et al. (刘少英等), 2020; Wilson and Mittermeier, 2019; Zhang et al., 2009;
Koopman, 1993; Zhu (朱斌良), 2008; Pan et al. (潘清华等), 2007; Wilson and Reeder, 2005; Wang (王应祥), 2003

165 / 印支鼠耳蝠

Myotis indochinensis
Son, Görföl, Francis, Motokawa, Estók, Endo,
Thong, Dang, Oshida and Csorba, 2013

· Indochina Mouse-eared Bat

▲ 分类地位 / Taxonomy

翼手目 Chiroptera / 蝙蝠科 Vespertilionidae / 鼠耳蝠属 *Myotis*

科建立者及其文献 / Family Authority
Gray, 1821

属建立者及其文献 / Genus Authority
Kaup, 1829

亚种 / Subspecies
无 None

模式标本产地 / Type Locality
越南
16°06′N, 107°19′E, Loa village, Dong Son commune, A Luoi District, Thua Thien-Hue
Province, Vietnam, 970 m a. s. l

王晓云 / 供图

▲ 其他名称 / Other Name(s)

其他中文名 / Other Chinese Name(s)
中印鼠耳蝠

其他英文名 / Other English Name(s)
无 None

同物异名 / Synonym(s)
无 None

▲ 形态及生境 / Morphology and Habitat

形态特征 / Morphological Characteristics
齿式：2.1.3.3/3.1.3.3=38。头体长 56~62 mm。前臂长 43.7~46.6 mm。
胫骨长 19~20 mm。后足长 8~10 mm。尾长 42~48 mm。头骨粗壮，
吻突宽短，覆盖稀疏毛发。耳长中等。耳屏长达耳长之半。后足长不
及胫长之半。背毛较腹毛色深，毛基偏黑，毛尖色浅。
Dental formula: 2.1.3.3/3.1.3.3=38. Head and body length 56-62 mm. Forearm length
43.7-46.6 mm. Tibia length 19-20 mm. Hind foot length 8-10 mm. Tail length 42-
48 mm. Skull stout. Snout short and wide, and it is covered with sparse hairs. Ears of
medium length. Tragus is half the length of the ear. Hind foot is less than half the length
of the tibia. Dorsal hairs are darker than the abdominal hairs, the hair bases are black,
and the hair tips are pale.

生境 / Habitat
热带和亚热带湿润山地森林
Tropical and subtropical moist montane forest

▲ 地理分布 / Geographic Distribution

国内分布 / Domestic Distribution
江西、广东
Jiangxi, Guangdong

全球分布 / World Distribution
中国、老挝、越南
China, Laos, Vietnam

生物地理界 / Biogeographic Realm
印度马来界 Indomalaya

WWF 生物群系 / WWF Biome
热带和亚热带湿润阔叶林
Tropical & Subtropical Moist Broadleaf Forests

动物地理分布型 / Zoogeographic Distribution Type
Wa

分布标注 / Distribution Note
非特有种 Non-Endemic

▲ 濒危状况 / Threatened Status

中国生物多样性红色名录等级 / CB RL Category (2021)
无危 LC

IUCN 红色名录 / IUCN Red List (2021)
无危 LC

威胁因子 / Threats
无 None

▲ 法律保护地位 / Legal Protection Status

国家重点保护野生动物等级 / Category of National Key Protected Wild Animals (2021)
未列入 Not listed

"三有" 名录 / TWIESSV (2023)
未列入 Not listed

CITES 附录等级 / CITES Appendix (2023)
未列入 Not listed

迁徙物种公约附录 / CMS Appendix (2020)
未列入 Not listed

保护行动 / Conservation Action
尚无保护行动 No conservation action so far

▲ 参考文献 / References

Jiang et al. (蒋志刚等), 2021; Burgin et al., 2020; IUCN, 2020; Liu et al. (刘少英等), 2020; Wilson and Mittermeier, 2019; Wang et al., 2017

166 / 伊氏鼠耳蝠

Myotis ikonnikovi Ognev, 1912

• Ikonnikov's Mouse-eared Bat

翼手目 Chiroptera / 蝙蝠科 Vespertilionidae / 鼠耳蝠属 *Myotis*

科建立者及其文献 / Family Authority
Gray, 1821

属建立者及其文献 / Genus Authority
Kaup, 1829

亚种 / Subspecies
无 None

模式标本产地 / Type Locality
俄罗斯
Russia, Primorsk. Krai (Ussuri Region), Dalnerechen Dist., Euseevka

▲ 其他名称 / Other Name(s)

其他中文名 / Other Chinese Name(s)
中印鼠耳蝠

其他英文名 / Other English Name(s)
Ikonnikov's Bat, Ikonnikov's Whiskered
Bat

同物异名 / Synonym(s)
无 None

▲ 形态及生境 / Morphology and Habitat

形态特征 / Morphological Characteristics
齿式：2.1.3.3/3.1.3.3=38。头体长 36~52 mm。耳长 11~13 mm。前臂长 30~36 mm。后足长 7~9 mm。尾长 30~38 mm。颅全长 12.5~13.8 mm。耳郭短，向前翻转几乎达到吻突尖端。翅膜附着在趾基部。尾翼外缘无梳状茸毛。背毛深棕色，腹侧表面棕色。
Dental formula: 2.1.3.3/3.1.3.3=38. Head and body length 36-52 mm. Ear length 11-13 mm. Forearm length 30-36 mm. Hind foot length 7-9 mm. Tail length 30-38 mm. Greatest skull length 12.5-13.8 mm. Ears short, when folded forward nearly reach rostrum tip and not distinctly beyond it. Wing membrane attached at the basal part of the toe. Exterior margins of uropatagium without comb-shaped fuzz. Dorsal hairs dark brown; ventral surface brown.

生境 / Habitat
洞穴、热带和亚热带湿润山地森林
Caves, Tropical and subtropical moist montane forest

▲ 地理分布 / Geographic Distribution

国内分布 / Domestic Distribution
内蒙古、黑龙江、吉林、辽宁、甘肃、陕西
Inner Mongolia, Heilongjiang, Jilin, Liaoning, Gansu, Shaanxi

全球分布 / World Distribution
中国、日本、哈萨克斯坦、朝鲜、蒙古国、俄罗斯
China, Japan, Kazakhstan, Democratic People's Republic of Korea, Mongolia,
Russia

生物地理界 / Biogeographic Realm
古北界 Palearctic

WWF 生物群系 / WWF Biome
温带阔叶和混交林
Temperate Broadleaf & Mixed Forests

动物地理分布型 / Zoogeographic Distribution Type
Wa

分布标注 / Distribution Note
非特有种 Non-Endemic

▲ 濒危状况 / Threatened Status

中国生物多样性红色名录等级 / CB RL Category (2021)
无危 LC

IUCN 红色名录 / IUCN Red List (2021)
无危 LC

威胁因子 / Threats
无 None

▲ 法律保护地位 / Legal Protection Status

国家重点保护野生动物等级 / Category of National Key Protected Wild Animals (2021)
未列入 Not listed

"三有" 名录 / TWIESSV (2023)
未列入 Not listed

CITES 附录等级 / CITES Appendix (2023)
未列入 Not listed

迁徙物种公约附录 / CMS Appendix (2020)
未列入 Not listed

保护行动 / Conservation Action
尚无保护行动 No conservation action so far

▲ 参考文献 / References

Jiang et al. (蒋志刚等), 2021; Burgin et al., 2020; IUCN, 2020; Wilson and Mittermeier, 2019; Wu and Pei (吴家炎和裴俊峰), 2011; Zhang et al., 2009; Gao (高志英), 2008; Pan et al. (潘清华等), 2007; Sun et al. (孙克萍等), 2006; Wilson and Reeder, 2005; Wang (王应祥), 2003; Zhang (张荣祖), 1997

167 / 华南水鼠耳蝠

Myotis laniger Peters, 1871

• Chinese Water Mouse-eared Bat

▲ 分类地位 / Taxonomy

翼手目 Chiroptera / 蝙蝠科 Vespertilionidae / 鼠耳蝠属 *Myotis*

科建立者及其文献 / Family Authority
Gray, 1821

属建立者及其文献 / Genus Authority
Kaup, 1829

亚种 / Subspecies
无 None

模式标本产地 / Type Locality
中国
China, Fujian, Amoy

余文华 / 供图

阙腾侯 / 供图

▲ 其他名称 / Other Name(s)

其他中文名 / Other Chinese Name(s)
水鼠耳蝠

其他英文名 / Other English Name(s)
Chinese Water Bat, Chinese Myotis,
Indochinese Myotis

同物异名 / Synonym(s)
无 None

▲ 形态及生境 Morphology and Habitat

形态特征 / Morphological Characteristics

齿式：2.1.3.3/3.1.3.3=38。头体长 40~42 mm。耳长 12~16 mm。前臂长 34~36 mm。后足长 8~11 mm。尾长 38~40 mm。后足长约为胫骨长的一半。背毛毛基黑色，毛尖深棕色。腹毛色较浅，毛基黑色，毛尖浅灰色。翼膜游离缘止于外趾基部。尾端突出股间膜。

Dental formula: 2.1.3.3/3.1.3.3=38. Head and body length 40-42 mm. Ear length 12-16 mm. Forearm length 34-36 mm. Hind foot length 8-11 mm. Tail length 38-40 mm. Hind foot is about 1/2 the length of the tibia. Hair bases black, tips dark brown in the back pelage. Abdominal pelage color is light, the hair bases are black, and the tips are light gray. Free margin of the pterygoid ends at the lateral base of the toe. Caudal end protrudes the interfemoral membrane.

生境 / Habitat
岩洞 Caves

▲ 地理分布 / Geographic Distribution

国内分布 / Domestic Distribution
山东、江苏、安徽、浙江、福建、江西、广东、河南、香港、海南、贵州、重庆、云南、西藏、四川、陕西、台湾
Shandong, Jiangsu, Anhui, Zhejiang, Fujian, Jiangxi, Guangdong, Henan, Hong Kong, Hainan, Guizhou, Chongqing, Yunnan, Tibet, Sichuan, Shaanxi, Taiwan

全球分布 / World Distribution
中国、印度、越南、老挝
China, India, Vietnam, Laos

生物地理界 / Biogeographic Realm
古北界 Palearctic

WWF 生物群系 / WWF Biome
热带和亚热带湿润阔叶林
Tropical & Subtropical Moist Broadleaf Forests

动物地理分布型 / Zoogeographic Distribution Type
We

分布标注 / Distribution Note
非特有种 Non-Endemic

▲ 濒危状况 / Threatened Status

中国生物多样性红色名录等级 / CB RL Category (2021)
无危 LC

IUCN 红色名录 / IUCN Red List (2021)
无危 LC

威胁因子 / Threats
未知 Unknown

▲ 法律保护地位 / Legal Protection Status

国家重点保护野生动物等级 / Category of National Key Protected Wild Animals (2021)
未列入 Not listed

"三有"名录 / TWIESSV (2023)
未列入 Not listed

CITES 附录等级 / CITES Appendix (2023)
未列入 Not listed

迁徙物种公约附录 / CMS Appendix (2020)
未列入 Not listed

保护行动 / Conservation Action
尚无保护行动 No conservation action so far

▲ 参考文献 / References

Jiang et al. (蒋志刚等), 2021; Burgin et al., 2020; IUCN, 2020; Wilson and Mittermeier, 2019; Pan et al. (潘清华等), 2007; Ruedi et al., 2015; Peng and Zhong (彭基泰和钟祥清), 2005; Wilson and Reeder, 2005

168 / 长指鼠耳蝠

Myotis longipes (Dobson, 1873)

• Kashmir Cave Mouse-eared Bat

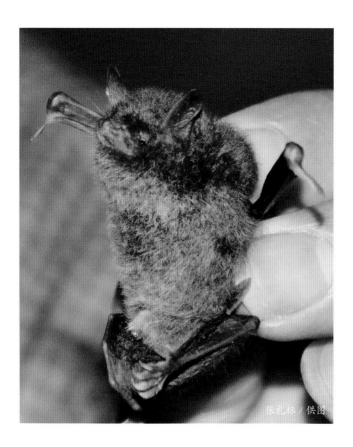

张礼标 / 供图

▲ 分类地位 / Taxonomy

翼手目 Chiroptera / 蝙蝠科 Vespertilionidae /
鼠耳蝠属 *Myotis*

科建立者及其文献 / Family Authority
Gray, 1821

属建立者及其文献 / Genus Authority
Kaup, 1829

亚种 / Subspecies
无 None

模式标本产地 / Type Locality
印度
India, Kashmir, Bhima Devi Caves 1,829 m

▲ 其他名称 / Other Name(s)

其他中文名 / Other Chinese Name(s)
长足鼠耳蝠

其他英文名 / Other English Name(s)
Kashmir Cave Bat

同物异名 / Synonym(s)
无 None

▲ 形态及生境 / Morphology and Habitat

形态特征 / Morphological Characteristics
齿式：2.1.3.3/3.1.3.3=38。头体长 43~46 mm。耳长 10~15 mm。前臂长 36~39 mm。后足长 9~10 mm。尾长 37~42 mm。头骨粗壮。耳壳狭长，耳屏长约为耳长之半。后足长超过胫长之半，爪长而粗壮。背毛黑色，毛尖淡灰色。腹毛黑或棕色，毛尖奶油白。翼狭长，翼膜附着于跖部末端。
Dental formula: 2.1.3.3/3.1.3.3=38. Head and body length 43-46 mm. Ear length 10-15 mm. Forearm length 36-39 mm. Hind foot length 9-10 mm. Tail length 37-42 mm. Skull robust. Auricular shell is long and narrow. Tragus is about half the length of the ear. Hind feet are longer than half of the tibia, and the claws are long and stout. Dorsal hairs are black with grayish tips. Abdominal hairs black or brown, tips cream white. Membrane wing is long and narrow, with the membrane attached to the end of the plantar.

生境 / Habitat
洞穴、人造建筑
Caves, man-made building

▲ 地理分布 / Geographic Distribution

国内分布 / Domestic Distribution
重庆、贵州、湖南、广东、广西
Chongqing, Guizhou, Hunan, Guangdong, Guangxi

全球分布 / World Distribution
阿富汗、巴基斯坦、中国、印度、尼泊尔
Afghanistan, Pakistan, China, India, Nepal

生物地理界 / Biogeographic Realm
古北界 Palearctic

WWF 生物群系 / WWF Biome
温带落叶和混交林
Temperate Broadleaf and Mixed Forests

动物地理分布型 / Zoogeographic Distribution Type
Wd

分布标注 / Distribution Note
非特有种 Non-Endemic

▲ 濒危状况 / Threatened Status

中国生物多样性红色名录等级 / CB RL Category (2021)
无危 LC

IUCN 红色名录 / IUCN Red List (2021)
数据缺乏 DD

威胁因子 / Threats
未知 Unknown

▲ 法律保护地位 / Legal Protection Status

国家重点保护野生动物等级 / Category of National Key Protected Wild Animals (2021)
未列入 Not listed

"三有" 名录 / TWIESSV (2023)
未列入 Not listed

CITES 附录等级 / CITES Appendix (2023)
未列入 Not listed

迁徙物种公约附录 / CMS Appendix (2020)
未列入 Not listed

保护行动 / Conservation Action
尚无保护行动 No conservation action so far

▲ 参考文献 / References

Jiang et al. (蒋志刚等), 2021; Burgin et al., 2020; IUCN, 2020; Liu et al. (刘少英等), 2020; Wilson and Mittermeier, 2019; Pan et al. (潘清华等), 2007; Wang (王应祥), 2003; Zhang (张荣祖), 1997

169 / 大趾鼠耳蝠

Myotis macrodactylus (Temminck, 1840)

• Big-footed Mouse-eared Bat

▲ 分类地位 / Taxonomy

翼手目 Chiroptera / 蝙蝠科 Vespertilionidae / 鼠耳蝠属 *Myotis*

科建立者及其文献 / Family Authority
Gray, 1821

属建立者及其文献 / Genus Authority
Kaup, 1829

亚种 / Subspecies
指名亚种 *M. m. macrodactylus* (Temminck, 1840)
吉林 Jilin

模式标本产地 / Type Locality
日本
Japan

郭东革 / 供图

▲ 其他名称 / Other Name(s)

其他中文名 / Other Chinese Name(s)
无 None

其他英文名 / Other English Name(s)
Nepalese Whiskered Bat, Whiskered Bat

同物异名 / Synonym(s)
无 None

▲ 形态及生境 / Morphology and Habitat

形态特征 / Morphological Characteristics
齿式：2.1.3.3/3.1.3.3=38。头体长 43~46 mm。前臂长 37~39 mm。尾长 38~43 m。体重 7~9 g。颅全长 14.5~15.3 mm。耳屏窄长，约为耳长的一半。背毛基部黑色，毛尖灰褐色。腹毛基部黑色，毛尖灰白色。翼膜起始于胫上离踝关节 3~5 mm 处，尾膜起始于踝关节。胫及附近尾膜着生毛发，尾尖略微超出尾膜。

Dental formula: 2.1.3.3/3.1.3.3=38. Head and body length 43-46 mm. Forearm length 37-39 mm. Tail length 38-43 mm. Body mass 7-9 g. Greatest skull length 14.5-15.3 mm. Tragus is narrow and long, about half the length of the ear. Bases of the dorsal hairs are black, and the tips are grayish-brown. Ventral hair bases are black, hair tips are grayish-white. Pterygoid membrane attached 3-5 mm above the tibial joint and the caudal membrane begins at the ankle joint. Shin and near the caudal membrane with hair. Tail tip slightly beyond the caudal membrane.

生境 / Habitat
洞穴、人造建筑、溪流边、淡水湖
Caves, man-made building, streams, freshwater lake

▲ 地理分布 / Geographic Distribution

国内分布 / Domestic Distribution
黑龙江、吉林
Heilongjiang, Jilin

全球分布 / World Distribution
中国、日本、朝鲜、韩国、俄罗斯
China, Japan, Democratic People's Republic of Korea, Republic of Korea, Russia

生物地理界 / Biogeographic Realm
古北界 Palearctic

WWF 生物群系 / WWF Biome
温带落叶和混交林
Temperate Broadleaf and Mixed Forests

动物地理分布型 / Zoogeographic Distribution Type
Xa

分布标注 / Distribution Note
非特有种 Non-Endemic

▲ 濒危状况 / Threatened Status

中国生物多样性红色名录等级 / CB RL Category (2021)
近危 NT

IUCN 红色名录 / IUCN Red List (2021)
无危 LC

威胁因子 / Threats
未知 Unknown

▲ 法律保护地位 / Legal Protection Status

国家重点保护野生动物等级 / Category of National Key Protected Wild Animals (2021)
未列入 Not listed

"三有"名录 / TWIESSV (2023)
未列入 Not listed

CITES 附录等级 / CITES Appendix (2023)
未列入 Not listed

迁徙物种公约附录 / CMS Appendix (2020)
未列入 Not listed

保护行动 / Conservation Action
在吉林省，针对该物种等建立了"集安市蝙蝠自然保护小区"，种群数量正在恢复
In Jilin Province, the "Ji'an Bat Small Reserve" has been established for several species including *Myotis macrodactylus*, and the population size of *Myotis macrodactylus* is recovering

▲ 参考文献 / References

Jiang et al. (蒋志刚等), 2021; Burgin et al., 2020; IUCN, 2020; Liu et al. (刘少英等), 2020; Wang et al., 2014; Wang (王红娜), 2010; Liu et al. (刘丰等), 2009; Luo et al. (罗金红等), 2009; Jiang et al. (江廷磊等), 2008; Pan et al. (潘清华等), 2007; Wang (王应祥), 2003

170 / 山地鼠耳蝠

Myotis montivagus (Dobson, 1874)

· Burmese Whiskered Mouse-eared Bat

▲ 分类地位 / Taxonomy

翼手目 Chiroptera / 蝙蝠科 Vespertilionidae / 鼠耳蝠属 *Myotis*

科建立者及其文献 / Family Authority
Gray, 1821

属建立者及其文献 / Genus Authority
Kaup, 1829

亚种 / Subspecies
指名亚种 *M. m. montivagus* (Dobson, 1874)
云南、福建和河北
Yunnan, Fujian and Hebei

模式标本产地 / Type Locality
中国
China, Yunnan, Hotha

张礼标 / 供图

▲ 其他名称 / Other Name(s)

其他中文名 / Other Chinese Name(s)
无 None

其他英文名 / Other English Name(s)
Large Brown Myotis

同物异名 / Synonym(s)
无 None

▲ 形态及生境 / Morphology and Habitat

形态特征 / Morphological Characteristics

齿式：2.1.3.3/3.1.3.3=38。头体长 56~62 mm。耳长 14~16 mm。前臂长 40~49 mm。后足长 9~10 mm。尾长 42~48 mm。耳短而钝，耳屏短。后足长短于胫长一半。背毛深棕色，毛基深色，毛尖浅色，腹部毛色与背毛相似，毛尖浅棕色。翼膜附于外趾基部。

Dental formula: 2.1.3.3/3.1.3.3=38. Head and body length 56-62 mm. Ear length 14-16 mm. Forearm length 40-49 mm. Hind foot length 9-10 mm. Tail length 42-48 mm. Ears short and blunt. Tragus short. Hind foot length less than half tibia length. Dorsal pelage dark brown, hairs with darker roots and paler tips. Ventral hair color similar to that on the back, with hair tips paler brown. Membrane wings attached to the base of toe.

生境 / Habitat

未知 Unknown

国内分布 / Domestic Distribution
云南、福建
Yunnan, Fujian

全球分布 / World Distribution
中国、印度、老挝、马来西亚、缅甸、越南
China, India, Laos, Malaysia, Myanmar, Vietnam

生物地理界 / Biogeographic Realm
印度马来界 Indomalaya

WWF 生物群系 / WWF Biome
热带和亚热带湿润阔叶林
Tropical & Subtropical Moist Broadleaf Forests

动物地理分布型 / Zoogeographic Distribution Type
Wa

分布标注 / Distribution Note
非特有种 Non-Endemic

▲ 濒危状况 / Threatened Status

中国生物多样性红色名录等级 / CB RL Category (2021)
无危 LC

IUCN 红色名录 / IUCN Red List (2021)
无危 LC

威胁因子 / Threats
无 None

▲ 法律保护地位 / Legal Protection Status

国家重点保护野生动物等级 / Category of National Key Protected Wild Animals (2021)
未列入 Not listed

"三有" 名录 / TWIESSV (2023)
未列入 Not listed

CITES 附录等级 / CITES Appendix (2023)
未列入 Not listed

迁徙物种公约附录 / CMS Appendix (2020)
未列入 Not listed

保护行动 / Conservation Action
尚无保护行动 No conservation action so far

▲ 参考文献 / References

Jiang et al. (蒋志刚等), 2021; Burgin et al., 2020; IUCN, 2020; Liu et al. (刘少英等), 2020; Wilson and Mittermeier, 2019; Goerfoel et al., 2013; Wilson and Reeder, 2005; Wang (王应祥), 2003

171 / 喜山鼠耳蝠

Myotis muricola (Gray, 1846)

• Nepalese Whiskered Mouse-eared Bat

▲ 分类地位 / Taxonomy

翼手目 Chiroptera / 蝙蝠科 Vespertilionidae / 鼠耳蝠属 *Myotis*

科建立者及其文献 / Family Authority
Gray, 1821

属建立者及其文献 / Genus Authority
Kaup, 1829

亚种 / Subspecies
川西亚种 *M. m. moupinensis* (Milne-Edwards,1872)
四川和云南
Sichuan and Yunnan
喜马拉雅亚种 *M. m. blanfordi* (Dobson, 1871)
西藏和云南
Tibet and Yunnan

模式标本产地 / Type Locality
尼泊尔
Nepal

余文华 / 供图

▲ 其他名称 / Other Name(s)

其他中文名 / Other Chinese Name(s)
无 None

其他英文名 / Other English Name(s)
Nepalese Whiskered Bat, Whiskered Bat

同物异名 / Synonym(s)
无 None

▲ 形态及生境 / Morphology and Habitat

形态特征 / Morphological Characteristics
齿式：2.1.3.3/3.1.3.3=38。头体长 41~47 mm。耳长 6~13 mm。前臂长 31~37 mm。后足长 4~7 mm。尾长 25~39 mm。耳长而尖，耳缘有明显的缺刻，耳屏相对较短，尖端弯向前。被毛厚，头及背部毛为暗褐色，毛尖黄棕色。腹部毛尖灰白色。翼膜终止于外趾基部，距细长，距缘膜狭窄或不显。

Dental formula: 2.1.3.3/3.1.3.3=38. Head and body length 41-47 mm. Ear length 6-13 mm. Forearm length 31-37 mm. Hind foot length 4-7 mm. Tail length 25-39 mm. Ears large and pointed, with distinctive and sharp notch near rear edge of concha, tragus relatively short and bend forward. Pelage thick and soft. Dorsal hairs black basally with paler tips. Ventral hair tips are slightly paler. Wing attached to the base of metatarsal. Hind foot length less than half tibia length.

生境 / Habitat
热带亚热带湿润低地森林、次生林、灌丛、乡村种植园、洞穴
Tropical subtropical moist lowland forest, secondary forest, shrubland, rural garden, caves

▲ 地理分布 / Geographic Distribution

国内分布 / Domestic Distribution
四川、云南、西藏、台湾
Sichuan, Yunnan, Tibet, Taiwan

全球分布 / World Distribution
阿富汗、不丹、文莱、柬埔寨、中国、印度、印度尼西亚、老挝、
马来西亚、缅甸、尼泊尔、巴基斯坦、菲律宾、新加坡、泰国、
越南
Afghanistan, Bhutan, Brunei, Cambodia, China, India, Indonesia, Laos, Malaysia,
Myanmar, Nepal, Pakistan, Philippines, Singapore, Thailand, Vietnam

生物地理界 / Biogeographic Realm
印度马来界、古北界、大洋洲界
Indomalaya, Palearctic, Oceanian

WWF 生物群系 / WWF Biome
热带和亚热带湿润阔叶林
Tropical & Subtropical Moist Broadleaf Forests

动物地理分布型 / Zoogeographic Distribution Type
Hm

分布标注 / Distribution Note
非特有种 Non-Endemic

▲ 濒危状况 / Threatened Status

中国生物多样性红色名录等级 / CB RL Category (2021)
近危 NT

IUCN 红色名录 / IUCN Red List (2021)
无危 LC

威胁因子 / Threats
森林砍伐、火灾、耕种
Logging, fire, farming

▲ 法律保护地位 / Legal Protection Status

国家重点保护野生动物等级 / Category of National Key Protected Wild Animals (2021)
未列入 Not listed

"三有" 名录 / TWIESSV (2023)
未列入 Not listed

CITES 附录等级 / CITES Appendix (2023)
未列入 Not listed

迁徙物种公约附录 / CMS Appendix (2020)
未列入 Not listed

保护行动 / Conservation Action
尚无保护行动 No conservation action so far

▲ 参考文献 / References

Jiang et al. (蒋志刚等), 2021; Burgin et al., 2020; IUCN, 2020; Wilson and Mittermeier, 2019; Wang et al. (王磊等), 2010; Smith and Xie, 2009; Pan et al. (潘清华等), 2007; Wilson and Reeder, 2005; Wang (王应祥), 2003

172 / 尼泊尔鼠耳蝠

Myotis nipalensis (Dobson, 1871)

· Nepal Mouse-eared Bat

▲ 分类地位 / Taxonomy

翼手目 Chiroptera / 蝙蝠科 Vespertilionidae / 鼠耳蝠属 *Myotis*

科建立者及其文献 / Family Authority
Gray, 1821

属建立者及其文献 / Genus Authority
Kaup, 1829

亚种 / Subspecies
无 None

模式标本产地 / Type Locality
尼泊尔
Nepal, Katmandu

Sanjan Thapa/ 供图

▲ 其他名称 / Other Name(s)

其他中文名 / Other Chinese Name(s)
无 None

其他英文名 / Other English Name(s)
无 None

同物异名 / Synonym(s)
无 None

▲ 形态及生境 / Morphology and Habitat

形态特征 / Morphological Characteristics

齿式：2.1.3.3/3.1.3.3=38。头体长 38~47 mm。耳长 12~14 mm。前臂长 34~37 mm。后足长 7~8 mm。尾长 32~40 mm。颅基长 12~13 mm。耳小，耳屏长而窄，仅耳的一半长。后足不到胫长的一半。背毛基深色，毛尖棕色，或红灰色。腹部毛基深灰色，毛尖浅灰色。翼膜附着在跖骨远端。Dental formula: 2.1.3.3/3.1.3.3=38. Head and body length 38-47 mm. Ear length 12-14 mm. Forearm length 34-37 mm. Hind foot length 7-8 mm. Tail length 32-40 mm. Ears small. Tragus long and narrow, half as long as the ear. Hind foot small, less than half tibia length. Dorsal pelage dark basally with brown tips, or sometimes more reddish-gray. Ventral pelage is also dark basally but with paler grayish tips. Wing membrane attached to the distal end of metatarsal.

生境 / Habitat
森林、灌丛、草地、荒漠、人造建筑、洞穴
Forest, shrubland, grassland, desert, man-made building, caves

▲ 地理分布 / Geographic Distribution

国内分布 / Domestic Distribution
新疆、甘肃、青海、西藏、陕西
Xinjiang, Gansu, Qinghai, Tibet, Shaanxi

全球分布 / World Distribution
阿富汗、亚美尼亚、阿塞拜疆、中国、印度、格鲁吉亚、伊朗，哈萨克斯坦、吉尔吉斯斯坦、尼泊尔、塔吉克斯坦、土库曼斯坦、乌兹别克斯坦、土耳其
Afghanistan, Armenia, Azerbaijan, China, India, Georgia, Iran, Kazakhstan, Kyrgyzstan, Nepal, Tajikistan, Turkey, Turkmenistan, Uzbekistan

生物地理界 / Biogeographic Realm
古北界 Palearctic

WWF 生物群系 / WWF Biome
温带针叶林
Temperate Conifer Forests

动物地理分布型 / Zoogeographic Distribution Type
Ud

分布标注 / Distribution Note
非特有种 Non-Endemic

▲ 濒危状况 / Threatened Status

中国生物多样性红色名录等级 / CB RL Category (2021)
数据缺乏 DD

IUCN 红色名录 / IUCN Red List (2021)
无危 LC

威胁因子 / Threats
未知 Unknown

▲ 法律保护地位 / Legal Protection Status

国家重点保护野生动物等级 / Category of National Key Protected Wild Animals (2021)
未列入 Not listed

"三有" 名录 / TWIESSV (2023)
未列入 Not listed

CITES 附录等级 / CITES Appendix (2023)
未列入 Not listed

迁徙物种公约附录 / CMS Appendix (2020)
未列入 Not listed

保护行动 / Conservation Action
尚无保护行动 No conservation action so far

▲ 参考文献 / References

Jiang et al. (蒋志刚等), 2021; Burgin et al., 2020; IUCN, 2020; Liu et al. (刘奇等), 2014; Hu et al. (胡一鸣等), 2014

173 / 北京鼠耳蝠

Myotis pequinius Thomas, 1908

• Peking Mouse-eared Bat

▲ 分类地位 / Taxonomy

翼手目 Chiroptera / 蝙蝠科 Vespertilionidae /
鼠耳蝠属 *Myotis*

科建立者及其文献 / Family Authority
Gray, 1821

属建立者及其文献 / Genus Authority
Kaup, 1829

亚种 / Subspecies
无 None

模式标本产地 / Type Locality
中国
China, Hopeh, 48 km W Peking, 183 m

▲ 其他名称 / Other Name(s)

其他中文名 / Other Chinese Name(s)
无 None

其他英文名 / Other English Name(s)
无 None

同物异名 / Synonym(s)
无 None

▲ 形态及生境 / Morphology and Habitat

形态特征 / Morphological Characteristics

齿式：2.1.3.3/3.1.3.3=38。头体长约47 mm。耳长18 mm。前臂长45~50 mm。后足长约9 mm。尾约42 mm。耳郭前折略超过吻端，耳屏相对较短，约为耳长之半。后足长度超过胫长的一半，被毛短而柔软。体背毛基暗褐，毛尖灰褐色，腹面毛基暗褐，毛尖灰白翼膜附着在脚踝处。外缘无毛。

Dental formula: 2.1.3.3/3.1.3.3=38. Head and body length 47 mm. Ear length 18 mm. Forearm length 45-50 mm. Hind foot length 9 mm. Tail length 42 mm. Ears do not go beyond rostrum tip, and tragus is relatively short, about half the length of ear. Length of hind foot more than half tibia length. Fur rather short and velvety. Dorsal hairs gray brown. Ventral surface off-white. Wing membrane attached at the ankle. Exterior margins of uropatagium are hairless.

生境 / Habitat
洞穴、森林
Caves, forest

▲ 地理分布 / Geographic Distribution

国内分布 / Domestic Distribution
北京、河北、四川、山东、河南、安徽、江苏
Beijing, Hebei, Sichuan, Shandong, Henan, Anhui, Jiangsu

全球分布 / World Distribution
中国 China

生物地理界 / Biogeographic Realm
古北界 Palearctic

WWF 生物群系 / WWF Biome
温带落叶和混交林
Temperate Broadleaf and Mixed Forests

动物地理分布型 / Zoogeographic Distribution Type
Ea

分布标注 / Distribution Note
特有种 Endemic

▲ 濒危状况 / Threatened Status

中国生物多样性红色名录等级 / CB RL Category (2021)
无危 LC

IUCN 红色名录 / IUCN Red List (2021)
无危 LC

威胁因子 / Threats
未知 Unknown

▲ 法律保护地位 / Legal Protection Status

国家重点保护野生动物等级 / Category of National Key Protected Wild Animals (2021)
未列入 Not listed

"三有"名录 / TWIESSV (2023)
未列入 Not listed

CITES 附录等级 / CITES Appendix (2023)
未列入 Not listed

迁徙物种公约附录 / CMS Appendix (2020)
未列入 Not listed

保护行动 / Conservation Action
尚无保护行动 No conservation action so far

▲ 参考文献 / References

Jiang et al. (蒋志刚等), 2021; Burgin et al., 2020; IUCN, 2020; Liu et al. (刘少英等), 2020; Wilson and Mittermeier, 2019; Wu and Pei (吴家炎和裴俊峰), 2011; Smith et al., 2009; Pan et al. (潘清华等), 2007; Wilson and Reeder, 2005; Wang (王应祥), 2003; Zhang (张荣祖), 1997

174 / 东亚水鼠耳蝠

Myotis petax Hollister, 1912

• Eastern Daubenton's Mouse-eared Bat

▲ 分类地位 / Taxonomy

翼手目 Chiroptera / 蝙蝠科 Vespertilionidae / 鼠耳蝠属 *Myotis*

科建立者及其文献 / Family Authority
Gray, 1821

属建立者及其文献 / Genus Authority
Kaup, 1829

亚种 / Subspecies
指名亚种 *M. p. petax* Hollister, 1912
吉林、内蒙古、湖南
Jilin, Inner Mongolia, Hunan

黑龙江亚种 *M. p. loukashkini* Shamei, 1942
黑龙江 Heilongjiang

模式标本产地 / Type Locality
俄罗斯
Chuiskaya Steppe (Altai, south of Western Siberia)

郭东革 / 供图

▲ 其他名称 / Other Name(s)

其他中文名 / Other Chinese Name(s)
无 None

其他英文名 / Other English Name(s)
Eastern Water Myotis, Sakhalin Myotis

同物异名 / Synonym(s)
无 None

▲ 形态及生境 / Morphology and Habitat

形态特征 / Morphological Characteristics
齿式：2.1.3.3/3.1.3.3=38。头体长 40~54 mm。前臂长 35~40 mm。尾长 27~48 mm。体重 5~15 g。翼展 240~275 mm。颅全长 14~15.7 mm。耳狭长，耳屏长约为耳长一半。向前折叠略超过吻尖，边缘有 3~5 个横纹。从基部逐渐变细变尖。被毛短、柔软。背毛灰褐色，腹毛灰白色。翼膜浅褐色，起始于跖骨中部，尾膜起始于胫基。

Dental formula: 2.1.3.3/3.1.3.3=38. Head and body length 40-54 mm. Forearm length 35-40 mm. Tail length 27-48 mm. Body mass 5-15 g. Wingspan of 240-275 mm. Greatest skull length 14-15.7 mm. Ears are long and narrow, and the tragus is about half the length of the ear, with 3-5 stripes at the edges. When folding forward, the ear just over the tip of the snout, Tapering from the base to a point. Pelage short and soft. Dorsal hairs are grayish-brown, and the belly hairs are grayish-white. Pterygoid membrane is light brown, beginning at the middle of the metatarsal bone. Caudal membrane beginning at the tibial base.

生境 / Habitat
洞穴、淡水湖、人工水域、人造建筑、森林、灌丛
Caves, freshwater lake, artificial-aquatic, man-made building, forest, shrubland

▲ 地理分布 / Geographic Distribution

国内分布 / Domestic Distribution
黑龙江、吉林、内蒙古、新疆、湖南
Heilongjiang, Jilin, Inner Mongolia, Xinjiang, Hunan

全球分布 / World Distribution
阿尔巴尼亚、安道尔、奥地利、白俄罗斯、比利时、保加利亚、中国、克罗地亚、捷克、丹麦、爱沙尼亚、芬兰、法国、德国、直布罗陀、希腊、匈牙利、印度、爱尔兰、意大利、日本、哈萨克斯坦、朝鲜、韩国、拉脱维亚、列支敦士登、立陶宛、卢森堡、马其顿、摩尔多瓦、摩纳哥、黑山、荷兰、挪威、波兰、葡萄牙、罗马尼亚、俄罗斯、圣马力诺、塞尔维亚、斯洛伐克、斯洛文尼亚、西班牙、瑞典、瑞士、土耳其、乌克兰、英国
Albania, Andorra, Austria, Belarus, Belgium, Bulgaria, China, Croatia, Czech, Denmark, Estonia, Finland, France, Germany, Gibraltar, Greece, Hungary, India, Ireland, Italy, Japan, Kazakhstan, Democratic People's Republic of Korea, Republic of Korea, Latvia, Liechtenstein, Lithuania, Luxembourg, Macedonia, Moldova, Monaco, Montenegro, Netherlands, Norway, Poland, Portugal, Romania, Russia, San Marino, Serbia, Slovakia, Slovenia, Spain, Sweden, Switzerland, Turkey, Ukraine, United Kingdom

生物地理界 / Biogeographic Realm
古北界 Palearctic

WWF 生物群系 / WWF Biome
温带阔叶和混交林、温带针叶林
Temperate Broadleaf & Mixed Forests, Temperate Conifer Forest

动物地理分布型 / Zoogeographic Distribution Type
Ub+Uc+Ud

分布标注 / Distribution Note
非特有种 Non-Endemic

▲ 濒危状况 / Threatened Status

中国生物多样性红色名录等级 / CB RL Category (2021)
数据缺乏 DD

IUCN 红色名录 / IUCN Red List (2021)
无危 LC

威胁因子 / Threats
未知 Unknown

▲ 法律保护地位 / Legal Protection Status

国家重点保护野生动物等级 / Category of National Key Protected Wild Animals (2021)
未列入 Not listed

"三有" 名录 / TWIESSV (2023)
未列入 Not listed

CITES 附录等级 / CITES Appendix (2023)
未列入 Not listed

迁徙物种公约附录 / CMS Appendix (2020)
未列入 Not listed

保护行动 / Conservation Action
尚无保护行动 No conservation action so far

▲ 参考文献 / References

Jiang et al. (蒋志刚等), 2021; Burgin et al., 2020; IUCN, 2020; Wilson and Mittermeier, 2019; Feng et al. (冯磊等), 2019; Wang et al. (王磊等), 2010; China Species Information Service, 2008; Jones et al., 2006; Woodman, 1993

175 / 大足鼠耳蝠

Myotis pilosus (Peters, 1869)

· Rickett's Big-footed Mouse-eared Bat

▲ 分类地位 / Taxonomy

翼手目 Chiroptera / 蝙蝠科 Vespertilionidae / 鼠耳蝠属 *Myotis*

科建立者及其文献 / Family Authority
Gray, 1821

属建立者及其文献 / Genus Authority
Kaup, 1829

亚种 / Subspecies
无 None

模式标本产地 / Type Locality
中国
China, Fukien (Fujian), Foochow

周佳俊 / 供图

吴毅 / 供图

▲ 其他名称 / Other Name(s)

其他中文名 / Other Chinese Name(s)
大足蝠、里氏大足蝠、
大脚鼠耳蝠

其他英文名 / Other English Name(s)
Rickett's Big-footed Bat

同物异名 / Synonym(s)
无 None

▲ 形态及生境 / Morphology and Habitat

形态特征 / Morphological Characteristics

齿式：2.1.3.3/3.1.3.3=38。头体长 65 mm。耳长 15~18 mm。前臂长 53~56 mm。后足长 15~17 mm。尾长 45~54 mm。颅全长 21 mm。耳相对较短，前折不达吻端。耳屏不及耳长之半。体毛呈绒毛状。背部毛基黑褐色，毛尖灰褐色。腹部毛基黑褐色，毛尖灰白色。曾采到白化个体。翼膜止于胫部中间。后足很发达，连爪长度几乎与胫长相当。尾末端一节尾椎从股间膜后缘穿出。

Dental formula: 2.1.3.3/3.1.3.3=38. Head and body length 65 mm. Ear length 15-18 mm. Forearm length 53-56 mm. Hind foot length 15-17 mm. Tail length 45-54 mm. Greatest skull length 21 mm. Ears short, folded forward not to reach the snout. Tragus is less than half the length of the ear. Body hairs are villous. Upper body hairs are light gray-brown with darker roots. Underparts are pale gray with darker roots. Albino individuals have been found. Pterygoid ends at the center or base of the shin. Hind feet with claws. The last vertebra of the tail penetrates through the posterior margin of the interfemoral membrane.

生境 / Habitat

次生林、喀斯特地貌洞穴、淡水湖
Secondary forest, karst caves, freshwater lake

▲ 地理分布 / Geographic Distribution

国内分布 / Domestic Distribution
北京、山西、陕西、重庆、云南、江苏、江西、山东、浙江、澳门、广东、海南、香港
Beijing, Shanxi, Shaanxi, Chongqing, Yunnan, Jiangsu, Jiangxi, Shandong, Zhejiang, Macao, Guangdong, Hainan, Hong Kong

全球分布 / World Distribution
中国、老挝、越南
China, Laos, Vietnam

生物地理界 / Biogeographic Realm
古北界、印度马来界
Palearctic, Indomalaya

WWF 生物群系 / WWF Biome
北方森林、针叶林
Boreal Forests, Taiga

动物地理分布型 / Zoogeographic Distribution Type
Ue

分布标注 / Distribution Note
非特有种 Non-Endemic

▲ 濒危状况 / Threatened Status

中国生物多样性红色名录等级 / CB RL Category (2021)
近危 NT

IUCN 红色名录 / IUCN Red List (2021)
近危 NT

威胁因子 / Threats
城市污水、工业废水
Urban waste water, industrial waste water

▲ 法律保护地位 / Legal Protection Status

国家重点保护野生动物等级 / Category of National Key Protected Wild Animals (2021)
未列入 Not listed

"三有"名录 / TWIESSV (2023)
未列入 Not listed

CITES 附录等级 / CITES Appendix (2023)
未列入 Not listed

迁徙物种公约附录 / CMS Appendix (2020)
未列入 Not listed

保护行动 / Conservation Action
尚无保护行动 No conservation action so far

▲ 参考文献 / References

Jiang et al. (蒋志刚等), 2021; Liu et al. (刘少英等), 2020; Wilson and Mittermeier, 2019; Pei and Feng (裴俊峰和冯祁君), 2014; Huang et al. (黄继展等), 2013; Luo et al. (罗丽等), 2011; Smith et al., 2009; Jiang et al. (江廷磊等), 2008; Pan et al. (潘清华等), 2007; Li et al. (李玉春等), 2006; Ma et al. (马杰等), 2005; Wilson and Reeder, 2005; Wang (王应祥), 2003; Feng et al. (冯江等), 2003; Liu et al. (刘少英等), 2001; Zhang (张荣祖等), 1997

176 / 渡濑氏鼠耳蝠

Myotis rufoniger (Tomats, 1858)

· Reddish-black Myotis

▲ 分类地位 / Taxonomy

翼手目 Chiroptera / 蝙蝠科 Vespertilionidae / 鼠耳蝠属 *Myotis*

科建立者及其文献 / Family Authority
Gray, 1821

属建立者及其文献 / Genus Authority
Kaup, 1829

亚种 / Subspecies
指名亚种 *M. r. rufoniger* (Tomats, 1858)
吉林、辽宁、江西、福建、安徽、上海、广西
Jilin, Liaoning, Jiangxi, Fujian, Anhui, Shanghai, Guangxi

台湾亚种 *M. r. watase* Kishida, 1924
台湾 Taiwan

模式标本产地 / Type Locality
中国
Shanghai, China

▲ 其他名称 / Other Name(s)

其他中文名 / Other Chinese Name(s)
无 None

其他英文名 / Other English Name(s)
Black-winged Myotis,
Red and Black Mouse-eared Bat

同物异名 / Synonym(s)
无 None

▲ 形态及生境 / Morphology and Habitat

形态特征 / Morphological Characteristics
齿式：2.1.3.3/3.1.3.3=38。头体长 47~55 mm。耳长 14~19 mm。前臂长 45~56 mm。胫骨长 21~25 mm。后足长 9~12 mm。尾长 45~55 mm。耳狭长，略呈卵圆形，耳屏狭长，约耳长之半，具一小基叶。背毛长而密，毛基部深褐色近黑色，毛中部浅黄色，毛端部红褐色；腹毛亮红褐色，耳橙红色，耳缘黑色，翼膜黑色，但掌指骨周缘处及近躯干处为橙棕色。股间膜橙棕色，具稀疏短毛。

Dental formula: 2.1.3.3/3.1.3.3=38. Head and body length 47-55 mm. Ear length 14-19 mm. Forearm length 45-56 mm. Tibia length 21-25 mm. Hind Foot length 9-12 mm. Ears long and narrow, slightly ovoid. Tragus narrow, about half the length of ear, with a small basal lobe. Dorsal hairs long and thick, with slate-gray base, yellowish middle band, and dark brown tips. Abdominal hairs and most of membrane are orange-yellow. Ears are dark orange with black margin. Wings are black, with orange broadly surrounding bones and near body. The interfemoral membrane is orange-yellow, with sparse short hairs.

生境 / Habitat
次生林、溪流边、洞穴、人造建筑
Secondary forest, streams, caves, man-made building

▲ 地理分布 / Geographic Distribution

国内分布 / Domestic Distribution
吉林、辽宁、陕西、重庆、四川、贵州、湖北、江西、河南、福建、
安徽、浙江、江苏、上海、广西、台湾
Jilin, Liaoning, Shaanxi, Chongqing, Sichuan, Guizhou, Hubei, Jiangxi, Henan,
Fujian, Anhui, Zhejiang, Jiangsu, Shanghai, Guangxi, Taiwan

全球分布 / World Distribution
中国、老挝、越南、朝鲜、日本
China, Laos, Vietnam, Democratic People's Republic of Korea, Japan

生物地理界 / Biogeographic Realm
古北界、印度马来界
Palearctic, Indomalaya

WWF 生物群系 / WWF Biome
热带和亚热带湿润阔叶林、温带阔叶和混交林
Tropical & Subtropical Moist Broadleaf Forests, Temperate Broadleaf & Mixed
Forests

动物地理分布型 / Zoogeographic Distribution Type
Uh

分布标注 / Distribution Note
非特有种 Non-Endemic

▲ 濒危状况 / Threatened Status

中国生物多样性红色名录等级 / CB RL Category (2021)
易危 VU

IUCN 红色名录 / IUCN Red List (2021)
近危 NT

威胁因子 / Threats
未知 Unknown

▲ 法律保护地位 / Legal Protection Status

国家重点保护野生动物等级 / Category of National Key Protected Wild Animals (2021)
未列入 Not listed

"三有" 名录 / TWIESSV (2023)
未列入 Not listed

CITES 附录等级 / CITES Appendix (2023)
未列入 Not listed

迁徙物种公约附录 / CMS Appendix (2020)
未列入 Not listed

保护行动 / Conservation Action
尚无保护行动 No conservation action so far

▲ 参考文献 / References

Jiang et al. (蒋志刚等), 2021; Burgin et al., 2020; IUCN, 2020; Liu et al. (刘少英等), 2020; Wilson and Mittermeier, 2019; Dang et al. (党飞红等),
2017; Lu et al., 2013

177 / 高颅鼠耳蝠

Myotis siligorensis (Horsfield, 1855)

· Himalayan Whiskered Mouse-eared Bat

▲ 分类地位 / Taxonomy

翼手目 Chiroptera / 蝙蝠科 Vespertilionidae / 鼠耳蝠属 *Myotis*

科建立者及其文献 / Family Authority
Gray, 1821

属建立者及其文献 / Genus Authority
Kaup, 1829

亚种 / Subspecies
福建亚种 *M. s. sawerbyi* Howell, 1926
福建和海南
Fujian and Hainan

模式标本产地 / Type Locality
尼泊尔
Nepal, Siligori

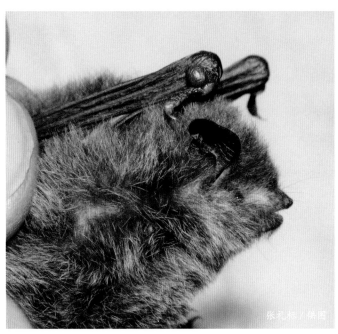
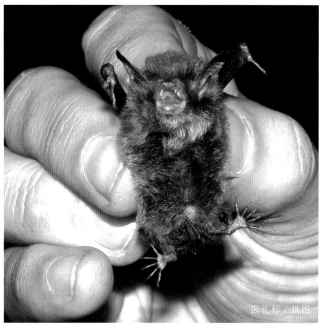

▲ 其他名称 / Other Name(s)

其他中文名 / Other Chinese Name(s)
无 None

其他英文名 / Other English Name(s)
Siliguri Bat, Small-toothed Myotis

同物异名 / Synonym(s)
无 None

▲ 形态及生境 / Morphology and Habitat

形态特征 / Morphological Characteristics
齿式：2.1.3.3/3.1.3.3=38。头体长 33.7~43.3 mm。前臂长 30~36 mm。后足长 5.9~10.8 mm。尾长 31.4~39.3 mm。颅全长约 13 mm。吻部被毛，具吻须。耳狭长，外耳缘微凹，耳背面无毛，耳屏直而细长，为耳长之半，顶端尖锐，基部具基叶。覆毛浓密，背毛深烟灰色，腹毛灰褐色，毛尖灰白色。翼膜黑色，附着于趾基，距有明显的距缘膜。

Dental formula: 2.1.3.3/3.1.3.3=38. Head and body length 33.7-43.3 mm. Ear length 8-13 mm. Forearm length 30-36 mm. Hind foot length 5.9-10.8 mm. Tail length 31.4-39.3 mm. Greatest skull length 13 mm. Snout hairy with whiskers. Ears long and narrow, the outer edges of the ears are slightly concave, the backs of the ears are hairless. Tragus is straight and long, the half of the ear length, the top is sharp, with a basal leaf. Pelage thick. Dorsal hairs deep smoke gray, abdominal hairs gray, hair tips pale. Pterygoid membrane is black, attached to the base of the toe.

生境 / Habitat
次生林、溪流边、洞穴、人造建筑
Secondary forest, streams, caves, man-made building

▲ 地理分布 / Geographic Distribution

国内分布 / Domestic Distribution
云南、海南、福建、广东
Yunnan, Hainan, Fujian, Guangdong

全球分布 / World Distribution
柬埔寨、中国、印度、印度尼西亚、老挝、马来西亚、缅甸、尼泊尔、越南
Cambodia, China, India, Indonesia, Laos, Malaysia, Myanmar, Nepal, Vietnam

生物地理界 / Biogeographic Realm
印度马来界 Indomalaya

WWF 生物群系 / WWF Biome
热带和亚热带湿润阔叶林
Tropical & Subtropical Moist Broadleaf Forests

动物地理分布型 / Zoogeographic Distribution Type
Wa

分布标注 / Distribution Note
非特有种 Non-Endemic

▲ 濒危状况 / Threatened Status

中国生物多样性红色名录等级 / CB RL Category (2021)
近危 NT

IUCN 红色名录 / IUCN Red List (2021)
近危 NT

威胁因子 / Threats
无 None

▲ 法律保护地位 / Legal Protection Status

国家重点保护野生动物等级 / Category of National Key Protected Wild Animals (2021)
未列入 Not listed

"三有" 名录 / TWIESSV (2023)
未列入 Not listed

CITES 附录等级 / CITES Appendix (2023)
未列入 Not listed

迁徙物种公约附录 / CMS Appendix (2020)
未列入 Not listed

保护行动 / Conservation Action
尚无保护行动 No conservation action so far

▲ 参考文献 / References

Jiang et al. (蒋志刚等), 2021; Wilson and Mittermeier, 2019; Xiao et al. (肖宁等), 2017; Liu et al. (刘志霄等), 2013; Lin et al. (林洪军等), 2012; Wei et al. (韦力等), 2011, 2006; Chen, 2009; Zhu (朱斌良), 2008; Pan et al. (潘清华等), 2007; Wilson and Reeder, 2005; Wang (王应祥), 2003; Zhang (张荣祖), 1997

178 / 台湾鼠耳蝠

Myotis taiwanensis Linde, 1908

• Taiwanese Mouse-eared Bat

▲ 分类地位 / Taxonomy

翼手目 Chiroptera / 蝙蝠科 Vespertilionidae / 鼠耳蝠属 *Myotis*

科建立者及其文献 / Family Authority
Gray, 1821

属建立者及其文献 / Genus Authority
Kaup, 1829

亚种 / Subspecies
无 None

模式标本产地 / Type Locality
中国
Takao, Anping, Taiwan, China

▲ 其他名称 / Other Name(s)

其他中文名 / Other Chinese Name(s)
无 None

其他英文名 / Other English Name(s)
无 None

同物异名 / Synonym(s)
无 None

▲ 形态及生境 / Morphology and Habitat

形态特征 / Morphological Characteristics

齿式：2.1.3.3/3.1.3.3=38。头体长 51.3 mm。耳长 11.3 mm。前臂长 39.9 mm。体重 5.65 g。颅全长 16.9 mm。耳屏短而直，不及耳郭高度 的一半。背毛暗黑色，略呈棕色调，腹毛比背毛浅，毛根黑色，毛尖白色。 Dental formula: 2.1.3.3/3.1.3.3=38. Head and body length 51.3 mm. Ear length 11.3 mm. Forearm length 39.9 mm. Body mass 5.65 g. Greatest length of skull 16.9 mm. Tragus straight and short, less than half the height of the pinna. Dorsal fur is dark black tinged with a brown tone. Ventral fur is lighter than the dorsal pelage, with hairs of black bases and white tips.

生境 / Habitat
未知 Unknown

▲ 地理分布 / Geographic Distribution

国内分布 / Domestic Distribution
台湾 Taiwan

全球分布 / World Distribution
中国 China

生物地理界 / Biogeographic Realm
印度马来界 Indomalaya

WWF 生物群系 / WWF Biome
热带和亚热带湿润阔叶林
Tropical & Subtropical Moist Broadleaf Forests

动物地理分布型 / Zoogeographic Distribution Type
J

分布标注 / Distribution Note
特有种 Endemic

▲ 濒危状况 / Threatened Status

中国生物多样性红色名录等级 / CB RL Category (2021)
近危 NT

IUCN 红色名录 / IUCN Red List (2021)
未评定 NE

威胁因子 / Threats
森林砍伐、耕种
Logging, farming

▲ 法律保护地位 / Legal Protection Status

国家重点保护野生动物等级 / Category of National Key Protected Wild Animals (2021)
未列入 Not listed

"三有" 名录 / TWIESSV (2023)
未列入 Not listed

CITES 附录等级 / CITES Appendix (2023)
未列入 Not listed

迁徙物种公约附录 / CMS Appendix (2020)
未列入 Not listed

保护行动 / Conservation Action
尚无保护行动 No conservation action so far

▲ 参考文献 / References

Jiang et al. (蒋志刚等), 2021; Burgin et al., 2020; IUCN, 2020; Liu et al. (刘少英等), 2020; Pan et al. (潘清华等), 2007; Wang(王应祥), 2003

179 / 宽吻鼠耳蝠

Submyotodon latirostris Kishida, 1932

• Taiwan Broad-muzzled
  Mouse-eared Bat

▲ 分类地位 / Taxonomy

翼手目 Chiroptera / 蝙蝠科 Vespertilionidae / 宽吻蝠属 *Submyotodon*

科建立者及其文献 / Family Authority
Gray, 1821

属建立者及其文献 / Genus Authority
Kaup, 1829

亚种 / Subspecies
无 None

模式标本产地 / Type Locality
不明
Unknown

▲ 其他名称 / Other Name(s)

其他中文名 / Other Chinese Name(s)
无 None

其他英文名 / Other English Name(s)
Broad-muzzled Bat,
Taiwan Broad-muzzled Myotis

同物异名 / Synonym(s)
无 None

▲ 形态及生境 / Morphology and Habitat

形态特征 / Morphological Characteristics

齿式：2.1.3.3/3.1.3.3=38。头体长 35~40.5 mm。前臂长 32~36 mm。尾长 35~38 mm。颅全长 12.25~13.37 mm。吻端突出。耳郭外缘有缺刻。耳长大于耳宽，顶端钝圆。耳屏相对较短，尖端略向前弯。背毛长而蓬松，毛基黑色、尖端咖啡色。腹毛灰黑色。翼膜宽长，左右延伸连接到脚趾基部。

Dental formula: 2.1.3.3/3.1.3.3=38. Head and body length 35-40.5 mm. Forearm length 32-36 mm. Tail length 35-38 mm. Greatest skull length 12.25-13.37 mm. Snout protrudes. Auricle is notched. Ear length are longer than the width of the ear and each ear has a blunt rounded tip. Tragus is relatively short and bent forward. Its pelage is long and shaggy. Dorsal hairs dense, with black hair bases and brown tips. Ventral hairs are gray-black. Pterygoid membrane is broad and long, extending left and right to the base of the toe.

生境 / Habitat

未知 Unknown

▲ 地理分布 / Geographic Distribution

国内分布 / Domestic Distribution
台湾 Taiwan

全球分布 / World Distribution
中国 China

生物地理界 / Biogeographic Realm
印度马来界 Indomalaya

WWF 生物群系 / WWF Biome
热带和亚热带湿润阔叶林
Tropical & Subtropical Moist Broadleaf Forests

动物地理分布型 / Zoogeographic Distribution Type
J

分布标注 / Distribution Note
特有种 Endemic

▲ 濒危状况 / Threatened Status

中国生物多样性红色名录等级 / CB RL Category (2021)
近危 NT

IUCN 红色名录 / IUCN Red List (2021)
无危 LC

威胁因子 / Threats
未知 Unknown

▲ 法律保护地位 / Legal Protection Status

国家重点保护野生动物等级 / Category of National Key Protected Wild Animals (2021)
未列入 Not listed

"三有"名录 / TWIESSV (2023)
未列入 Not listed

CITES 附录等级 / CITES Appendix (2023)
未列入 Not listed

迁徙物种公约附录 / CMS Appendix (2020)
未列入 Not listed

保护行动 / Conservation Action
尚无保护行动 No conservation action so far

▲ 参考文献 / References

Jiang et al. (蒋志刚等), 2021; Wilson and Mittermeier, 2019; Han et al., 2010; Lee et al., 2007; Lin et al., 2002

180 / 东亚伏翼

Pipistrellus abramus (Temminck, 1838)

• Japanese Pipistrelle

▲ 分类地位 / Taxonomy

翼手目 Chiroptera / 蝙蝠科 Vespertilionidae / 伏翼属 *Pipistrellus*

科建立者及其文献 / Family Authority
Gray, 1821

属建立者及其文献 / Genus Authority
Kaup, 1829

亚种 / Subspecies
无 None

模式标本产地 / Type Locality
日本
Japan, Kyushu, Nagasaki

周佳俊 / 供图

▲ 其他名称 / Other Name(s)

其他中文名 / Other Chinese Name(s)
日本伏翼、小伏翼

其他英文名 / Other English Name(s)
Japanese House Bat

同物异名 / Synonym(s)
无 None

▲ 形态及生境 / Morphology and Habitat

形态特征 / Morphological Characteristics

齿式：2.1.2.3/3.1.2.3=34。头体长 38~60 mm。耳长 8~13 mm。前臂长 31~36 mm。后足长 6~10 mm。尾长 29~45 mm。颅全长 12.2~13.4 mm。耳屏小，端部钝圆，外缘基部有凹缺，向前微弯。第 5 指比第 3 或第 4 指长，背毛深褐色，腹毛灰褐色，毛色有变异。后足短小。翼膜止于趾基。股间膜呈锥状。

Dental formula: 2.1.2.3/3.1.2.3=34. Head and body length 38-60 mm. Ear length 8-13 mm. Forearm length 31-36 mm. Hind foot length 6-10 mm. Tail length 29-45 mm. Greatest skull length 12.2-13.4 mm. Tragus small, blunt round tip, outer margin at the base concave, slightly curved forward. The 5th finger is longer than the 3rd or 4th fingers. Dorsal hairs dark brown, and abdominal hairs taupe. Hair color varies. Hind feet short. Pterygoid membrane terminates at the toe base. Interfemoral membrane apex shaped.

生境 / Habitat

人造建筑、森林
Man-made building, forest

▲ 地理分布 / Geographic Distribution

国内分布 / Domestic Distribution

内蒙古、黑龙江、吉林、辽宁、河北、天津、山东、山西、陕西、甘肃、西藏、浙江、江苏、安徽、重庆、湖北、湖南、四川、贵州、云南、海南、福建、台湾、广东、广西、香港、澳门

Inner Mongolia, Heilongjiang, Jilin, Liaoning, Hebei, Tianjin, Shandong, Shanxi, Shaanxi, Gansu, Tibet, Zhejiang, Jiangsu, Anhui, Chongqing, Hubei, Hunan, Sichuan, Guizhou, Yunnan, Hainan, Fujian, Taiwan, Guangxi, Guangdong, Hong Kong, Macao

全球分布 / World Distribution

中国、日本、朝鲜、韩国、老挝、缅甸、越南
China, Japan, Democratic People's Republic of Korea, Republic of Korea, Laos, Myanmar, Vietnam

生物地理界 / Biogeographic Realm

古北界、印度马来界
Palearctic, Indomalaya

WWF 生物群系 / WWF Biome

热带和亚热带湿润阔叶林、温带阔叶和混交林
Tropical & Subtropical Moist Broadleaf Forests, Temperate Broadleaf & Mixed Forests

动物地理分布型 / Zoogeographic Distribution Type

Ug

分布标注 / Distribution Note

非特有种 Non-Endemic

▲ 濒危状况 / Threatened Status

中国生物多样性红色名录等级 / CB RL Category (2021)

无危 LC

IUCN 红色名录 / IUCN Red List (2021)

无危 LC

威胁因子 / Threats

无 None

▲ 法律保护地位 / Legal Protection Status

国家重点保护野生动物等级 / Category of National Key Protected Wild Animals (2021)

未列入 Not listed

"三有"名录 / TWIESSV (2023)

未列入 Not listed

CITES 附录等级 / CITES Appendix (2023)

未列入 Not listed

迁徙物种公约附录 / CMS Appendix (2020)

未列入 Not listed

保护行动 / Conservation Action

尚无保护行动 No conservation action so far

▲ 参考文献 / References

Jiang et al. (蒋志刚等), 2021; Wilson and Mittermeier, 2019; Ruedi et al., 2013; Pan et al. (潘清华等), 2007; Wilson and Reeder, 2005; Wang (王应祥), 2003

181 / 阿拉善伏翼

Pipistrellus alaschanicus (Bobrinskii, 1926)

· Alashanian Pipistrelle

▲ 分类地位 / Taxonomy

翼手目 Chiroptera / 蝙蝠科 Vespertilionidae / 伏翼属 *Pipistrellus*

科建立者及其文献 / Family Authority
Gray, 1821

属建立者及其文献 / Genus Authority
Kaup, 1829

亚种 / Subspecies
无 None

模式标本产地 / Type Locality
斯里兰卡
Sri Lanka, Trincomalee

▲ 其他名称 / Other Name(s)

其他中文名 / Other Chinese Name(s)
无 None

其他英文名 / Other English Name(s)
Black Gilded Pipistrelle

同物异名 / Synonym(s)
无 None

▲ 形态及生境 / Morphology and Habitat

形态特征 / Morphological Characteristics
齿式：2.1.2.3/3.1.2.3=34。颅全长 14 mm。头体长 38 mm。耳长 3 mm。前臂长 36~38 mm。尾长 39~40 mm。体型小。耳屏狭窄，基叶区别明确。背毛浅棕色，毛尖较淡。腹毛较淡，但毛基深棕色。尾尖稍伸出于尾膜外。
Dental formula: 2.1.2.3/3.1.2.3=34. Greatest skull length 14 mm. The head and body length 38 mm. Ear length 3mm. The forearm 36-38 mm long. Tail length 39-40 mm. A small Pipistrellus. Tragus narrow, distinct basal lobe. Dorsal hairs light brown with lighter tip. Abdominal hairs are lighter with dark brown base. The caudal tip extends slightly beyond the caudate membrane.

生境 / Habitat
洞穴 Caves

▲ 地理分布 / Geographic Distribution

国内分布 / Domestic Distribution

内蒙古、甘肃、宁夏、四川、陕西、河南、江苏、湖北、湖南、贵州、安徽、山东、河北、北京、山西、辽宁、吉林和黑龙江
Inner Mongolia, Gansu, Ningxia, Sichuan, Shaanxi, Henan, Jiangsu, Hubei, Hunan, Guizhou, Anhui, Shandong, Hebei, Beijing, Shanxi, Liaoning, Jilin and Heilongjiang

全球分布 / World Distribution

蒙古国、俄罗斯、朝鲜、日本
Mongolia, Russia, Democratic People's Republic of Korea, Japan

生物地理界 / Biogeographic Realm

古北界 Palearctic

WWF 生物群系 / WWF Biome

沙漠和干燥的灌木地温带阔叶和混交林
Deserts & Xeric Shrublands, Temperate Broadleaf & Mixed Forests

动物地理分布型 / Zoogeographic Distribution Type

Xb

分布标注 / Distribution Note

非特有种 Non-Endemic

▲ 濒危状况 / Threatened Status

中国生物多样性红色名录等级 / CB RL Category (2021)

未评估 NE

IUCN 红色名录 / IUCN Red List (2021)

无危 LC

威胁因子 / Threats

未知 Unknown

▲ 法律保护地位 / Legal Protection Status

国家重点保护野生动物等级 / Category of National Key Protected Wild Animals (2021)

未列入 Not listed

"三有" 名录 / TWIESSV (2023)

未列入 Not listed

CITES 附录等级 / CITES Appendix (2023)

未列入 Not listed

迁徙物种公约附录 / CMS Appendix (2020)

未列入 Not listed

保护行动 / Conservation Action

尚无保护行动 No conservation action so far

▲ 参考文献 / References

Burgin et al., 2020; IUCN, 2020; Wilson and Mittermeier, 2019; Zhang (张劲硕), 2010; Pan et al. (潘清华等), 2007

182 / 锡兰伏翼

Pipistrellus ceylonicus (Kelaart, 1852)

· Kelaart's Pipistrelle

▲ 分类地位 / Taxonomy

翼手目 Chiroptera / 蝙蝠科 Vespertilionidae / 伏翼属 *Pipistrellus*

科建立者及其文献 / Family Authority
Gray, 1821

属建立者及其文献 / Genus Authority
Kaup, 1829

亚种 / Subspecies
越北亚种 *P. c. raptor* Thomas, 1904
广西、广东
Guangxi, Guangdong
海南亚种 *P. c. tongfangensis* Wang, 1964
海南 Hainan

模式标本产地 / Type Locality
斯里兰卡
Sri Lanka, Trincomalee

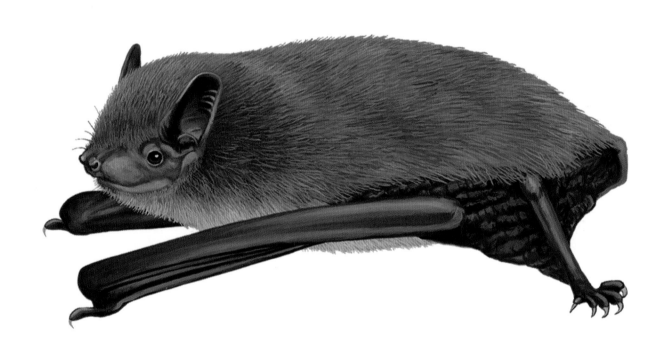

▲ 其他名称 / Other Name(s)

其他中文名 / Other Chinese Name(s)
无 None

其他英文名 / Other English Name(s)
Black Gilded Pipistrelle

同物异名 / Synonym(s)
无 None

▲ 形态及生境 / Morphology and Habitat

形态特征 / Morphological Characteristics
齿式：2.1.2.3/3.1.2.3=34。头体长 45~64 mm。耳长 9~14 mm。前臂长 33~42 mm。后足长 6~11 mm。尾长 30~45 mm。体重 9~10 g。颅全长 14.7~15.8 mm。毛发深棕色或棕色，毛发尖端无灰色。
Dental formula: 2.1.2.3/3.1.2.3=34. Head and body length 45-64 mm. Ear length 9-14 mm. Forearm length 33-42 mm. Hind foot length 6-11 mm. Tail length 30-45 mm. Body mass 9-10 g. Greatest skull length 14.7-15.8 mm. Hair color dark brown or brown, and hairs without gray tips.

生境 / Habitat
热带和亚热带湿润山地森林、洞穴、人造建筑
Tropical and subtropical moist montane forest, caves, man-made building

▲ 地理分布 / Geographic Distribution

国内分布 / Domestic Distribution
海南、广西、广东
Hainan, Guangxi, Guangdong

全球分布 / World Distribution
孟加拉国、中国、印度、印度尼西亚、马来西亚、缅甸、巴基斯坦、
斯里兰卡、越南
Bangladesh, China, India, Indonesia, Malaysia, Myanmar, Pakistan, Sri Lanka,
Vietnam

生物地理界 / Biogeographic Realm
印度马来界、大洋洲界
Indomalaya, Oceanian

WWF 生物群系 / WWF Biome
热带和亚热带湿润阔叶林
Tropical & Subtropical Moist Broadleaf Forests

动物地理分布型 / Zoogeographic Distribution Type
Wb

分布标注 / Distribution Note
非特有种 Non-Endemic

▲ 濒危状况 / Threatened Status

中国生物多样性红色名录等级 / CB RL Category (2021)
无危 LC

IUCN 红色名录 / IUCN Red List (2021)
无危 LC

威胁因子 / Threats
无 None

▲ 法律保护地位 / Legal Protection Status

国家重点保护野生动物等级 / Category of National Key Protected Wild Animals (2021)
未列入 Not listed

"三有"名录 / TWIESSV (2023)
未列入 Not listed

CITES 附录等级 / CITES Appendix (2023)
未列入 Not listed

迁徙物种公约附录 / CMS Appendix (2020)
未列入 Not listed

保护行动 / Conservation Action
尚无保护行动 No conservation action so far

▲ 参考文献 / References

Jiang et al. (蒋志刚等), 2021; Burgin et al., 2020; IUCN, 2020; Liu et al. (刘少英等), 2020; Wilson and Mittermeier, 2019; Huang et al. (黄继展等), 2013; Cen and Peng (岑业文和彭红元), 2010; Wei et al., 2010; Wu et al., 2009; James et al., 2008; Niu (牛红星), 2008; Pan et al. (潘清华等), 2007; Wu et al. (吴毅等), 2007; Wilson and Reeder, 2005; Wang (王应祥), 2003; Zhang (张荣祖), 1997

183 / 印度伏翼

Pipistrellus coromandra (Gray, 1838)

- Coromandel Pipistrelle

翼手目 Chiroptera / 蝙蝠科 Vespertilionidae / 伏翼属 *Pipistrellus*

科建立者及其文献 / Family Authority
Gray, 1821

属建立者及其文献 / Genus Authority
Kaup, 1829

亚种 / Subspecies
无 None

模式标本产地 / Type Locality
印度
India, Coromandel Coast, Pondicherry

吴毅 / 供图

▲ 其他名称 / Other Name(s)

其他中文名 / Other Chinese Name(s)
无 None

其他英文名 / Other English Name(s)
Coromandel Pipistrelle, Little Indian Bat

同物异名 / Synonym(s)
无 None

▲ 形态及生境 / Morphology and Habitat

形态特征 / Morphological Characteristics
齿式：2.1.2.3/3.1.2.3=34。头体长 34~49 mm。耳长 7~14 mm。前臂长 25~35 mm。后足长 3~8 mm。尾长 22~39 mm。颅全长 10.6~11.9 mm。体重 4~6 g。背部毛皮呈中棕色到深棕色，腹面毛色较浅，毛尖米褐色或肉桂棕色，毛根深色。耳壳和翼膜呈棕色至深棕色，基本裸露，在靠近身体和尾巴的尾膜上下有少许毛发。

Dental formula: 2.1.2.3/3.1.2.3=34. Head and body length 34-49 mm. Ear length 7-14 mm. Forearm length 25-35 mm. Hind foot length 3-8 mm. Tail length 22-39 mm. Greatest skull length 10.6-11.9 mm. Dorsal pelage uniformly mid-to dark brown; ventral surface paler, hairs with beige-brown or cinnamon-brown tips, and dark roots. Ears and membranes mid-to dark brown and essentially naked, with a few hairs on the uropatagium next to the body and tail, above and below.

生境 / Habitat
人造建筑、森林、耕地
Man-made building, forest, arable land

▲ 地理分布 / Geographic Distribution

国内分布 / Domestic Distribution
西藏、广东
Tibet, Guangdong

全球分布 / World Distribution
阿富汗、孟加拉国、不丹、柬埔寨、中国、印度、老挝、缅甸、尼泊尔、巴基斯坦、斯里兰卡、越南
Afghanistan, Bangladesh, Bhutan, Cambodia, China, India, Laos, Myanmar, Nepal, Pakistan, Sri Lanka, Vietnam

生物地理界 / Biogeographic Realm
印度马来界、古北界、大洋洲界
Indomalaya, Palearctic, Oceanian

WWF 生物群系 / WWF Biome
热带和亚热带湿润阔叶林
Tropical & Subtropical Moist Broadleaf Forests

动物地理分布型 / Zoogeographic Distribution Type
O01

分布标注 / Distribution Note
非特有种 Non-Endemic

▲ 濒危状况 / Threatened Status

中国生物多样性红色名录等级 / CB RL Category (2021)
无危 LC

IUCN 红色名录 / IUCN Red List (2021)
无危 LC

威胁因子 / Threats
无 None

▲ 法律保护地位 / Legal Protection Status

国家重点保护野生动物等级 / Category of National Key Protected Wild Animals (2021)
未列入 Not listed

"三有"名录 / TWIESSV (2023)
未列入 Not listed

CITES 附录等级 / CITES Appendix (2023)
未列入 Not listed

迁徙物种公约附录 / CMS Appendix (2020)
未列入 Not listed

保护行动 / Conservation Action
尚无保护行动 No conservation action so far

▲ 参考文献 / References

Jiang et al. (蒋志刚等), 2021; Burgin et al., 2020; IUCN, 2020; Wilson and Mittermeier, 2019; Pan et al. (潘清华等), 2007; Wilson and Reeder, 2005; Wang (王应祥), 2003; Zhang (张荣祖), 1997

184 / 爪哇伏翼

Pipistrellus javanicus (Gray, 1838)

· Javan Pipistrelle

▲ 分类地位 / Taxonomy

翼手目 Chiroptera / 蝙蝠科 Vespertilionidae / 伏翼属 *Pipistrellus*

科建立者及其文献 / Family Authority
Gray, 1821

属建立者及其文献 / Genus Authority
Kaup, 1829

亚种 / Subspecies
无 None

模式标本产地 / Type Locality
印度尼西亚
Indonesia, Java

余文华 / 供图

▲ 其他名称 / Other Name(s)

其他中文名 / Other Chinese Name(s)
无 None

其他英文名 / Other English Name(s)
无 None

同物异名 / Synonym(s)
无 None

▲ 形态及生境 / Morphology and Habitat

形态特征 / Morphological Characteristics

齿式：2.1.2.3/3.1.2.3=34。头体长 40~55 mm。耳长 5~15 mm。前臂长 30~36 mm。后足长 3~8 mm。尾长 26~40 mm。吻部宽扁、短窄。无眶上突，矢状嵴和人字嵴不明显。耳壳短，外缘基部具凸形突叶。耳壳、耳屏黑色。耳屏高度约为耳壳之半，外缘基部具凹形缺刻。体背烟褐色或深褐色。背色红棕色到深棕色，毛尖色浅发白。腹毛浅淡，毛基黑褐色，毛尖淡黄灰色或淡黄红色，翼膜和深棕色，无毛。翼膜止于趾基。后足较小，超过胫骨长的二分之一。

Dental formula: 2.1.2.3/3.1.2.3=34. Head and body length 40-55 mm. Ear length 5-15 mm. Forearm length 30-36 mm. Hind foot length 3-8 mm. Tail length 26-40 mm. The snout is wide and flat, short and narrow. There is no supraorbital process, and the sagittal crest and herringoid crest are not obvious. Auricle is short, with convex lobes at the base of the outer margin. Auricle and tragus are black and brown. Tragus is about half the height of the auricle, with a concave notch at the base of the outer margin. Dorsal coloration reddish brown to darker brown, sometimes with frosting from paler-tipped hairs. Hairs on belly are light, and the hair bases are black-brown, and the hair tips are yellowish gray or yellowish red. Wings and tail membranes are darker brown and essentially hairless. Pterygoid membrane terminates at the toe base. Hind feet are small, but more than 1/2 the length of the tibia.

生境 / Habitat

森林、耕地、种植园、城市、人造建筑
Forest, arable land, plantations, urban area, man-made building

▲ 地理分布 / Geographic Distribution

国内分布 / Domestic Distribution
西藏、云南
Tibet, Yunnan

全球分布 / World Distribution
阿富汗、孟加拉国、文莱、柬埔寨、中国、印度、印度尼西亚、老挝、马来西亚、缅甸、尼泊尔、巴基斯坦、菲律宾、新加坡、泰国、越南
Afghanistan, Bangladesh, Brunei, Cambodia, China, India, Indonesia, Laos, Malaysia, Myanmar, Nepal, Pakistan, Philippines, Singapore, Thailand, Vietnam

生物地理界 / Biogeographic Realm
古北界 Palearctic

WWF 生物群系 / WWF Biome
热带和亚热带湿润阔叶林
Tropical & Subtropical Moist Broadleaf Forests

动物地理分布型 / Zoogeographic Distribution Type
Wc

分布标注 / Distribution Note
非特有种 Non-Endemic

▲ 濒危状况 / Threatened Status

中国生物多样性红色名录等级 / CB RL Category (2021)
近危 NT

IUCN 红色名录 / IUCN Red List (2021)
无危 LC

威胁因子 / Threats
森林砍伐、耕种
Logging, farming

▲ 法律保护地位 / Legal Protection Status

国家重点保护野生动物等级 / Category of National Key Protected Wild Animals (2021)
未列入 Not listed

"三有" 名录 / TWIESSV (2023)
未列入 Not listed

CITES 附录等级 / CITES Appendix (2023)
未列入 Not listed

迁徙物种公约附录 / CMS Appendix (2020)
未列入 Not listed

保护行动 / Conservation Action
尚无保护行动 No conservation action so far

▲ 参考文献 / References

Jiang et al. (蒋志刚等), 2021; Burgin et al., 2020; IUCN, 2020; Liu et al. (刘少英等), 2020; Wilson and Mittermeier, 2019; Pan et al. (潘清华等), 2007; Wilson and Reeder, 2005; Gu et al. (谷晓明等), 2003; Wilson and Reeder, 2005; Wang (王应祥), 2003; Zhang (张荣祖), 1997; Hu and Wu (胡锦矗和吴毅), 1993

185 | 棒茎伏翼

Pipistrellus paterculus Thomas, 1915

• Mount Popa Pipistrelle

▲ 分类地位 / Taxonomy

翼手目 Chiroptera / 蝙蝠科 Vespertilionidae / 伏翼属 *Pipistrellus*

科建立者及其文献 / Family Authority
Gray, 1821

属建立者及其文献 / Genus Authority
Kaup, 1829

亚种 / Subspecies
云南亚种 *P. p. yunnanensis* (Wang, 1982)
云南 Yunnan

模式标本产地 / Type Locality
缅甸
Burma (Myanmar), Mt. Popa

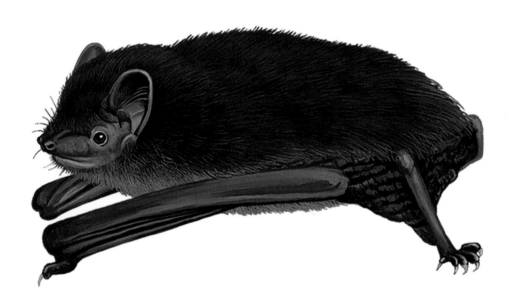

▲ 其他名称 / Other Name(s)

其他中文名 / Other Chinese Name(s)
无 None

其他英文名 / Other English Name(s)
Paternal Pipistrelle

同物异名 / Synonym(s)
无 None

▲ 形态及生境 / Morphology and Habitat

形态特征 / Morphological Characteristics
齿式：2.1.2.3/3.1.2.3=34。头体长 42~48 mm。耳长 10~13 mm。前臂长 29~34 mm。后足长 6~7 mm。尾长 31~38 mm。颅全长 13~14 mm。背部深棕色，腹部双色，发梢颜色较浅。尾膜近端三分之一被覆短毛。
Dental formula: 2.1.2.3/3.1.2.3=34. Head and body length 42-48 mm. Ear length 10-13 mm. Forearm length 29-34 mm. Hind foot length 6-7 mm. Tail length 31-38 mm. Greatest skull length 13-14 mm. Dorsal hairs are dark brown. Ventral bicolored, with paler hair tips. Uropatagium covered with hairs for proximal one-third.

生境 / Habitat
森林、种植园
Forest, plantations

▲ 地理分布 / Geographic Distribution

国内分布 / Domestic Distribution
云南、贵州
Yunnan, Guizhou

全球分布 / World Distribution
中国、印度、老挝、缅甸、越南
China, India, Laos, Myanmar, Vietnam

生物地理界 / Biogeographic Realm
印度马来界、古北界
Indomalaya, Palearctic

WWF 生物群系 / WWF Biome
热带和亚热带湿润阔叶林
Tropical & Subtropical Moist Broadleaf Forests

动物地理分布型 / Zoogeographic Distribution Type
Wb

分布标注 / Distribution Note
非特有种 Non-Endemic

▲ 濒危状况 / Threatened Status

中国生物多样性红色名录等级 / CB RL Category (2021)
无危 LC

IUCN 红色名录 / IUCN Red List (2021)
无危 LC

威胁因子 / Threats
无 None

▲ 法律保护地位 / Legal Protection Status

国家重点保护野生动物等级 / Category of National Key Protected Wild Animals (2021)
未列入 Not listed

"三有" 名录 / TWIESSV (2023)
未列入 Not listed

CITES 附录等级 / CITES Appendix (2023)
未列入 Not listed

迁徙物种公约附录 / CMS Appendix (2020)
未列入 Not listed

保护行动 / Conservation Action
尚无保护行动 No conservation action so far

▲ 参考文献 / References

Jiang et al. (蒋志刚等), 2021; Burgin et al., 2020; IUCN, 2020; Wilson and Mittermeier, 2019; Smith et al., 2009; Pan et al. (潘清华等), 2007; Wilson and Reeder, 2005; Wang (王应祥), 2003; Zhang (张荣祖), 1997

186 / 普通伏翼

Pipistrellus pipistrellus Schreber, 1774

• Common Pipistrelle

张礼标 / 供图

翼手目 Chiroptera / 蝙蝠科 Vespertilionidae /
伏翼属 *Pipistrellus*

科建立者及其文献 / Family Authority
Gray, 1821

属建立者及其文献 / Genus Authority
Kaup, 1829

亚种 / Subspecies
无 None

模式标本产地 / Type Locality
法国
France

▲ 其他名称 / Other Name(s)

其他中文名 / Other Chinese Name(s)
无 None

其他英文名 / Other English Name(s)
Small House Bat

同物异名 / Synonym(s)
无 None

▲ 形态及生境 / Morphology and Habitat

形态特征 / Morphological Characteristics
齿 式：2.1.2.3/3.1.2.3=34。头 体 长 40~48 mm。耳 长
10~12 mm。前 臂 长 30~32 mm。后 足 长 6~7 mm。尾
长 29~35 mm。体重 4~5 g。颅全长 10.5 mm。头骨狭
窄。吻突狭长，中央具浅槽。前颌骨和上颌骨愈合。
无矢状嵴，人字嵴不明显。吻鼻部腺体明显。耳短宽，
前缘稍外凸，顶端钝，后缘具一小凹刻。耳屏接近耳
高之半。阴茎骨小，前端双叉。毛被浅黑灰色，体侧
较浅。头部和背部浅黄到栗褐色，毛基乌黑。喉部和
腹部毛基乌黑，毛尖色浅。翼膜和尾膜深褐色，无毛。
Dental formula: 2.1.2.3/3.1.2.3=34. Head and body length 40-
48 mm. Ear length 10-12 mm. Forearm length 30-32 mm. Hind
foot length 6-7 mm. Tail length 29-35 mm. Body mass 4-5 g.
Greatest skull length 10.5 mm. The skull is narrow. Proboscis
is long and narrow with a shallow groove in the center. The
anterior and maxillary bones are healed. There is no sagittal crest
and the lambdoid crest is not prominent. Rostral nasal glands
are conspicuous. Ears short and broad, anterior margin slightly
convex, apex blunt, posterior margin with a small indentation.
The tragus is about half the height of the ear. Baculum small,
anterior bifid. Pelage color blackish gray, paler near sides of the
body. Ventral part of hairs jet black. Wing and tail membranes
are dark brown without any hair on them.

生境 / Habitat
人造建筑、森林
Man-made building, forest

▲ 地理分布 / Geographic Distribution

国内分布 / Domestic Distribution

陕西、新疆、江西、云南、台湾、四川、重庆、山东、广西、广东、福建、浙江、澳门

Shaanxi, Xinjiang, Jiangxi, Yunnan, Taiwan, Sichuan, Chongqing, Shandong, Guangxi, Guangdong, Fujian, Zhejiang, Macao

全球分布 / World Distribution

阿富汗、阿尔巴尼亚、阿尔及利亚、安道尔、亚美尼亚、奥地利、阿塞拜疆、白俄罗斯、比利时、波斯尼亚和黑塞哥维那、保加利亚、中国、克罗地亚、塞浦路斯、捷克、丹麦、爱沙尼亚、芬兰、法国、格鲁吉亚、德国、直布罗陀、希腊、梵蒂冈、匈牙利、印度、伊朗、爱尔兰、以色列、意大利、哈萨克斯坦、拉脱维亚、黎巴嫩、利比亚、列支敦士登、立陶宛、卢森堡、马其顿、马耳他、摩尔多瓦、摩纳哥、黑山、摩洛哥、缅甸、荷兰、挪威、巴基斯坦、波兰、葡萄牙、罗马尼亚、俄罗斯、圣马力诺、塞尔维亚、斯洛伐克、斯洛文尼亚、西班牙、瑞典、瑞士、突尼斯、土耳其、乌克兰、英国

Afghanistan, Albania, Algeria, Andorra, Armenia, Austria, Azerbaijan, Belarus, Belgium, Bosnia and Herzegovina, Bulgaria, China, Croatia, Cyprus, Czech, Denmark, Estonia, Finland, France, Georgia, Germany, Gibraltar, Greece, Vatican, Hungary, India, Iran, Ireland, Israel, Italy, Kazakhstan, Latvia, Lebanon, Libya, Liechtenstein, Lithuania, Luxembourg, Macedonia, Malta, Moldova, Monaco, Montenegro, Morocco, Myanmar, Netherlands, Norway, Pakistan, Poland, Portugal, Romania, Russia, San Marino, Serbia, Slovakia, Slovenia, Spain, Sweden, Switzerland, Tunisia, Turkey, Ukraine, United Kingdom

生物地理界 / Biogeographic Realm

印度马来界、古北界
Indomalaya, Palearctic

WWF 生物群系 / WWF Biome

温带阔叶和混交林、温带针叶林
Temperate Broadleaf & Mixed Forests, Temperate Conifer Forest

动物地理分布型 / Zoogeographic Distribution Type

We

分布标注 / Distribution Note

非特有种 Non-Endemic

▲ 濒危状况 / Threatened Status

中国生物多样性红色名录等级 / CB RL Category (2021)

无危 LC

IUCN 红色名录 / IUCN Red List (2021)

无危 LC

威胁因子 / Threats

无 None

▲ 法律保护地位 / Legal Protection Status

国家重点保护野生动物等级 / Category of National Key Protected Wild Animals (2021)

未列入 Not listed

"三有"名录 / TWIESSV (2023)

未列入 Not listed

CITES 附录等级 / CITES Appendix (2023)

未列入 Not listed

迁徙物种公约附录 / CMS Appendix (2020)

未列入 Not listed

保护行动 / Conservation Action

尚无保护行动 No conservation action so far

▲ 参考文献 / References

Jiang et al. (蒋志刚等), 2021; Burgin et al., 2020; IUCN, 2020; Wilson and Mittermeier, 2019; Zhou and Yang (周江和杨天友), 2009, 2010; Smith and Xie, 2009; Pan et al. (潘清华等), 2007; Wilson and Reeder, 2005; Wang (王应祥), 2003; Zhang (张荣祖), 1997; Hu and Wu (胡锦矗和吴毅), 1993

187 / 小伏翼

Pipistrellus tenuis (Temminck, 1840)

· Least Pipistrelle

张礼标 / 供图

▲ 分类地位 / Taxonomy

翼手目 Chiroptera / 蝙蝠科 Vespertilionidae /
伏翼属 *Pipistrellus*

科建立者及其文献 / Family Authority
Gray, 1821

属建立者及其文献 / Genus Authority
Kaup, 1829

亚种 / Subspecies
无 None

模式标本产地 / Type Locality
印度尼西亚
Indonesia, Sumatra

▲ 其他名称 / Other Name(s)

其他中文名 / Other Chinese Name(s)
侏伏翼

其他英文名 / Other English Name(s)
Indian Pygmy Bat

同物异名 / Synonym(s)
无 None

▲ 形态及生境 / Morphology and Habitat

形态特征 / Morphological Characteristics
齿式：2.1.2.3/3.1.2.3=34。头体长 33~45 mm。尾长 20~35 mm。前臂长 25~31 mm。后足长 4~6 mm。体重 3.2~4.2 g。颅全长约 12 mm。吻狭窄，耳郭和耳屏相对本属其他种类较短。吻鼻和面部微具短毛。鼻略呈管状，鼻孔向两外侧倾斜。耳屏长小于耳长之半，顶端钝圆。额毛青黑色。颈背、背部和臀部被毛深褐色，毛尖微棕。腹毛多浅灰色。翼膜止于趾基，有距缘膜。

Dental formula: 2.1.2.3/3.1.2.3=34. Head and body length 33-45 mm. Ear length 5-11 mm. Forearm length 25-31 mm. Hind foot length 4-6 mm. Tail length 20-35 mm. Body mass 3.2-4.2 g. Greatest skull length 9-11 mm. Snout is narrow, and the ears and tragus are short relative to other species of this genus. Snout and face are slightly covered with short hairs. Nose is slightly tubular and the nostrils slant laterally. Tragus with a blunt round tip which is less than half the length of the ear. Hairs on the forehead are bluish-black. Neck, back, and buttocks are covered with dark brown hair with a light brown tip. Ventral hairs are lighter gray. Pterygium terminates at the toe base and has a distal membrane.

生境 / Habitat
人造建筑、森林、城市、次生林
Man-made building, forest, urban area, secondary forest

▲ 地理分布 / Geographic Distribution

国内分布 / Domestic Distribution
浙江、福建、广西、广东、海南、四川、贵州、云南、重庆
Zhejiang, Fujian, Guangxi, Guangdong, Hainan, Sichuan, Guizhou, Yunnan, Chongqing

全球分布 / World Distribution
阿富汗、孟加拉国、柬埔寨、中国、圣诞岛、可可岛、印度、印度尼西亚、老挝、马来西亚、缅甸、尼泊尔、巴基斯坦、菲律宾、斯里兰卡、泰国、东帝汶、越南
Afghanistan, Bangladesh, Cambodia, China, Christmas Island, Coco Island, India, Indonesia, Laos, Malaysia, Myanmar, Nepal, Pakistan, Philippines, Sri Lanka, Thailand, Timor-Leste, Vietnam

生物地理界 / Biogeographic Realm
印度马来界、古北界、大洋洲界
Indomalaya, Palearctic, Oceanian

WWF 生物群系 / WWF Biome
热带和亚热带湿润阔叶林
Tropical & Subtropical Moist Broadleaf Forests

动物地理分布型 / Zoogeographic Distribution Type
We

分布标注 / Distribution Note
非特有种 Non-Endemic

▲ 濒危状况 / Threatened Status

中国生物多样性红色名录等级 / CB RL Category (2021)
近危 NT

IUCN 红色名录 / IUCN Red List (2021)
无危 LC

威胁因子 / Threats
未知 Unknown

▲ 法律保护地位 / Legal Protection Status

国家重点保护野生动物等级 / Category of National Key Protected Wild Animals (2021)
未列入 Not listed

"三有" 名录 / TWIESSV (2023)
未列入 Not listed

CITES 附录等级 / CITES Appendix (2023)
未列入 Not listed

迁徙物种公约附录 / CMS Appendix (2020)
未列入 Not listed

保护行动 / Conservation Action
尚无保护行动 No conservation action so far

▲ 参考文献 / References

Jiang et al. (蒋志刚等), 2021; Burgin et al., 2020; IUCN, 2020; Liu et al. (刘少英等), 2020; Wilson and Mittermeier, 2019; Zhou et al. (周江等), 2011; Pan et al. (潘清华等), 2007; Wilson and Reeder, 2005; International Co mmission on Zoological Nomenclature, 2003; Wang (王应祥), 2003

188 / 大黑伏翼

Arielulus circumdatus (Temminck, 1840)

• Bronze Sprite

张礼标 / 供图

▲ 分类地位 / Taxonomy

翼手目 Chiroptera / 蝙蝠科 Vespertilionidae /
金背伏翼属 *Arielulus*

科建立者及其文献 / Family Authority
Gray, 1821

属建立者及其文献 / Genus Authority
Hill & Harrison, 1987

亚种 / Subspecies
无 None

模式标本产地 / Type Locality
印度尼西亚
Indonesia, Java, Tapos

▲ 其他名称 / Other Name(s)

其他中文名 / Other Chinese Name(s)
无 None

其他英文名 / Other English Name(s)
Black-gilded Pipistrelle

同物异名 / Synonym(s)
无 None

▲ 形态及生境 / Morphology and Habitat

形态特征 / Morphological Characteristics
齿式：2.1.2.3/3.1.2.3=34。头体长 46~58 mm。耳长 15 mm。前臂长 39~44 mm。后足长 10 mm。尾长 40 mm。颅全长 15~16 mm。头骨额凹显著。吻突宽短。脑颅圆滑，颧弓粗，无矢状嵴。耳中等，耳尖钝圆，深棕黑色。耳屏宽阔，端部钝或稍尖，前端稍凸，后部稍凹。吻鼻部被稀疏绒毛。背毛黑色，毛尖带橙色，头和背有橙色光泽。腹面棕色，毛略有双色，毛基比毛尖稍深。拇指连爪长 8~9 mm。第五指长等于或超过第四掌骨加第一指节的总长度。

Dental formula: 2.1.2.3/3.1.2.3=34. Head and body length 46-58 mm. Ear length 15 mm. Forearm length 39-44 mm. Hind foot length 10 mm. Tail length 40 mm. Greatest skull length 15-16 mm. Forehead of the skull is obviously concave. Muzzle is broad and short. Craniocranial smooth, thick zygomatic arch, no sagittal ridge. Medium ears, tip blunt round, deep brown-black, tragus wide, and blunt end or slightly pointed, slightly convex front end, slightly concave rear. Snout is sparsely furred. Dorsal hairs are black with orange tips, and the head and back have an orange sheen. Ventral hairs are brown, and the hairs are slightly bicolor, and the bases are slightly darker than the tips. Thumb and claw length 8-9 mm. 5th finger length is equal to or more than the total length of the 4th metacarpal plus the 1st knuckle.

生境 / Habitat
未知 Unknown

▲ 地理分布 / Geographic Distribution

国内分布 / Domestic Distribution
云南、广东
Yunnan, Guangdong

全球分布 / World Distribution
柬埔寨、中国、印度、印度尼西亚、马来西亚、缅甸、尼泊尔、泰国、
越南
Cambodia, China, India, Indonesia, Malaysia, Myanmar, Nepal, Thailand,
Vietnam

生物地理界 / Biogeographic Realm
古北界 Palearctic

WWF 生物群系 / WWF Biome
热带和亚热带湿润阔叶林
Tropical & Subtropical Moist Broadleaf Forests

动物地理分布型 / Zoogeographic Distribution Type
Wa

分布标注 / Distribution Note
非特有种 Non-Endemic

▲ 濒危状况 / Threatened Status

中国生物多样性红色名录等级 / CB RL Category (2021)
易危 VU

IUCN 红色名录 / IUCN Red List (2021)
无危 LC

威胁因子 / Threats
未知 Unknown

▲ 法律保护地位 / Legal Protection Status

国家重点保护野生动物等级 / Category of National Key Protected Wild Animals (2021)
未列入 Not listed

"三有" 名录 / TWIESSV (2023)
未列入 Not listed

CITES 附录等级 / CITES Appendix (2023)
未列入 Not listed

迁徙物种公约附录 / CMS Appendix (2020)
未列入 Not listed

保护行动 / Conservation Action
尚无保护行动 No conservation action so far

▲ 参考文献 / References

Jiang et al. (蒋志刚等), 2021; Burgin et al., 2020; IUCN, 2020; Wilson and Mittermeier, 2019; Pan et al. (潘清华等), 2007; Wilson and Reeder, 2005; Wang (王应祥), 2003; Zhang (张荣祖), 1997

189 / 茶褐伏翼

Hypsugo affinis Dobson, 1871

· Chocolate Pipistrelle

▲ 分类地位 / Taxonomy

翼手目 Chiroptera / 蝙蝠科 Vespertilionidae / 高级伏翼属 *Hypsugo*

科建立者及其文献 / Family Authority
Gray, 1821

属建立者及其文献 / Genus Authority
Troughton, 1943

亚种 / Subspecies
无 None

模式标本产地 / Type Locality
缅甸
Burma (Myanmar), Bhamo

Sanjan Thapa / 供图

▲ 其他名称 / Other Name(s)

其他中文名 / Other Chinese Name(s)
无 None

其他英文名 / Other English Name(s)
Grizzled Pipistrelle, Highland Pipistrelle

同物异名 / Synonym(s)
无 None

▲ 形态及生境 / Morphology and Habitat

形态特征 / Morphological Characteristics

齿式：2.1.2.3/3.1.2.3=34。头体长 43~51 mm。耳长 12~15 mm。前臂长 38~42 mm。后足长 7~8 mm。尾长 30~41 mm。颅全长 14.1 mm。吻突延长，睛后突明显。耳壳宽长，耳尖钝圆，耳屏短，外缘有凹形小缺刻。后足小。背毛棕色，毛尖灰色，腹毛浅灰白色。翼膜、耳和吻鼻部裸露部位黑褐色。翼膜游离缘终止于趾骨基部。

Dental formula: 2.1.2.3/3.1.2.3=34. Head and body length 43-51 mm. Forearm length 38-42 mm. Hind foot length 7-8 mm. Tail length 30-41 mm. Greatest skull length 14.1 mm. Ear length 12-15 mm. Rostrum process is prolonged and the posterior orbital process is obvious. Auricle is wide and long. Tips of the ears are blunt and round, and the tragus is short, and the outer edge has a small concave notch. Hind feet small. Dorsal hairs are brown with gray tips and the abdomen is pale grayish-white. Naked parts of pterygium, ears, and snout are dark brown. Free margin of the pterygium terminates at the base of the phalanges.

生境 / Habitat
人造建筑、森林
Man-made building, forest

▲ 地理分布 / Geographic Distribution

国内分布 / Domestic Distribution
西藏、云南、广西
Tibet, Yunnan, Guangxi

全球分布 / World Distribution
中国、印度、缅甸、尼泊尔、斯里兰卡
China, India, Myanmar, Nepal, Sri Lanka

生物地理界 / Biogeographic Realm
印度马来界、古北界
Indomalaya, Palearctic

WWF 生物群系 / WWF Biome
热带和亚热带湿润阔叶林
Tropical & Subtropical Moist Broadleaf Forests

动物地理分布型 / Zoogeographic Distribution Type
Wb

分布标注 / Distribution Note
非特有种 Non-Endemic

▲ 濒危状况 / Threatened Status

中国生物多样性红色名录等级 / CB RL Category (2021)
无危 LC

IUCN 红色名录 / IUCN Red List (2021)
无危 LC

威胁因子 / Threats
无 None

▲ 法律保护地位 / Legal Protection Status

国家重点保护野生动物等级 / Category of National Key Protected Wild Animals (2021)
未列入 Not listed

"三有" 名录 / TWIESSV (2023)
未列入 Not listed

CITES 附录等级 / CITES Appendix (2023)
未列入 Not listed

迁徙物种公约附录 / CMS Appendix (2020)
未列入 Not listed

保护行动 / Conservation Action
尚无保护行动 No conservation action so far

▲ 参考文献 / References

Jiang et al. (蒋志刚等), 2021; Burgin et al., 2020; IUCN, 2020; Liu et al. (刘少英等), 2020; Wilson and Mittermeier, 2019; Wang (王应祥), 2003; Corbet and Hill, 1992

190 / 卡氏伏翼

Hypsugo cadornae Thomas, 1916

• Cadorna's pipistrelle

▲ 分类地位 / Taxonomy

翼手目 Chiroptera / 蝙蝠科 Vespertilionidae / 高级伏翼属 *Hypsugo*

科建立者及其文献 / Family Authority
Gray, 1821

属建立者及其文献 / Genus Authority
Troughton, 1943

亚种 / Subspecies
无 None

模式标本产地 / Type Locality
印度
India, Darjeeling, Pashok, 3,500 ft. (1,067 m)

▲ 其他名称 / Other Name(s)

其他中文名 / Other Chinese Name(s)
无 None

其他英文名 / Other English Name(s)
Thomas's Pipistrelle

同物异名 / Synonym(s)
无 None

▲ 形态及生境 / Morphology and Habitat

形态特征 / Morphological Characteristics

齿式：2.1.2.3/3.1.2.3=34。头体长 40~47 mm。耳长 11~14 mm。前臂长 32~37 mm。后足长 5~7 mm。尾长 36~45 mm。第五掌骨几乎与第三和第四掌骨一样长。吻突短、窄。鼻缺刻宽而圆。颅骨饱满而圆，中间有浅矢状嵴。吻突和颅骨之间，额部凹陷缺如。颌下毛发为金黄色。背毛色深黑棕色，毛尖金黄褐色。腹面毛色棕色，毛尖灰淡黄色。

Dental formula: 2.1.2.3/3.1.2.3=34. Head and body length 40-47 mm. Ear length 11-14 mm. Forearm length 32-37 mm. Hind foot length 5-7 mm. Tail length 36-45 mm. The fifth metacarpal is nearly as long as the third and fourth metacarpal. Rostrum is short and narrow. Nasal notch is broad and round. Braincase is full and rounded with a shallow sagittal crest in the midpart. Cranial profile lacked a frontal depression between the rostrum and braincase. Golden brown hairs under the jaw. Dorsal hair color dark brown with golden brown hair tip. Abdominal hair color is brown, with a gray light-yellow hair tip.

生境 / Habitat
栖息于建筑物、也见于棕榈科植物枯叶内
Roosting in artificial buildings, also seen in dried leaves of palmae

▲ 地理分布 / Geographic Distribution

国内分布 / Domestic Distribution
云南 Yunnan

全球分布 / World Distribution
中国、印度、老挝、缅甸、泰国、越南
China, India, Laos, Myanmar, Thailand, Vietnam

生物地理界 / Biogeographic Realm
印度马来界 Indomalaya

WWF 生物群系 / WWF Biome
热带和亚热带湿润阔叶林
Tropical & Subtropical Moist Broadleaf Forests

动物地理分布型 / Zoogeographic Distribution Type
Wa

分布标注 / Distribution Note
非特有种 Non-Endemic

▲ 濒危状况 / Threatened Status

中国生物多样性红色名录等级 / CB RL Category (2021)
未评定 NE

IUCN 红色名录 / IUCN Red List (2021)
无危 LC

威胁因子 / Threats
未知 Unknown

▲ 法律保护地位 / Legal Protection Status

国家重点保护野生动物等级 / Category of National Key Protected Wild Animals (2021)
未列入 Not listed

"三有"名录 / TWIESSV (2023)
未列入 Not listed

CITES 附录等级 / CITES Appendix (2023)
未列入 Not listed

迁徙物种公约附录 / CMS Appendix (2020)
未列入 Not listed

保护行动 / Conservation Action
尚无保护行动 No conservation action so far

▲ 参考文献 / References

Xie et al., 2021; Burgin et al., 2020; IUCN, 2020

191 / 大灰伏翼

Hypsugo mordax (Peters, 1866)

· Pungent Pipistrelle

▲ 分类地位 / Taxonomy

翼手目 Chiroptera / 蝙蝠科 Vespertilionidae / 高级伏翼属 *Hypsugo*

科建立者及其文献 / Family Authority
Gray, 1821

属建立者及其文献 / Genus Authority
Troughton, 1943

亚种 / Subspecies
无 None

模式标本产地 / Type Locality
印度尼西亚
Indonesia, Java

▲ 其他名称 / Other Name(s)

其他中文名 / Other Chinese Name(s)
无 None

其他英文名 / Other English Name(s)
无 None

同物异名 / Synonym(s)
无 None

▲ 形态及生境 / Morphology and Habitat

形态特征 / Morphological Characteristics

齿式：2.1.2.3/3.1.2.3=34。头体长 47~56 mm。耳长 14~16 mm。前臂长 40~42 mm。后足长 6~9 mm。尾长 37~42 mm。耳屏中等长度。背毛锈棕色，发尖颜色较浅。腹毛黑棕色，发尖淡褐色。肛门区域淡褐黄色。翼膜附在脚趾基部。尾尖突出于尾膜外。

Dental formula: 2.1.2.3/3.1.2.3=34. Head and body length 47-56 mm. Ear length 14-16 mm. Forearm length 40-42 mm. Hind foot length 6-9 mm. Tail length 37-42 mm. Dorsum rust brown with paler hair tips. Venter black-brown with pale brown hair tips. Anal region pale brownish-yellow, Tragus medium-long. Wing membrane attached to the base of toes. Tail projects beyond uropatagium.

生境 / Habitat
未知 Unknown

▲ 地理分布 / Geographic Distribution

国内分布 / Domestic Distribution
云南 Yunnan

全球分布 / World Distribution
中国、印度尼西亚
China, Indonesia

生物地理界 / Biogeographic Realm
印度马来界 Indomalaya

WWF 生物群系 / WWF Biome
热带和亚热带湿润阔叶林
Tropical & Subtropical Moist Broadleaf Forests

动物地理分布型 / Zoogeographic Distribution Type
Wa

分布标注 / Distribution Note
非特有种 Non-Endemic

▲ 濒危状况 / Threatened Status

中国生物多样性红色名录等级 / CB RL Category (2021)
近危 NT

IUCN 红色名录 / IUCN Red List (2021)
数据缺乏 DD

威胁因子 / Threats
未知 Unknown

▲ 法律保护地位 / Legal Protection Status

国家重点保护野生动物等级 / Category of National Key Protected Wild Animals (2021)
未列入 Not listed

"三有"名录 / TWIESSV (2023)
未列入 Not listed

CITES 附录等级 / CITES Appendix (2023)
未列入 Not listed

迁徙物种公约附录 / CMS Appendix (2020)
未列入 Not listed

保护行动 / Conservation Action
尚无保护行动 No conservation action so far

▲ 参考文献 / References

Jiang et al. (蒋志刚等), 2021; Burgin et al., 2020; IUCN, 2020; Liu et al. (刘少英等), 2020; Pan et al. (潘清华等), 2007; Wang (王应祥), 2003; Csorba and Lee, 1999

192 / 灰伏翼

Hypsugo pulveratus Peters, 1871

• Chinese Pipistrelle

张礼标 / 供图

翼手目 Chiroptera / 蝙蝠科 Vespertilionidae / 高级伏翼属 *Hypsugo*

科建立者及其文献 / Family Authority
Gray, 1821

属建立者及其文献 / Genus Authority
Troughton, 1943

亚种 / Subspecies
无 None

模式标本产地 / Type Locality
中国
China, Fukien, Amoy

▲ 其他名称 / Other Name(s)

其他中文名 / Other Chinese Name(s)
无 None

其他英文名 / Other English Name(s)
无 None

同物异名 / Synonym(s)
无 None

▲ 形态及生境 / Morphology and Habitat

形态特征 / Morphological Characteristics
齿式：2.1.2.3/3.1.2.3=34。头体长 40~48 mm。前臂长 33~37 mm。后足长 6~8 mm。尾长 33~39 mm。颅全长 14~15 mm。脑颅部凸出，后部宽。眶上突发达。吻突长。耳狭窄，耳缘泛白，耳屏高为耳长之的三分之一。背毛色暗，近乎浅黑棕色，毛尖浅金黄褐色。腹面毛色相对较淡，近乎棕色，毛尖有点灰白。翼膜附着于趾基。
Dental formula: 2.1.2.3/3.1.2.3=34. Head and body length 40-48 mm. Forearm length 33-37 mm. Hind foot length 6-8 mm. Tail length 33-39 mm. Greatest skull length 14-15 mm. Ear length 12-14 mm. Craniocranial bulge, wide posterior. Orbital burst up. Rostrum long. Ears are narrow, and the fringe of the ears are white, and tragus height is 1/3 of the length of the ear. Dorsal hairs are almost blackish-brown with slight golden-brown tips. Hairs on the ventral side are relatively light brown with grayish tips. Pterygoid membrane is attached to the toe base.

生境 / Habitat
森林、人造建筑
Forest, man-made building

▲ 地理分布 / Geographic Distribution

国内分布 / Domestic Distribution
陕西、上海、江苏、安徽、湖南、四川、重庆、贵州、云南、广东、海南、福建、香港
Shaanxi, Shanghai, Jiangsu, Anhui, Hunan, Sichuan, Chongqing, Guizhou, Yunnan, Guangdong, Hainan, Fujian, Hong Kong

全球分布 / World Distribution
中国、老挝、缅甸、泰国、越南
China, Laos, Myanmar, Thailand, Vietnam

生物地理界 / Biogeographic Realm
印度马来界、古北界
Indomalaya, Palearctic

WWF 生物群系 / WWF Biome
热带和亚热带湿润阔叶林、温带阔叶和混交林
Tropical & Subtropical Moist Broadleaf Forests, Temperate Broadleaf & Mixed Forests

动物地理分布型 / Zoogeographic Distribution Type
We

分布标注 / Distribution Note
非特有种 Non-Endemic

▲ 濒危状况 / Threatened Status

中国生物多样性红色名录等级 / CB RL Category (2021)
近危 NT

IUCN 红色名录 / IUCN Red List (2021)
无危 LC

威胁因子 / Threats
未知 Unknown

▲ 法律保护地位 / Legal Protection Status

国家重点保护野生动物等级 / Category of National Key Protected Wild Animals (2021)
未列入 Not listed

"三有" 名录 / TWIESSV (2023)
未列入 Not listed

CITES 附录等级 / CITES Appendix (2023)
未列入 Not listed

迁徙物种公约附录 / CMS Appendix (2020)
未列入 Not listed

保护行动 / Conservation Action
尚无保护行动 No conservation action so far

▲ 参考文献 / References

Jiang et al. (蒋志刚等), 2021; Burgin et al., 2020; IUCN, 2020; Liu et al. (刘少英等), 2020; Wilson and Mittermeier, 2019; Pan et al. (潘清华等), 2007; Wilson and Reeder, 2005; Wang (王应祥), 2003; Zhang (张荣祖), 1997

193 / 萨氏伏翼

Hypsugo savii Bonaparte, 1837

· Savi's Pipistrelle

翼手目 Chiroptera / 蝙蝠科 Vespertilionidae / 高级伏翼属 *Hypsugo*

科建立者及其文献 / Family Authority
Gray, 1821

属建立者及其文献 / Genus Authority
Kolenati, 1856

亚种 / Subspecies
无 None

模式标本产地 / Type Locality
意大利
Italy, Pisa

▲ 其他名称 / Other Name(s)

其他中文名 / Other Chinese Name(s)
无 None

其他英文名 / Other English Name(s)
无 None

同物异名 / Synonym(s)
无 None

▲ 形态及生境 / Morphology and Habitat

形态特征 / Morphological Characteristics

齿式：2.1.2.3/3.1.2.3=34。头体长 47~60 mm。耳长 10~14 mm。前臂长 32~36 mm。后足长 6~8 mm。尾长 30~35 mm。颅全长 13.1~14.2 mm。耳和耳屏宽而短。翼膜附脚跟。尾距尾鞘后缘约 5 mm。背毛深棕色。腹侧毛发灰褐色。

Dental formula: 2.1.2.3/3.1.2.3=34. Head and body length 47-60 mm. Ear length 10-14 mm. Forearm length 32-36 mm. Tail length 30-35 mm. Hind foot length 6-8 mm. Greatest skull length 13.1-14.2 mm. Ears and tragus wide and short. Wing membrane attaches to the heel. Tail extends about 5 mm from the posterior margin of the uropatagium. Dorsal hairs are dark brown. Ventral surface taupe.

生境 / Habitat

人造建筑、森林、湿地、牧场、内陆岩石区域
Man-made building, forest, wetlands (inland), pastureland, inland rocky areas

▲ 地理分布 / Geographic Distribution

国内分布 / Domestic Distribution

新疆 Xinjiang

全球分布 / World Distribution

阿富汗、阿尔巴尼亚、阿尔及利亚、安道尔、奥地利、阿塞拜疆、孟加拉国、波斯尼亚和黑塞哥维那、保加利亚、中国、克罗地亚、塞浦路斯、捷克、法国、格鲁吉亚、直布罗陀、希腊、梵蒂冈、匈牙利、印度、伊朗、伊拉克、以色列、意大利、日本、哈萨克斯坦、朝鲜、韩国、黎巴嫩、摩纳哥、黑山、摩洛哥、巴勒斯坦、葡萄牙、俄罗斯、圣马力诺、塞尔维亚、斯洛伐克、斯洛文尼亚、西班牙、瑞士、叙利亚、突尼斯、土耳其、土库曼斯坦、乌克兰

Afghanistan, Albania, Algeria, Andorra, Austria, Azerbaijan, Bangladesh, Bosnia and Herzegovina, Bulgaria, China, Croatia, Cyprus, Czech, France, Georgia, Gibraltar, Greece, Vatican, Hungary, India, Iran, Iraq, Israel, Italy, Japan, Kazakhstan, Democratic People's Republic of Korea, Republic of Korea, Lebanon, Monaco, Montenegro, Morocco, Palestine, Portugal, Russia, San Marino, Serbia, Slovakia, Slovenia, Spain, Switzerland, Syria, Tunisia, Turkey, Turkmenistan, Ukraine

生物地理界 / Biogeographic Realm

古北界 Palearctic

WWF 生物群系 / WWF Biome

温带阔叶和混交林、温带针叶林
Temperate Broadleaf and Mixed Forests, Temperate Conifer Forest

动物地理分布型 / Zoogeographic Distribution Type

Pa

分布标注 / Distribution Note

非特有种 Non-Endemic

▲ 濒危状况 / Threatened Status

中国生物多样性红色名录等级 / CB RL Category (2021)

近危 NT

IUCN 红色名录 / IUCN Red List (2021)

无危 LC

威胁因子 / Threats

未知 Unknown

▲ 法律保护地位 / Legal Protection Status

国家重点保护野生动物等级 / Category of National Key Protected Wild Animals (2021)
未列入 Not listed

"三有"名录 / TWIESSV (2023)
未列入 Not listed

CITES 附录等级 / CITES Appendix (2023)
未列入 Not listed

迁徙物种公约附录 / CMS Appendix (2020)
未列入 Not listed

保护行动 / Conservation Action
尚无保护行动 No conservation action so far

▲ 参考文献 / References

Jiang et al. (蒋志刚等), 2021; Burgin et al., 2020; IUCN, 2020; Liu et al. (刘少英等), 2020; Wilson and Mittermeier, 2019; Pan et al. (潘清华等), 2007; Wang (王应祥), 2003; Zhang (张荣祖), 1997

194 / 南蝠

Ia io Thomas, 1902

· Great Evening Bat

▲ 分类地位 / Taxonomy

翼手目 Chiroptera / 蝙蝠科 Vespertilionidae / 南蝠属 *Ia*

科建立者及其文献 / Family Authority
Gray, 1821

属建立者及其文献 / Genus Authority
Thomas, 1902

亚种 / Subspecies
无 None

模式标本产地 / Type Locality
中国
China, Hupeh, Chungyang

汤亮 / 供图

▲ 其他名称 / Other Name(s)

其他中文名 / Other Chinese Name(s)
大夜蝠、大油蝠

其他英文名 / Other English Name(s)
无 None

同物异名 / Synonym(s)
无 None

▲ 形态及生境 / Morphology and Habitat

形态特征 / Morphological Characteristics

头体长 89~104 mm。耳长 22~29 mm。前臂长 71~80 mm。后足长 13~18 mm。尾长 61~83 mm。颅全长 27 mm。面颊裸露无毛，两耳前端内侧有密毛。耳郭略呈三角形，耳屏较短，内弯，端部钝圆肾形。下颌有一簇硬毛，其下有颌下腺开孔。足连爪的长度超过胫长之半。背毛烟褐色，毛基深褐色，毛尖灰褐色，腹毛略浅。

Head and body length 89-104 mm. Ear length 22-29 mm. Forearm length 71-80 mm. Hind foot length 13-18 mm. Tail length 61-83 mm. Greatest skull length 27 mm. Cheeks are glabrous. Front and inner sides of the ears are densely hairy. Auricle is slightly triangular, the tragus is short, inward-curved, and the ends are in kidney shape, blunt and round. Lower jaw has a tuft of bristles and a submandibular gland opening under it. Feet and claws are more than half the length of the shins. Dorsal hairs smoke brown, basal part of hairs dark brown, hair tips grayish brown, abdominal hairs slightly shallow in color.

生境 / Habitat
洞穴、森林 Caves, forest

▲ 地理分布 / Geographic Distribution

国内分布 / Domestic Distribution
江苏、安徽、湖北、湖南、江西、四川、云南、贵州、广东
Jiangsu, Anhui, Hubei, Hunan, Jiangxi, Sichuan, Yunnan, Guizhou, Guangdong

全球分布 / World Distribution
中国、印度、老挝、尼泊尔、泰国、越南
China, India, Laos, Nepal, Thailand, Vietnam

生物地理界 / Biogeographic Realm
印度马来界、古北界
Indomalaya, Palearctic

WWF 生物群系 / WWF Biome
热带和亚热带湿润阔叶林
Tropical & Subtropical Moist Broadleaf Forests

动物地理分布型 / Zoogeographic Distribution Type
Wc

分布标注 / Distribution Note
非特有种 Non-Endemic

▲ 濒危状况 / Threatened Status

中国生物多样性红色名录等级 / CB RL Category (2021)
近危 NT

IUCN 红色名录 / IUCN Red List (2021)
无危 LC

威胁因子 / Threats
未知 Unknown

▲ 法律保护地位 / Legal Protection Status

国家重点保护野生动物等级 / Category of National Key Protected Wild Animals (2021)
未列入 Not listed

"三有" 名录 / TWIESSV (2023)
未列入 Not listed

CITES 附录等级 / CITES Appendix (2023)
未列入 Not listed

迁徙物种公约附录 / CMS Appendix (2020)
未列入 Not listed

保护行动 / Conservation Action
尚无保护行动 No conservation action so far

▲ 参考文献 / References

Jiang et al. (蒋志刚等), 2021; Burgin et al., 2020; IUCN, 2020; Wilson and Mittermeier, 2019; Chen et al. (陈毅等), 2013; Han and He (韩宝银和贺红早), 2012

195 / 双色蝙蝠

Vespertilio murinus Linnaeus, 1758

· Particoloured Bat

▲ 分类地位 / Taxonomy

翼手目 Chiroptera / 蝙蝠科 Vespertilionidae / 蝙蝠属 *Vespertilio*

科建立者及其文献 / Family Authority
Gray, 1821

属建立者及其文献 / Genus Authority
Linnaeus, 1758

亚种 / Subspecies
指名亚种 *V. m. murinus* Linnaeus, 1758
内蒙古、甘肃和新疆
Inner Mongolia, Gansu and Xinjiang
乌苏里亚种 *V. m. ussuriensis* Wallin, 1969
内蒙古和黑龙江
Inner Mongolia and Heilongjiang

模式标本产地 / Type Locality
瑞典
Sweden. Baage (2001) indicated that the type locality is probably near Uppsala, Central Sweden

▲ 其他名称 / Other Name(s)

其他中文名 / Other Chinese Name(s)
无 None

其他英文名 / Other English Name(s)
Eurasian Particolored Bat

同物异名 / Synonym(s)
无 None

▲ 形态及生境 / Morphology and Habitat

形态特征 / Morphological Characteristics
齿式：2.1.1.3/3.1.2.3=32。头体长 55~66 mm。耳长 14~16 mm。前臂长 41~46 mm。后足长 8~10 mm。尾长 40~48 mm。颅全长 16.5 mm。背毛和腹毛基部黑棕色。毛尖呈白色。腹毛白色尖端较背毛明显，侧腹毛和喉毛均为白色。
Dental formula: 2.1.1.3/3.1.2.3=32. Head and body length 55-66 mm. Ear length 14-16 mm. Forearm length 41-46 mm. Hind foot length 8-10 mm. Tail length 40-48 mm. Greatest skull length 16.5 mm. Basal part of dorsal and ventral hairs are blackish brown. Hair tips white. White tips of ventral hairs are more distinct than those of dorsal hairs. Hairs on the lateral venter and throat are all white.

生境 / Habitat
森林、草地、耕地、人造建筑
Forest, grassland, arable land, man-made building

国内分布 / Domestic Distribution

黑龙江、内蒙古、新疆、甘肃

Heilongjiang, Inner Mongolia, Xinjiang, Gansu

全球分布 / World Distribution

阿富汗、阿尔巴尼亚、亚美尼亚、奥地利、阿塞拜疆、白俄罗斯、比利时、保加利亚、中国、克罗地亚、捷克、丹麦、爱沙尼亚、芬兰、法国、格鲁吉亚、德国、希腊、匈牙利、伊朗、意大利、日本、朝鲜、拉脱维亚、列支敦士登、立陶宛、卢森堡、马其顿、摩尔多瓦、蒙古国、黑山、荷兰、挪威、波兰、罗马尼亚、俄罗斯、塞尔维亚、斯洛伐克、斯洛文尼亚、瑞典、瑞士、土耳其、土库曼斯坦、乌克兰、英国、乌兹别克斯坦

Afghanistan, Albania, Armenia, Austria, Azerbaijan, Belarus, Belgium, Bulgaria, China, Croatia, Czech, Denmark, Estonia, Finland, France, Georgia, Germany, Greece, Hungary, Iran, Italy, Japan, Democratic People's Republic of Korea, Latvia, Liechtenstein, Lithuania, Luxembourg, Macedonia, Moldova, Mongolia, Montenegro, Netherlands, Norway, Poland, Romania, Russia, Serbia, Slovakia, Slovenia, Sweden, Switzerland, Turkey, Turkmenistan, Ukraine, United Kingdom, Uzbekistan

生物地理界 / Biogeographic Realm

古北界 Palearctic

WWF 生物群系 / WWF Biome

北方森林 / 针叶林

Boreal Forests / Coniferous Forests

动物地理分布型 / Zoogeographic Distribution Type

Ub

分布标注 / Distribution Note

非特有种 Non-Endemic

▲ 濒危状况 / Threatened Status

中国生物多样性红色名录等级 / CB RL Category (2021)

无危 LC

IUCN 红色名录 / IUCN Red List (2021)

无危 LC

威胁因子 / Threats

未知 Unknown

▲ 法律保护地位 / Legal Protection Status

国家重点保护野生动物等级 / Category of National Key Protected Wild Animals (2021)

未列入 Not listed

"三有"名录 / TWIESSV (2023)

未列入 Not listed

CITES 附录等级 / CITES Appendix (2023)

未列入 Not listed

迁徙物种公约附录 / CMS Appendix (2020)

未列入 Not listed

保护行动 / Conservation Action

尚无保护行动 No conservation action so far

▲ 参考文献 / References

Jiang et al. (蒋志刚等), 2021; Burgin et al., 2020; IUCN, 2020; Wilson and Mittermeier, 2019; Smith et al., 2009; Pan et al. (潘清华等), 2007; Wilson and Reeder, 2005; Wang (王应祥), 2003; Zhang(张荣祖), 1997

196 / 东方蝙蝠

Vespertilio sinensis (Peters, 1880)

- Asian Particolored Bat

▲ 分类地位 / Taxonomy

翼手目 Chiroptera / 蝙蝠科 Vespertilionidae / 蝙蝠属 *Vespertilio*

科建立者及其文献 / Family Authority
Gray, 1821

属建立者及其文献 / Genus Authority
Linnaeus, 1758

亚种 / Subspecies
无 None

模式标本产地 / Type Locality
中国
China, Peking (Beijing)

江廷磊 / 供图

▲ 其他名称 / Other Name(s)

其他中文名 / Other Chinese Name(s)
东亚蝙蝠

其他英文名 / Other English Name(s)
无 None

同物异名 / Synonym(s)
无 None

▲ 形态及生境 / Morphology and Habitat

形态特征 / Morphological Characteristics
齿式：2.1.1.3/3.1.2.3=32。头体长 58~80 mm。耳长 14~21 mm。前臂长 43~55 mm。后足长 9~16 mm。尾长 34~54 mm。颅全长 17.3~18 mm。体重 12~19 g。耳短而宽，略呈三角形。背毛的毛基棕褐色，毛尖淡棕褐色，腹毛的毛基淡褐色，毛尖灰褐色。尾发达。翼膜起始于趾基部，距缘膜较狭呈小弧形。体毛延伸至股间膜。

Dental formula: 2.1.1.3/3.1.2.3=32. Head and body length 58-80 mm. Ear length 14-21 mm. Forearm length 43-55 mm. Hind foot length 9-16 mm. Tail length 34-54 mm. Greatest skull length 17.3-18 mm. Body mass 12-19 g. Big eyes. Ears are short and broad, slightly triangular. Basal parts of dorsal hairs brown, and hair tips light tan. Basal parts of hairs on the belly are gray-brown and yellow-brown at the hair tips. Tail is well developed, protruding the interfemoral membrane about 3 mm. Pterygoid membrane starts from the base of the toe and is narrow from the membrane margin, showing a small arc. Body hairs extend to the interfemoral membrane.

生境 / Habitat
洞穴、草原、沙漠
Caves, grassland, desert

▲ 地理分布 / Geographic Distribution

国内分布 / Domestic Distribution
山西、广西、北京、内蒙古、黑龙江、江西、湖北、湖南、四川、陕西、甘肃、台湾、福建、河南、辽宁、山东、云南、浙江、天津、重庆
Shanxi, Guangxi, Beijing, Inner Mongolia, Heilongjiang, Jiangxi, Hubei, Hunan, Sichuan, Shaanxi, Gansu, Taiwan, Fujian, Henan, Liaoning, Shandong, Yunnan, Zhejiang, Tianjin, Chongqing

全球分布 / World Distribution
中国、日本、朝鲜、韩国、蒙古国、俄罗斯
China, Japan, Democratic People's Republic of Korea, Republic of Korea, Mongolia, Russia

生物地理界 / Biogeographic Realm
印度马来界 Indomalaya

WWF 生物群系 / WWF Biome
温带草原、温带针叶林
Temperate Grasslands, Temperate Conifer Forest

动物地理分布型 / Zoogeographic Distribution Type
Ub

分布标注 / Distribution Note
非特有种 Non-Endemic

▲ 濒危状况 / Threatened Status

中国生物多样性红色名录等级 / CB RL Category (2021)
无危 LC

IUCN 红色名录 / IUCN Red List (2021)
无危 LC

威胁因子 / Threats
未知 Unknown

▲ 法律保护地位 / Legal Protection Status

国家重点保护野生动物等级 / Category of National Key Protected Wild Animals (2021)
未列入 Not listed

"三有"名录 / TWIESSV (2023)
未列入 Not listed

CITES 附录等级 / CITES Appendix (2023)
未列入 Not listed

迁徙物种公约附录 / CMS Appendix (2020)
未列入 Not listed

保护行动 / Conservation Action
尚无保护行动 No conservation action so far

▲ 参考文献 / References

Jiang et al. (蒋志刚等), 2021; Burgin et al., 2020; IUCN, 2020; Liu et al. (刘少英等), 2020; Wilson and Mittermeier, 2019; Pan et al. (潘清华等), 2007; Wu et al. (武明录等), 2006; Wilson and Reeder, 2005; Wang (王应祥), 2003; Zhang (张荣祖), 1997; Rydell and Baagoe, 1994; Wilson and Reeder, 1993

197 / 戈壁北棕蝠

Eptesicus gobiensis Bobrinski, 1926

· Gobi Big Brown Bat

▲ 分类地位 / Taxonomy

翼手目 Chiroptera / 蝙蝠科 Vespertilionidae / 棕蝠属 *Eptesicus*

科建立者及其文献 / Family Authority
Gray, 1821

属建立者及其文献 / Genus Authority
Rafinesque, 1820

亚种 / Subspecies
无 None

模式标本产地 / Type Locality
蒙古国
Mongolia, Gobi Altai Mtns, Burchastei-tala

陈文杰 / 供图

▲ 其他名称 / Other Name(s)

其他中文名 / Other Chinese Name(s)
无 None

其他英文名 / Other English Name(s)
无 None

同物异名 / Synonym(s)
无 None

▲ 形态及生境 / Morphology and Habitat

形态特征 / Morphological Characteristics

齿式：2.1.1.3/3.1.2.3=32。头体长 57~65 mm。前臂长 38~42 mm。上颌内切牙明显高于外切牙。背毛红棕色，毛基部深棕色。腹部被毛棕白色。
Dental formula: 2.1.1.3/3.1.2.3=32. Head and body length 57-65 mm. Forearm length of 38-42 mm. The inner incisors of the upper jaw are distinctly higher than the outer ones. Dorsal hairs are reddish-brown, with the bases of the hairs being dark brown. Hairs on the underparts are brownish white.

生境 / Habitat
洞穴、草原、沙漠
Caves, grassland, desert

▲ 地理分布 / Geographic Distribution

国内分布 / Domestic Distribution
西藏、新疆、内蒙古、甘肃、宁夏、青海
Tibet, Xinjiang, Inner Mongolia, Gansu, Ningxia, Qinghai

全球分布 / World Distribution
中国、阿富汗、印度、哈萨克斯坦、吉尔吉斯斯坦、蒙古国、巴基斯坦、
俄罗斯、叙利亚、塔吉克斯坦、土库曼斯坦、乌兹别克斯坦
China, Afghanistan, India, Kazakhstan, Kyrgyzstan, Mongolia, Pakistan, Russia,
Syrian, Tajikistan, Turkmenistan, Uzbekistan

生物地理界 / Biogeographic Realm
古北界 Palearctic

WWF 生物群系 / WWF Biome
沙漠和干旱灌木地
Deserts & Xeric Shrublands

动物地理分布型 / Zoogeographic Distribution Type
Da

分布标注 / Distribution Note
非特有种 Non-Endemic

▲ 濒危状况 / Threatened Status

中国生物多样性红色名录等级 / CB RL Category (2021)
未评定 NE

IUCN 红色名录 / IUCN Red List (2021)
无危 LC

威胁因子 / Threats
未知 Unknown

▲ 法律保护地位 / Legal Protection Status

国家重点保护野生动物等级 / Category of National Key Protected Wild Animals (2021)
未列入 Not listed

"三有"名录 / TWIESSV (2023)
未列入 Not listed

CITES 附录等级 / CITES Appendix (2023)
未列入 Not listed

迁徙物种公约附录 / CMS Appendix (2020)
未列入 Not listed

保护行动 / Conservation Action
尚无保护行动 No conservation action so far

▲ 参考文献 / References

Burgin et al., 2020; IUCN, 2020; Liu et al. (刘少英等), 2020; Huang and Liu, 2019; Wilson and Mittermeier, 2019

198 / 北棕蝠

Eptesicus nilssonii
(Keyserling & Blasius, 1839)

· Northern Serotine

▲ 分类地位 / Taxonomy

翼手目 Chiroptera / 蝙蝠科 Vespertilionidae / 棕蝠属 *Eptesicus*

科建立者及其文献 / Family Authority
Gray, 1821

属建立者及其文献 / Genus Authority
Rafinesque, 1820

亚种 / Subspecies
指名亚种 *E. n. nilssonii* (Keyserling & Blasius, 1839)
黑龙江、山东、吉林、内蒙古、北京、河北、新疆
Heilongjiang, Shandong, Jilin, Inner Mongolia, Beijing, Hebei, Xinjiang

模式标本产地 / Type Locality
瑞典
Sweden

▲ 其他名称 / Other Name(s)

其他中文名 / Other Chinese Name(s)
无 None

其他英文名 / Other English Name(s)
Northern Bat

同物异名 / Synonym(s)
无 None

▲ 形态及生境 / Morphology and Habitat

形态特征 / Morphological Characteristics
齿式：2.1.1.3/3.1.2.3=32。头体长 54~64 mm。耳长 13~18 mm。前臂长 37~44 mm。后足长 10~12 mm。尾长 35~50 mm。颅全长 14~16 mm。耳壳肉质，短圆。耳屏短而圆，向内弯曲。背毛暗棕色或带黑色，点缀具金色毛尖的毛。腹部毛黄棕色。翅长而宽，翅尖短而圆。距骨延伸至尾包膜长度的一半。尾略超出尾鞘。翼膜附着在脚趾的底部。
Dental formula: 2.1.1.3/3.1.2.3=32. Head and body length 54-64 mm. Ear length 13-18 mm. Forearm length 37-44 mm. Hind foot length 10-12 mm. Tail length 35-50 mm. Greatest skull length 14-16 mm. Ears short, round, and fleshy. Tragus is short and round, curving inward. Dorsal hairs are dark brown or blackish, interspersed with some golden-tipped hairs. Venter hairs are yellowish brown. Wings long and broad, with short, rounded tips. Calcar extends half the length of uropatagium. Tail extends slightly beyond the uropatagium. Wing membrane attaches at the base of the toe.

生境 / Habitat
森林、人造建筑
Forest, man-made building

▲ 地理分布 / Geographic Distribution

国内分布 / Domestic Distribution

黑龙江、山东、吉林、内蒙古、北京、河北、河南、山东、陕西、山西、宁夏、甘肃、青海、西藏、新疆
Heilongjiang, Shandong, Jilin, Inner Mongolia, Beijing, Hebei, Henan, Shandong, Shaanxi, Shanxi, Ningxia, Gansu, Qinghai, Tibet, Xinjiang

全球分布 / World Distribution

奥地利、阿塞拜疆、白俄罗斯、比利时、保加利亚、中国、克罗地亚、捷克、丹麦、爱沙尼亚、芬兰、法国、格鲁吉亚、德国、匈牙利、伊朗、伊拉克、意大利、日本、朝鲜、韩国、吉尔吉斯斯坦、拉脱维亚、列支敦士登、立陶宛、卢森堡、蒙古国、荷兰、挪威、波兰、罗马尼亚、俄罗斯、斯洛伐克、斯洛文尼亚、瑞典、瑞士、土耳其、乌克兰
Austria, Azerbaijan, Belarus, Belgium, Bulgaria, China, Croatia, Czech, Denmark, Estonia, Finland, France, Georgia, Germany, Hungary, Iran, Islamic Republic of, Iraq, Italy, Japan, Democratic People's Republic of Korea, Republic of Korea, Kyrgyzstan, Latvia, Liechtenstein, Lithuania, Luxembourg, Mongolia, Netherlands, Norway, Poland, Romania, Russia, Slovakia, Slovenia, Sweden, Switzerland, Turkey, Ukraine

生物地理界 / Biogeographic Realm
古北界 Palearctic

WWF 生物群系 / WWF Biome
温带草原、热带稀树草原和灌木地
Temperate Grasslands, Savannas & Shrublands

动物地理分布型 / Zoogeographic Distribution Type
Da

分布标注 / Distribution Note
非特有种 Non-Endemic

▲ 濒危状况 / Threatened Status

中国生物多样性红色名录等级 / CB RL Category (2021)
无危 LC

IUCN 红色名录 / IUCN Red List (2021)
无危 LC

威胁因子 / Threats
未知 Unknown

▲ 法律保护地位 / Legal Protection Status

国家重点保护野生动物等级 / Category of National Key Protected Wild Animals (2021)
未列入 Not listed

"三有"名录 / TWIESSV (2023)
未列入 Not listed

CITES 附录等级 / CITES Appendix (2023)
未列入 Not listed

迁徙物种公约附录 / CMS Appendix (2020)
未列入 Not listed

保护行动 / Conservation Action
尚无保护行动 No conservation action so far

▲ 参考文献 / References

Jiang et al. (蒋志刚等), 2021; Burgin et al., 2020; IUCN, 2020; Liu et al. (刘少英等), 2020; Wilson and Mittermeier, 2019; Peng and Zhong (彭基泰和钟祥清), 2005

199 / 东方棕蝠

Eptesicus pachyomus (Tomes, 1857)

• Oriental Serotine

翼手目 Chiroptera / 蝙蝠科 Vespertilionidae / 棕蝠属 *Eptesicus*

科建立者及其文献 / Family Authority
Gray, 1821

属建立者及其文献 / Genus Authority
Rafinesque, 1820

亚种 / Subspecies
安氏亚种 *E. p. andersoni* Dobson, 1871
云南 Yunnan

帕氏亚种 *E. p. pallens* Miller, 1911
甘肃 Gansu

台湾亚种 *E. p. horikawai* Kishida, 1924
台湾 Taiwan

模式标本产地 / Type Locality
印度
Rajputana, India

▲ 其他名称 / Other Name(s)

其他中文名 / Other Chinese Name(s)
无 None

其他英文名 / Other English Name(s)
Mouse-like Serotine, Thick-muzzled Serotine

同物异名 / Synonym(s)
Eptesicus erotinus pachyomus

▲ 形态及生境 / Morphology and Habitat

形态特征 / Morphological Characteristics

齿式：2.1.1.3/3.1.2.3=32。以前被认为是大棕蝠 *E. serotinus* 的亚种。Juste et al.（2013）将其提升为种。根据 2018 年 KFBG 测量的一具标本，耳长 14.3 mm，前臂长 51.1 mm，头体长 77.1 mm，尾长 62 mm，胫骨长 21.9 mm，后足长 10.8 mm。背毛黄棕色，尖端白色，有光泽，肩部毛色深棕色，胸部和腹部被覆灰褐色绒毛。翅膜发达，黑褐色，止于趾的基部。尾膜为三角形，被白色的毛所覆盖。尾膜从尾部略微突出。该物种以蝴蝶和飞蛾为食，通常独自或成群栖息在阴暗区域，如岩石和建筑物的裂缝。偶尔也栖息在洞穴的前部。

Dental formula: 2.1.1.3/3.1.2.3=32. This species was formerly considered a subspecies of Common Serotine, *E. serotinus*. It has now been raised to species status by Juste et al. (2013) by comparing Cytb and RAG2 genes. According to a specimen measure by Kadoorie Farm and Botanic Garden (KFBG) in 2018, ear length 14.3 mm, forearm length 51.1 mm, head and body length 77.1 mm, tail length 62 mm, tib 21.9 mm, hindfoot 10.8 mm. Dorsal hairs brown, shiny, and chest and abdomen by yellow brown fluff. Wing membrane developed, dark brown, stops at the base of the toe. The caudal membrane is triangular. The caudal membrane protrudes slightly from the end of the tail. The species roosts in shaded areas such as cracks in rocks and buildings. It also occasionally roosts in the frontal part of caves. It generally roosts alone or in small groups and it feeds on butterflies and moths.

生境 / Habitat

森林、灌木丛洞穴和地下的栖息地
Forest, shrubland, caves and subterranean habitat

▲ 地理分布 / Geographic Distribution

国内分布 / Domestic Distribution
黑龙江、吉林、辽宁、内蒙古、河北、北京、天津、山东、山西、江西、福建、台湾、湖北、湖南、贵州、云南、四川、陕西、甘肃、宁夏、重庆、香港
Heilongjiang, Jilin, Liaoning, Inner Mongolia, Hebei, Beijing, Tianjin, Shandong, Jiangxi, Fujian, Taiwan, Hubei, Hunan, Guizhou, Yunnan, Sichuan, Shaanxi, Gansu, Ningxia, Chongqing, Hong Kong

全球分布 / World Distribution
中国、印度、伊朗、老挝、缅甸、巴基斯坦、泰国、越南
China, India, Iran, Laos, Myanmar, Pakistan, Thailand, Vietnam

生物地理界 / Biogeographic Realm
古北界、印度马来界
Palearctic, Indomalaya

WWF 生物群系 / WWF Biome
温带阔叶和混交林、山地草原和灌丛、热带和亚热带湿润阔叶林
Temperate Broadleaf & Mixed Forests, Montane Grasslands & Shrublands, Tropical & Subtropical Moist Broadleaf Forests

动物地理分布型 / Zoogeographic Distribution Type
We

分布标注 / Distribution Note
非特有种 Non-Endemic

▲ 濒危状况 / Threatened Status

中国生物多样性红色名录等级 / CB RL Category (2021)
未评定 NE

IUCN 红色名录 / IUCN Red List (2021)
无危 LC

威胁因子 / Threats
未知 Unknown

▲ 法律保护地位 / Legal Protection Status

国家重点保护野生动物等级 / Category of National Key Protected Wild Animals (2021)
未列入 Not listed

"三有" 名录 / TWIESSV (2023)
未列入 Not listed

CITES 附录等级 / CITES Appendix (2023)
未列入 Not listed

迁徙物种公约附录 / CMS Appendix (2020)
未列入 Not listed

保护行动 / Conservation Action
尚无保护行动 No conservation action so far

▲ 参考文献 / References

Wei et al. (魏辅文等), 2021; Burgin et al., 2020; IUCN, 2020; Srinivasulu et al., 2019; Wilson and Mittermeier, 2019; Kadoorie Farm and Botanic Garden, 2018; Juste et al., 2013

200 / 肥耳棕蝠

Eptesicus pachyotis (Dobson, 1871)

• Thick-eared Bat

翼手目 Chiroptera / 蝙蝠科 Vespertilionidae / 棕蝠属 *Eptesicus*

科建立者及其文献 / Family Authority
Gray, 1821

属建立者及其文献 / Genus Authority
Rafinesque, 1820

亚种 / Subspecies
无 None

模式标本产地 / Type Locality
印度
India, Assam (Meghalaya), Khasi Hills

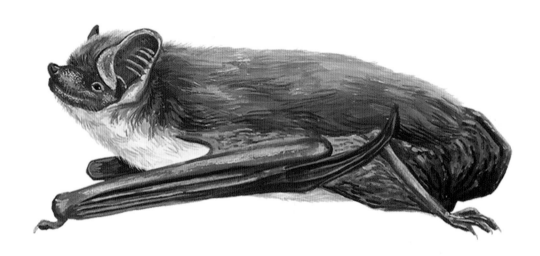

▲ 其他名称 / Other Name(s)

其他中文名 / Other Chinese Name(s)
无 None

其他英文名 / Other English Name(s)
Thick-eared Serotine

同物异名 / Synonym(s)
无 None

▲ 形态及生境 / Morphology and Habitat

形态特征 / Morphological Characteristics
齿式：2.1.1.3/3.1.2.3=32。头体长 55~56 mm。耳长 13~14 mm。前臂长 38~40 mm。后足长 8~9 mm。尾长 40~41 mm。颅全长 21 mm。耳三角形，具圆形尖，肉质、相当厚。耳屏短而圆，具内弯。头部扁平。吻部短。翅膜附在脚趾底部。背毛颜色深棕色。腹毛较浅。

Dental formula: 2.1.1.3/3.1.2.3=32. Head and body length 55-56 mm. Ear length 13-14 mm. Forearm length 38-40 mm. Hind foot length 8-9 mm. Tail length 40-41 mm. Greatest skull length 21 mm. Ears triangular, quite thick and fleshy with a rounded tip. Tragus short and round, with inward curve. Head flat. Muzzle short. Wings attached to the base of toes. Dorsal hair color dark brown. Ventral hairs are paler.

生境 / Habitat
森林 Forest

▲ 地理分布 / Geographic Distribution

国内分布 / Domestic Distribution
宁夏、四川、西藏、甘肃、青海、内蒙古
Ningxia, Sichuan, Tibet, Gansu, Qinghai, Inner Mongolia

全球分布 / World Distribution
中国、孟加拉国、缅甸、泰国、印度
China, Bangladesh, Myanmar, Thailand, India

生物地理界 / Biogeographic Realm
古北界 Palearctic

WWF 生物群系 / WWF Biome
温带阔叶和混交林，热带和亚热带湿润阔叶林
Temperate Broadleaf & Mixed Forests; Tropical & Subtropical Moist Broadleaf Forests

动物地理分布型 / Zoogeographic Distribution Type
Pf

分布标注 / Distribution Note
非特有种 Non-Endemic

▲ 濒危状况 / Threatened Status

中国生物多样性红色名录等级 / CB RL Category (2021)
无危 LC

IUCN 红色名录 / IUCN Red List (2021)
无危 LC

威胁因子 / Threats
未知 Unknown

▲ 法律保护地位 / Legal Protection Status

国家重点保护野生动物等级 / Category of National Key Protected Wild Animals (2021)
未列入 Not listed

"三有"名录 / TWIESSV (2023)
未列入 Not listed

CITES 附录等级 / CITES Appendix (2023)
未列入 Not listed

迁徙物种公约附录 / CMS Appendix (2020)
未列入 Not listed

保护行动 / Conservation Action
尚无保护行动 No conservation action so far

▲ 参考文献 / References

Jiang et al. (蒋志刚等), 2021; Burgin et al., 2020; IUCN, 2020; Wilson and Mittermeier, 2019; Smith et al., 2009; Pan et al. (潘清华等), 2007; Wilson and Reeder, 2005; Wang (王应祥), 2003; Rydell, 1993

201 / 大山蝠

Nyctalus aviator (Thomas, 1911)

· Birdlike Noctule

翼手目 Chiroptera / 蝙蝠科 Vespertilionidae / 山蝠属 *Nyctalus*

科建立者及其文献 / Family Authority
Gray, 1821

属建立者及其文献 / Genus Authority
Bowditch, 1825

亚种 / Subspecies
无 None

模式标本产地 / Type Locality
日本
Japan, Honshu, Tokyo

▲ 其他名称 / Other Name(s)

其他中文名 / Other Chinese Name(s)
无 None

其他英文名 / Other English Name(s)
Japanese Large Noctule

同物异名 / Synonym(s)
无 None

▲ 形态及生境 / Morphology and Habitat

形态特征 / Morphological Characteristics

齿式：2.1.2.3/3.1.2.3=34。头体长 80~106 mm。耳长 16~23 mm。前臂长 58~64 mm。后足长 12~17 mm。尾长 80~106 mm。口鼻部宽，眼睛和鼻孔之间有腺体。耳壳短而厚，耳屏肾形。拇指短，具强爪。第五指最短，第三指最长。毛发深黄棕色，浓密柔软。翼膜附在脚踝上。尾尖未延伸超过尾膜。

Dental formula: 2.1.2.3/3.1.2.3=34. Head and body length 80-106 mm. Ear length 16-23 mm. Forearm length 58-64 mm. Hind foot length 12-17 mm. Tail length 80-106 mm. Muzzle broad, with gland between eye and nostril. Tragus short and stubby. The tragus is kidney shaped. Thumb short, with strong claw. Fifth digit shortest, third-longest. Hair color deep yellowish-brown, dense and velvety. Wing membrane attached to the ankle. Tail barely extends past uropatagium.

生境 / Habitat

亚热带湿润山地森林
Subtropical moist montane forest

▲ 地理分布 / Geographic Distribution

国内分布 / Domestic Distribution
吉林、黑龙江、浙江、安徽、河南、上海、贵州
Jilin, Heilongjiang, Zhejiang, Anhui, Henan, Shanghai, Guizhou

全球分布 / World Distribution
中国、日本、朝鲜、韩国
China, Japan, Democratic People's Republic of Korea, Republic of Korea

生物地理界 / Biogeographic Realm
古北界 Palearctic

WWF 生物群系 / WWF Biome
温带阔叶和混交林
Temperate Broadleaf & Mixed Forests

动物地理分布型 / Zoogeographic Distribution Type
Ea

分布标注 / Distribution Note
非特有种 Non-Endemic

▲ 濒危状况 / Threatened Status

中国生物多样性红色名录等级 / CB RL Category (2021)
近危 NT

IUCN 红色名录 / IUCN Red List (2021)
近危 NT

威胁因子 / Threats
人类活动干扰
Human disturbance

▲ 法律保护地位 / Legal Protection Status

国家重点保护野生动物等级 / Category of National Key Protected Wild Animals (2021)
未列入 Not listed

"三有" 名录 / TWIESSV (2023)
未列入 Not listed

CITES 附录等级 / CITES Appendix (2023)
未列入 Not listed

迁徙物种公约附录 / CMS Appendix (2020)
未列入 Not listed

保护行动 / Conservation Action
尚无保护行动 No conservation action so far

▲ 参考文献 / References

Jiang et al. (蒋志刚等), 2021; Burgin et al., 2020; IUCN, 2020; Wilson and Mittermeier, 2019; Smith and Xie, 2009; Zhou and Yang (周江和杨天友), 2012; Wang et al. (王志伟等), 2010; Zhu et al. (朱旭等), 2009; Mayer et al., 2007; Pan et al. (潘清华等), 2007; Wilson and Reeder, 2005; Wang (王应祥), 2003

202 / 褐山蝠

Nyctalus noctula (Schreber, 1774)

· Noctule

邢睿 / 供图

▲ 分类地位 / Taxonomy

翼手目 Chiroptera / 蝙蝠科 Vespertilionidae /
山蝠属 *Nyctalus*

科建立者及其文献 / Family Authority
Gray, 1821

属建立者及其文献 / Genus Authority
Bowditch, 1825

亚种 / Subspecies
新疆亚种 *N. n. meklenburzcui* Kuzyakin, 1934
新疆 Xinjiang

模式标本产地 / Type Locality
法国
France

▲ 其他名称 / Other Name(s)

其他中文名 / Other Chinese Name(s)
无 None

其他英文名 / Other English Name(s)
Noctule Bat

同物异名 / Synonym(s)
无 None

▲ 形态及生境 / Morphology and Habitat

形态特征 / Morphological Characteristics
齿式：2.1.1.3/3.1.2.3=32。头体长 60~82 mm。
耳长 16~21 mm。前臂长 47~60 mm。后足长
12~14 mm。尾长 41~61 mm。颅全长 17~19
mm。耳宽短，耳外缘延伸向下达口角下，耳
屏明显，内缘弯转呈肾形。第五指长度略超过
第三或第四掌骨之长。背毛深褐色，腹毛色浅。
背腹侧毛沿翼膜和尾间膜向外延伸，从前臂到
第五掌骨基部密布绒毛。

Dental formula: 2.1.1.3/3.1.2.3=32. Head and body
length 60-82 mm. Ear length 16-21 mm. Forearm length
47-60 mm. Hind foot length 12-14 mm. Tail length 41-
61 mm. Greatest skull length 17-19 mm. Ears are wide
and short and the outer edges of the ears extend to
the lower corner of the mouth. Tragus is obvious, and
the inner edge curves in the shape of a kidney. The 5th
finger is slightly longer than the 3rd or 4th metacarpal.
Dorsal hairs range from golden to dark brown and
ventral hairs are usually pale brown. Dorsal ventral hairs
extend outward along the pterygous membrane and
intercaudal membrane. From the forearm to the base of
the 5th metacarpal bone covered with villi.

生境 / Habitat
森林、人造建筑、城市
Forest, man-made building, urban Area

▲ 地理分布 / Geographic Distribution

国内分布 / Domestic Distribution

新疆、河北、北京、天津、山东、陕西、甘肃
Xinjiang, Hebei, Beijing, Tianjin, Shandong, Shaanxi, Gansu

全球分布 / World Distribution

阿尔巴尼亚、安道尔、亚美尼亚、奥地利、阿塞拜疆、白俄罗斯、比利时、波斯尼亚和黑塞哥维那、保加利亚、中国、克罗地亚、塞浦路斯、捷克、丹麦、爱沙尼亚、芬兰、法国、格鲁吉亚、德国、希腊、梵蒂冈、匈牙利、印度、伊朗、伊拉克、以色列、意大利、哈萨克斯坦、拉脱维亚、黎巴嫩、列支敦士登、立陶宛、卢森堡、马其顿、马来西亚、马耳他、摩纳哥、黑山、摩洛哥、缅甸、荷兰、挪威、阿曼、巴基斯坦、波兰、罗马尼亚、俄罗斯、圣马力诺、塞尔维亚、斯洛伐克、斯洛文尼亚、西班牙、瑞典、瑞士、土耳其、乌克兰、英国、越南

Albania, Andorra, Armenia, Austria, Azerbaijan, Belarus, Belgium, Bosnia and Herzegovina, Bulgaria, China, Croatia, Cyprus, Czech, Denmark, Estonia, Finland, France, Georgia, Germany, Greece, Vatican, Hungary, India, Iran, Iraq, Israel, Italy, Kazakhstan, Latvia, Lebanon, Liechtenstein, Lithuania, Luxembourg, Macedonia, Malaysia, Malta, Monaco, Montenegro, Morocco, Myanmar, Netherlands, Norway, Oman, Pakistan, Poland, Romania, Russia, San Marino, Serbia, Slovakia, Slovenia, Spain, Sweden, Switzerland, Turkey, Ukraine, United Kingdom, Vietnam

生物地理界 / Biogeographic Realm

古北界 Palearctic

WWF 生物群系 / WWF Biome

温带草原、温带阔叶和混交林
Temperate Grasslands, Temperate Broadleaf & Mixed Forest

动物地理分布型 / Zoogeographic Distribution Type

U01

分布标注 / Distribution Note

非特有种 Non-Endemic

▲ 濒危状况 / Threatened Status

中国生物多样性红色名录等级 / CB RL Category (2021)

近危 NT

IUCN 红色名录 / IUCN Red List (2021)

无危 LC

威胁因子 / Threats

人类活动干扰、城市建设
Human disturbance, urban development

▲ 法律保护地位 / Legal Protection Status

国家重点保护野生动物等级 / Category of National Key Protected Wild Animals (2021)

未列入 Not listed

"三有"名录 / TWIESSV (2023)

未列入 Not listed

CITES 附录等级 / CITES Appendix (2023)

未列入 Not listed

迁徙物种公约附录 / CMS Appendix (2020)

未列入 Not listed

保护行动 / Conservation Action

尚无保护行动 No conservation action so far

▲ 参考文献 / References

Jiang et al. (蒋志刚等), 2021; Burgin et al., 2020; IUCN, 2020; Liu et al. (刘少英等), 2020; Wilson and Mittermeier, 2019; Smith and Xie, 2009; Pan et al. (潘清华等), 2007; Wang (王应祥), 2003; Zhang (张荣祖), 1997

203 / 中华山蝠

Nyctalus plancyi Gerbe, 1880

· Chinese Noctule

▲ 分类地位 / Taxonomy

翼手目 Chiroptera / 蝙蝠科 Vespertilionidae / 山蝠属 *Nyctalus*

科建立者及其文献 / Family Authority
Gray, 1821

属建立者及其文献 / Genus Authority
Bowditch, 1825

亚种 / Subspecies
无 None

模式标本产地 / Type Locality
中国
China, Beijing

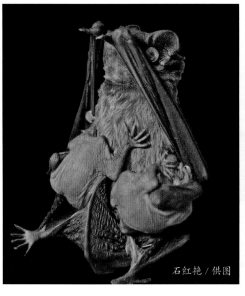

石红艳 / 供图

吴毅 / 供图

▲ 其他名称 / Other Name(s)

其他中文名 / Other Chinese Name(s)
绒山蝠

其他英文名 / Other English Name(s)
Chinese Mountain Bat,
Mountain Noctule, Villus Noctule

同物异名 / Synonym(s)
无 None

▲ 形态及生境 / Morphology and Habitat

形态特征 / Morphological Characteristics

齿式：2.1.1.3/3.1.2.3=32。头体长 65~75 mm。耳长 15~18 mm。前臂长 47~56 mm。后足长 10~11 mm。尾长 36~52 mm。颅全长 17~19 mm。吻鼻部裸露无毛。耳短呈钝三角形。耳壳后缘延伸至下颌角后方。耳屏短阔，似半月形。翼狭长，翼膜起于跖部。背毛深褐或棕褐色，具光泽。上臂至膝及股间膜近体部均具短褐毛。胸部被毛呈沙灰色，毛基暗褐色，体后部毛色较浅，前臂、第五掌骨附近以及体侧至膝部，股间膜近体部被覆厚密黄褐色短毛。

Dental formula: 2.1.1.3/3.1.2.3=32. Head and body length 65-75 mm. Ear length 15-18 mm. Forearm length 47-56 mm. Hind foot length 10-11 mm. Tail length 36-52 mm. Greatest skull length 17-19 mm. Snout nose is bare and hairless. Ears are short and blunt triangular. Posterior edge of the auricle extends to the rear of the jaw. Dorsal hairs dark brown or tan, shiny. Upper arm to the knee and the proximal part of the femoral membrane has short brown hairs. Chest hairs are sand gray, the hair bases are dark brown. Hair color of the rear part of the body is lighter. Forearm, near the 5th metacarpal and the side of the body to the knee, and to the interfemoral membrane near the body is covered with thick yellowish-brown short hairs.

生境 / Habitat

人造建筑、洞穴、森林
Man-made building, caves, forest

▲ 地理分布 / Geographic Distribution

国内分布 / Domestic Distribution

辽宁、吉林、北京、山东、山西、陕西、甘肃、浙江、安徽、湖北、湖南、四川、重庆、贵州、云南、福建、台湾、广东、广西、江西、香港
Liaoning, Jilin, Beijing, Shandong, Shanxi, Shaanxi, Gansu, Zhejiang, Anhui, Hubei, Hunan, Sichuan, Chongqing, Guizhou, Yunnan, Fujian, Taiwan, Guangdong, Guangxi, Jiangxi, Hong Kong

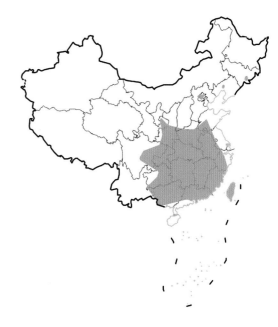

全球分布 / World Distribution
中国 China

生物地理界 / Biogeographic Realm
印度马来界 Indomalaya

WWF 生物群系 / WWF Biome
热带和亚热带湿润阔叶林
Tropical & Subtropical Moist Broadleaf Forests

动物地理分布型 / Zoogeographic Distribution Type
Ea

分布标注 / Distribution Note
特有种 Endemic

▲ 濒危状况 / Threatened Status

中国生物多样性红色名录等级 / CB RL Category (2021)
无危 LC

IUCN 红色名录 / IUCN Red List (2021)
无危 LC

威胁因子 / Threats
未知 Unknown

▲ 法律保护地位 / Legal Protection Status

国家重点保护野生动物等级 / Category of National Key Protected Wild Animals (2021)
未列入 Not listed

"三有"名录 / TWIESSV (2023)
未列入 Not listed

CITES 附录等级 / CITES Appendix (2023)
未列入 Not listed

迁徙物种公约附录 / CMS Appendix (2020)
未列入 Not listed

保护行动 / Conservation Action
尚无保护行动 No conservation action so far

▲ 参考文献 / References

Jiang et al. (蒋志刚等), 2021; Burgin et al., 2020; IUCN, 2020; Liu et al. (刘少英等), 2020; Wilson and Mittermeier, 2019; Pan et al. (潘清华等), 2007; Wilson and Reeder, 2005; Wang (王应祥), 2003; Zheng and Li (郑生武和李保国), 1999

204 / 华南扁颅蝠

Tylonycteris fulvida Temminck, 1840

· Lesser Bamboo Bat

翼手目 Chiroptera / 蝙蝠科 Vespertilionidae / 扁颅蝠属 *Tylonycteris*

科建立者及其文献 / Family Authority
Gray, 1821

属建立者及其文献 / Genus Authority
Peters, 1872

亚种 / Subspecies
无 None

模式标本产地 / Type Locality
缅甸
From the late Major Berdmore, of Schwe Gyen, in the valley of the Sitang river, Tenasserim Provinces

▲ 其他名称 / Other Name(s)

其他中文名 / Other Chinese Name(s)
扁颅蝠

其他英文名 / Other English Name(s)
Amber Bamboo Bat

同物异名 / Synonym(s)
无 None

▲ 形态及生境 / Morphology and Habitat

形态特征 / Morphological Characteristics
颅全长 11~12 mm。耳长 5~8 mm。前臂长 25~28 mm。后足长 4~6 mm。鼻孔呈管状向两侧突出。背毛基部呈金黄色，头颈部被毛毛尖呈褐色，身体下部呈浅金黄色。耳近三角形，裸露无毛，顶部圆，后缘有褶皱。耳屏短而钝。翅膜是深棕色的。足部与脚底裸露无毛，呈棕色。
Greatest skull length is 11-12 mm. Ear length is 5-8 mm. Forearm is 25-28 mm. Hind foot length is 4-6 mm. Nostrils are tubular and protrude laterally. Bases of the dorsal hairs are golden yellow, and the tips of the hairs on the head and neck are brown. Hairs on the lower part of the body is light golden yellow. Ears subtriangular, bare and glabrous, rounded apically, and trailing margin wrinkled. Tragus is short and blunt. The membrane is dark brown. Feet and soles bare and glabrous, brown in color.

生境 / Habitat
竹林、洞穴
Bamboo grove, caves

▲ 地理分布 / Geographic Distribution

国内分布 / Domestic Distribution
广东、广西、贵州、云南、香港、澳门
Guangdong, Guangxi, Guizhou, Yunnan, Hong Kong, Macao

全球分布 / World Distribution
孟加拉国、文莱、柬埔寨、中国、印度、印度尼西亚、老挝、马来西亚、缅甸、菲律宾、泰国、越南
Bangladesh, Brunei, Cambodia, China, India, Indonesia, Laos, Malaysia, Myanmar, Philippines, Thailand, Vietnam

生物地理界 / Biogeographic Realm
印度马来界 Indomalaya

WWF 生物群系 / WWF Biome
热带和亚热带湿润阔叶林
Tropical & Subtropical Moist Broadleaf Forests

动物地理分布型 / Zoogeographic Distribution Type
Sb

分布标注 / Distribution Note
非特有种 Non-Endemic

▲ 濒危状况 / Threatened Status

中国生物多样性红色名录等级 / CB RL Category (2021)
无危 LC

IUCN 红色名录 / IUCN Red List (2021)
无危 LC

威胁因子 / Threats
伐竹、耕种
Bamboo logging, farming

▲ 法律保护地位 / Legal Protection Status

国家重点保护野生动物等级 / Category of National Key Protected Wild Animals (2021)
未列入 Not listed

"三有"名录 / TWIESSV (2023)
未列入 Not listed

CITES 附录等级 / CITES Appendix (2023)
未列入 Not listed

迁徙物种公约附录 / CMS Appendix (2020)
未列入 Not listed

保护行动 / Conservation Action
尚无保护行动 No conservation action so far

▲ 参考文献 / References

Jiang et al. (蒋志刚等), 2021; Burgin et al., 2020; IUCN, 2020; Wilson and Mittermeier, 2019; Zhu et al. (朱光剑等), 2008; Wang and Xie (汪松和解焱), 2004

205 / 小扁颅蝠

Tylonycteris pygmaeus
Feng, Li & Wang, 2008

• Minimum Bamboo Bat

翼手目 Chiroptera / 蝙蝠科 Vespertilionidae / 扁颅蝠属 *Tylonycteris*

科建立者及其文献 / Family Authority
Gray, 1821

属建立者及其文献 / Genus Authority
Peters, 1872

亚种 / Subspecies
无 None

模式标本产地 / Type Locality
中国
Damenglong (100°42'E, 21°35'N), Jinghong County, Xishuangbanna Prefecture, Yunnan, China

▲ 其他名称 / Other Name(s)

其他中文名 / Other Chinese Name(s)
无 None

其他英文名 / Other English Name(s)
无 None

同物异名 / Synonym(s)
无 None

▲ 形态及生境 / Morphology and Habitat

形态特征 / Morphological Characteristics
齿式：2.1.1.3/3.1.2.3=32。头体长 23.5~27.7 mm。前臂长 23.9~25.6 mm。体重 2.5~3.5 g。头宽平扁。吻短而宽。耳屏短，端部钝圆。背部毛色深棕色，腹毛浅棕色。后足第一趾基部及趾间具盘状肉垫。
Dental formula: 2.1.1.3/3.1.2.3=32. Head and body length 23.5-27.7 mm. Forearm length 23.9-25.6 mm. Body mass 2.5-3.5 g. Braincase broad and flat. Muzzle short and wide. Tragus short, top blunt round. Dark brown hairs on the back and light brown hairs on the abdomen. The base and intertoe of the first toe of the hind foot have dish-shaped pads.

生境 / Habitat
竹林 Bamboo grove

▲ 地理分布 / Geographic Distribution

国内分布 / Domestic Distribution
云南 Yunnan

全球分布 / World Distribution
中国 China

生物地理界 / Biogeographic Realm
印度马来界 Indomalaya

WWF 生物群系 / WWF Biome
热带和亚热带湿润阔叶林
Tropical & Subtropical Moist Broadleaf Forests

动物地理分布型 / Zoogeographic Distribution Type
Sb

分布标注 / Distribution Note
特有种 Endemic

▲ 濒危状况 / Threatened Status

中国生物多样性红色名录等级 / CB RL Category (2021)
数据缺乏 DD

IUCN 红色名录 / IUCN Red List (2021)
未评定 NE

威胁因子 / Threats
未知 Unknown

▲ 法律保护地位 / Legal Protection Status

国家重点保护野生动物等级 / Category of National Key Protected Wild Animals (2021)
未列入 Not listed

"三有"名录 / TWIESSV (2023)
未列入 Not listed

CITES 附录等级 / CITES Appendix (2023)
未列入 Not listed

迁徙物种公约附录 / CMS Appendix (2020)
未列入 Not listed

保护行动 / Conservation Action
尚无保护行动 No conservation action so far

▲ 参考文献 / References

Jiang et al. (蒋志刚等), 2021; Burgin et al., 2020; IUCN, 2020; Feng et al. (冯庆等), 2006

206 / 托京褐扁颅蝠

Tylonycteris tonkinensis
Tu, Csorba, Ruedi & Hassanin, 2017

· Tonkin Greater Bamboo Bat

▲ 分类地位 / Taxonomy

翼手目 Chiroptera / 蝙蝠科 Vespertilionidae / 扁颅蝠属 *Tylonycteris*

科建立者及其文献 / Family Authority
Gray, 1821

属建立者及其文献 / Genus Authority
Peters, 1872

亚种 / Subspecies
无 None

模式标本产地 / Type Locality
越南
Vietnam: Copia Nature Reserve, Co Ma co mmune, Thuan Chau District, Son La Province, Malaysia, Borneo, Sarawak, Upper Sarawak

▲ 其他名称 / Other Name(s)

其他中文名 / Other Chinese Name(s)
无 None

其他英文名 / Other English Name(s)
Tonkin Bamboo Bat

同物异名 / Synonym(s)
无 None

▲ 形态及生境 / Morphology and Habitat

形态特征 / Morphological Characteristics

齿式：2.1.1.3/3.1.2.3=32。最大颅骨长 11.9~12.6 mm。耳呈三角形，尖端粗而圆。耳屏短而钝。前臂长 25.1~27.8 mm。头部背侧扁平。毛色相对多变，背毛基部或多或少有金红色，背毛尖端深棕色，背毛底部浅金棕色。翅膜深棕色。拇指基部和后脚底有肉垫。

Teeth formula: 2.1.1.3/3.1.2.3=32. Greatest skull length 11.9-12.6 mm. The ears are triangular with thick, rounded tips. The tragus is short and blunt. The forearms are 25.1-27.8 mm long. The head is flat on the back. The pelage color is relatively variable, more or less golden-red at the base of the dorsum, dark brown at the tips of the dorsal hairs, and light golden-brown at the bases of the dorsal hairs. Wing membrane dark brown. There are pads at the base of the thumb and on the sole of the hind foot.

生境 / Habitat

竹林 Bamboo grove

▲ 地理分布 / Geographic Distribution

国内分布 / Domestic Distribution
福建、江西、香港、广东、海南、广西、贵州、四川
Fujian, Jiangxi, Hong Kong, Guangdong, Hainan, Guangxi, Guizhou, Sichuan

全球分布 / World Distribution
中国、越南、老挝
China, Vietnam, Laos

生物地理界 / Biogeographic Realm
印度马来界 Indomalaya

WWF 生物群系 / WWF Biome
热带和亚热带湿润阔叶林
Tropical & Subtropical Moist Broadleaf Forests

动物地理分布型 / Zoogeographic Distribution Type
Wa

分布标注 / Distribution Note
非特有种 Non-Endemic

▲ 濒危状况 / Threatened Status

中国生物多样性红色名录等级 / CB RL Category (2021)
未评定 NE

IUCN 红色名录 / IUCN Red List (2021)
未评定 NE

威胁因子 / Threats
未知 Unknown

▲ 法律保护地位 / Legal Protection Status

国家重点保护野生动物等级 / Category of National Key Protected Wild Animals (2021)
未列入 Not listed

"三有"名录 / TWIESSV (2023)
未列入 Not listed

CITES 附录等级 / CITES Appendix (2023)
未列入 Not listed

迁徙物种公约附录 / CMS Appendix (2020)
未列入 Not listed

保护行动 / Conservation Action
尚无保护行动 No conservation action so far

▲ 参考文献 / References

Liang et al. (梁晓玲等), 2021; Yu et al. (余文华等), 2021; Wei et al. (魏辅文等), 2021; Schoch et al., 2020; Tu et al., 2017; Huang et al., 2014

207 / 北京宽耳蝠

Barbastella beijingensis
Zhang, Han, Jones, Lin, Zhang, Zhu,
Huang & Zhang, 2007

· Beijing Barbastelle

▲ 分类地位 / Taxonomy

翼手目 Chiroptera / 蝙蝠科 Vespertilionidae / 宽耳蝠属 *Barbastella*

科建立者及其文献 / Family Authority
Gray, 1821

属建立者及其文献 / Genus Authority
Gray, 1821

亚种 / Subspecies
无 None

模式标本产地 / Type Locality
中国
San-Qing Cave (39^{0}45'N, 115^{0}45'E) at Wang-Lao-Pu Village, Fangshan District,
southwest Beijing, China

▲ 其他名称 / Other Name(s)

其他中文名 / Other Chinese Name(s)
无 None

其他英文名 / Other English Name(s)
无 None

同物异名 / Synonym(s)
无 None

▲ 形态及生境 / Morphology and Habitat

形态特征 / Morphological Characteristics

齿式：2.1.2.3/3.1.2.3=34。前臂长 41~46 mm。体重 10.5~13.9 g。本属中体型相对较大的一种。背毛为暗黑色，毛尖为棕灰色。腹毛颜色较浅。耳郭形态和回声定位频率与东方宽耳蝠不同。

Dental formula: 2.1.2.3/3.1.2.3=34. Forearm length 41-46 mm. Body mass of 10.5-13.9 g. A relatively large member of the genus *Barbastella*. Dorsal hairs are dark black with brown-gray tips. Ventral hairs are lighter colored. Shape of its ear and the frequency of its echolocation calls is distinct from those of its closest relatives, the Asian barbastelle *Barbastella leucomelas*.

生境 / Habitat

洞穴 Caves

▲ 地理分布 / Geographic Distribution

国内分布 / Domestic Distribution
北京 Beijing

全球分布 / World Distribution
中国 China

生物地理界 / Biogeographic Realm
古北界 Palearctic

WWF 生物群系 / WWF Biome
热带和亚热带湿润阔叶林
Tropical & Subtropical Moist Broadleaf Forests

动物地理分布型 / Zoogeographic Distribution Type
Ea

分布标注 / Distribution Note
特有种 Endemic

▲ 濒危状况 / Threatened Status

中国生物多样性红色名录等级 / CB RL Category (2021)
数据缺乏 DD

IUCN 红色名录 / IUCN Red List (2021)
未评定 NE

威胁因子 / Threats
未知 Unknown

▲ 法律保护地位 / Legal Protection Status

国家重点保护野生动物等级 / Category of National Key Protected Wild Animals (2021)
未列入 Not listed

"三有" 名录 / TWIESSV (2023)
未列入 Not listed

CITES 附录等级 / CITES Appendix (2023)
未列入 Not listed

迁徙物种公约附录 / CMS Appendix (2020)
未列入 Not listed

保护行动 / Conservation Action
尚无保护行动 No conservation action so far

▲ 参考文献 / References

Jiang et al. (蒋志刚等), 2021; Burgin et al., 2020; IUCN, 2020; Liu et al. (刘少英等), 2020; Wilson and Mittermeier, 2019; Zhang et al., 2007

208 / 亚洲宽耳蝠

Barbastella leucomelas
(Cretzschmar, 1826)

• Eastern Barbastelle

翼手目 Chiroptera / 蝙蝠科 Vespertilionidae / 宽耳蝠属 *Barbastella*

科建立者及其文献 / Family Authority
Gray, 1821

属建立者及其文献 / Genus Authority
Gray, 1821

亚种 / Subspecies
无 None

模式标本产地 / Type Locality
埃及
Egypt, Sinai

吴毅 / 供图

▲ 其他名称 / Other Name(s)

其他中文名 / Other Chinese Name(s)
东方宽耳蝠、阔耳蝠、亚洲阔耳蝠

其他英文名 / Other English Name(s)
Asian Barbastelle

同物异名 / Synonym(s)
无 None

▲ 形态及生境 / Morphology and Habitat

形态特征 / Morphological Characteristics
齿式：2.1.2.3/3.1.2.3=34。头体长 47~51 mm。耳长 15~17 mm。前臂长 38~45 mm。后足长 7~8 mm。尾长 40~47 mm。颅全长 14~16 mm。耳宽大。耳屏呈狭长三角形，左右耳郭在额顶部相连。鼻孔朝上。尾长。背毛黑棕色，毛尖灰色或米黄色。腹毛色与背毛相近，毛尖浅白色；股间膜近腹侧处覆毛白色。股间膜发达，呈盾形。

Dental formula: 2.1.2.3/3.1.2.3=34. Head and body length 47-51 mm. Ear length 15-17 mm. Forearm length 38-45 mm. Hind foot length 7-8 mm. Tail length 40-47 mm. Greatest skull length 14-16 mm. Ears wide, tragus long triangle. The left and right auricles are almost joined at the top of the forehead. Nostrils up. The fifth finger is longer than the metacarpal of the third and fourth fingers. Dorsal hairs are dark brown with gray or beige tips. The abdominal hair color is similar to the back, and the hair tips are light white. The proximal ventral lining of the interfemoral membrane is white. The interfemoral membrane is well developed and is shield-shaped.

生境 / Habitat
热带和亚热带湿润山地森林、洞穴、人造建筑
Tropical and subtropical moist montane forest, caves, man-made building

▲ 地理分布 / Geographic Distribution

国内分布 / Domestic Distribution
内蒙古、新疆、陕西、甘肃、青海、四川、云南、重庆、台湾、河南、
湖南、江西
Inner Mongolia, Xinjiang, Shaanxi, Gansu, Qinghai, Sichuan, Yunnan,
Chongqing, Taiwan, Henan, Hunan, Jiangxi

全球分布 / World Distribution
阿富汗、亚美尼亚、阿塞拜疆、不丹、中国、埃及、厄立特里亚、
格鲁吉亚、印度、伊朗、以色列、日本、吉尔吉斯斯坦、尼泊尔、
巴基斯坦、俄罗斯、沙特阿拉伯、塔吉克斯坦、土库曼斯坦、乌
兹别克斯坦
Afghanistan, Armenia, Azerbaijan, Bhutan, China, Egypt, Eritrea, Georgia, India,
Iran, Israel, Japan, Kyrgyzstan, Nepal, Pakistan, Russia, Saudi Arabia, Tajikistan,
Turkmenistan, Uzbekistan

生物地理界 / Biogeographic Realm
古北界 Palearctic

WWF 生物群系 / WWF Biome
热带和亚热带湿润阔叶林、温带针叶林
Tropical & Subtropical Moist Broadleaf Forests, Temperate Conifer Forests

动物地理分布型 / Zoogeographic Distribution Type
We

分布标注 / Distribution Note
非特有种 Non-Endemic

▲ 濒危状况 / Threatened Status

中国生物多样性红色名录等级 / CB RL Category (2021)
易危 VU

IUCN 红色名录 / IUCN Red List (2021)
无危 LC

威胁因子 / Threats
砍伐 Logging & wood harvesting

▲ 法律保护地位 / Legal Protection Status

国家重点保护野生动物等级 / Category of National Key Protected Wild Animals (2021)
未列入 Not listed

"三有" 名录 / TWIESSV (2023)
未列入 Not listed

CITES 附录等级 / CITES Appendix (2023)
未列入 Not listed

迁徙物种公约附录 / CMS Appendix (2020)
未列入 Not listed

保护行动 / Conservation Action
尚无保护行动 No conservation action so far

▲ 参考文献 / References

Jiang et al. (蒋志刚等), 2021; Liu et al. (刘少英等), 2020; Zhang et al. (张翰博等), 2020; Wilson and Mittermeier, 2019; Pan et al. (潘清华等),
2007; Zhang et al., 2007; Wilson and Reeder, 2005; Wang (王应祥), 2003; Zhang (张荣祖), 1997

209 / 斑蝠

Scotomanes ornatus (Blyth, 1851)

• Harlequin Bat

▲ 分类地位 / Taxonomy

翼手目 Chiroptera / 蝙蝠科 Vespertilionidae / 斑蝠属 *Scotomanes*

科建立者及其文献 / Family Authority
Gray, 1821

属建立者及其文献 / Genus Authority
Dobson, 1875

亚种 / Subspecies
华南亚种 *S. o. sinensis* Thomas, 1921
安徽、福建、广东、广西、海南、湖南、四川和贵州
Anhui, Fujian, Guangdong, Guangxi, Hainan, Hunan, Sichuan and Guizhou

模式标本产地 / Type Locality
印度
India, Assam, Khasi Hills, Cherrapunji

周佳俊 / 供图

吴毅 / 供图

▲ 其他名称 / Other Name(s)

其他中文名 / Other Chinese Name(s)
花蝠、印度斑蝠、中华斑蝠

其他英文名 / Other English Name(s)
无 None

同物异名 / Synonym(s)
无 None

▲ 形态及生境 / Morphology and Habitat

形态特征 / Morphological Characteristics
齿式：1.1.1.3/3.1.2.3=30。头体长 64~85 mm。耳长 19~23 mm。前臂长 50~62 mm。后足长 12~15 mm。尾长 52~66 mm。颅全长 20 mm。耳呈椭圆形。耳屏短，端部钝，外缘弧形内缘直。体毛具光泽。背部褐棕色，中央有白色条纹，头顶和两肩后有白色斑。颈部和腹部中央有"V"字形棕褐色斑纹。翼膜黑褐色，翼膜止于趾基部。股间膜淡棕褐色。
Dental formula: 2.1.1.3/3.1.2.3=32. Head and body length 64-85 mm. Ear length 19-23 mm. Forearm length 50-62 mm. Hind foot length 12-15 mm. Tail length 52-66 mm. Greatest skull length 20 mm. Ears are oval. Tragus short, blunt end, outer margin arc and inner margin straight. Body hairs are shiny. Dorsal hairs are brown with a white stripe in the center, the top of the head and behind the shoulders are white spots, and the neck and abdomen have "V" shaped tan markings in the center. Pterygoid dark brown, and pterygoid at the base of the toe. Interfemoral membrane is light brown.

生境 / Habitat
森林、洞穴
Forest, caves

▲ 地理分布 / Geographic Distribution

国内分布 / Domestic Distribution
安徽、浙江、湖南、四川、贵州、云南、广西、广东、海南、福建、
重庆
Anhui, Zhejiang, Hunan, Sichuan, Guizhou, Yunnan, Guangxi, Guangdong,
Hainan, Fujian, Chongqing

全球分布 / World Distribution
孟加拉国、中国、印度、老挝、缅甸、尼泊尔、泰国、越南
Bangladesh, China, India, Laos, Myanmar, Nepal, Thailand, Vietnam

生物地理界 / Biogeographic Realm
印度马来界 Indomalaya

WWF 生物群系 / WWF Biome
热带和亚热带湿润阔叶林
Tropical & Subtropical Moist Broadleaf Forests

动物地理分布型 / Zoogeographic Distribution Type
Wc

分布标注 / Distribution Note
非特有种 Non-Endemic

▲ 濒危状况 / Threatened Status

中国生物多样性红色名录等级 / CB RL Category (2021)
近危 NT

IUCN 红色名录 / IUCN Red List (2021)
无危 LC

威胁因子 / Threats
森林砍伐、耕种
Logging, farming

▲ 法律保护地位 / Legal Protection Status

国家重点保护野生动物等级 / Category of National Key Protected Wild Animals (2021)
未列入 Not listed

"三有"名录 / TWIESSV (2023)
未列入 Not listed

CITES 附录等级 / CITES Appendix (2023)
未列入 Not listed

迁徙物种公约附录 / CMS Appendix (2020)
未列入 Not listed

保护行动 / Conservation Action
尚无保护行动 No conservation action so far

▲ 参考文献 / References

Jiang et al. (蒋志刚等), 2021; Burgin et al., 2020; IUCN, 2020; Wilson and Mittermeier, 2019; Smith et al., 2009; Pan et al. (潘清华等), 2007; Wilson and Reeder, 2005; Wang (王应祥), 2003; Lin et al., 2002; Zhang (张荣祖), 1997; Ding and Wang (丁铁明和王作义), 1989; Liang et al. (梁仁济等), 1983

210 / 大黄蝠

Scotophilus heathi (Horsfield, 1831)

· Greater Asiatic Yellow House Bat

▲ 分类地位 / Taxonomy

翼手目 Chiroptera / 蝙蝠科 Vespertilionidae / 黄蝠属 *Scotophilus*

科建立者及其文献 / Family Authority
Gray, 1821

属建立者及其文献 / Genus Authority
Leach, 1821

亚种 / Subspecies
华南亚种 *S. h. belangeri* (Geoffroy, 1834)
云南、广西、湖南、广东和福建
Yunnan, Guangxi, Hunan, Guangdong and Fujian

海南亚种 *S. h. insularis* J. Allen, 1906
海南 Hainan

模式标本产地 / Type Locality
印度
India, Madras

吴毅 / 供图

▲ 其他名称 / Other Name(s)

其他中文名 / Other Chinese Name(s)
黄蝠

其他英文名 / Other English Name(s)
Common Yellow Bat

同物异名 / Synonym(s)
无 None

▲ 形态及生境 / Morphology and Habitat

形态特征 / Morphological Characteristics
齿式：2.1.1.3/3.1.2.3=32。头体长 67~93 mm。耳长 13~21 mm。前臂长 55~66 mm。后足长 9~15 mm。尾长 43~71 mm。颅全长 21~26 mm。头部粗壮，耳短，端部圆，耳屏呈弯月形。后足大，距缘膜狭长。尾端从股间膜后缘稍微穿出。背毛呈栗褐色，腹部毛土黄色。
Dental formula: 2.1.1.3/3.1.2.3=32. Head and body length 67-93 mm. Ear length 13-21 mm. Forearm length 55-66 mm. Hind Foot length 9-15 mm. Tail length 43-71 mm. Greatest skull length 21-26 mm. Head is stout. Ears are short, with round end. Tragus is crescent. Hind foot is large, long, and narrow from the margin membrane. Caudal end passes slightly through the posterior margin of the interfemoral membrane. Dorsal hairs are thick and long, and are yellowish-brown. Abdominal hairs are yellow and bright in color.

生境 / Habitat
城市、人造建筑、森林
Urban area, man-made building, forest

▲ 地理分布 / Geographic Distribution

国内分布 / Domestic Distribution
云南、湖南、广东、广西、福建、海南
Yunnan, Hunan, Guangdong, Guangxi, Fujian, Hainan

全球分布 / World Distribution
阿富汗、孟加拉国、柬埔寨、中国、印度、印度尼西亚、老挝、缅甸、
尼泊尔、巴基斯坦、菲律宾、斯里兰卡、泰国、越南
Afghanistan, Bangladesh, Cambodia, China, India, Indonesia, Laos, Myanmar,
Nepal, Pakistan, Philippines, Sri Lanka, Thailand, Vietnam

生物地理界 / Biogeographic Realm
印度马来界 Indomalaya

WWF 生物群系 / WWF Biome
热带和亚热带湿润阔叶林
Tropical & Subtropical Moist Broadleaf Forests

动物地理分布型 / Zoogeographic Distribution Type
Yb

分布标注 / Distribution Note
非特有种 Non-Endemic

▲ 濒危状况 / Threatened Status

中国生物多样性红色名录等级 / CB RL Category (2021)
无危 LC

IUCN 红色名录 / IUCN Red List (2021)
无危 LC

威胁因子 / Threats
无 None

▲ 法律保护地位 / Legal Protection Status

国家重点保护野生动物等级 / Category of National Key Protected Wild Animals (2021)
未列入 Not listed

"三有" 名录 / TWIESSV (2023)
未列入 Not listed

CITES 附录等级 / CITES Appendix (2023)
未列入 Not listed

迁徙物种公约附录 / CMS Appendix (2020)
未列入 Not listed

保护行动 / Conservation Action
尚无保护行动 No conservation action so far

▲ 参考文献 / References

Jiang et al. (蒋志刚等), 2021; Burgin et al., 2020; IUCN, 2020; Liu et al. (刘少英等), 2020; Smith et al., 2009; Pan et al. (潘清华等), 2007; Wilson and Reeder, 2005; Wang (王应祥), 2003; Zhang (张荣祖), 1997

211 / 小黄蝠

Scotophilus kuhlii Leach, 1821

· Lesser Asiatic Yellow House Bat

▲ 分类地位 / Taxonomy

翼手目 Chiroptera / 蝙蝠科 Vespertilionidae / 黄蝠属 *Scotophilus*

科建立者及其文献 / Family Authority
Gray, 1821

属建立者及其文献 / Genus Authority
Leach, 1821

亚种 / Subspecies
华南亚种 *S. k. swinhoei* (Byth, 1896)
福建、广东、香港和广西
Fujian, Guangdong, Hong Kong and Guangxi

泰国亚种 *S. k. gairdneri* Kross, 1917
云南 Yunnan

海南亚种 *S. k. consobrinus* J. Allen, 1906
海南和台湾
Hainan and Taiwan

模式标本产地 / Type Locality
印度
India

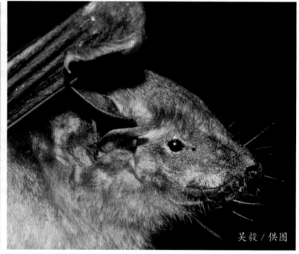

吴毅 / 供图

吴毅 / 供图

▲ 其他名称 / Other Name(s)

其他中文名 / Other Chinese Name(s)
高颡蝠、库氏黄蝠、中黄蝠、
高头蝠

其他英文名 / Other English Name(s)
Lesser Asian House Bat

同物异名 / Synonym(s)
无 None

▲ 形态及生境 / Morphology and Habitat

形态特征 / Morphological Characteristics
齿式：2.1.1.3/3.1.2.3=32。头体长 60~78 mm。耳长 9~17 mm。前臂长
44~55 mm。后足长 8~13 mm。尾长 40~65 mm。颅全长 16~20 mm。吻
鼻部宽且两侧肿胀。鼻孔稍微向外突出。耳小，无毛，具横向皱褶，具
弧形弯曲的耳屏。毛被短而紧密。背部呈栗黄色，腹毛土浅黄色。
Dental formula: 2.1.1.3/3.1.2.3=32. Head and body length 60-78 mm. Ear length 9-17
mm. Forearm length 44-55 mm. Hind foot length 8-13 mm. Tail length 40-65 mm.
Greatest skull length 16-20 mm. Snout is wide and swollen on both sides, almost bare.
Nostrils protrude slightly. Ears small, glabrous, with transverse folds and antitragus.
Tragus is arcuate-curved. Hairs are short and close. Dorsal hairs are chestnut brown.
Ventral hairs are light soil yellow.

生境 / Habitat
森林、次生林、城市、人造建筑
Forest, secondary forest, urban area, man-made building

▲ 地理分布 / Geographic Distribution

国内分布 / Domestic Distribution
广西、广东、云南、海南、福建、台湾
Guangxi, Guangdong, Yunnan, Hainan, Fujian, Taiwan

全球分布 / World Distribution
孟加拉国、柬埔寨、中国、印度、印度尼西亚、老挝、马来西亚、
缅甸、巴基斯坦、菲律宾、斯里兰卡、泰国、东帝汶、越南
Bangladesh, Cambodia, China, India, Indonesia, Laos, Malaysia, Myanmar,
Pakistan, Philippines, Sri Lanka, Thailand, Timor-Leste, Vietnam

生物地理界 / Biogeographic Realm
印度马来界、大洋洲界
Indomalaya, Oceanian

WWF 生物群系 / WWF Biome
热带和亚热带湿润阔叶林
Tropical & Subtropical Moist Broadleaf Forests

动物地理分布型 / Zoogeographic Distribution Type
Wa

分布标注 / Distribution Note
非特有种 Non-Endemic

▲ 濒危状况 / Threatened Status

中国生物多样性红色名录等级 / CB RL Category (2021)
无危 LC

IUCN 红色名录 / IUCN Red List (2021)
无危 LC

威胁因子 / Threats
未知 Unknown

▲ 法律保护地位 / Legal Protection Status

国家重点保护野生动物等级 / Category of National Key Protected Wild Animals (2021)
未列入 Not listed

"三有"名录 / TWIESSV (2023)
未列入 Not listed

CITES 附录等级 / CITES Appendix (2023)
未列入 Not listed

迁徙物种公约附录 / CMS Appendix (2020)
未列入 Not listed

保护行动 / Conservation Action
尚无保护行动 No conservation action so far

▲ 参考文献 / References

Jiang et al. (蒋志刚等), 2021; Burgin et al., 2020; IUCN, 2020; Liu et al. (刘少英等), 2020; Wilson and Mittermeier, 2019; Smith and Xie, 2009; Pan et al. (潘清华等), 2007; Yu et al., 2012; Wilson and Reeder, 2005; Wang (王应祥), 2003; Zhang (张荣祖), 1997

212 / 环颈蝠

Thainycteris aureocollaris
Kock & Storch, 1996

· Collared Sprite

▲ 分类地位 / Taxonomy

翼手目 Chiroptera / 蝙蝠科 Vespertilionidae / 金颈蝠属 *Thainycteris*

科建立者及其文献 / Family Authority
Gray, 1821

属建立者及其文献 / Genus Authority
Kock and Storch, 1996

亚种 / Subspecies
无 None

模式标本产地 / Type Locality
泰国
Thailand, Chiang Mai

吴毅 / 供图

▲ 其他名称 / Other Name(s)

其他中文名 / Other Chinese Name(s)
环颈伏翼

其他英文名 / Other English Name(s)
Collared Pipistrelle, Siam Goldnecklet,
Thai Golden-throated Bat

同物异名 / Synonym(s)
无 None

▲ 形态及生境 / Morphology and Habitat

形态特征 / Morphological Characteristics
齿式：2.1.2.3/3.1.2.3=34。前臂长 47~52 mm。后足长 9~12 mm。吻部短宽，被稀毛。矢状嵴弱，眶上嵴发达。背毛毛基偏黑，毛尖赭黄色；腹毛毛尖略显白。耳基部至喉部有明显赭黄色项圈。

Dental formula: 2.1.2.3/3.1.2.3=34. Forearm length 47-52 mm. Hind foot length 9-12 mm. Snout is short and broad, with sparse hairs. Sagittal crest is weak and the supraorbital crest is well developed. Dorsal hairs slant black at the base, with ochre yellow hair tips. Abdominal hair tips are slightly white. There is a clear ochre yellow collar from ear base to the throat.

生境 / Habitat
森林 Forest

▲ 地理分布 / Geographic Distribution

国内分布 / Domestic Distribution
云南、贵州、福建
Yunnan, Guizhou, Fujian

全球分布 / World Distribution
中国、老挝、泰国、越南
China, Laos, Thailand, Vietnam

生物地理界 / Biogeographic Realm
古北界、印度马来界
Palearctic, Indomalaya

WWF 生物群系 / WWF Biome
热带和亚热带湿润阔叶林
Tropical & Subtropical Moist Broadleaf Forests

动物地理分布型 / Zoogeographic Distribution Type
Wa

分布标注 / Distribution Note
非特有种 Non-Endemic

▲ 濒危状况 / Threatened Status

中国生物多样性红色名录等级 / CB RL Category (2021)
无危 LC

IUCN 红色名录 / IUCN Red List (2021)
无危 LC

威胁因子 / Threats
未知 Unknown

▲ 法律保护地位 / Legal Protection Status

国家重点保护野生动物等级 / Category of National Key Protected Wild Animals (2021)
未列入 Not listed

"三有"名录 / TWIESSV (2023)
未列入 Not listed

CITES 附录等级 / CITES Appendix (2023)
未列入 Not listed

迁徙物种公约附录 / CMS Appendix (2020)
未列入 Not listed

保护行动 / Conservation Action
尚无保护行动 No conservation action so far

▲ 参考文献 / References

Jiang et al. (蒋志刚等), 2021; Xie et al., 2021; Burgin et al., 2020; IUCN, 2020; Wilson and Mittermeier, 2019; Guo et al. (郭伟健等), 2017

213 / 黄颈蝠

Thainycteris torquatus Csorba & Lee, 1999

· Necklace Sprite

▲ 分类地位 / Taxonomy

翼手目 Chiroptera / 蝙蝠科 Vespertilionidae / 金颈蝠属 *Thainycteris*

科建立者及其文献 / Family Authority
Gray, 1821

属建立者及其文献 / Genus Authority
Hill & Harrison, 1987

亚种 / Subspecies
无 None

模式标本产地 / Type Locality
中国
China, Taiwan, Taichung County, Wu-ling Farm; 1,800 m; 24°4'N, 121°8'E

周政翰 / 供图

▲ 其他名称 / Other Name(s)

其他中文名 / Other Chinese Name(s)
黄喉黑伏翼

其他英文名 / Other English Name(s)
Arielulus torquatus

同物异名 / Synonym(s)
无 None

▲ 形态及生境 / Morphology and Habitat

形态特征 / Morphological Characteristics
齿式：2.1.2.3/3.1.2.3=34。前臂长 43~46 mm。颅全长 16~17 mm。口吻短而宽。耳三角形，黑色，没有白色边缘。耳屏短而弯曲。喉部有明亮的赭色斑块。背毛黑色，毛尖青铜色。腹毛色深色，毛尖银色。尾膜近端一半被毛。侧膜止于第五趾基部。

Dental formula: 2.1.2.3/3.1.2.3=34. Forearm length 43-46 mm. Greatest skull length 16-17 mm. Muzzle short and broad. Ears triangular, black, without pale edging. Tragus short and curved. Throat with bright ochraceous collar. Ventral pelage is dark with silver-colored tips. Proximal half of uropatagium furred. Plagiopatagium attaches at the base of the fifth toe.

生境 / Habitat
未知 Unknown

▲ 地理分布 / Geographic Distribution

国内分布 / Domestic Distribution
台湾 Taiwan

全球分布 / World Distribution
中国 China

生物地理界 / Biogeographic Realm
印度马来界 Indomalaya

WWF 生物群系 / WWF Biome
热带和亚热带湿润阔叶林
Tropical & Subtropical Moist Broadleaf Forests

动物地理分布型 / Zoogeographic Distribution Type
J

分布标注 / Distribution Note
特有种 Endemic

▲ 濒危状况 / Threatened Status

中国生物多样性红色名录等级 / CB RL Category (2021)
易危 VU

IUCN 红色名录 / IUCN Red List (2021)
无危 LC

威胁因子 / Threats
住宅区及商业发展、公路铁路、耕种
Residential and commercial development, roads, railroads, farming

▲ 法律保护地位 / Legal Protection Status

国家重点保护野生动物等级 / Category of National Key Protected Wild Animals (2021)
未列入 Not listed

"三有" 名录 / TWIESSV (2023)
未列入 Not listed

CITES 附录等级 / CITES Appendix (2023)
未列入 Not listed

迁徙物种公约附录 / CMS Appendix (2020)
未列入 Not listed

保护行动 / Conservation Action
尚无保护行动 No conservation action so far

▲ 参考文献 / References

Jiang et al. (蒋志刚等), 2021; Burgin et al., 2020; IUCN, 2020; Liu et al. (刘少英等), 2020; Pan et al. (潘清华等), 2007; Wilson and Reeder, 2005; Wang (王应祥), 2003; Hill and Harrison, 1986

214 / 大耳蝠

Plecotus auritus Linnaeus, 1758

• Brown Big-eared Bat

▲ 分类地位 / Taxonomy

翼手目 Chiroptera / 蝙蝠科 Vespertilionidae / 大耳蝠属 *Plecotus*

科建立者及其文献 / Family Authority
Gray, 1821

属建立者及其文献 / Genus Authority
E. Geoffroy, 1818

亚种 / Subspecies
无 None

模式标本产地 / Type Locality
瑞典
Sweden

冯利民 / 供图

▲ 其他名称 / Other Name(s)

其他中文名 / Other Chinese Name(s)
大耳蝠、兔耳蝠、兔蝠

其他英文名 / Other English Name(s)
Common Long-eared Bat

同物异名 / Synonym(s)
无 None

▲ 形态及生境 / Morphology and Habitat

形态特征 / Morphological Characteristics
齿式：2.1.2.3/3.1.3.3=36。头体长 40~45 mm。耳长 39~41 mm。前臂长 36~46 mm。后足长 7~8 mm。尾长 48~50 mm。颅全长 16.5~18.8 mm。耳郭几乎和头体一样长。在飞行时，耳保持竖立并向前延伸。休息时，耳就像公羊的角一样向旁边蜷曲。在冬眠期间，会收起耳郭，仅长长的尖耳可见。鼻孔微朝上。尾长达到或超过头体长。背毛暗褐色。腹毛灰白色，毛尖浅褐色，毛基黑色。翼膜灰褐色。

Dental formula: 2.1.2.3/3.1.3.3=36. Head and body length 40-45 mm. Ear length 39-41 mm. Forearm length 36-46 mm. Hind foot length 7-8 mm. Tail length 48-50 mm. Greatest skull length 16.5-18.8 mm. Ears are almost as long as the body. While flying, big-eared bats keep their ears fully erected and extended forward. At rest, the bats' ears curl sideways like the horns of a ram. During hibernation, the bats fold back their ears, leaving only long, pointed caps. Nostrils are angled slightly upward. Tail reaches or exceeds the head and body length. Dorsal hairs are dark brown. Abdominal hairs grayish-white, with light brown hair tips and black bases. Pterygoid membrane is grayish-brown.

生境 / Habitat
森林、乡村种植园、人造建筑、洞穴
Forest, rural garden, man-made building, caves

▲ 地理分布 / Geographic Distribution

国内分布 / Domestic Distribution
黑龙江、吉林、内蒙古、甘肃、河北、山西、陕西
Heilongjiang, Jilin, Inner Mongolia, Gansu, Hebei, Shanxi, Shaanxi

全球分布 / World Distribution
中国、阿尔巴尼亚、安道尔、奥地利、阿塞拜疆、白俄罗斯、比利时、波斯尼亚和黑塞哥维那、保加利亚、克罗地亚、捷克、丹麦、爱沙尼亚、芬兰、法国、格鲁吉亚、德国、希腊、匈牙利、伊朗、爱尔兰、意大利、哈萨克斯坦、拉脱维亚、列支敦士登、立陶宛、卢森堡、摩尔多瓦、摩纳哥、黑山、荷兰、挪威、波兰、葡萄牙、罗马尼亚、俄罗斯、圣马力诺、塞尔维亚、斯洛伐克、斯洛文尼亚、西班牙、瑞典、瑞士、土耳其、乌克兰、英国

China, Albania, Andorra, Austria, Azerbaijan, Belarus, Belgium, Bosnia and Herzegovina, Bulgaria, Croatia, Czech, Denmark, Estonia, Finland, France, Georgia, Germany, Greece, Hungary, Iran, Ireland, Italy, Kazakhstan, Latvia, Liechtenstein, Lithuania, Luxembourg, Moldova, Monaco, Montenegro, Netherlands, Norway, Poland, Portugal, Romania, Russia, San Marino, Serbia, Slovakia, Slovenia, Spain, Sweden, Switzerland, Turkey, Ukraine, United Kingdom

生物地理界 / Biogeographic Realm
古北界 Palearctic

WWF 生物群系 / WWF Biome
温带阔叶和混交林
Temperate Broadleaf & Mixed Forests

动物地理分布型 / Zoogeographic Distribution Type
Ub

分布标注 / Distribution Note
非特有种 Non-Endemic

▲ 濒危状况 / Threatened Status

中国生物多样性红色名录等级 / CB RL Category (2021)
无危 LC

IUCN 红色名录 / IUCN Red List (2021)
无危 LC

威胁因子 / Threats
未知 Unknown

▲ 法律保护地位 / Legal Protection Status

国家重点保护野生动物等级 / Category of National Key Protected Wild Animals (2021)
未列入 Not listed

"三有"名录 / TWIESSV (2023)
未列入 Not listed

CITES 附录等级 / CITES Appendix (2023)
未列入 Not listed

迁徙物种公约附录 / CMS Appendix (2020)
未列入 Not listed

保护行动 / Conservation Action
尚无保护行动 No conservation action so far

▲ 参考文献 / References

Jiang et al. (蒋志刚等), 2021; Burgin et al., 2020; IUCN, 2020; Liu et al. (刘少英等), 2020; Wilson and Mittermeier, 2019; Zhu et al. (朱光剑等), 2008; Pan et al. (潘清华等), 2007; Wilson and Reeder, 2005; Wang (王应祥), 2003; Zhang (张荣祖), 1997

215 / 灰大耳蝠

Plecotus austriacus (Fischer, 1829)

• Gray Big-eared Bat

裴俊峰 / 供图

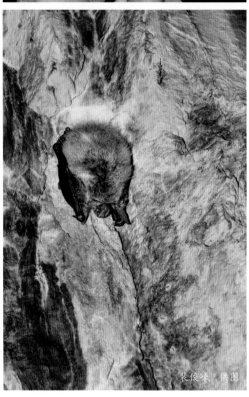

裴俊峰 / 供图

▲ 分类地位 / Taxonomy

翼手目 Chiroptera / 蝙蝠科 Vespertilionidae / 大耳蝠属 *Plecotus*

科建立者及其文献 / Family Authority
Gray, 1821

属建立者及其文献 / Genus Authority
E. Geoffroy, 1818

亚种 / Subspecies
四川亚种 *P. a. ariel* Thomas, 1911
四川和青海
Sichuan and Qinghai

克什米尔亚种 *P. a. wardi* Thomas, 1911
西藏 Tibet

阿拉善亚种 *P. a. kozlovi* Borbinskii, 1926
内蒙古、青海、甘肃和宁夏
Inner Mongolia, Qinghai, Gansu and Ningxia

新疆亚种 *P. a. mordax* Thomas, 1926
新疆 Xinjiang

模式标本产地 / Type Locality
奥地利
Austria, Vienna

▲ 其他名称 / Other Name(s)

其他中文名 / Other Chinese Name(s)
灰长耳蝠

其他英文名 / Other English Name(s)
Lesser Asiatic Yellow House Bat

同物异名 / Synonym(s)
无 None

▲ 形态及生境 / Morphology and Habitat

形态特征 / Morphological Characteristics
齿式：2.1.2.3/3.1.3.3=36。头体长 41~58 mm。耳长 37~42 mm。前臂长 37~45 mm。后足长 7~10 mm。尾长 37~55 mm。耳长达到或超过头体长。耳屏向前折超过吻端。耳壳内侧有皱褶。鼻孔朝上。尾长，全部包在股间膜内，或仅尾端少许穿出。背毛灰褐色。腹毛灰白色，毛尖浅褐色，毛基深棕色或浅黄色。翼膜深褐色。
Dental formula: 2.1.2.3/3.1.3.3=36. Head and body length 41-58 mm. Ear length 37-42 mm. Forearm length 37-45 mm. Hind foot length 7-10 mm. Tail length 37-55 mm. Ears equal to or longer than the length of the head and body length. Auricle is elliptical and folded forward over the snout. Inner part of the auricle is wrinkled. Nostrils up. Tail is long, but all enclosed in the interfemoral membrane or only a little through the tail. Dorsal hairs grayish-brown. Abdominal hairs grayish-white with light brown hair tips and dark brown or faint yellow bases. Pterygoid membrane is dark brown.

生境 / Habitat
耕地、人造建筑、洞穴
Arable land, man-made building, caves

▲ 地理分布 / Geographic Distribution

国内分布 / Domestic Distribution
新疆、四川、青海、西藏、内蒙古、甘肃、宁夏、陕西
Xinjiang, Sichuan, Qinghai, Tibet, Inner Mongolia, Gansu, Ningxia, Shaanxi

全球分布 / World Distribution
中国、阿尔巴尼亚、安道尔、奥地利、比利时、波斯尼亚和黑塞哥维那、保加利亚、克罗地亚、捷克、法国、德国、直布罗陀、希腊、匈牙利、意大利、列支敦士登、卢森堡、马其顿、马耳他、毛里塔尼亚、摩尔多瓦、摩纳哥、黑山、荷兰、波兰、葡萄牙、罗马尼亚、圣马力诺、塞尔维亚、斯洛伐克、斯洛文尼亚、西班牙、瑞典、瑞士、土耳其、乌克兰、英国
China, Albania, Andorra, Austria, Belgium, Bosnia and Herzegovina, Bulgaria, Croatia, Czech, France, Germany, Gibraltar, Greece, Hungary, Italy, Liechtenstein, Luxembourg, Macedonia, Malta, Mauritania, Moldova, Monaco, Montenegro, Netherlands, Poland, Portugal, Romania, San Marino, Serbia, Slovakia, Slovenia, Spain, Sweden, Switzerland, Turkey, Ukraine, United Kingdom

生物地理界 / Biogeographic Realm
古北界 Palearctic

WWF 生物群系 / WWF Biome
温带阔叶和混交林
Temperate Broadleaf & Mixed Forests

动物地理分布型 / Zoogeographic Distribution Type
Ue

分布标注 / Distribution Note
非特有种 Non-Endemic

▲ 濒危状况 / Threatened Status

中国生物多样性红色名录等级 / CB RL Category (2021)
近危 NT

IUCN 红色名录 / IUCN Red List (2021)
无危 LC

威胁因子 / Threats
森林砍伐、耕种
Logging, farming

▲ 法律保护地位 / Legal Protection Status

国家重点保护野生动物等级 / Category of National Key Protected Wild Animals (2021)
未列入 Not listed

"三有" 名录 / TWIESSV (2023)
未列入 Not listed

CITES 附录等级 / CITES Appendix (2023)
未列入 Not listed

迁徙物种公约附录 / CMS Appendix (2020)
未列入 Not listed

保护行动 / Conservation Action
尚无保护行动 No conservation action so far

▲ 参考文献 / References

Jiang et al. (蒋志刚等), 2021; Burgin et al., 2020; IUCN, 2020; Wilson and Mittermeier, 2019; Pan et al. (潘清华等), 2007; Peng and Zhong (彭基泰和钟祥清), 2005; Wilson and Reeder, 2005; Wang (王应祥), 2003; Zhang (张荣祖), 1997; Wilson and Reeder, 1993

216 / 奥氏长耳蝠

Plecotus ognevi (Kishida, 1927)

· Ognev's Long-eared Bat

▲ 分类地位 / Taxonomy

翼手目 Chiroptera / 蝙蝠科 Vespertilionidae / 大耳蝠属 *Plecotus*

科建立者及其文献 / Family Authority
Gray, 1821

属建立者及其文献 / Genus Authority
E. Geoffroy, 1818

亚种 / Subspecies
无 None

模式标本产地 / Type Locality
俄罗斯
Sakhalin, Russia

▲ 其他名称 / Other Name(s)

其他中文名 / Other Chinese Name(s)
高颅蝠、库氏黄蝠、中黄蝠、
高头蝠

其他英文名 / Other English Name(s)
Siberian Long-eared Bat

同物异名 / Synonym(s)
Plecotus auritus ognevi

▲ 形态及生境 / Morphology and Habitat

形态特征 / Morphological Characteristics
齿式：2.1.2.3/3.1.3.3=36。头体长 48~50 mm。耳长 31~33 mm。前臂长
40~42 mm。后足长约 8 mm。尾长 44~46 mm。与褐大耳蝠 *P. auritus* 相似，
曾被视作为其的一个亚种。前臂比褐大耳蝠略长。
Dental formula: 2.1.2.3/3.1.3.3=36. Head and body length 48-50 mm. Ear length 31-
33 mm. Forearm length 40-42 mm. Hind foot length 8 mm. Tail length 44-46 mm.
The morphology similar to Brown Long-eared Bat (*P. auritus*) but forearm is slightly
larger.

生境 / Habitat
森林、灌木林、草原、洞穴和地下生境
Forest, shrubland, grassland, caves and subterranean habitat

▲ 地理分布 / Geographic Distribution

国内分布 / Domestic Distribution
黑龙江、吉林、内蒙古、辽宁、河北
Heilongjiang, Jilin, Inner Mongolia, Liaoning, Hebei

全球分布 / World Distribution
中国、哈萨克斯坦、朝鲜、韩国、蒙古国、俄罗斯
China, Kazakhstan, Democratic People's Republic of Korea, Republic of Korea, Mongolia, Russia

生物地理界 / Biogeographic Realm
古北界 Palearctic

WWF 生物群系 / WWF Biome
温带阔叶和混交林
Temperate Broadleaf & Mixed Forests

动物地理分布型 / Zoogeographic Distribution Type
Ue

分布标注 / Distribution Note
非特有种 Non-Endemic

▲ 濒危状况 / Threatened Status

中国生物多样性红色名录等级 / CB RL Category (2021)
未评估 NE

IUCN 红色名录 / IUCN Red List (2021)
无危 LC

威胁因子 / Threats
未知 Unknown

▲ 法律保护地位 / Legal Protection Status

国家重点保护野生动物等级 / Category of National Key Protected Wild Animals (2021)
未列入 Not listed

"三有"名录 / TWIESSV (2023)
未列入 Not listed

CITES 附录等级 / CITES Appendix (2023)
未列入 Not listed

迁徙物种公约附录 / CMS Appendix (2020)
未列入 Not listed

保护行动 / Conservation Action
尚无保护行动 No conservation action so far

▲ 参考文献 / References

Wei et al. (魏辅文等), 2021; Burgin et al., 2020; IUCN, 2020; Wilson and Mittermeier, 2019

217 / 台湾大耳蝠

Plecotus taivanus Yoshiyuki, 1991

• Taiwan Long-eared Bat

▲ 分类地位 / Taxonomy

翼手目 Chiroptera / 蝙蝠科 Vespertilionidae / 大耳蝠属 *Plecotus*

科建立者及其文献 / Family Authority
Gray, 1821

属建立者及其文献 / Genus Authority
E. Geoffroy, 1818

亚种 / Subspecies
无 None

模式标本产地 / Type Locality
中国
Anma Mountain of Taichung, Wuling area and Cuifeng, Meifeng of Nantou

周政翰／供图

▲ 其他名称 / Other Name(s)

其他中文名 / Other Chinese Name(s)
台湾长耳蝠

其他英文名 / Other English Name(s)
Taiwan Big-eared Bat,
Taiwanese Long-eared Bat

同物异名 / Synonym(s)
无 None

▲ 形态及生境 / Morphology and Habitat

形态特征 / Morphological Characteristics
齿式：2.1.2.3/3.1.3.3=36。头体长 38~40 mm。耳长 36~39 mm。前臂长 37~38 mm。后足长 9~11 mm。尾长 47~49 mm。颅全长 15~16 mm。吻端突出，鼻孔大。耳长大于耳宽，顶端钝圆。耳屏长约为耳长一半。头部褐色，背毛基部为深棕色到黑色，顶端黄褐色。腹毛灰白色。翼膜宽圆，延伸到趾基部。

Dental formula: 2.1.2.3/3.1.3.3=36. Head and body length 38-40 mm. Ear length 36-39 mm. Forearm length 37-38 mm. Hind foot length 9-11 mm. Tail length 47-49 mm. Greatest skull length 15-16 mm. Snout is prominent and the nostrils are large. Head is brown. Bases of dorsal hairs are dark brown to black and the hair tips are tawny. Abdominal hairs are grayish-white. Ear length is longer than the width of the ear and tip blunt rounded. Tragus is about half the length of the ear. Pterygoid is broadly circular, extending to the base of the toe.

生境 / Habitat
森林、洞穴、人造建筑
Forest, caves, man-made building

▲ 地理分布 / Geographic Distribution

国内分布 / Domestic Distribution
台湾 Taiwan

全球分布 / World Distribution
中国 China

生物地理界 / Biogeographic Realm
印度马来界 Indomalaya

WWF 生物群系 / WWF Biome
热带和亚热带湿润阔叶林
Tropical & Subtropical Moist Broadleaf Forests

动物地理分布型 / Zoogeographic Distribution Type
Sa

分布标注 / Distribution Note
特有种 Endemic

▲ 濒危状况 / Threatened Status

中国生物多样性红色名录等级 / CB RL Category (2021)
近危 NT

IUCN 红色名录 / IUCN Red List (2021)
近危 NT

威胁因子 / Threats
未知 Unknown

▲ 法律保护地位 / Legal Protection Status

国家重点保护野生动物等级 / Category of National Key Protected Wild Animals (2021)
未列入 Not listed

"三有"名录 / TWIESSV (2023)
未列入 Not listed

CITES 附录等级 / CITES Appendix (2023)
未列入 Not listed

迁徙物种公约附录 / CMS Appendix (2020)
未列入 Not listed

保护行动 / Conservation Action
尚无保护行动 No conservation action so far

▲ 参考文献 / References

Jiang et al. (蒋志刚等), 2021; Burgin et al., 2020; IUCN, 2020; Liu et al. (刘少英等), 2020; Wilson and Mittermeier, 2019; Pan et al. (潘清华等), 2007; Wang (王应祥), 2003; Yoshiyuki, 1995; Wilson and Reeder, 1993

218 / 金芒管鼻蝠

Harpiola isodon
Kuo, Fang, Csorba & Lee, 2006

· Taiwan Tube-Nosed Bat

▲ 分类地位 / Taxonomy

翼手目 Chiroptera / 蝙蝠科 Vespertilionidae / 金芒蝠属 *Harpiola*

科建立者及其文献 / Family Authority
Gray, 1821

属建立者及其文献 / Genus Authority
Thomas, 1915

亚种 / Subspecies
无 None

模式标本产地 / Type Locality
中国
Taiwan, China

周政翰 / 供图

▲ 其他名称 / Other Name(s)

其他中文名 / Other Chinese Name(s)
无 None

其他英文名 / Other English Name(s)
Golden-haired Tube-nosed Bat

同物异名 / Synonym(s)
无 None

▲ 形态及生境 / Morphology and Habitat

形态特征 / Morphological Characteristics

齿式：2.1.2.3/3.1.2.3=34。头体长 45~47 mm。前臂长 30~36.5 mm。尾长 30~33 mm。雌性较雄性大。头部褐色。吻端明显突出。鼻孔延长成短管状。耳郭相对于体型短。耳长大于耳宽，顶端钝圆。耳屏长而窄。面部被毛暗褐色，背毛为褐色，有金属光泽，腹毛灰白色。翼膜宽圆，连接到趾基部。尾膜及足上覆盖细密毛发。

Dental formula: 2.1.2.3/3.1.2.3=34. Head and body length 45-47 mm. Forearm length 30-36.5 mm. Tail length 30-33 mm. Female is larger than the male. Head is brown. Snout conspicuous protrudes. Nostrils lengthen into short tubes. Pinna is short relative to the body. Ear length is larger than the ear width. Ear tip is blunt and round. Tragus is long and narrow. Facial hairs are dark brown. Dorsal hairs are brown with a metallic luster. Abdominal hairs are grayish-white. Pterygoid membrane is broadly circular and connected to the base of the toe. Tail membrane and feet are covered with fine and dense hairs.

生境 / Habitat
洞穴、森林
Caves, forest

▲ 地理分布 / Geographic Distribution

国内分布 / Domestic Distribution
台湾 Taiwan

全球分布 / World Distribution
中国、越南
China, Vietnam

生物地理界 / Biogeographic Realm
印度马来界 Indomalaya

WWF 生物群系 / WWF Biome
热带和亚热带湿润阔叶林
Tropical & Subtropical Moist Broadleaf Forests

动物地理分布型 / Zoogeographic Distribution Type
J

分布标注 / Distribution Note
非特有种 Non-Endemic

▲ 濒危状况 / Threatened Status

中国生物多样性红色名录等级 / CB RL Category (2021)
未评定 NE

IUCN 红色名录 / IUCN Red List (2021)
无危 LC

威胁因子 / Threats
未知 Unknown

▲ 法律保护地位 / Legal Protection Status

国家重点保护野生动物等级 / Category of National Key Protected Wild Animals (2021)
未列入 Not listed

"三有"名录 / TWIESSV (2023)
未列入 Not listed

CITES 附录等级 / CITES Appendix (2023)
未列入 Not listed

迁徙物种公约附录 / CMS Appendix (2020)
未列入 Not listed

保护行动 / Conservation Action
尚无保护行动 No conservation action so far

▲ 参考文献 / References

Burgin et al., 2020; IUCN, 2020; Liu et al. (刘少英等), 2020; Kuo and Huang, 2020; Wilson and Mittermeier, 2019; Kuo et al., 2006

219 / 金管鼻蝠

Murina aurata Milne-Edwards, 1872

· Little Tube-nosed Bat

▲ 分类地位 / Taxonomy

翼手目 Chiroptera / 蝙蝠科 Vespertilionidae / 管鼻蝠属 *Murina*

科建立者及其文献 / Family Authority
Gray, 1821

属建立者及其文献 / Genus Authority
Gray, 1842

亚种 / Subspecies
指名亚种 *M. a. aurata* Milne-Edwards, 1872
四川、甘肃
Sichuan, Gansu

缅甸亚种 *M. a. feai* Thomas, 1891
西藏、云南
Tibet, Yunnan

模式标本产地 / Type Locality
中国四川
China, Szechwan, Moupin

关毅 / 供图

▲ 其他名称 / Other Name(s)

其他中文名 / Other Chinese Name(s)
小管鼻蝠

其他英文名 / Other English Name(s)
Tibetan Tube-nosed Bat

同物异名 / Synonym(s)
无 None

▲ 形态及生境 / Morphology and Habitat

形态特征 / Morphological Characteristics

齿式：2.1.2.3/3.1.2.3=34。头体长 33~35 mm。耳长 10~12 mm。前臂长 28~32 mm。后足长 7~8 mm。尾长 29~31 mm。颅全长 15.6~15.9 mm。吻部狭长，中线具一浅凹槽。鼻孔呈管状，上翘。耳壳后缘无凹刻，耳屏狭窄，末端尖细。胫骨短。后足小，趾细长，爪小，弯曲尖锐。背毛毛基灰黑色，毛尖为金色，腹毛毛尖灰白色。翼膜宽，无毛。肋膜、股间膜、尾和后足被覆稀疏长毛。

Dental formula: 2.1.2.3/3.1.2.3=34. Head and body length 33-35 mm. Ear length 10-12 mm. Forearm length 28-32 mm. Hind foot length 7-8 mm. Tail length 29-31 mm. Greatest skull length 15.6-15.9 mm. Snout long and narrow, and midline of the snout with a shallow groove. Nostrils are tubular and upturned. Posterior margin of the auricle is not concave, the tragus is narrow, and the terminal tip is tapering. Short shins. Hind feet small, long toes, small claws, sharp bending. Dorsal hairs with gray-black basal part and golden tips. Ventral hair tips gray-white. Pterygum broad, glabrous. Pleura, interfemoral membrane, tail, and hind feet are covered with thin long hairs.

生境 / Habitat

森林 Forest

▲ 地理分布 / Geographic Distribution

国内分布 / Domestic Distribution
甘肃、西藏、四川、云南、贵州、广东、黑龙江、吉林
Gansu, Tibet, Sichuan, Yunnan, Guizhou, Guangdong, Heilongjiang, Jilin

全球分布 / World Distribution
中国、印度、老挝、缅甸、尼泊尔、泰国、越南
China, India, Laos, Myanmar, Nepal, Thailand, Vietnam

生物地理界 / Biogeographic Realm
古北界、印度马来界
Palearctic, Indomalaya

WWF 生物群系 / WWF Biome
热带和亚热带湿润阔叶林、温带阔叶和混交林
Tropical & Subtropical Moist Broadleaf Forests, Temperate Broadleaf & Mixed
Forests

动物地理分布型 / Zoogeographic Distribution Type
Hm

分布标注 / Distribution Note
非特有种 Non-Endemic

▲ 濒危状况 / Threatened Status

中国生物多样性红色名录等级 / CB RL Category (2021)
近危 NT

IUCN 红色名录 / IUCN Red List (2021)
无危 LC

威胁因子 / Threats
未知 Unknown

▲ 法律保护地位 / Legal Protection Status

国家重点保护野生动物等级 / Category of National Key Protected Wild Animals (2021)
未列入 Not listed

"三有" 名录 / TWIESSV (2023)
未列入 Not listed

CITES 附录等级 / CITES Appendix (2023)
未列入 Not listed

迁徙物种公约附录 / CMS Appendix (2020)
未列入 Not listed

保护行动 / Conservation Action
尚无保护行动 No conservation action so far

▲ 参考文献 / References

Jiang et al. (蒋志刚等), 2021; Burgin et al., 2020; IUCN, 2020; Liu et al. (刘少英等), 2020; Wilson and Mittermeier, 2019; Pan et al. (潘清华等), 2007; Wilson and Reeder, 2005; Wang (王应祥), 2003; Zhang (张荣祖), 1997

220 / 黄胸管鼻蝠

Murina bicolor
Kuo, Fang, Csorba & Lee, 2009

• Yellow-chested Tube-nosed Bat

▲ 分类地位 / Taxonomy

翼手目 Chiroptera / 蝙蝠科 Vespertilionidae / 管鼻蝠属 *Murina*

科建立者及其文献 / Family Authority
Gray, 1821

属建立者及其文献 / Genus Authority
Gray, 1842

亚种 / Subspecies
无 None

模式标本产地 / Type Locality
中国
China, Taiwan, Nantou County, Renai Township, Taroko National Park, Hehuanshan, 3,020 m

▲ 其他名称 / Other Name(s)

其他中文名 / Other Chinese Name(s)
无 None

其他英文名 / Other English Name(s)
Bicolored Tube-nosed Bat

同物异名 / Synonym(s)
无 None

▲ 形态及生境 / Morphology and Habitat

形态特征 / Morphological Characteristics
齿式：2.1.2.3/3.1.2.3=34。头体长 47~50 mm。前臂长 36~42 mm。尾长 40~45 mm。头部褐色。吻端突出。鼻孔延长成短管状。耳郭具有缺刻。耳长大于耳宽，顶端钝圆。耳屏长而窄。头部褐色。背部毛发褐色或灰褐色，夹杂红褐色的长毛。腹毛黄色或白色。翼膜宽圆，连接到趾基部。尾膜及足上均有细毛。

Dental formula: 2.1.2.3/3.1.2.3=34. Head and body length 47-50 mm. Forearm length 36-42 mm. Tail length 40-45 mm. Head hairs are brown. Snout protrudes. Nostrils are lengthened into short tubes. Auricle has notches. Ear length is longer than the ear width, and ear tip is blunt round. Tragus is long and narrow. Head brown. Dorsal hairs brown or grayish brown, mixed with reddish-brown long hairs. Ventral hairs are yellow or white. Pterygoid membrane is broadly circular and connected to the base of the toe. Tail membrane and feet are covered with fine hairs.

生境 / Habitat
森林、人造建筑
Forest, man-made building

▲ 地理分布 / Geographic Distribution

国内分布 / Domestic Distribution
台湾 Taiwan

全球分布 / World Distribution
中国 China

生物地理界 / Biogeographic Realm
印度马来界 Indomalaya

WWF 生物群系 / WWF Biome
热带和亚热带湿润阔叶林
Tropical & Subtropical Moist Broadleaf Forests

动物地理分布型 / Zoogeographic Distribution Type
Sa

分布标注 / Distribution Note
特有种 Endemic

▲ 濒危状况 / Threatened Status

中国生物多样性红色名录等级 / CB RL Category (2021)
数据缺乏 DD

IUCN 红色名录 / IUCN Red List (2021)
无危 LC

威胁因子 / Threats
未知 Unknown

▲ 法律保护地位 / Legal Protection Status

国家重点保护野生动物等级 / Category of National Key Protected Wild Animals (2021)
未列入 Not listed

"三有"名录 / TWIESSV (2023)
未列入 Not listed

CITES 附录等级 / CITES Appendix (2023)
未列入 Not listed

迁徙物种公约附录 / CMS Appendix (2020)
未列入 Not listed

保护行动 / Conservation Action
尚无保护行动 No conservation action so far

▲ 参考文献 / References

Jiang et al. (蒋志刚等), 2021; Burgin et al., 2020; IUCN, 2020; Liu et al. (刘少英等), 2020; Wilson and Mittermeier, 2019; Kuo et al., 2009

221 / 金毛管鼻蝠

Murina chrysochaetes Eger & Lim, 2011

· Golden-haired Tube-nosed Bat

▲ 分类地位 / Taxonomy

翼手目 Chiroptera / 蝙蝠科 Vespertilionidae / 管鼻蝠属 *Murina*

科建立者及其文献 / Family Authority
Gray, 1821

属建立者及其文献 / Genus Authority
Gray, 1842

亚种 / Subspecies
无 None

模式标本产地 / Type Locality
中国
Diding Headwater Forest Nature Preserve (formerly known as Jing Xi County Provincial Nature Reserve), Jing Xi County, Guangxi Zhuang Autonomous Region, China

▲ 其他名称 / Other Name(s)

其他中文名 / Other Chinese Name(s)
无 None

其他英文名 / Other English Name(s)
无 None

同物异名 / Synonym(s)
无 None

▲ 形态及生境 / Morphology and Habitat

形态特征 / Morphological Characteristics
齿式：2.1.2.3/3.1.2.3=34。头体长 38~50 mm。耳长 12~15 mm。前臂长 30~35 mm。后足长 7~10 mm。尾长 32~42 mm。鼻孔管状。耳短而圆。前脑壳呈球状。上唇有突出的毛发边缘。背毛基部深棕色，毛尖烟熏棕色。腹毛浅棕色。拇指较长。尾短于头体长。足小，有茸毛。爪长、锋利。后脚背部被覆短茸毛。翼膜半透明。

Dental formula: 2.1.2.3/3.1.2.3=34. Head and body length 38-50 mm. Ear length 12-15 mm. Forearm length 30-35 mm. Hind foot length 7-10 mm. Tail length 32-42 mm. Tubular nostrils short, round ears. Anterior braincase is ballooned. Upper lip possesses a protruding fringe of hairs. Thumbs relatively long. Bases of dorsal hairs dark brown, and tips smoky brown. Ventral surface light brown. Tail is shorter than the length of head and body. Feet are hairy, small, and the claws are relatively long and sharp. Uropatagium with hairs. Back of hind foot also with short hairs. Wing membrane is semi-transparent.

生境 / Habitat
森林、喀斯特地貌
Forest, karst landscape

▲ 地理分布 / Geographic Distribution

国内分布 / Domestic Distribution
广西、贵州、四川、广东、云南
Guangxi, Guizhou, Sichuan, Guangdong, Yunnan

全球分布 / World Distribution
中国 China

生物地理界 / Biogeographic Realm
印度马来界 Indomalaya

WWF 生物群系 / WWF Biome
热带和亚热带湿润阔叶林
Tropical & Subtropical Moist Broadleaf Forests

动物地理分布型 / Zoogeographic Distribution Type
Sb

分布标注 / Distribution Note
特有种 Endemic

▲ 濒危状况 / Threatened Status

中国生物多样性红色名录等级 / CB RL Category (2021)
数据缺乏 DD

IUCN 红色名录 / IUCN Red List (2021)
未评定 NE

威胁因子 / Threats
未知 Unknown

▲ 法律保护地位 / Legal Protection Status

国家重点保护野生动物等级 / Category of National Key Protected Wild Animals (2021)
未列入 Not listed

"三有"名录 / TWIESSV (2023)
未列入 Not listed

CITES 附录等级 / CITES Appendix (2023)
未列入 Not listed

迁徙物种公约附录 / CMS Appendix (2020)
未列入 Not listed

保护行动 / Conservation Action
尚无保护行动 No conservation action so far

▲ 参考文献 / References

Jiang et al. (蒋志刚等), 2021; Zhong et al. (钟韦凌等), 2021; Burgin et al., 2020; IUCN, 2020; Liu et al. (刘少英等), 2020; Wilson and Mittermeier, 2019; Eger & Lim, 2011

222 / 圆耳管鼻蝠

Murina cyclotis Dobson, 1872

• Round-eared Tube-nosed Bat

▲ 分类地位 / Taxonomy

翼手目 Chiroptera / 蝙蝠科 Vespertilionidae / 管鼻蝠属 *Murina*

科建立者及其文献 / Family Authority
Gray, 1821

属建立者及其文献 / Genus Authority
Gray, 1842

亚种 / Subspecies
指名亚种 *M. c. cyclotis* Dobson, 1872
江西和海南
Jiangxi and Hainan

模式标本产地 / Type Locality
印度
India, Darjeeling

余文华 / 供图

▲ 其他名称 / Other Name(s)

其他中文名 / Other Chinese Name(s)
无 None

其他英文名 / Other English Name(s)
无 None

同物异名 / Synonym(s)
无 None

▲ 形态及生境 / Morphology and Habitat

形态特征 / Morphological Characteristics
齿式：2.1.2.3/3.1.2.3=34。头体长 38~50 mm。耳长 12~15 mm。前臂长 30~35 mm。后足长 7~10 mm。尾长 32~42 mm。体重 9~12 g。鼻部管状，有纵沟。耳壳近圆形，耳屏外缘薄，尖端圆锥状。背毛赤褐色，毛基多为棕褐色。股部、胫部和尾基部多毛，腹面几近裸露。足背毛厚，且超出脚趾。尾尖游离。

Dental formula: 2.1.2.3/3.1.2.3=34. Head and body length 38-50 mm. Ear length 12-15 mm. Forearm length 30-35 mm. Hind foot length 7-10 mm. Tail length 32-42 mm. Body mass 9-12 g. Nose is tubular with longitudinal furrows. Auricle is nearly round. Outer margin of the tragus is thin and the tip is conical. Dorsal hairs fresh russet, and basal part of the hairs are brown. Thighs, shins, and tail base are hairy. Ventral surface is nearly bare. Hairs on the back of the feet are thick and extend beyond the toes. Tail tip-free.

生境 / Habitat
洞穴、内陆岩石区域、森林
Caves, inland rocky area, forest

▲ 地理分布 / Geographic Distribution

国内分布 / Domestic Distribution
江西、海南、广东、贵州、广西、云南
Jiangxi, Hainan, Guangdong, Guizhou, Guangxi, Yunnan

全球分布 / World Distribution
文莱、柬埔寨、中国、印度、老挝、马来西亚、缅甸、尼泊尔、菲律宾、
斯里兰卡、泰国、越南
Brunei, Cambodia, China, India, Laos, Malaysia, Myanmar, Nepal, Philippines,
Sri Lanka, Thailand, Vietnam

生物地理界 / Biogeographic Realm
印度马来界 Indomalaya

WWF 生物群系 / WWF Biome
热带和亚热带湿润阔叶林
Tropical & Subtropical Moist Broadleaf Forests

动物地理分布型 / Zoogeographic Distribution Type
Sa

分布标注 / Distribution Note
非特有种 Non-Endemic

▲ 濒危状况 / Threatened Status

中国生物多样性红色名录等级 / CB RL Category (2021)
近危 NT

IUCN 红色名录 / IUCN Red List (2021)
无危 LC

威胁因子 / Threats
森林砍伐、耕种、洞穴开发
Logging, farming, cave exploitation

▲ 法律保护地位 / Legal Protection Status

国家重点保护野生动物等级 / Category of National Key Protected Wild Animals (2021)
未列入 Not listed

"三有"名录 / TWIESSV (2023)
未列入 Not listed

CITES 附录等级 / CITES Appendix (2023)
未列入 Not listed

迁徙物种公约附录 / CMS Appendix (2020)
未列入 Not listed

保护行动 / Conservation Action
尚无保护行动 No conservation action so far

▲ 参考文献 / References

Jiang et al. (蒋志刚等), 2021; Zhou et al. (周全等), 2021; Li et al. (李彦男等), 2020; Liu et al. (刘少英等), 2020; Wilson and Mittermeier, 2019;
Francis & Eger, 2012; Eger and Lim, 2011; Pan et al. (潘清华等), 2007; Wilson and Reeder, 2005; Wang (王应祥), 2003; Zhang (张荣祖), 1997

223 / 艾氏管鼻蝠

Murina eleryi
Furey, Thong, Bates & Csorba, 2009

· Elery's Tube-nosed Bat

▲ 分类地位 / Taxonomy

翼手目 Chiroptera / 蝙蝠科 Vespertilionidae / 管鼻蝠属 *Murina*

科建立者及其文献 / Family Authority
Gray, 1821

属建立者及其文献 / Genus Authority
Gray, 1842

亚种 / Subspecies
无 None

模式标本产地 / Type Locality
越南
Kim Hy Commune, Na Ri District of Kim Hy Nature Reserve, Bac Kan Province, Vietnam, 22°16.392'N, 106°03.427'E, 525 m a. s. l

▲ 其他名称 / Other Name(s)

其他中文名 / Other Chinese Name(s)
无 None

其他英文名 / Other English Name(s)
无 None

同物异名 / Synonym(s)
无 None

▲ 形态及生境 / Morphology and Habitat

形态特征 / Morphological Characteristics

齿式：2.1.1.3/3.1.2.3=32。前臂长 30~33 mm。耳短宽。耳屏端部尖细。鼻孔呈管状。背毛暗黄褐色，毛尖金黄色。腹毛灰白色，无细长毛。翼膜前臂处和股间膜覆有黄色短毛。足背腹短毛均为黑色。翼膜止于趾基外缘。股间膜始于距基部，尾端从股间膜后缘穿出约 1 mm。

Dental formula: 2.1.1.3/3.1.2.3=32. Forearm length 30-33 mm. Ears short wide. Tragus apex tapering. Nostrils are tubular. Dorsal hairs are dark tawny, hair tips are golden in tone. Ventral hairs gray, no fine long hair. Pterygium forearm and interfemoral membrane are covered with yellow short hairs. Black short hairs on upper and underparts of feet. Pterygium terminates at the outer edge of the toe base. Interfemoral membrane begins from the base. Tail penetrates about 1 mm from the posterior edge of the interfemoral membrane.

生境 / Habitat
森林、喀斯特地貌
Forest, karst landscape

▲ 地理分布 / Geographic Distribution

国内分布 / Domestic Distribution
湖南、贵州、广西、广东、海南
Hunan, Guizhou, Guangxi, Guangdong, Hainan

全球分布 / World Distribution
中国、越南
China, Vietnam

生物地理界 / Biogeographic Realm
印度马来界 Indomalaya

WWF 生物群系 / WWF Biome
热带和亚热带湿润阔叶林
Tropical & Subtropical Moist Broadleaf Forests

动物地理分布型 / Zoogeographic Distribution Type
Wb

分布标注 / Distribution Note
非特有种 Non-Endemic

▲ 濒危状况 / Threatened Status

中国生物多样性红色名录等级 / CB RL Category (2021)
近危 NT

IUCN 红色名录 / IUCN Red List (2021)
未评定 NE

威胁因子 / Threats
森林砍伐、耕种、洞穴开发
Logging, farming, cave exploitation

▲ 法律保护地位 / Legal Protection Status

国家重点保护野生动物等级 / Category of National Key Protected Wild Animals (2021)
未列入 Not listed

"三有"名录 / TWIESSV (2023)
未列入 Not listed

CITES 附录等级 / CITES Appendix (2023)
未列入 Not listed

迁徙物种公约附录 / CMS Appendix (2020)
未列入 Not listed

保护行动 / Conservation Action
尚无保护行动 No conservation action so far

▲ 参考文献 / References

Jiang et al. (蒋志刚等), 2021; Burgin et al., 2020; IUCN, 2020; Liu et al. (刘少英等), 2020; Hu et al. (胡宜峰等), 2019; Liu et al. (刘志霄等), 2014; Furey et al., 2009

224 / 梵净山管鼻蝠

Murina fanjingshanensis
He, Xiao & Zhou, 2016

· Fanjingshan Tube-nosed Bat

▲ 分类地位 / Taxonomy

翼手目 Chiroptera / 蝙蝠科 Vespertilionidae / 管鼻蝠属 *Murina*

科建立者及其文献 / Family Authority
Gray, 1821

属建立者及其文献 / Genus Authority
Gray, 1842

亚种 / Subspecies
无 None

模式标本产地 / Type Locality
中国贵州
In an abandoned gold mine of Luojiawan village, Wuluo Town, Songtao County (1069m, 28°01'09.77"N, 108°45' 37.34"E), located in the east of the FNNR in Guizhou Province, China

吴毅 / 供图

▲ 其他名称 / Other Name(s)

其他中文名 / Other Chinese Name(s)
无 None

其他英文名 / Other English Name(s)
无 None

同物异名 / Synonym(s)
无 None

▲ 形态及生境 / Morphology and Habitat

形态特征 / Morphological Characteristics

齿式：2.1.2.3/3.1.2.3=34。前臂长 40~42 mm。头骨隆起。矢状嵴和人字嵴不明显。鼻吻部宽。鼻部短管状，向两侧突出。鼻间沟明显。其后方有疣粒突起。额弓细长。鼻吻部及下唇均具毛。背毛棕黄，腹毛污白。翼膜黑褐色，股间膜及后足密生棕黄色长毛，无距缘膜，距尖端。尾椎突出尾膜。

Dental formula: 2.1.2.3/3.1.2.3=34. Forearm length 40-42 mm. Skull hump-shaped. Sagittal crest and herringbone crest are not obvious. Nose snout is wide. Nose is short tubular, protruding to both sides. Nasal groove is obvious and the rear has warts. Frontal arch is slender. Nose snout and lower lip hairy. Dorsal hairs are brown and the belly hairs are dirty white. Pterygoid membrane is dark brown. Interfemoral membrane and the posterior foot have densely brownish-yellow long hairs. Talar membrane absent. Coccygeal vertebra protruding caudal membrane.

生境 / Habitat
森林、喀斯特地貌
Forest, karst landscape

▲ 地理分布 / Geographic Distribution

国内分布 / Domestic Distribution
贵州 、湖南
Guizhou, Hunan

全球分布 / World Distribution
中国 China

生物地理界 / Biogeographic Realm
古北界 Palearctic

WWF 生物群系 / WWF Biome
热带和亚热带湿润阔叶林
Tropical & Subtropical Moist Broadleaf Forests

动物地理分布型 / Zoogeographic Distribution Type
Sd

分布标注 / Distribution Note
特有种 Endemic

▲ 濒危状况 / Threatened Status

中国生物多样性红色名录等级 / CB RL Category (2021)
数据缺乏 DD

IUCN 红色名录 / IUCN Red List (2021)
未评定 NE

威胁因子 / Threats
未知 Unknown

▲ 法律保护地位 / Legal Protection Status

国家重点保护野生动物等级 / Category of National Key Protected Wild Animals (2021)
未列入 Not listed

"三有"名录 / TWIESSV (2023)
未列入 Not listed

CITES 附录等级 / CITES Appendix (2023)
未列入 Not listed

迁徙物种公约附录 / CMS Appendix (2020)
未列入 Not listed

保护行动 / Conservation Action
尚无保护行动 No conservation action so far

▲ 参考文献 / References

Jiang et al. (蒋志刚等), 2021; Liu et al. (刘少英等), 2020; Wilson and Mittermeier, 2019; Huang et al., 2018; He et al., 2015

225 / 菲氏管鼻蝠

Murina feae Thomas, 1891

· Murina feae

翼手目 Chiroptera / 蝙蝠科 Vespertilionidae / 管鼻蝠属 *Murina*

科建立者及其文献 / Family Authority
Gray, 1821

属建立者及其文献 / Genus Authority
Gray, 1842

亚种 / Subspecies
无 None

模式标本产地 / Type Locality
缅甸
Biapo, Burma

余文华 / 供图

▲ 其他名称 / Other Name(s)

其他中文名 / Other Chinese Name(s)
无 None

其他英文名 / Other English Name(s)
无 None

同物异名 / Synonym(s)
无 None

▲ 形态及生境 / Morphology and Habitat

形态特征 / Morphological Characteristics

齿式: 2.1.2.3/3.1.2.3=34。前臂长约 29 mm。颅全长 15~16 mm。头骨小，吻突窄，凹陷发达。矢状嵴与人字嵴不明显。颧弓相对发达。鼻孔前端突出延长成短管状。鼻吻部深灰褐色。耳圆无缺刻，耳屏长度约为耳长之半。背毛基部黑色、中部浅灰色，毛尖深灰褐色。腹毛基部黑色或灰黑色，毛尖银白色。后足背面及股间膜覆盖稀疏的灰褐色毛发。股间膜呈浅灰褐色。膜边缘的毛向后延伸游离可见。翼膜延至第一趾中部。

Dental formula: 2.1.2.3/3.1.2.3=34. Forearm length 29 mm. Skull is small. Greatest skull length 15-16 mm. Muzzle is narrow and the depression is well developed. Sagittal crest and the herringbone crest are not obvious. Zygomatic arch is relatively developed. Nose tip protrudes into a short tube. Snout is dark grayish brown. Ears are round and intact. Tragus is about half the length of the ear. Bases of the dorsal hairs are black, light gray in the middle, and the tips are dark grayish-brown. Ventral hair bases are black or grayish black, and hair tips silvery white. Dorsal part of the hind foot and the interfemoral membrane are covered with sparse grayish-brown hairs. Interfemoral membrane is light grayish-brown. Hairs at the membrane margin extend freely backward and are visible. Pterygium extends to the middle of the first toe.

生境 / Habitat
森林、喀斯特地貌
Forest, karst landscape

▲ 地理分布 / Geographic Distribution

国内分布 / Domestic Distribution
贵州、江西、广东、广西
Guizhou, Jiangxi, Guangdong, Guangxi

全球分布 / World Distribution
中国、柬埔寨、老挝、缅甸、泰国、越南
China, Cambodia, Laos, Myanmar, Thailand, Vietnam

生物地理界 / Biogeographic Realm
印度马来界 Indomalaya

WWF 生物群系 / WWF Biome
热带和亚热带湿润阔叶林
Tropical & Subtropical Moist Broadleaf Forests

动物地理分布型 / Zoogeographic Distribution Type
Wb

分布标注 / Distribution Note
非特有种 Non-Endemic

▲ 濒危状况 / Threatened Status

中国生物多样性红色名录等级 / CB RL Category (2021)
未评定 NE

IUCN 红色名录 / IUCN Red List (2021)
无危 LC

威胁因子 / Threats
未知 Unknown

▲ 法律保护地位 / Legal Protection Status

国家重点保护野生动物等级 / Category of National Key Protected Wild Animals (2021)
未列入 Not listed

"三有" 名录 / TWIESSV (2023)
未列入 Not listed

CITES 附录等级 / CITES Appendix (2023)
未列入 Not listed

迁徙物种公约附录 / CMS Appendix (2020)
未列入 Not listed

保护行动 / Conservation Action
尚无保护行动 No conservation action so far

▲ 参考文献 / References

Liu et al. (刘少英等), 2020; Francis et al., 2012; Wu et al. (吴梦柳等), 2017

226 / 暗色管鼻蝠

Murina fusca Sowerby, 1922

• Dusky Tube-nosed Bat

翼手目 Chiroptera / 蝙蝠科 Vespertilionidae / 管鼻蝠属 *Murina*

科建立者及其文献 / Family Authority
Gray, 1821

属建立者及其文献 / Genus Authority
Gray, 1842

亚种 / Subspecies
无 None

模式标本产地 / Type Locality
中国
Imienpo area, Kirin, Manchuria, China

▲ 其他名称 / Other Name(s)

其他中文名 / Other Chinese Name(s)
无 None

其他英文名 / Other English Name(s)
Delicate Tube-nosed Bat, Slender Tube-nosed Bat, Taiwanese Little Tube-nosed Bat

同物异名 / Synonym(s)
无 None

▲ 形态及生境 / Morphology and Habitat

形态特征 / Morphological Characteristics
齿式：2.1.2.3/3.1.2.3=34。头体长 58 mm。耳长 18 mm。前臂长 40 mm。后脚长 8 mm。尾长 34 mm。颅骨结实，但缺少矢状嵴。枕骨向后突出不明显。背毛土棕色，中间杂有较长的白色刚毛，几乎呈灰色。腹侧毛色苍白。背毛延伸到翼膜基部、尾翼基部、拇指和足。
Dental formula: 2.1.2.3/3.1.2.3=34. Head and body length 58 mm, ear length 18 mm, forearm length 40 mm, hindfoot length 8 mm, tail length 34 mm. Skull is robust but lacks of the sagittal crest. Occipital does not noticeably project backward. Dorsal dust brown with longer whitish guard hairs interspersed, giving an almost grayish tinge. Ventral paler. Dorsal fur extends onto the base of wing, uropatagium, thumbs and feet.

生境 / Habitat
未知 Unknown

▲ 地理分布 / Geographic Distribution

国内分布 / Domestic Distribution
黑龙江 Heilongjiang

全球分布 / World Distribution
中国 China

生物地理界 / Biogeographic Realm
古北界 Palearctic

WWF 生物群系 / WWF Biome
温带阔叶和混交林
Temperate Broadleaf & Mixed Forests

动物地理分布型 / Zoogeographic Distribution Type
E

分布标注 / Distribution Note
特有种 Endemic

▲ 濒危状况 / Threatened Status

中国生物多样性红色名录等级 / CB RL Category (2021)
未评定 NE

IUCN 红色名录 / IUCN Red List (2021)
数据缺乏 DD

威胁因子 / Threats
未知 Unknown

▲ 法律保护地位 / Legal Protection Status

国家重点保护野生动物等级 / Category of National Key Protected Wild Animals (2021)
未列入 Not listed

"三有" 名录 / TWIESSV (2023)
未列入 Not listed

CITES 附录等级 / CITES Appendix (2023)
未列入 Not listed

迁徙物种公约附录 / CMS Appendix (2020)
未列入 Not listed

保护行动 / Conservation Action
尚无保护行动 No conservation action so far

▲ 参考文献 / References

Burgin et al., 2020; Wilson and Mittermeier, 2019; Wu and Yu, 2018

227 / 姬管鼻蝠

Murina gracilis
Kuo, Fang, Csorba & Lee, 2009

• Taiwanese Little Tube-nosed Bat

周政翰 / 供图

▲ 其他名称 / Other Name(s)

其他中文名 / Other Chinese Name(s)
无 None

其他英文名 / Other English Name(s)
Delicate Tube-nosed Bat, Slender Tube-
nosed Bat

同物异名 / Synonym(s)
无 None

▲ 形态及生境 / Morphology and Habitat

形态特征 / Morphological Characteristics
齿式：2.1.2.3/3.1.2.3=34。头体长 33~45 mm。前臂长 27~33 mm。尾长
26~36 mm。吻端突出。鼻孔延长成短管状。耳郭短。耳长大于耳宽，顶
端钝圆。背毛夹杂金属光泽，其余毛发基部黑色，尖端为浅褐色。腹毛
灰白色。翼膜宽圆，连接到趾基部。尾膜及足被覆细密的毛。
Dental formula: 2.1.2.3/3.1.2.3=34. Head and body length 33-45 mm. Forearm length
27-33 mm. Tail length 26-36 mm. Snout protrudes and the nostrils lengthen into
short tubes. Auricle is short. Ear length is larger than ear width. Ear tip is blunt and
round. Dorsal hairs with metallic luster, and the rest of the body hairs have black bases
and light brown tips. Ventral hairs are grayish-white. Pterygoid membrane is broadly
circular and connected to the base of the toe. Tail membrane and feet are covered with
dense fine hairs.

生境 / Habitat
森林、喀斯特地貌
Forest, karst landscape

▲ 地理分布 / Geographic Distribution

国内分布 / Domestic Distribution
台湾 Taiwan

全球分布 / World Distribution
中国 China

生物地理界 / Biogeographic Realm
印度马来界 Indomalaya

WWF 生物群系 / WWF Biome
热带和亚热带湿润阔叶林
Tropical & Subtropical Moist Broadleaf Forests

动物地理分布型 / Zoogeographic Distribution Type
Sa

分布标注 / Distribution Note
特有种 Endemic

▲ 濒危状况 / Threatened Status

中国生物多样性红色名录等级 / CB RL Category (2021)
未评定 NE

IUCN 红色名录 / IUCN Red List (2021)
无危 LC

威胁因子 / Threats
未知 Unknown

▲ 法律保护地位 / Legal Protection Status

国家重点保护野生动物等级 / Category of National Key Protected Wild Animals (2021)
未列入 Not listed

"三有" 名录 / TWIESSV (2023)
未列入 Not listed

CITES 附录等级 / CITES Appendix (2023)
未列入 Not listed

迁徙物种公约附录 / CMS Appendix (2020)
未列入 Not listed

保护行动 / Conservation Action
尚无保护行动 No conservation action so far

▲ 参考文献 / References

Liu et al. (刘少英等), 2020; Francis et al., 2012; Wu et al. (吴梦柳等), 2017

228 / 哈氏管鼻蝠

Murina harrisoni Csorba & Bates, 2005

· Harrison's Tube-nosed Bat

▲ 分类地位 / Taxonomy

翼手目 Chiroptera / 蝙蝠科 Vespertilionidae / 管鼻蝠属 *Murina*

科建立者及其文献 / Family Authority
Gray, 1821

属建立者及其文献 / Genus Authority
Gray, 1842

亚种 / Subspecies
无 None

模式标本产地 / Type Locality
柬埔寨
Tuk Chehn, Kirirom National Park, Kompong Speu Province, Cambodia, 11°29.611'N, 104°12.746'E

余文华 / 供图

▲ 其他名称 / Other Name(s)

其他中文名 / Other Chinese Name(s)
无 None

其他英文名 / Other English Name(s)
无 None

同物异名 / Synonym(s)
无 None

▲ 形态及生境 / Morphology and Habitat

形态特征 / Morphological Characteristics
齿式：2.1.2.3/3.1.2.3=34。前臂长 36 mm。体重 5 g。鼻部管状，向两侧突出。耳壳后缘无微刻。耳屏高度约为耳长之半。背毛红棕色，毛尖略黑。尾膜上部密生毛，最后一节尾椎骨露出尾膜。侧翼膜止于脚趾基部。
Dental formula: 2.1.2.3/3.1.2.3=34. Forearm length 36 mm. Body mass 5 g. Nose is tubular and protruding laterally. Posterior margin of the auricle is not marked, and the tragus is about half the length of the ear. Dorsal hairs are reddish-brown with slightly black tips. Upper part of the caudal membrane is densely hairy. Last vertebra is exposed to the caudal membrane. Lateral membrane stops at the base of the toe.

生境 / Habitat
森林、次生林
Forest, secondary forest

▲ 地理分布 / Geographic Distribution

国内分布 / Domestic Distribution
江西、广东、云南、海南
Jiangxi, Guangdong, Yunnan, Hainan

全球分布 / World Distribution
中国、柬埔寨
China, Cambodia

生物地理界 / Biogeographic Realm
印度马来界 Indomalaya

WWF 生物群系 / WWF Biome
热带和亚热带湿润阔叶林
Tropical & Subtropical Moist Broadleaf Forests

动物地理分布型 / Zoogeographic Distribution Type
Wb

分布标注 / Distribution Note
非特有种 Non-Endemic

▲ 濒危状况 / Threatened Status

中国生物多样性红色名录等级 / CB RL Category (2021)
数据缺乏 DD

IUCN 红色名录 / IUCN Red List (2021)
数据缺乏 DD

威胁因子 / Threats
砍伐 Logging

▲ 法律保护地位 / Legal Protection Status

国家重点保护野生动物等级 / Category of National Key Protected Wild Animals (2021)
未列入 Not listed

"三有"名录 / TWIESSV (2023)
未列入 Not listed

CITES 附录等级 / CITES Appendix (2023)
未列入 Not listed

迁徙物种公约附录 / CMS Appendix (2020)
未列入 Not listed

保护行动 / Conservation Action
尚无保护行动 No conservation action so far

▲ 参考文献 / References

Jiang et al. (蒋志刚等), 2021; Zhang et al. (张欣等), 2021; Liu et al. (刘少英等), 2020; Chen et al.(陈子禧等), 2018; Wu et al.(吴毅等), 2017; Csorba & Bates, 2005; Wu et al., 2011

229 / 东北管鼻蝠

Murina hilgendorfi Peters, 1880

· Greater Tube-nosed Bat

翼手目 Chiroptera / 蝙蝠科 Vespertilionidae / 管鼻蝠属 *Murina*

科建立者及其文献 / Family Authority
Gray, 1821

属建立者及其文献 / Genus Authority
Gray, 1842

亚种 / Subspecies
韩国亚种 *Mu. h. ognevilnner* (Peters, 1880)
黑龙江和内蒙古
Heilongjiang and Inner Mongolia

模式标本产地 / Type Locality
日本
Japan, near Tokyo, Yedo

▲ 其他名称 / Other Name(s)

其他中文名 / Other Chinese Name(s)
暗色管鼻蝠

其他英文名 / Other English Name(s)
Hilgendorf's Tube-nosed Bat

同物异名 / Synonym(s)
无 None

▲ 形态及生境 / Morphology and Habitat

形态特征 / Morphological Characteristics
齿式：2.1.2.3/3.1.2.3=34。头体长 46~70 mm。耳长 14~20 mm。前臂长 40~45 mm。后足长 10~15 mm。尾长 32~45 mm。颅全长 16~19 mm。体型大。耳边缘上有两个凹刻。耳屏向外偏转。毛皮柔软，有光泽。背毛从基部到毛尖有4条色带，依次为暗灰褐色、浅灰褐色、橄榄色到橘色和金黄色。腹毛基部深棕色，毛尖银色。尾膜、体侧膜和拇指背面被毛。
Dental formula: 2.1.2.3/3.1.2.3=34. Head and body length 46-70 mm. Ear length 14-20 mm. Forearm length 40-45 mm. Hind foot length 10-15 mm. Tail length 32-45 mm. Greatest skull length 16-19 mm. Large body size. Ear with two concavities on margin. Tragus deflects outward. Hairs are soft and shiny. Dorsal hairs have four color bands from the base to the tip, in order: dark grayish-brown, light grayish-brown, olive to orange and golden. Ventral hairs are dark brown at the hair bases and silver at the tips. Dorsal surfaces of uropatagium, plagiopatagium, and thumb are covered with hairs.

生境 / Habitat
洞穴、森林、人造建筑
Caves, forest, man-made building

▲ 地理分布 / Geographic Distribution

国内分布 / Domestic Distribution
黑龙江、内蒙古
Heilongjiang, Inner Mongolia

全球分布 / World Distribution
中国、日本、朝鲜、韩国、俄罗斯、哈萨克斯坦、蒙古国
China, Japan, Democratic People's Republic of Korea, Republic of Korea, Russia, Kazakhstan, Mongolia

生物地理界 / Biogeographic Realm
古北界 Palearctic

WWF 生物群系 / WWF Biome
北方森林 / 针叶林
Boreal Forests / Coniferous Forests

动物地理分布型 / Zoogeographic Distribution Type
Ma

分布标注 / Distribution Note
非特有种 Non-Endemic

▲ 濒危状况 / Threatened Status

中国生物多样性红色名录等级 / CB RL Category (2021)
无危 LC

IUCN 红色名录 / IUCN Red List (2021)
数据缺乏 DD

威胁因子 / Threats
未知 Unknown

▲ 法律保护地位 / Legal Protection Status

国家重点保护野生动物等级 / Category of National Key Protected Wild Animals (2021)
未列入 Not listed

"三有"名录 / TWIESSV (2023)
未列入 Not listed

CITES 附录等级 / CITES Appendix (2023)
未列入 Not listed

迁徙物种公约附录 / CMS Appendix (2020)
未列入 Not listed

保护行动 / Conservation Action
尚无保护行动 No conservation action so far

▲ 参考文献 / References

Jiang et al. (蒋志刚等), 2021; Burgin et al., 2020; IUCN, 2020; Liu et al. (刘少英等), 2020; Wilson and Mittermeier, 2019; Pan et al. (潘清华等), 2007; Csorba and Bates, 2005; Wilson and Reeder, 2005; Wang (王应祥), 2003

230 / 中管鼻蝠

Murina huttoni (Peters, 1872)

· Hutton's Tube-nosed Bat

▲ 分类地位 / Taxonomy

翼手目 Chiroptera / 蝙蝠科 Vespertilionidae / 管鼻蝠属 *Murina*

科建立者及其文献 / Family Authority
Gray, 1821

属建立者及其文献 / Genus Authority
Gray, 1842

亚种 / Subspecies
福建亚种 *M. h. rubella* Thomas, 1914
福建、江西和广西
Fujian, Jiangxi and Guangxi

模式标本产地 / Type Locality
印度
India, Uttar Pradesh, Kumaon, Dehra Dun

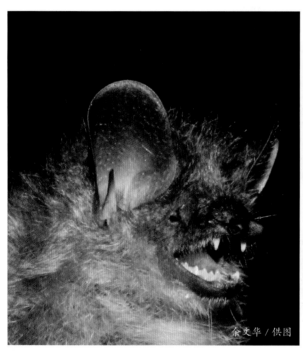

余文华 / 供图

吴毅 / 供图

▲ 其他名称 / Other Name(s)

其他中文名 / Other Chinese Name(s)
无 None

其他英文名 / Other English Name(s)
White-bellied Tube-nosed Bat

同物异名 / Synonym(s)
无 None

▲ 形态及生境 / Morphology and Habitat

形态特征 / Morphological Characteristics
齿式：2.1.2.3/3.1.2.3=34。头体长 47~50 mm。耳长 16~18 mm。前臂长 29~38 mm。后足长 6~10 mm。尾长 31~39 mm。颅全长 15~18 mm。鼻部管状，向两侧突出。耳短而窄。耳后缘凸起，无凹刻。体毛长而柔软，背毛为棕红褐色，腹毛苍白，背部与腹部毛色差异不明显。尾膜和足有毛。
Dental formula: 2.1.2.3/3.1.2.3=34. Head and body length 47-50 mm. Ear length 16-18 mm. Forearm length 29-38 mm. Hind foot length 6-10 mm. Tail length 31-39 mm. Greatest skull length 15-18 mm. Nose is tubular and protruding laterally. Ears are short and narrow. Rear margin of the ear is raised without concave. Body is covered with long and soft hairs. Dorsal hairs are reddish-brown. Ventral hairs are pale. Dorsal and ventral and abdominal hair color difference is not obvious. Tail membrane and feet are hairy.

生境 / Habitat
森林、种植园
Forest, plantations

▲ 地理分布 / Geographic Distribution

国内分布 / Domestic Distribution
广西、广东、福建、湖北、江西、浙江
Guangxi, Guangdong, Fujian, Hubei, Jiangxi, Zhejiang

全球分布 / World Distribution
中国、印度、老挝、马来西亚、缅甸、尼泊尔、巴基斯坦、泰国、
越南
China, India, Laos, Malaysia, Myanmar, Nepal, Pakistan, Thailand, Vietnam

生物地理界 / Biogeographic Realm
印度马来界、古北界
Indomalaya, Palearctic

WWF 生物群系 / WWF Biome
热带和亚热带湿润阔叶林
Tropical & Subtropical Moist Broadleaf Forests

动物地理分布型 / Zoogeographic Distribution Type
Wb

分布标注 / Distribution Note
非特有种 Non-Endemic

▲ 濒危状况 / Threatened Status

中国生物多样性红色名录等级 / CB RL Category (2021)
无危 LC

IUCN 红色名录 / IUCN Red List (2021)
数据缺乏 DD

威胁因子 / Threats
未知 Unknown

▲ 法律保护地位 / Legal Protection Status

国家重点保护野生动物等级 / Category of National Key Protected Wild Animals (2021)
未列入 Not listed

"三有"名录 / TWIESSV (2023)
未列入 Not listed

CITES 附录等级 / CITES Appendix (2023)
未列入 Not listed

迁徙物种公约附录 / CMS Appendix (2020)
未列入 Not listed

保护行动 / Conservation Action
尚无保护行动 No conservation action so far

▲ 参考文献 / References

Jiang et al. (蒋志刚等), 2021; Burgin et al., 2020; IUCN, 2020; Liu et al. (刘少英等), 2020; Wilson and Mittermeier, 2019; Smith and Xie, 2009; Pan et al. (潘清华等), 2007; Wilsn and Reeder, 2005; Wang (王应祥), 2003; Zhang (张荣祖), 1997

231 / 锦蟊管鼻蝠

Murina jinchui
W.-H. Yu, Csorba and Y. Wu, 2020

• Jinchu Tube-nosed Bat

▲ 分类地位 / Taxonomy

翼手目 Chiroptera / 蝙蝠科 Vespertilionidae / 管鼻蝠属 *Murina*

科建立者及其文献 / Family Authority
Gray, 1821

属建立者及其文献 / Genus Authority
Gray, 1842

亚种 / Subspecies
无 None

模式标本产地 / Type Locality
中国
Sichuan, China

余文华 / 供图

▲ 其他名称 / Other Name(s)

其他中文名 / Other Chinese Name(s)
无 None

其他英文名 / Other English Name(s)
无 None

同物异名 / Synonym(s)
无 None

▲ 形态及生境 / Morphology and Habitat

形态特征 / Morphological Characteristics

齿式: 2.1.2.3/3.1.2.3=34。前臂长 35 mm。颅全长 15~17 mm。吻突较发达。矢状嵴与人字嵴不明显。颧弓纤弱。鼻孔前端突出延长成短管状，鼻吻部黑色。耳小且圆，无缺刻，耳屏尖长。背毛基部黑色，中部棕灰色，毛尖深褐色，杂有暖灰色长毛。腹毛毛基黑色，毛尖呈冷灰色，杂有金色毛。前臂和掌骨无毛，短拇指背表面有金色毛。后肢和尾膜深棕色。翼膜延至第一趾基部。

Dental formula: 2.1.2.3/3.1.2.3=34. Forearm 35 mm. Greatest skull length 15-17 mm. Muzzle is well developed. Sagittal crest and the herringbone crest are not obvious. Zygomatic arch is delicate. Nose tip protrudes and extends into short tubes. Snout is black. Ears are small and round, without a notch. Tragus tip is long. Basal parts of dorsal hairs are black, and the middle parts are brown and gray, and the tips are dark brown, mixed with warm gray long hairs. Ventral hair bases are black, hair tips are cold gray, mixed with golden hairs. Forearms and metacarpal glabrous. Golden hairs on the dorsal surface of short thumb. Hind limbs and caudal membrane are dark brown. Pterygium extends to the base of the first toe.

生境 / Habitat
森林 Forest

▲ 地理分布 / Geographic Distribution

国内分布 / Domestic Distribution
四川 Sichuan

全球分布 / World Distribution
中国 China

生物地理界 / Biogeographic Realm
古北界 Palearctic

WWF 生物群系 / WWF Biome
热带和亚热带湿润阔叶林
Tropical & Subtropical Moist Broadleaf Forests

动物地理分布型 / Zoogeographic Distribution Type
Sb

分布标注 / Distribution Note
特有种 Endemic

▲ 濒危状况 / Threatened Status

中国生物多样性红色名录等级 / CB RL Category (2021)
数据缺乏 DD

IUCN 红色名录 / IUCN Red List (2021)
未评定 NE

威胁因子 / Threats
未知 Unknown

▲ 法律保护地位 / Legal Protection Status

国家重点保护野生动物等级 / Category of National Key Protected Wild Animals (2021)
未列入 Not listed

"三有"名录 / TWIESSV (2023)
未列入 Not listed

CITES 附录等级 / CITES Appendix (2023)
未列入 Not listed

迁徙物种公约附录 / CMS Appendix (2020)
未列入 Not listed

保护行动 / Conservation Action
尚无保护行动 No conservation action so far

▲ 参考文献 / References

Jiang et al. (蒋志刚等), 2021; Liu et al. (刘少英等), 2020; Yu et al., 2020

232 / 白腹管鼻蝠

Murina leucogaster Milne-Edwards, 1872

• Rufous Tube-nosed Bat

▲ 分类地位 / Taxonomy

翼手目 Chiroptera / 蝙蝠科 Vespertilionidae / 管鼻蝠属 *Murina*

科建立者及其文献 / Family Authority

Gray, 1821

属建立者及其文献 / Genus Authority

Gray, 1842

亚种 / Subspecies

指名亚种 *M. l. leucogaster* Milne-Edwards, 1872

北京、广西、贵州、云南、西藏、四川、福建、陕西、山西、吉林、辽宁、内蒙古和黑龙江

Beijing, Guangxi, Guizhou, Yunnan, Tibet, Sichuan, Fujian, Shaanxi, Shanxi, Jilin, Liaoning, Inner Mongolia and Heilongjiang

模式标本产地 / Type Locality

中国四川

China, Szechwan, Moupin Dist

江廷磊 / 供图

▲ 其他名称 / Other Name(s)

其他中文名 / Other Chinese Name(s)

无 None

其他英文名 / Other English Name(s)

Greater Tube-nosed Bat

同物异名 / Synonym(s)

无 None

▲ 形态及生境 / Morphology and Habitat

形态特征 / Morphological Characteristics

齿式: 2.1.2.3/3.1.2.3=34。头体长 47~49 mm。耳长 14~15 mm。前臂长 40~43 mm。后足长 9~10 mm。尾长 35~45 mm。颅全长 18~19 mm。头部褐色，吻端突出。耳长略大于耳宽，顶端钝圆。耳屏长而窄，基部有缺刻。鼻孔延长呈短管状。面部暗褐色，背毛红褐色或灰褐色，腹部灰白色。尾膜及足上均覆有细而密的毛。不同种群个体体毛颜色有变化。翼膜宽圆，连接到趾基部。

Dental formula: 2.1.2.3/3.1.2.3=34. Head and body length 47-49 mm. Ear length 14-15 mm. Forearm length 40-43 mm. Hind foot length 9-10 mm. Tail length 35-45 mm. Greatest skull length 18-19 mm. Head is brown and the snout is prominent. Ear length is slightly longer than ear's width, with a blunt rounded tip. Tragus is long and narrow with a notch at the base. Nostrils are elongated into short tubes. Dark brown face. Dorsal hairs reddish-brown or grayish-brown. Ventral hairs are gray. Tail membrane and feet are covered with dense fine hairs. Hair color varies among different geographical areas. Pterygoid membrane is broadly circular and connected to the base of the toe.

生境 / Habitat

洞穴、森林、人造建筑

Caves, forest, man-made building

▲ 地理分布 / Geographic Distribution

国内分布 / Domestic Distribution
北京、广西、贵州、云南、西藏、四川、福建、陕西、山西、吉林、
辽宁、内蒙古和黑龙江
Beijing, Guangxi, Guizhou, Yunnan, Tibet, Sichuan, Fujian, Shaanxi, Shanxi, Jilin,
Liaoning, Inner Mongolia and Heilongjiang

全球分布 / World Distribution
不丹、中国、印度、日本、尼泊尔、泰国
Bhutan, China, India, Japan, Nepal, Thailand

生物地理界 / Biogeographic Realm
古北界、印度马来界
Palearctic, Indomalaya

WWF 生物群系 / WWF Biome
热带和亚热带湿润阔叶林
Tropical & Subtropical Moist Broadleaf Forests

动物地理分布型 / Zoogeographic Distribution Type
We

分布标注 / Distribution Note
非特有种 Non-Endemic

▲ 濒危状况 / Threatened Status

中国生物多样性红色名录等级 / CB RL Category (2021)
无危 LC

IUCN 红色名录 / IUCN Red List (2021)
数据缺乏 DD

威胁因子 / Threats
未知 Unknown

▲ 法律保护地位 / Legal Protection Status

国家重点保护野生动物等级 / Category of National Key Protected Wild Animals (2021)
未列入 Not listed

"三有"名录 / TWIESSV (2023)
未列入 Not listed

CITES 附录等级 / CITES Appendix (2023)
未列入 Not listed

迁徙物种公约附录 / CMS Appendix (2020)
未列入 Not listed

保护行动 / Conservation Action
尚无保护行动 No conservation action so far

▲ 参考文献 / References

Jiang et al. (蒋志刚等), 2021; Wilson and Mittermeier, 2019; Pan et al. (潘清华等), 2007; Wilson and Reeder, 2005; Wang(王应祥), 2003;
Zhang (张荣祖), 1997

233 / 荔波管鼻蝠

Murina liboensis
X. Zeng, J. Chen, H.-Q. Deng, N. Xiao
& J. Zhou, 2018

· Libo Tube-nosed Bat

▲ 分类地位 / Taxonomy

翼手目 Chiroptera / 蝙蝠科 Vespertilionidae / 管鼻蝠属 *Murina*

科建立者及其文献 / Family Authority
Gray, 1821

属建立者及其文献 / Genus Authority
Gray, 1842

亚种 / Subspecies
无 None

模式标本产地 / Type Locality
中国贵州
Guizhou, China

▲ 其他名称 / Other Name(s)

其他中文名 / Other Chinese Name(s)
无 None

其他英文名 / Other English Name(s)
无 None

同物异名 / Synonym(s)
无 None

▲ 形态及生境 / Morphology and Habitat

形态特征 / Morphological Characteristics
齿式：2.1.2.3/3.1.2.3=34。头体长 37.7~45.5 mm。前臂长度 28.1~32.6 mm。尾长 31~35 mm。体重 4.5 ~5.5 g。鼻宽，鼻间凹入沟明显。颧骨弓纤细，中间有明显隆起，颧骨弓向后逐渐变薄。耳郭卵圆形，耳郭前缘呈钝圆形，耳郭后缘无凹痕。背毛蓬松，呈褐灰色，尖端金黄色。被毛延伸到尾膜和后脚。腹部被毛较短，基部浅灰黑色，尖端灰色，在靠近下巴和腹部的一侧尖端有更多的棕色。体侧膜插入在第一趾的二分之一处。

Dental formula: 2.1.2.3/3.1.2.3=34. Head and body length 37.7-45.5 mm. Forearm length 28.1-32.6 mm. Tail length 31-35 mm. Body mass 4.5-5.5 g. Snout is wide, recessed into the recession between the nasal notably. Zygomatic arch is slender, has a distinct rise in the middle, followed by the zygomatic arch gradually becoming thinner. Ears are ovate-orbicular, and the anterior border of the ear is a dull circle and without an indentation on the posterior border of the pinna. Dorsal pelage is fluffy with brownish-grey and the tip is golden yellow. Pelage extends onto the tail membrane and the hind feet. Ventral pelage is relatively short, pale grayish-black basally and ash grey at the tips but browner near the chin and on the side of the abdomen. Plagiopatagium inserts at 1/2 of the 1st toe .

生境 / Habitat
森林、喀斯特地貌
Forest, karst landscape

▲ 地理分布 / Geographic Distribution

国内分布 / Domestic Distribution
贵州 Guizhou

全球分布 / World Distribution
中国 China

生物地理界 / Biogeographic Realm
古北界 Palearctic

WWF 生物群系 / WWF Biome
热带和亚热带湿润阔叶林
Tropical & Subtropical Moist Broadleaf Forests

动物地理分布型 / Zoogeographic Distribution Type
Sb

分布标注 / Distribution Note
特有种 Endemic

▲ 濒危状况 / Threatened Status

中国生物多样性红色名录等级 / CB RL Category (2021)
数据缺乏 DD

IUCN 红色名录 / IUCN Red List (2021)
未评定 NE

威胁因子 / Threats
未知 Unknown

▲ 法律保护地位 / Legal Protection Status

国家重点保护野生动物等级 / Category of National Key Protected Wild Animals (2021)
未列入 Not listed

"三有"名录 / TWIESSV (2023)
未列入 Not listed

CITES 附录等级 / CITES Appendix (2023)
未列入 Not listed

迁徙物种公约附录 / CMS Appendix (2020)
未列入 Not listed

保护行动 / Conservation Action
尚无保护行动 No conservation action so far

▲ 参考文献 / References

Jiang et al. (蒋志刚等), 2021; Burgin et al., 2020; IUCN, 2020; Zeng et al., 2018

234 / 罗蕾莱管鼻蝠

Murina lorelieae Eger & Lim, 2011

· Lorelie's Tube-nosed Bat

翼手目 Chiroptera / 蝙蝠科 Vespertilionidae / 管鼻蝠属 *Murina*

科建立者及其文献 / Family Authority
Gray, 1821

属建立者及其文献 / Genus Authority
Gray, 1842

亚种 / Subspecies
无 None

模式标本产地 / Type Locality
中国
Diding Headwater Forest Nature Preserve (formerly known as Jing Xi County Provincial Nature Reserve), Jing Xi County, Guangxi Zhuang Autonomous Region, China. 23°07'12"N, 105°57'36" E, elevation 978 m above sea level

李锋 / 供图

▲ 其他名称 / Other Name(s)

其他中文名 / Other Chinese Name(s)
无 None

其他英文名 / Other English Name(s)
无 None

同物异名 / Synonym(s)
无 None

▲ 形态及生境 / Morphology and Habitat

形态特征 / Morphological Characteristics
齿式：2.1.2.3/3.1.2.3=34。头体长 59~61 mm。耳长 17~19 mm。前臂长 32.57 mm。后足长 10~11 mm。尾长 32~36 mm。头骨具矢状嵴和人字嵴。鼻吻部中间有凹槽。脑颅表面光滑。鼻部延长呈管状。吻端较尖。耳郭大而圆，无缺刻。背毛呈红褐色，毛基黑色，中段为浅灰色。头部毛发毛尖呈红色，肩至尾部背毛毛尖为褐色。腹毛毛基棕黑色，毛尖灰白色，其间杂有长 6~7 mm 的白色针毛。尾膜背面边缘有红色毛发，尾膜边缘有灰黑色的短毛。股间膜与后足第一趾相连接。

Dental formula: 2.1.2.3/3.1.2.3=34. Head and body length 59-61 mm. Ear length 17-19 mm. Forearm length 32.57 mm. Hind foot length 10-11 mm. Tail length 32-36 mm. The skull has a sagittal ridge and a herringbone ridge. There's a groove in the middle of the snout. The cranial surface is smooth. The nose is prolonged and tubular. The rostral end is pointed. The auricle is big, round, and flawless. The dorsal hairs are reddish-brown, with a black base and a light gray midsection. The tips of the head hairs are red, and the tips of the hairs on the shoulder to the tail are brown. The ventral hair bases are brown-black, and the tips are off-white, interspersed with white acicular hairs 6-7 mm long. Red hairs on the back edge of the caudal membrane. Caudal membrane edge has short grey-black hairs. The interfemoral membrane is connected with the first toe of the hind foot.

生境 / Habitat

森林 Forest

▲ **地理分布 / Geographic Distribution**

国内分布 / Domestic Distribution
云南、广西
Yunnan, Guangxi

全球分布 / World Distribution
中国 China

生物地理界 / Biogeographic Realm
印度马来界、古北界
Indomalaya, Palearctic

WWF 生物群系 / WWF Biome
热带和亚热带湿润阔叶林
Tropical & Subtropical Moist Broadleaf Forests

动物地理分布型 / Zoogeographic Distribution Type
Sa

分布标注 / Distribution Note
特有种 Endemic

▲ **濒危状况 / Threatened Status**

中国生物多样性红色名录等级 / CB RL Category (2021)
数据缺乏 DD

IUCN 红色名录 / IUCN Red List (2021)
数据缺乏 DD

威胁因子 / Threats
砍伐 Logging

▲ **法律保护地位 / Legal Protection Status**

国家重点保护野生动物等级 / Category of National Key Protected Wild Animals (2021)
未列入 Not listed

"三有" 名录 / TWIESSV (2023)
未列入 Not listed

CITES 附录等级 / CITES Appendix (2023)
未列入 Not listed

迁徙物种公约附录 / CMS Appendix (2020)
未列入 Not listed

保护行动 / Conservation Action
尚无保护行动 No conservation action so far

▲ **参考文献 / References**

Jiang et al. (蒋志刚等), 2021; Liu et al. (刘少英等), 2020; Li et al. (黎舫等), 2017

235 / 台湾管鼻蝠

Murina puta Kishida, 1924

• Taiwanese Tube-nosed Bat

▲ 分类地位 / Taxonomy

翼手目 Chiroptera / 蝙蝠科 Vespertilionidae / 管鼻蝠属 *Murina*

科建立者及其文献 / Family Authority
Gray, 1821

属建立者及其文献 / Genus Authority
Gray, 1842

亚种 / Subspecies
无 None

模式标本产地 / Type Locality
中国
China, Taiwan, Chang Hua, Erh-Shui

周政翰 / 供图

▲ 其他名称 / Other Name(s)

其他中文名 / Other Chinese Name(s)
无 None

其他英文名 / Other English Name(s)
无 None

同物异名 / Synonym(s)
无 None

▲ 形态及生境 / Morphology and Habitat

形态特征 / Morphological Characteristics

齿式：2.1.2.3/3.1.2.3=34。头体长 59~61 mm。耳长 17~19 mm。前臂长 33~39 mm。后足长 10~11 mm。尾长 32~36 mm。吻端突出，鼻孔延长成短管状。耳郭较大，耳长大于耳宽，顶端钝圆。耳屏长而窄。头部毛为褐色，背毛浅褐色或灰褐色，腹毛灰白色到灰褐色。翼膜宽圆，延伸连接到趾基部。尾膜及足上均覆盖有细密的短毛。

Dental formula: 2.1.2.3/3.1.2.3=34. Head and body length 59-61 mm. Ear length 17-19 mm. Forearm length 33-39 mm. Hind foot length 10-11 mm. Tail length 32-36 mm. Snout protrude. Nostrils are lengthened into short tubes. Auricle is larger. Ear length is larger than ear width. Ear tips are blunt and round. Tragus is long and narrow. Hairs on the head are brown. Dorsal hairs are light brown or grayish brown. Ventral hairs are grayish to grayish brown. Pterygium is broadly circular and extends to the base of the toe. Tail membrane and feet are covered with dense fine hairs.

生境 / Habitat
亚热带湿润阔叶林
Temperate forest

▲ 地理分布 / Geographic Distribution

国内分布 / Domestic Distribution
台湾 Taiwan

全球分布 / World Distribution
中国 China

生物地理界 / Biogeographic Realm
印度马来界 Indomalaya

WWF 生物群系 / WWF Biome
热带和亚热带湿润阔叶林
Tropical & Subtropical Moist Broadleaf Forests

动物地理分布型 / Zoogeographic Distribution Type
Sa

分布标注 / Distribution Note
特有种 Endemic

▲ 濒危状况 / Threatened Status

中国生物多样性红色名录等级 / CB RL Category (2021)
近危 NT

IUCN 红色名录 / IUCN Red List (2021)
数据缺乏 DD

威胁因子 / Threats
砍伐 Logging

▲ 法律保护地位 / Legal Protection Status

国家重点保护野生动物等级 / Category of National Key Protected Wild Animals (2021)
未列入 Not listed

"三有"名录 / TWIESSV (2023)
未列入 Not listed

CITES 附录等级 / CITES Appendix (2023)
未列入 Not listed

迁徙物种公约附录 / CMS Appendix (2020)
未列入 Not listed

保护行动 / Conservation Action
尚无保护行动 No conservation action so far

▲ 参考文献 / References

Jiang et al. (蒋志刚等), 2021; Burgin et al., 2020; IUCN, 2020; Liu et al. (刘少英等), 2020; Wilson and Mittermeier, 2019; Eger and Lim, 2011

236 / 隐姬管鼻蝠

Murina recondita
Kuo, Fang, Csorba & Lee, 2009

· Faint-golden Little Tube-nosed Bat

▲ 分类地位 / Taxonomy

翼手目 Chiroptera / 蝙蝠科 Vespertilionidae / 管鼻蝠属 *Murina*

科建立者及其文献 / Family Authority
Gray, 1821

属建立者及其文献 / Genus Authority
Gray, 1842

亚种 / Subspecies
无 None

模式标本产地 / Type Locality
中国
China, Taiwan, Hualien County, Jhuosi Township, Rueisuei logging road, 1,300 m, 23°29'N, 121°16'E

周政翰 / 供图

▲ 其他名称 / Other Name(s)

其他中文名 / Other Chinese Name(s)
无 None

其他英文名 / Other English Name(s)
Dun Tube-nosed Bat, Faint-colored Tube-nosed Bat

同物异名 / Synonym(s)
无 None

▲ 形态及生境 / Morphology and Habitat

形态特征 / Morphological Characteristics

齿式：2.1.2.3/3.1.2.3=34。头体长 34~44 mm。前臂长 31~35 mm。尾长 27~39 mm。吻端明显突出。鼻孔延长成短管状。头部毛为褐色，背毛金黄色，杂有金属光泽。腹毛基部黑色，毛尖灰白色，毛厚而柔软。耳郭相对于体型较短，耳长大于耳宽，顶端钝圆。耳屏窄而尖长。翼膜宽圆，向后延伸连接到趾基部。尾膜及足上均被覆有细密短毛。

Dental formula: 2.1.2.3/3.1.2.3=34. Head and body length 34-44 mm. Forearm length 31-35 mm. Tail length 27-39 mm. Snout is prominent. Nostrils are elongated into short tubes. Head hairs are brown. Body hairs are thick and soft. Dorsal hairs are golden yellow, and the miscellany interspersed with hairs of metallic luster. Ventral hair bases are black and the hair tips are gray. Pinna is short relative to the body. Ear length is longer than ear width, and ear tip is blunt and rounded. Tragus is narrow and tapered. Pterygium is broadly rounded and extends back to the base of the toe. Tail membrane and feet are covered with dense fine hairs.

生境 / Habitat
未知 Unknown

▲ 地理分布 / Geographic Distribution

国内分布 / Domestic Distribution
台湾 Taiwan

全球分布 / World Distribution
中国 China

生物地理界 / Biogeographic Realm
印度马来界 Indomalaya

WWF 生物群系 / WWF Biome
热带和亚热带湿润阔叶林
Tropical & Subtropical Moist Broadleaf Forests

动物地理分布型 / Zoogeographic Distribution Type
Sa

分布标注 / Distribution Note
特有种 Endemic

▲ 濒危状况 / Threatened Status

中国生物多样性红色名录等级 / CB RL Category (2021)
数据缺乏 DD

IUCN 红色名录 / IUCN Red List (2021)
数据缺乏 DD

威胁因子 / Threats
砍伐 Logging

▲ 法律保护地位 / Legal Protection Status

国家重点保护野生动物等级 / Category of National Key Protected Wild Animals (2021)
未列入 Not listed

"三有"名录 / TWIESSV (2023)
未列入 Not listed

CITES 附录等级 / CITES Appendix (2023)
未列入 Not listed

迁徙物种公约附录 / CMS Appendix (2020)
未列入 Not listed

保护行动 / Conservation Action
尚无保护行动 No conservation action so far

▲ 参考文献 / References

Jiang et al. (蒋志刚等), 2021; Burgin et al., 2020; IUCN, 2020; Liu et al. (刘少英等), 2020; Wilson and Mittermeier, 2019

237 | 榕江管鼻蝠

Murina rongjiangensis
J. Chen, T. Liu, H. -Q. Deng, N. Xiao &
J. Zhou, 2017

· Rongjiang Tube-nosed Bat

翼手目 Chiroptera / 蝙蝠科 Vespertilionidae / 管鼻蝠属 *Murina*

科建立者及其文献 / Family Authority
Gray, 1821

属建立者及其文献 / Genus Authority
Gray, 1842

亚种 / Subspecies
无 None

模式标本产地 / Type Locality
中国
Guizhou, China

▲ 其他名称 / Other Name(s)

其他中文名 / Other Chinese Name(s)
无 None

其他英文名 / Other English Name(s)
无 None

同物异名 / Synonym(s)
无 None

▲ 形态及生境 / Morphology and Habitat

形态特征 / Morphological Characteristics
齿式：2.1.2.3/3.1.2.3=34。前臂长 30~34 mm。尾长 32~39 mm。颅全长
15.7~16.3 mm。颅骨呈半圆形。喙坚实。腭前微缺。喉部周围毛橙色，
背部毛绒毛状，呈棕色。胸腹部毛为黄色。腹部中央有白色短毛。体侧
膜在靠近第一个脚趾的三分之一处插入。
Dental formula: 2.1.2.3/3.1.2.3=34. Forearm length 30-34 mm, Tail length 32-39 mm.
Greatest skull length 15.7-16.3 mm. Skull is domed, and the rostrum is robust, and
the pre-palatal is emarginate. Dorsum is fluffy with brown color, and the ventral fur
is yellow on the chest. Orange hairs around the throat, and short white hairs on the
center of the abdominal. The insertion point of the plagiopatagium is close to 1/3 of
the 1st toe .

生境 / Habitat
森林、喀斯特地貌
Forest, karst landscape

▲ **地理分布 / Geographic Distribution**

国内分布 / Domestic Distribution
贵州 Guizhou

全球分布 / World Distribution
中国 China

生物地理界 / Biogeographic Realm
古北界 Palearctic

WWF 生物群系 / WWF Biome
热带和亚热带湿润阔叶林
Tropical & Subtropical Moist Broadleaf Forests

动物地理分布型 / Zoogeographic Distribution Type
Sb

分布标注 / Distribution Note
特有种 Endemic

▲ **濒危状况 / Threatened Status**

中国生物多样性红色名录等级 / CB RL Category (2021)
数据缺乏 DD

IUCN 红色名录 / IUCN Red List (2021)
未评定 DD

威胁因子 / Threats
未知 Unknown

▲ **法律保护地位 / Legal Protection Status**

国家重点保护野生动物等级 / Category of National Key Protected Wild Animals (2021)
未列入 Not listed

"三有" 名录 / TWIESSV (2023)
未列入 Not listed

CITES 附录等级 / CITES Appendix (2023)
未列入 Not listed

迁徙物种公约附录 / CMS Appendix (2020)
未列入 Not listed

保护行动 / Conservation Action
尚无保护行动 No conservation action so far

▲ **参考文献 / References**

Jiang et al. (蒋志刚等), 2021; Burgin et al., 2020; Chen et al., 2017

238 / 水甫管鼻蝠

Murina shuipuensis Eger & Lim, 2011

• Shuipu's Tube-nosed Bat

▲ 分类地位 / Taxonomy

翼手目 Chiroptera / 蝙蝠科 Vespertilionidae / 管鼻蝠属 *Murina*

科建立者及其文献 / Family Authority
Gray, 1821

属建立者及其文献 / Genus Authority
Gray, 1842

亚种 / Subspecies
无 None

模式标本产地 / Type Locality
中国
Shuipu Village, Yuping Town, Libo County, Guizhou, China. 25°28'59"N, 107°52'54"E, elevation 650 m above sea level

王晓云 / 供图   王晓云 / 供图

▲ 其他名称 / Other Name(s)

其他中文名 / Other Chinese Name(s)
无 None

其他英文名 / Other English Name(s)
无 None

同物异名 / Synonym(s)
无 None

▲ 形态及生境 / Morphology and Habitat

形态特征 / Morphological Characteristics
齿式：2.1.2.3/3.1.2.3=34。头体长 72 mm。耳长 11 mm。前臂长 30 mm。尾长 32 mm。体重 4 g。口鼻部黑色、狭长，鼻孔管状、突出。眼睛小。耳短而圆，耳后缘中间有一凹槽。背毛基部灰色，中部棕黄色，尖端黑色，杂有金色长毛。腹部被毛为橙黄色。腹面有成排白色小斑点，斑点上有白色毛。膜翅附着在大脚趾基部。足小，覆盖着毛发。尾较长，尾背面密布金褐色毛。

Dental formula: 2.1.2.3/3.1.2.3=34. Head and body length 72 mm. Ears length 11 mm. Forearm length 30 mm. Tail length 32 mm. Body mass 4 g. Snout black, long and narrow. Tubular nostrils prominent. Eyes small. Ears are short and round with a notch in the middle of the trailing edge. Dorsal hairs are gray at the base, brown at the middle part, and black at the tip, mixed with long golden hairs. Underbelly pelage is orange-yellow. There are rows of small white spots with white hairs on the abdomen. The membranous wing is attached to the base of the big toe. The feet are small and covered with hairs. Tail is long and covered with golden-brown hairs on the back.

生境 / Habitat
森林、喀斯特地貌
Forest, karst landscape

▲ 地理分布 / Geographic Distribution

国内分布 / Domestic Distribution
贵州、江西、广东
Guizhou, Jiangxi, Guangdong

全球分布 / World Distribution
中国 China

生物地理界 / Biogeographic Realm
印度马来界、古北界
Indomalaya, Palearctic

WWF 生物群系 / WWF Biome
热带和亚热带湿润阔叶林
Tropical & Subtropical Moist Broadleaf Forests

动物地理分布型 / Zoogeographic Distribution Type
J

分布标注 / Distribution Note
特有种 Endemic

▲ 濒危状况 / Threatened Status

中国生物多样性红色名录等级 / CB RL Category (2021)
数据缺乏 DD

IUCN 红色名录 / IUCN Red List (2021)
数据缺乏 DD

威胁因子 / Threats
未知 Unknown

▲ 法律保护地位 / Legal Protection Status

国家重点保护野生动物等级 / Category of National Key Protected Wild Animals (2021)
未列入 Not listed

"三有"名录 / TWIESSV (2023)
未列入 Not listed

CITES 附录等级 / CITES Appendix (2023)
未列入 Not listed

迁徙物种公约附录 / CMS Appendix (2020)
未列入 Not listed

保护行动 / Conservation Action
尚无保护行动 No conservation action so far

▲ 参考文献 / References

Jiang et al. (蒋志刚等), 2021; Liu et al. (刘少英等), 2020; Wilson and Mittermeier, 2019; Wang et al. (王晓云等), 2016; Eger & Lim, 2011

239 / 乌苏里管鼻蝠

Murina ussuriensis Ognev, 1913

· Ussuri Tube-nosed Bat

▲ 分类地位 / Taxonomy

翼手目 Chiroptera / 蝙蝠科 Vespertilionidae /
管鼻蝠属 *Murina*

科建立者及其文献 / Family Authority
Gray, 1821

属建立者及其文献 / Genus Authority
Gray, 1842

亚种 / Subspecies
无 None

模式标本产地 / Type Locality
俄罗斯
Russia, SE Siberia, Ussuri, Imansky distr., Evseevka

▲ 其他名称 / Other Name(s)

其他中文名 / Other Chinese Name(s)
无 None

其他英文名 / Other English Name(s)
Lesser Tube-nosed Bat, Ussurian Tube-nosed Bat

同物异名 / Synonym(s)
无 None

▲ 形态及生境 / Morphology and Habitat

形态特征 / Morphological Characteristics
齿式： 2.1.2.3/3.1.2.3=34。头体长 40 mm。耳
长 13 mm。前臂长 27 mm。后足长 10 mm。尾
长 25 mm。颅全长 15 mm。体型小。毛皮短而
柔软，毛基部深色，中间带灰色，毛尖红棕色。
腹部毛发白，略带灰色。尾膜、腿和后脚背部
多毛。

Dental formula: 2.1.2.3/3.1.2.3=34. Head and body
length 40 mm. Ear length 13 mm. Forearm length
27 mm. Hind foot length 10 mm. Tail length 25 mm.
Greatest skull length 15 mm. Size small. Pelage short and
soft. Hairs are dark at the base, with a pale intermediate
band and reddish-brown tip. Ventral hairs are paler with
grayish cast. Dorsal parts of the uropatagium, legs and
hind feet are hairy.

生境 / Habitat
森林、洞穴
Forest, caves

▲ 地理分布 / Geographic Distribution

国内分布 / Domestic Distribution
吉林、黑龙江
Jilin, Heilongjiang

全球分布 / World Distribution
中国、日本、朝鲜、俄罗斯
China, Japan, Democratic People's Republic of Korea, Russia

生物地理界 / Biogeographic Realm
古北界 Palearctic

WWF 生物群系 / WWF Biome
温带阔叶和混交林
Temperate Broadleaf & Mixed Forests

动物地理分布型 / Zoogeographic Distribution Type
Xa

分布标注 / Distribution Note
非特有种 Non-Endemic

▲ 濒危状况 / Threatened Status

中国生物多样性红色名录等级 / CB RL Category (2021)
数据缺乏 DD

IUCN 红色名录 / IUCN Red List (2021)
数据缺乏 DD

威胁因子 / Threats
未知 Unknown

▲ 法律保护地位 / Legal Protection Status

国家重点保护野生动物等级 / Category of National Key Protected Wild Animals (2021)
未列入 Not listed

"三有"名录 / TWIESSV (2023)
未列入 Not listed

CITES 附录等级 / CITES Appendix (2023)
未列入 Not listed

迁徙物种公约附录 / CMS Appendix (2020)
未列入 Not listed

保护行动 / Conservation Action
尚无保护行动 No conservation action so far

▲ 参考文献 / References

Jiang et al. (蒋志刚等), 2021; Burgin et al., 2020; IUCN, 2020; Wilson and Mittermeier, 2019; Eger and Lim, 2011; Pan et al. (潘清华等), 2007; Wang (王应祥), 2003

240 / 毛翼蝠

Harpiocephalus harpia (Temminck, 1840)

· Lesser Hairy-Winged Bat

▲ 分类地位 / Taxonomy

翼手目 Chiroptera / 蝙蝠科 Vespertilionidae / 毛翼蝠属 *Harpiocephalus*

科建立者及其文献 / Family Authority
Gray, 1821

属建立者及其文献 / Genus Authority
Gray, 1842

亚种 / Subspecies
无 None

模式标本产地 / Type Locality
印度尼西亚
Indonesia, Java, NE side of Mt. Gede

周佳俊 / 供图

周佳俊 / 供图

▲ 其他名称 / Other Name(s)

其他中文名 / Other Chinese Name(s)
无 None

其他英文名 / Other English Name(s)
无 None

同物异名 / Synonym(s)
无 None

▲ 形态及生境 / Morphology and Habitat

形态特征 / Morphological Characteristics

齿式：2.1.2.3/3.1.2.3=34。头体长 60~75 mm。耳长 17~18 mm。前臂长 44~50 mm。后足长 11~14 mm。尾 40~50 mm。颅全长 23 mm。鼻部前端呈短管状。耳壳卵圆形，耳屏披针形，较长，且有一基凹。后足相对较短。背部毛基黄褐色，毛尖褐栗色；翼膜淡黑褐色；后足、股间膜及尾膜密生黄褐色细毛。

Dental formula: 2.1.2.3/3.1.2.3=34. Head and body length 60-75 mm. Ear length 17-18 mm. Forearm length 44-50 mm. Hind foot length 11-14 mm. Greatest skull length 23 mm. Ears are round. Tragus long, with a basal notch. Nostrils protuberant. Pelage thick and soft. Dorsal hairs orange-brown. Ventral hairs are light brown. Hind legs, wing membrane, and uropatagium are covered with hairs in part.

生境 / Habitat
常绿阔叶林
Evergreen broad-leaved forest

▲ 地理分布 / Geographic Distribution

国内分布 / Domestic Distribution
贵州、云南、湖北、湖南、福建、江西、台湾、浙江、广东、广西、
海南、四川
Guizhou, Yunnan, Hubei, Hunan, Fujian, Jiangxi, Taiwan, Zhejiang, Guangdong,
Guangxi, Hainan, Sichuan

全球分布 / World Distribution
中国、马来西亚、泰国、缅甸、新几内亚、澳大利亚
China, Malaysia, Thailand, Myanmar, New Guinea, Australia

生物地理界 / Biogeographic Realm
印度马来界 Indomalaya

WWF 生物群系 / WWF Biome
热带和亚热带湿润阔叶林
Tropical & Subtropical Moist Broadleaf Forests

动物地理分布型 / Zoogeographic Distribution Type
Wc

分布标注 / Distribution Note
非特有种 Non-Endemic

▲ 濒危状况 / Threatened Status

中国生物多样性红色名录等级 / CB RL Category (2021)
近危 NT

IUCN 红色名录 / IUCN Red List (2021)
数据缺乏 DD

威胁因子 / Threats
砍伐 Logging

▲ 法律保护地位 / Legal Protection Status

国家重点保护野生动物等级 / Category of National Key Protected Wild Animals (2021)
未列入 Not listed

"三有"名录 / TWIESSV (2023)
未列入 Not listed

CITES 附录等级 / CITES Appendix (2023)
未列入 Not listed

迁徙物种公约附录 / CMS Appendix (2020)
未列入 Not listed

保护行动 / Conservation Action
尚无保护行动 No conservation action so far

▲ 参考文献 / References

Jiang et al. (蒋志刚等), 2021; Burgin et al., 2020; IUCN, 2020; Liu et al. (刘少英等), 2020; Smith and Xie, 2009; Pan et al. (潘清华等), 2007;
Wilson and Reeder, 2005; Si mmons, 2005; Wang (王应祥), 2003; Zhang (张荣祖), 1997; Maeda, 1980

241 / 暗褐彩蝠

Kerivoula furva
Kuo, Soisook, Ho & Rossiter, 2017

• Leaf-roosting Bat

翼手目 Chiroptera / 蝙蝠科 Vespertilionidae / 彩蝠属 *Kerivoula*

科建立者及其文献 / Family Authority
Gray, 1821

属建立者及其文献 / Genus Authority
Gray, 1842

亚种 / Subspecies
无 None

模式标本产地 / Type Locality
中国
China: Taiwan, Yilan County, Yuanshan Township, 3 km East of Shuanglianpi, 24°45.21'N, 121°39.63'E, 180 m a.s.l.

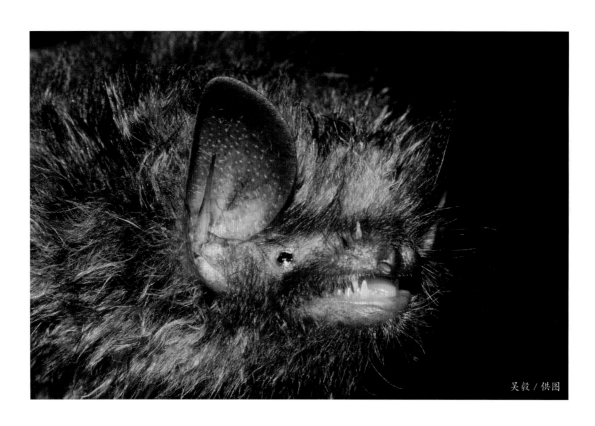

吴毅 / 供图

▲ 其他名称 / Other Name(s)

其他中文名 / Other Chinese Name(s)
无 None

其他英文名 / Other English Name(s)
无 None

同物异名 / Synonym(s)
无 None

▲ 形态及生境 / Morphology and Habitat

形态特征 / Morphological Characteristics
前臂长 30~34 mm。颅全长 14~15 mm。耳郭漏斗状，耳屏略呈披针形。背毛毛色存在个体差异，从深褐色至深灰色，毛基部一般为黑色。腹毛基部为黑色，尖端稍白而略带灰褐色。翼膜和尾间膜呈灰色，被覆稀疏浅毛。翼膜后缘末端延伸至后足第一趾基部，尾间膜前缘末端止于脚踝。
Forearm length 30-34 mm. Greatest skull length 14-15 mm. Auricle funnel-shaped. Tragus slightly lanceolate. Dorsal coat is nearly gray, and the bases of the hairs are black and the tips are dark gray. Ventral hairs are black at the bases and slightly white and grayish brown at the tips. Pterygoid and intercaudal membranes are gray, covered with sparse hairs. Posterior edge of the pterygium extends to the base of the first toe of the hind foot. Anterior edge of the intercaudal membrane terminates at the ankle.

生境 / Habitat
常绿阔叶林
Evergreen broad-leaved forest

▲ 地理分布 / Geographic Distribution

国内分布 / Domestic Distribution
云南、重庆、四川、湖南、江西、福建、台湾、广东、广西、海南
Yunnan, Chongqing, Sichuan, Hunan, Jiangxi, Fujian, Taiwan, Guangdong,
Guangxi, Hainan

全球分布 / World Distribution
中国、印度、老挝、缅甸、尼泊尔、巴基斯坦、越南
China, India, Laos, Myanmar, Nepal, Pakistan, Vietnam

生物地理界 / Biogeographic Realm
印度马来界 Indomalaya

WWF 生物群系 / WWF Biome
热带和亚热带湿润阔叶林
Tropical & Subtropical Moist Broadleaf Forests

动物地理分布型 / Zoogeographic Distribution Type
Wb

分布标注 / Distribution Note
非特有种 Non-Endemic

▲ 濒危状况 / Threatened Status

中国生物多样性红色名录等级 / CB RL Category (2021)
未评定 NE

IUCN 红色名录 / IUCN Red List (2021)
未评定 NE

威胁因子 / Threats
未知 Unknown

▲ 法律保护地位 / Legal Protection Status

国家重点保护野生动物等级 / Category of National Key Protected Wild Animals (2021)
未列入 Not listed

"三有"名录 / TWIESSV (2023)
未列入 Not listed

CITES 附录等级 / CITES Appendix (2023)
未列入 Not listed

迁徙物种公约附录 / CMS Appendix (2020)
未列入 Not listed

保护行动 / Conservation Action
尚无保护行动 No conservation action so far

▲ 参考文献 / References

Burgin et al., 2020; IUCN, 2020; Liu et al. (刘少英等), 2020; Yu et al., 2018

242 / 彩蝠

Kerivoula picta (Pallas, 1767)

· Painted Woolly Bat

翼手目 Chiroptera / 蝙蝠科 Vespertilionidae / 彩蝠属 *Kerivoula*

科建立者及其文献 / Family Authority

Gray, 1821

属建立者及其文献 / Genus Authority

Gray, 1842

亚种 / Subspecies

华南亚种 *K. p. bellissima* Thomas, 1906
贵州、福建、广东、广西和海南
Guizhou, Fujian, Guangdong, Guangxi and Hainan

模式标本产地 / Type Locality

印度尼西亚
Indonesia, Molucca Isls, Ternate Isl. See discussion in Corbet and Hill (1992)

吴毅 / 供图

▲ 其他名称 / Other Name(s)

其他中文名 / Other Chinese Name(s)
无 None

其他英文名 / Other English Name(s)
Painted Bat

同物异名 / Synonym(s)
无 None

▲ 形态及生境 / Morphology and Habitat

形态特征 / Morphological Characteristics

头体长 40~48 mm。耳长 13~16 mm。前臂长 31~38 mm。后足长 4~8 mm。尾长 43~48 mm。颅全长 15 mm。体重 8~10 g。耳壳大，耳基部管状，耳内缘凸起，耳屏细长呈披针形。背毛橙黄色，腹毛色淡。前臂、掌和指部为橙色。翼短而宽。指间翼膜为黑褐色。翼膜与趾基相连。足背被覆黑色短毛。

Head and body length 40-48 mm. Ear length 13-16 mm. Forearm length 31-38 mm. Hind foot length 4-8 mm. Tail length 43-48 mm. Greatest skull length 15 mm. Body mass 8-10 g. Auricle is large. Bases of the ears are tubular, and the inner edges of the ears are raised, and the tragus is long and lanceolate. Dorsal hairs orange-yellow. Ventral hair color light. Forearms, palms, and fingers are orange. Wings are short and wide. Interphalangeal membrane is dark brown. Pterygoid membrane is attached to the toe base. The 5th metacarpal is longer than the 3rd and 4th metacarpal. Foot back is covered with black short hairs.

生境 / Habitat

未知 Unknown

▲ 地理分布 / Geographic Distribution

国内分布 / Domestic Distribution
广东、广西、海南、福建、贵州
Guangdong, Guangxi, Hainan, Fujian, Guizhou

全球分布 / World Distribution
孟加拉国、柬埔寨、中国、印度、印度尼西亚、老挝、马来西亚、
缅甸、尼泊尔、斯里兰卡、泰国、越南
Bangladesh, Cambodia, China, India, Indonesia, Laos, Malaysia, Myanmar, Nepal, Sri Lanka, Thailand, Vietnam

生物地理界 / Biogeographic Realm
印度马来界、大洋洲界
Indomalaya, Oceanian

WWF 生物群系 / WWF Biome
热带和亚热带湿润阔叶林
Tropical & Subtropical Moist Broadleaf Forests

动物地理分布型 / Zoogeographic Distribution Type
Wa

分布标注 / Distribution Note
非特有种 Non-Endemic

▲ 濒危状况 / Threatened Status

中国生物多样性红色名录等级 / CB RL Category (2021)
濒危 EN

IUCN 红色名录 / IUCN Red List (2021)
数据缺乏 DD

威胁因子 / Threats
森林砍伐、耕种
Logging, farming

▲ 法律保护地位 / Legal Protection Status

国家重点保护野生动物等级 / Category of National Key Protected Wild Animals (2021)
未列入 Not listed

"三有" 名录 / TWIESSV (2023)
未列入 Not listed

CITES 附录等级 / CITES Appendix (2023)
未列入 Not listed

迁徙物种公约附录 / CMS Appendix (2020)
未列入 Not listed

保护行动 / Conservation Action
尚无保护行动 No conservation action so far

▲ 参考文献 / References

Jiang et al. (蒋志刚等), 2021; Liu et al. (刘少英等), 2020; Wilson and Mittermeier, 2019; Pan et al. (潘清华等), 2007; Wilson and Reeder, 2005; Wang (王应祥), 2003; Zhang (张荣祖), 1997

243 / 亚洲长翼蝠

Miniopterus fuliginosus Hodgson, 1835

· Eastern Long-fingered Bat

▲ 分类地位 / Taxonomy

翼手目 Chiroptera / 长翼蝠科 Miniopteridae / 长翼蝠属 *Miniopterus*

科建立者及其文献 / Family Authority
Dobson, 1875

属建立者及其文献 / Genus Authority
Bonaparte, 1837

亚种 / Subspecies
无 None

模式标本产地 / Type Locality
尼泊尔
Nepal

周佳俊 / 供图

▲ 其他名称 / Other Name(s)

其他中文名 / Other Chinese Name(s)
普通长翼蝠、长指蝠、褶翅蝠

其他英文名 / Other English Name(s)
Asian Long-winged Bat, Schreiber's Bent-winged Bat

同物异名 / Synonym(s)
无 None

▲ 形态及生境 / Morphology and Habitat

形态特征 / Morphological Characteristics

齿式：2.1.2.3/3.1.3.3=36。头体长 67~78 mm。耳长 12~14 mm。前臂长 47~50 mm。后足长 9~12 mm。尾长 50~62 mm。颅全长 16~17 mm。耳短圆，耳屏长度不及耳长之半，端部钝圆，稍向内弯。体被丝绒状短毛，背毛深褐色，腹毛色淡，毛基鼠灰色。第三指的第二指节为第一指节的 3 倍长，且在静止时呈倒折状态。翼狭长。尾长等于或大于头体长，股间膜呈锥状。

Dental formula: 2.1.2.3/3.1.3.3=36. Head and body length 67-78 mm. Ear length 12-14 mm. Forearm length 47-50 mm. Hind foot length 9-12 mm. Tail length 50-62 mm. Greatest skull length 16-17 mm. Ears are short and round, tragus less than half the length of the ear, tips obtuse, and slightly curved inwards. Body is covered with velvety short hairs. Dorsal hairs are dark brown. Ventral hairs are light-colored with a mouse gray hair bases. The second knuckle of the third finger is three times as long as the first knuckle and is inverted when at rest. Wing is long and narrow. Tail length is equal to or greater than the head and body length, all enclosed in the interfemoral membrane which is conical.

生境 / Habitat
洞穴、耕地、亚热带热带严重退化森林
Caves, arable land, subtropical tropical degraded forest

▲ 地理分布 / Geographic Distribution

国内分布 / Domestic Distribution
云南、四川、重庆、贵州、广西、海南、广东、福建、台湾、浙江、安徽、陕西、河南、河北、北京
Yunnan, Sichuan, Chongqing, Guizhou, Guangxi, Hainan, Guangdong, Fujian, Taiwan, Zhejiang, Anhui, Shaanxi, Henan, Hebei, Beijing

全球分布 / World Distribution
中国、阿富汗、印度、日本、朝鲜、韩国、缅甸、尼泊尔、巴基斯坦、斯里兰卡、越南
China, Afghanistan, India, Japan, Democratic People's Republic of Korea, Republic of Korea, Nepal, Pakistan, Sri Lana, Vietnam

生物地理界 / Biogeographic Realm
古北界 Palearctic

WWF 生物群系 / WWF Biome
热带和亚热带湿润阔叶林
Tropical & Subtropical Moist Broadleaf Forests

动物地理分布型 / Zoogeographic Distribution Type
U

分布标注 / Distribution Note
非特有种 Non-Endemic

▲ 濒危状况 / Threatened Status

中国生物多样性红色名录等级 / CB RL Category (2021)
近危 NT

IUCN 红色名录 / IUCN Red List (2021)
无危 LC

威胁因子 / Threats
森林砍伐、耕种、火灾、采矿、采石、旅游
Logging, farming, fire, mining, quarrying, tourism

▲ 法律保护地位 / Legal Protection Status

国家重点保护野生动物等级 / Category of National Key Protected Wild Animals (2021)
未列入 Not listed

"三有" 名录 / TWIESSV (2023)
未列入 Not listed

CITES 附录等级 / CITES Appendix (2023)
未列入 Not listed

迁徙物种公约附录 / CMS Appendix (2020)
未列入 Not listed

保护行动 / Conservation Action
尚无保护行动 No conservation action so far

▲ 参考文献 / References

Jiang et al. (蒋志刚等), 2021; Burgin et al., 2020; IUCN, 2020; Wilson and Mittermeier, 2019; Wang (王应祥), 2003; Wilson and Reeder, 1993

244 / 几内亚长翼蝠

Miniopterus magnater Sanborn, 1931

· Western Long-fingered Bat

翼手目 Chiroptera / 长翼蝠科 Miniopteridae / 长翼蝠属 *Miniopterus*

科建立者及其文献 / Family Authority
Miller-Butterworth et al., 2007

属建立者及其文献 / Genus Authority
Bonaparte, 1837

亚种 / Subspecies
无 None

模式标本产地 / Type Locality
巴布亚新几内亚
Papua New Guinea, E Sepik, Marienberg

▲ 其他名称 / Other Name(s)

其他中文名 / Other Chinese Name(s)
无 None

其他英文名 / Other English Name(s)
Large Bent-winged Bat,
Western Bent-winged Bat

同物异名 / Synonym(s)
无 None

▲ 形态及生境 / Morphology and Habitat

形态特征 / Morphological Characteristics
齿式：2.1.2.3/3.1.3.3=36。头体长 58~75 mm。耳长 11~17 mm。前臂长
47~54 mm。后足长 9~13 mm。尾长 52~64 mm。体型大，体背部毛皮长、
软，黑棕色。腹部毛色深棕色，毛尖色浅。
Dental formula: 2.1.2.3/3.1.3.3=36. Head and body length 58-75 mm. Ear length 11-
17 mm. Forearm length 47-54 mm. Hind foot length 9-13 mm. Tail length 52-64 mm.
Body size large. Dorsal pelage long, soft, and blackish brown in color. Ventral pelage
dark brown, with paler hair tips.

生境 / Habitat
洞穴、森林、人工环境
Caves, forest, artificial environment

▲ 地理分布 / Geographic Distribution

国内分布 / Domestic Distribution
香港、海南
Hong Kong, Hainan

全球分布 / World Distribution
中国、印度尼西亚、巴布亚新几内亚
China, Indonesia, Papua New Guinea

生物地理界 / Biogeographic Realm
印度马来界、大洋洲界
Indomalaya, Oceanian

WWF 生物群系 / WWF Biome
热带和亚热带湿润阔叶林
Tropical & Subtropical Moist Broadleaf Forests

动物地理分布型 / Zoogeographic Distribution Type
Wa

分布标注 / Distribution Note
非特有种 Non-Endemic

▲ 濒危状况 / Threatened Status

中国生物多样性红色名录等级 / CB RL Category (2021)
近危 NT

IUCN 红色名录 / IUCN Red List (2021)
无危 LC

威胁因子 / Threats
森林砍伐、耕种、火灾、采矿、采石、旅游
Logging, farming, fire, mining, quarrying, tourism

▲ 法律保护地位 / Legal Protection Status

国家重点保护野生动物等级 / Category of National Key Protected Wild Animals (2021)
未列入 Not listed

"三有"名录 / TWIESSV (2023)
未列入 Not listed

CITES 附录等级 / CITES Appendix (2023)
未列入 Not listed

迁徙物种公约附录 / CMS Appendix (2020)
未列入 Not listed

保护行动 / Conservation Action
尚无保护行动 No conservation action so far

▲ 参考文献 / References

Jiang et al. (蒋志刚等), 2021; Burgin et al., 2020; IUCN, 2020; Liu et al. (刘少英等), 2020; Wilson and Mittermeier, 2019; Pan et al. (潘清华等), 2007; Wilson and Reeder, 2005; Wang (王应祥), 2003; Tan (谭邦杰), 1992

245 / 南长翼蝠

Miniopterus pusillus Dobson, 1876

· Small Long-fingered Bat

翼手目 Chiroptera / 长翼蝠科 Miniopteridae / 长翼蝠属 *Miniopterus*

科建立者及其文献 / Family Authority
Miller-Butterworth et al., 2007

属建立者及其文献 / Genus Authority
Bonaparte, 1837

亚种 / Subspecies
无 None

模式标本产地 / Type Locality
印度
India, Nicobar Isls (NW of Sumatra)

刘少英 / 供图

▲ 其他名称 / Other Name(s)

其他中文名 / Other Chinese Name(s)
无 None

其他英文名 / Other English Name(s)
Nicobar Bent-winged Bat, Nicobar Long-fingered Bat, Small Bent-winged Bat

同物异名 / Synonym(s)
无 None

▲ 形态及生境 / Morphology and Habitat

形态特征 / Morphological Characteristics

齿式：2.1.2.3/3.1.3.3=36。头体长 45~48 mm。耳长 10~11 mm。前臂长 39~42 mm。后足长 7~8 mm。尾长 40~48 mm。颅全长 14~15 mm。口鼻部较长。耳壳短而宽，耳屏长而细。背部毛发深褐色，腹部毛发棕色。翅膀长、窄。稀疏被毛延伸到尾膜。

Dental formula: 2.1.2.3/3.1.3.3=36. Head and body length 45-48 mm. Ear length 10-11 mm. Forearm length 39-42 mm. Hind foot length 7-8 mm. Tail length 40-48 mm. Greatest skull length 14-15 mm. Muzzle relatively long. Auricular shell is short and wide. Tragus is long and thin. Dorsal hairs are dark brown and the belly hairs are brown. Wing is long and narrow. Sparses hairs extend onto uropatagium.

生境 / Habitat
洞穴、森林、人工环境
Caves, forest, artificial environment

▲ 地理分布 / Geographic Distribution

国内分布 / Domestic Distribution
广东、云南、海南、香港、澳门、福建
Guangdong, Yunnan, Hainan, Hong Kong, Macao, Fujian

全球分布 / World Distribution
中国、印度、印度尼西亚、老挝、缅甸、尼泊尔、泰国、越南
China, India, Indonesia, Laos, Myanmar, Nepal, Thailand, Vietnam

生物地理界 / Biogeographic Realm
印度马来界、大洋洲界
Indomalaya, Oceanian

WWF 生物群系 / WWF Biome
热带和亚热带湿润阔叶林
Tropical & Subtropical Moist Broadleaf Forests

动物地理分布型 / Zoogeographic Distribution Type
Wb

分布标注 / Distribution Note
非特有种 Non-Endemic

▲ 濒危状况 / Threatened Status

中国生物多样性红色名录等级 / CB RL Category (2021)
近危 NT

IUCN 红色名录 / IUCN Red List (2021)
无危 LC

威胁因子 / Threats
未知 Unknown

▲ 法律保护地位 / Legal Protection Status

国家重点保护野生动物等级 / Category of National Key Protected Wild Animals (2021)
未列入 Not listed

"三有"名录 / TWIESSV (2023)
未列入 Not listed

CITES 附录等级 / CITES Appendix (2023)
未列入 Not listed

迁徙物种公约附录 / CMS Appendix (2020)
未列入 Not listed

保护行动 / Conservation Action
尚无保护行动 No conservation action so far

▲ 参考文献 / References

Jiang et al. (蒋志刚等), 2021; Burgin et al., 2020; IUCN, 2020; Liu et al. (刘少英等), 2020; Wilson and Mittermeier, 2019; Smith et al., 2009; Pan et al. (潘清华等), 2007; Wang (王应祥), 2003; Zhang (张荣祖), 1997